CONTEMPORARY MATHEMATICS

715

Advances in Rings and Modules

Sergio R. López-Permouth
Jae Keol Park
S. Tariq Rizvi
Cosmin S. Roman
Editors

AMS AMERICAN MATHEMATICAL SOCIETY

2010 *Mathematics Subject Classification.* Primary 16Dxx, 16Sxx, 16Wxx, 16Exx, 16Gxx, 16Nxx, 16Pxx, 16Uxx, 06Cxx, 15Bxx.

Library of Congress Cataloging-in-Publication Data

Names: López-Permouth, S. R. (Sergio R.), 1957– editor.

Title: Advances in rings and modules / Sergio R. López-Permouth [and three others], editors. Description: Providence, Rhode Island : American Mathematical Society, [2018] | Series: Contemporary mathematics ; volume 715 | Includes bibliographical references.

Identifiers: LCCN 2018007403 | ISBN 9781470435554 (alk. paper)

Subjects: LCSH: Rings (Algebra) | Modules (Algebra) | AMS: Associative rings and algebras – Modules, bimodules and ideals – Modules, bimodules and ideals. msc | Associative rings and algebras – Rings and algebras arising under various constructions – Rings and algebras arising under various constructions. msc | Associative rings and algebras – Rings and algebras with additional structure – Rings and algebras with additional structure. msc | Associative rings and algebras – Homological methods – Homological methods. msc | Associative rings and algebras – Representation theory of rings and algebras – Representation theory of rings and algebras. msc | Associative rings and algebras – Radicals and radical properties of rings – Radicals and radical properties of rings. msc | Associative rings and algebras – Chain conditions, growth conditions, and other forms of finiteness – Chain conditions, growth conditions, and other forms of finiteness. msc | Associative rings and algebras – Conditions on elements – Conditions on elements. msc | Order, lattices, ordered algebraic structures – Modular lattices, complemented lattices – Modular lattices, complemented lattices. msc | Linear and multilinear algebra; matrix theory – Special matrices – Special matrices. msc

Classification: LCC QA247 .A2824 2018 | DDC 512/.46–dc23 LC record available at https://lccn.loc.gov/2018007403

Contemporary Mathematics ISSN: 0271-4132 (print); ISSN: 1098-3627 (online)

DOI: https://doi.org/10.1090/conm/715

∞ The paper used in this book is acid-free and falls within the guidelines
established to ensure permanence and durability.
Visit the AMS home page at https://www.ams.org/

10 9 8 7 6 5 4 3 2 1   23 22 21 20 19 18

This volume is dedicated to the memory of Professor Bruno J. Müller
for his numerous contributions to the theory of Rings and Modules.

Photo courtesy of Hermine Vasarinsh

Bruno J. Müller (1934–2014)

Contents

Preface

Bruno J. Müller was a dear colleague and a mathematical inspiration to each one of us. He was a leader in the field and a wise mentor and advisor to his doctoral students. Most of all, he was a dear friend, one that has been sorely missed since he passed away on October 27, 2014.

Based at McMaster University, Hamilton, Canada, and specializing in the theories of rings and modules, Bruno was a world-renowned mathematician. Among other topics, Bruno did pioneering work on Localizations, Duality, and on the Theory of Continuous and Discrete Modules.

He was an author of the first ever book (published by Cambridge University Press) detailing the latest research developments in the theory of continuous, discrete and extending modules. Since its publication in 1990, that seminal book became one of the most cited works in the area and continues today to be used by researchers around the world. Professor Müller contributed fundamental research papers in many areas of algebra and impacted the discipline through his many PhD students at McMaster University.

Bruno was passionate about mathematics; he loved rings and modules and other algebraic structures. To recognize his contributions to Algebra, this volume collects a series of contributions from members of the community presented here in remembrance of Bruno and as a tribute to his memory. We are thankful to all who have participated for their enthusiastic embrace of the project and for their valuable contributions.

All papers in the volume are peer-refereed as usual. We express our gratitude to all referees who helped in the review of the papers. The topics in the assortment of the papers of this volume include Almost Self-Injective Modules, Commutators and Anti-Commutators of Idempotents, Derivations, Differential Polynomial Rings, Distributive Hierarchies of Binary Operations, Generalizations of Injectivity, HNP (Hereditary Noetherian Prime) Rings, Intrinsic Extensions of Rings, Leavitt Path Algebras, Module Hulls, Module Invariants, Reduced Ring Orders, Tilting Cotorsion Pairs, Twisted Invariants of Graded Algebras, and Universal Localizations.

As algebra continues to grow at a rapid pace, we hope the interesting results presented here will fuel and inspire young researchers for continued growth. We are sure Bruno would have enjoyed this volume and hope that you, the reader, will do the same.

As always, we are indebted to the staff of the American Mathematical Society for their outstanding work. Special thanks are due, in particular, to Ms. Christine Thivierge for her tireless efforts to see this project to conclusion. We are thankful to production editor Mike Saitas for his dedication and professional efforts to produce the nice final form of the volume.

Sergio R. López-Permouth
Jae Keol Park
S. Tariq Rizvi
Cosmin S. Roman

Contemporary Mathematics
Volume **715**, 2018
https://doi.org/10.1090/conm/715/14400

Chains of semiprime and prime ideals in Leavitt path algebras

G. Abrams, B. Greenfeld, Z. Mesyan, and K. M. Rangaswamy

ABSTRACT. Semiprime ideals of an arbitrary Leavitt path algebra L are described in terms of their generators. This description is then used to show that the semiprime ideals form a complete sublattice of the lattice of ideals of L, and they enjoy a certain gap property identified by Kaplansky in prime spectra of commutative rings. It is also shown that the totally ordered sets that can be realized as prime spectra of Leavitt path algebras are precisely those that have greatest lower bounds on subchains and enjoy the aforementioned gap property. Finally, it is shown that a conjecture of Kaplansky regarding identifying von Neumann regular rings via their prime factors holds for Leavitt path algebras.

1. Introduction

Kaplansky included the following results in two of his monographs from the 1970s, the first of which he attributed to R. Hamsher. (See [**11**, page 1] and [**10**, Theorem 11], respectively.)

Theorem (K-1): A commutative ring R is von Neumann regular if and only if R is semiprime and every prime factor ring of R is von Neumann regular.

Theorem (K-2): If $P \subsetneq Q$ are prime ideals of a commutative unital ring R, then there are prime ideals P', Q' of R such that $P \subseteq P' \subsetneq Q' \subseteq Q$ and Q' covers P', that is, there are no prime ideals of R strictly between P' and Q'.

Kaplansky took an interest in extending both theorems to noncommutative rings. In each case, his questions inspired substantial new results in the literature. Our goal in this note is to investigate whether Theorems (K-1) and (K-2) hold for Leavitt path algebras.

In [**11**] Kaplansky conjectured that Theorem (K-1) extends to all noncommutative rings. While this turned out to not be the case, J. Fisher and R. Snider [**7**] proved that Kaplansky's conjecture holds precisely when R satisfies the additional condition that the union of any ascending chain of semiprime ideals of R is again a semiprime ideal. (Recall that an ideal I of a ring R is said to be *semiprime* if

2010 *Mathematics Subject Classification.* 16W10, 16D25.

Key words and phrases. Leavitt path algebra, semiprime ideal, prime ideal, Kaplansky conjecture.

The first author is partially supported by a Simons Foundation Collaboration Grants for Mathematicians Award #208941.

I is the intersection of a collection of prime ideals of R; equivalently, whenever $A^k \subseteq I$ implies $A \subseteq I$, for any ideal A of R and positive integer k.) In view of the connection between semiprime ideals and Kaplansky's conjecture indicated in the result of Fisher and Snider, we investigate semiprime ideals in the context of Leavitt path algebras (Section 3). We first characterize the semiprime ideals of a Leavitt path algebra $L := L_K(E)$ for an arbitrary graph E and field K. Specifically, we characterize such ideals I in terms of their generators, namely, the vertices inside I and polynomials over K, with square-free factors, evaluated at cycles without exits. The semiprime ideals of Leavitt path algebras turn out to be much more well-behaved than the prime ideals. For instance, the semiprime ideals of L form a complete sublattice of the ideal lattice of L. As a consequence, we show that the above conjecture of Kaplansky regarding Theorem (K-1) holds for Leavitt path algebras. We also show that every ideal of L is semiprime if and only if the graph E satisfies Condition (K), and that every semiprime ideal of L is prime if and only if the prime ideals of L form a chain under set inclusion.

In [**10**] Kaplansky asked whether the property of the prime ideals of R in Theorem (K-2), together with the existence of least upper bounds and greatest lower bounds for chains in Spec(R), the prime spectrum of R, characterizes Spec(R) as a partially ordered set. While this also turned out not to be the case, it opened the door for serious investigations of various aspects of Spec(R) as a partially ordered set (see, e.g., [**9, 12, 13, 16, 17**]). Recently, it was shown in [**3**] that a wide class of partially ordered sets can be realized as prime spectra of Leavitt path algebras. In particular, every partially ordered set $(P, \leq)$ that satisfies the property in Theorem (K-2), has greatest lower bounds for chains, and enjoys an additional property (for every downward directed subset S of P and every $p \in P \setminus S$ such that p is greater than the greatest lower bound of S, there exists $s \in S$ such that $p > s$) can be realized as the prime spectrum of a Leavitt path algebra. However, for noncommutative rings, it is still an open question whether Theorem (K-2) holds. In Section 4, we continue to investigate Theorem (K-2) in the context of Leavitt path algebras. We say that a ring R satisfies the *Kaplansky Property* (KAP, for short) for a class $\mathcal{F}$ of ideals in R, if every pair $P, Q \in \mathcal{F}$ satisfies the condition in Theorem (K-2). The properties of semiprime ideals mentioned above enable us to show that the Kaplansky Property always holds for semiprime ideals in a Leavitt path algebra. We also show that a totally ordered set can be realized as the prime spectrum of a Leavitt path algebra if and only if the set has greatest lower bounds for all subchains and satisfies the Kaplansky Property.

In the final section (Section 5) we consider ideals which are not prime but which are unions of ascending chains of prime ideals. Following [**8**], such ideals are called *union-prime ideals.* They are of interest to us since their presence hinders the Kaplansky Property from holding in the prime spectrum of a ring. We first show that the union-prime ideals in a Leavitt path algebra are all graded, and hence always semiprime. To measure the extent to which the union-prime ideals are prevalent in a Leavitt path algebra $L_K(E)$, we consider the union-prime index $P^{\Uparrow}$ introduced in [**8**], and show that $P^{\Uparrow}(L_K(E)) \leq |E^0|$, where E^0 is the set of vertices of the graph E. Finally, an example is constructed to show that for every infinite cardinal κ there is a directed graph E_κ such that $P^{\Uparrow}(L_K(E_\kappa)) = \kappa$.

2. Preliminaries

We give below a short outline of some of the needed basic concepts and results about ordered sets and Leavitt path algebras. For general terminology and facts about Leavitt path algebras, we refer the reader to [**1**, **14**].

2.1. Partially ordered sets. A *preordered set* $(P, \leq)$ is a set P together with a binary relation $\leq$ which is reflexive and transitive. If in addition $\leq$ is antisymmetric, then $(P, \leq)$ is called a *partially ordered set* (often shortened to *poset*). If a poset $(P, \leq)$ has the property that $p \leq q$ or $q \leq p$ for all $p, q \in P$, then $(P, \leq)$ is called a *totally ordered set*, or a *chain*.

Let $(P, \leq)$ be a partially ordered set. An element x of P is called a *least* element of P if $x \leq y$ for all $y \in P$, and x is a *greatest* element of P if $y \leq x$ for all $y \in P$. P is *downward directed* if it is nonempty, and for all $p, q \in P$ there exists $r \in P$ such that $p \geq r$ and $q \geq r$.

For any poset $(P, \leq)$ and subset S of P, by restriction $(S, \leq)$ is a poset. A *lower bound* for $S \subseteq P$ is an element x of P such that $x \leq s$ for all $s \in S$. A *greatest lower bound* of S is a lower bound x of S with the additional property that $y \leq x$ for every $y \in P$ having $y \leq s$ for all $s \in S$. *(Least) upper bounds* are defined analogously.

The set of the integers will be denoted by $\mathbb{Z}$, the set of the natural numbers (including 0) will be denoted by $\mathbb{N}$, and the cardinality of a set X will be denoted by $|X|$.

2.2. Directed Graphs. A *(directed) graph* $E = (E^0, E^1, r, s)$ consists of two sets E^0 and E^1 together with maps $r, s : E^1 \to E^0$. The elements of E^0 are called *vertices* and the elements of E^1 *edges*. All the graphs E that we consider (excepting when specifically stated) are arbitrary in the sense that no restriction is placed either on the number of vertices in E or on the number of edges emitted by a single vertex.

A vertex $v \in E^0$ is called a *sink* if it emits no edges, and v is called a *regular vertex* if it emits a nonempty finite set of edges. An *infinite emitter* is a vertex which emits infinitely many edges. A graph without infinite emitters is said to be *row-finite*. For each $e \in E^1$, we call e^* a *ghost edge*. We let $r(e^*)$ denote $s(e)$, and we let $s(e^*)$ denote $r(e)$. A *path* μ of length $n > 0$, denoted $|\mu| = n$, is a finite sequence of edges $\mu = e_1 e_2 \cdots e_n$ with $r(e_i) = s(e_{i+1})$ for all $i = 1, \dots, n-1$. In this case $\mu^* = e_n^* \cdots e_2^* e_1^*$ is the corresponding *ghost path*. A vertex is considered to be a path of length 0. The set of all vertices on a path μ is denoted by μ^0.

A path $\mu = e_1 \cdots e_n$ in E is *closed* if $r(e_n) = s(e_1)$, in which case μ is said to be *based* at the vertex $s(e_1)$. A closed path μ as above is called *simple* provided it does not pass through its base more than once, i.e., $s(e_i) \neq s(e_1)$ for all $i = 2, \dots, n$. The closed path μ is called a *cycle* if it does not pass through any of its vertices twice, that is, if $s(e_i) \neq s(e_j)$ for every $i \neq j$. A cycle consisting of just one edge is called a *loop*. An *exit* for a path $\mu = e_1 \cdots e_n$ is an edge f that satisfies $s(f) = s(e_i)$ for some i, but where $f \neq e_i$. The graph E is said to satisfy *Condition (L)* if every cycle in E has an exit in E. The graph E is said to satisfy *Condition (K)* if any vertex on a simple closed path μ is also the base for a simple closed path γ different from μ. If E has no cycles, then it is called *acyclic*.

If there is a path in E from a vertex u to a vertex v, then we write $u \geq v$. It is easy to see that $(E^0, \leq)$ is a preordered set. A nonempty subset D of E^0 is said to be *downward directed* if for any $u, v \in D$, there exists a $w \in D$ such that $u \geq w$ and $v \geq w$. A subset H of E^0 is called *hereditary* if, whenever $v \in H$ and $w \in E^0$ satisfy $v \geq w$, then $w \in H$. A set $H \subseteq E^0$ is *saturated* if $r(s^{-1}(v)) \subseteq H$ implies that $v \in H$, for any regular vertex v.

2.3. Leavitt Path Algebras. Given a (nonempty) graph E and a field K, the *Leavitt path algebra* $L_K(E)$ is defined to be the K-algebra generated by $\{v : v \in E^0\} \cup \{e, e^* : e \in E^1\}$, subject to the following relations:

(V) $vv = v$ and $vw = 0$ if $v \neq w$, for all $v, w \in E^0$;
(E1) $s(e)e = e = er(e)$ for all $e \in E^1$;
(E2) $r(e)e^* = e^* = e^*s(e)$ for all $e \in E^1$;
(CK-1) $e^*e = r(e)$ and $e^*f = 0$ if $e \neq f$, for all $e, f \in E^1$;
(CK-2) $v = \sum_{e \in s^{-1}(v)} ee^*$ for every regular $v \in E^0$.

We note that every element of $L_K(E)$ can be expressed in the form $\sum_{i=1}^n k_i\alpha_i\beta_i^*$ for some $k_i \in K$ and paths α_i, β_i in E, and that sums of distinct vertices in E^0 form a set of local units for $L_K(E)$. Moreover, every Leavitt path algebra $L_K(E)$ is $\mathbb{Z}$-graded. Specifically, $L_K(E) = \bigoplus_{n \in \mathbb{Z}} L_n$, where

$$L_n = \Big\{ \sum_i k_i\alpha_i\beta_i^* \in L : \ |\alpha_i| - |\beta_i| = n \Big\}.$$

This grading is induced by defining, for all $v \in E^0$ and $e \in E^1$, $\deg(v) = 0$, $\deg(e) = 1$, $\deg(e^*) = -1$. Here the L_n are abelian subgroups satisfying $L_mL_n \subseteq L_{m+n}$ for all $m, n \in \mathbb{Z}$. An ideal I of $L_K(E)$ is said to be a *graded ideal* if $I = \bigoplus_{n \in \mathbb{Z}}(I \cap L_n)$. Equivalently, if $a \in I$ and $a = a_{i_1} + \cdots + a_{i_m}$ is a graded sum with $a_{i_k} \in L_{i_k}$ for all $k = 1, \ldots, m$, then $a_{i_k} \in I$ for all k.

We shall be using some of the results from [**2**, **4**, **18**], where the graphs that the authors consider are assumed to be countable. We point out, however, that the results in these papers hold for arbitrary graphs, with no restriction on the number of vertices or the number of edges emitted by any vertex. In fact, these results without any restriction on the size of the graph are proved in [**1**]. For the convenience of the reader, we shall still give references to [**2**, **4**, **18**] in addition to [**1**].

Given a graph E, a *breaking vertex* of a hereditary saturated subset $H \subseteq E^0$ is an infinite emitter $w \in E^0 \backslash H$ with the property that $0 < |s^{-1}(w) \cap r^{-1}(E^0 \backslash H)| < \infty$. The set of all breaking vertices of H is denoted by B_H. For any $v \in B_H$, v^H denotes the element $v - \sum_{s(e)=v,\, r(e) \notin H} ee^*$. Given a hereditary saturated subset H and $S \subseteq B_H$, (H, S) is called an *admissible pair*, and the ideal of $L_K(E)$ generated by $H \cup \{v^H : v \in S\}$ is denoted by $I(H, S)$. It was shown in [**18**, Theorem 5.7] (see also [**1**, Theorem 2.5.8]) that the graded ideals of $L_K(E)$ are precisely the ideals of the form $I(H, S)$. Moreover, given two admissible pairs (H_1, S_1) and (H_2, S_2), setting $(H_1, S_1) \leq (H_2, S_2)$ whenever $H_1 \subseteq H_2$ and $S_1 \subseteq H_2 \cup S_2$, defines a partial order on the set of all admissible pairs of $L_K(E)$. The map $(H, S) \mapsto I(H, S)$ gives a one-to-one order-preserving correspondence between the poset of admissible pairs and the set of all graded ideals of $L_K(E)$, ordered by inclusion. In [**18**, Theorem 5.7] (see also [**1**, Theorem 2.4.15]) it was also shown that $L_K(E)/I(H, S) \cong L_K(E \backslash (H, S))$ for any admissible pair (H, S). Here $E \backslash (H, S)$ is a *quotient graph of* E where

$$(E \backslash (H, S))^0 = (E^0 \backslash H) \cup \{v' : v \in B_H \backslash S\}$$

and

$$(E\backslash(H,S))^1 = \{e \in E^1 : r(e) \notin H\} \cup \{e' : e \in E^1 \text{ with } r(e) \in B_H\backslash S\},$$

and r, s are extended to $(E\backslash(H,S))^0$ by setting $s(e') = s(e)$ and $r(e') = r(e)'$.

We conclude this section by recalling the following three results, since we shall refer to them frequently in our investigation.

PROPOSITION 2.1 (well-known; see, e.g., Proposition 3 in [**3**]). *Let R be any ring, and let $\{P_\lambda : \lambda \in \Lambda\}$ be a downward directed set (under set inclusion) of prime ideals of R. Then $I = \bigcap_{\lambda\in\Lambda} P_\lambda$ is a prime ideal of R.*

THEOREM 2.2 (Theorem 4 in [**15**]). *Let E be a graph, K a field, and $L := L_K(E)$. Also, let I be a non-graded ideal of L, and set $H = I \cap E^0$ and $S = \{u \in B_H : u^H \in I\}$. Then $I = I(H,S) + \sum_{i\in X}\langle f_i(c_i)\rangle$ where X is an index set, each c_i is a cycle without exits in $E\backslash(H,S)$, and each $f_i(x) \in K[x]$ is a polynomial with a nonzero constant term.*

Here $I(H,S)$ is called the graded part *of the ideal I, and is denoted by* $\mathrm{gr}(I)$.

In a graph E, a cycle $c = e_1e_2\cdots e_n$ is said to be *WK* (for "without Condition (K)") if no vertex along c is the source of another distinct cycle in E. (In this context we view c and any "shift" $e_i\cdots e_ne_1\cdots e_{i-1}$ of c as being the same cycle.)

THEOREM 2.3 (Theorem 3.12 in [**14**]). *Let E be a graph, K a field, I a proper ideal of $L_K(E)$, and $H = I \cap E^0$. Then I is a prime ideal if and only if I satisfies one of the following conditions.*

1. *$I = \langle H \cup \{v^H : v \in B_H\}\rangle$ and $E^0 \setminus H$ is downward directed.*
2. *$I = \langle H \cup \{v^H : v \in B_H \setminus \{u\}\}\rangle$ for some $u \in B_H$ and $E^0 \setminus H = \{v \in E^0 : v \geq u\}$.*
3. *$I = \langle H \cup \{v^H : v \in B_H\} \cup \{f(c)\}\rangle$ where c is a WK cycle having source $u \in E^0$, $E^0 \setminus H = \{v \in E^0 : v \geq u\}$, and $f(x)$ is an irreducible polynomial in $K[x, x^{-1}]$.*

3. Semiprime Ideals

In this section, we give a complete description of the semiprime ideals in a Leavitt path algebra $L := L_K(E)$, of an arbitrary graph E over a field K, in terms their generators. We show that the semiprime ideals of L form a complete sublattice of the lattice of two-sided ideals of L. In particular, the union of an ascending chain of semiprime ideals is always semiprime. We also describe when every ideal in L is semiprime and when every semiprime ideal is prime.

REMARK 3.1. We note that every nonzero ideal I in the principal ideal domain $K[x, x^{-1}]$ is generated by a polynomial $f(x) \in K[x]$ with a nonzero constant term. For, suppose that $I = \langle g(x, x^{-1})\rangle$, where $g(x, x^{-1}) = \sum_{i=m}^{n} a_i x^i$ for some $m \leq n$ $(m, n \in \mathbb{Z})$ and $a_i \in K$, with $a_m \neq 0 \neq a_n$. Then

$$g(x, x^{-1}) = x^m \sum_{i=0}^{n-m} b_i x^i = x^m f(x),$$

where $b_i = a_{i+m}$ and $b_0 = a_m \neq 0$. Since x^m is a unit in $K[x, x^{-1}]$, we have $I = \langle f(x)\rangle$, where $f(x)$ is a polynomial in $K[x]$ with a nonzero constant term.

Recall that an element r of a commutative unital ring R is called *irreducible* if r is a nonunit, and r cannot be written as $r = st$ with both s, t nonunits in R. It follows from the previous observation (and the fact that the units in $K[x, x^{-1}]$ are of the form kx^m for $k \in K \setminus \{0\}$ and $m \in \mathbb{Z}$) that the irreducible polynomials in $K[x, x^{-1}]$ are precisely those of the form $x^m f(x)$ for some $m \in \mathbb{Z}$, where $f(x)$ is an irreducible polynomial in $K[x]$.

We record a well-known fact which will be useful in the proof Theorem 3.3.

LEMMA 3.2. *Let K be a field and A an ideal of $K[x, x^{-1}]$, and write $A = \langle f(x) \rangle$ for some $f(x) \in K[x, x^{-1}]$. Then A is semiprime if and only if $f(x)$ is a product of distinct irreducible polynomials.*

PROOF. If $f(x) = p_1(x) \cdots p_n(x)$ is a product of distinct irreducible polynomials, then it is easy to see that $\langle f(x) \rangle = \langle p_1(x) \rangle \cap \cdots \cap \langle p_n(x) \rangle$, and since each $\langle p_i(x) \rangle$ is a prime ideal, A is semiprime.

Conversely, suppose that $A = \langle f(x) \rangle$, where $f(x) = p(x)^k g(x)$, $p(x)$ is an irreducible polynomial, $k \geq 2$, and $\gcd(p(x), g(x)) = 1$. Then $\langle p(x)g(x) \rangle^k \subseteq A$, but $\langle p(x)g(x) \rangle \nsubseteq A$, so that A is not semiprime. $\square$

Next, observe that any graded ideal $I(H, S)$ of a Leavitt path algebra L is semiprime, since $L/I(H, S) \cong L_K(E \backslash (H, S))$ is a Leavitt path algebra, and hence has zero prime radical. (It is well-known that the Jacobson radical, and hence also the prime radical, of any Leavitt path algebra is zero; see [**2**, Proposition 6.3] or [**1**, Proposition 2.3].) But not all semiprime ideals need to be graded ideals. For example, let E be the graph having one vertex v and one loop e, with $s(e) = v = r(e)$, pictured below.

Now $K[x, x^{-1}] \cong L_K(E)$, via the map induced by sending $1 \mapsto v$, $x \mapsto e$, and $x^{-1} \mapsto e^*$. Then the natural $\mathbb{Z}$-grading of $K[x, x^{-1}]$ induced by integral powers of x coincides with the grading of $L_K(E)$ described above. Under this grading $\{0\}$ is the only proper graded ideal of $K[x, x^{-1}]$, since integral powers of x are units in $K[x, x^{-1}]$, and hence every nonzero proper ideal of $K[x, x^{-1}]$, and particularly every nonzero semiprime ideal, is non-graded.

The next result gives a complete description of all the semiprime ideals of $L_K(E)$ in terms of their generators.

THEOREM 3.3. *Let E be a graph, K a field, and $L := L_K(E)$. Also let A be an ideal of L, with $A \cap E^0 = H$ and $S = \{v \in B_H : v^H \in A\}$. Then the following are equivalent.*

(1) *A is a semiprime ideal.*
(2) *$A = I(H, S) + \sum_{i \in X} \langle f_i(c_i) \rangle$ where X is a possibly empty index set, each c_i is a cycle without exits in $E \backslash (H, S)$, and each $f_i(x)$ is a polynomial with nonzero constant term, which is a product of distinct irreducible polynomials in $K[x, x^{-1}]$.*

PROOF. We first observe that if A is a non-graded ideal, then by Theorem 2.2, $A = I(H, S) + \sum_{i \in X} \langle f_i(c_i) \rangle$, where X is a nonempty index set, each $f_i(x) \in K[x]$ has nonzero constant term, and each c_i is a cycle without exits in $E \backslash (H, S)$, based at a vertex v_i. Moreover, we may assume that $c_i^0 \neq c_j^0$ for $i \neq j$. In $L/I(H, S)$, as noted in [**4**, Proposition 3.5] (see also [**1**, Theorem 2.7.3]), the ideal M generated by $\{c_i^0 : i \in X\}$ decomposes as $M = \bigoplus_{i\in X} M_i$, where $M_i = \langle c_i^0\rangle \cong M_{\Lambda_i}(K[x,x^{-1}])$ is the ring of $\Lambda_i \times \Lambda_i$ matrices with at most finitely many nonzero entries, and Λ_i is the set of all paths that end at v_i but do not include the entire c_i. Moreover, given two paths $r, p \in \Lambda_i$, the isomorphism $\theta_i : M_i \to M_{\Lambda_i}(K[x,x^{-1}])$ takes $rc_i^k p^*$ to $x^k e_{rp}$ for all $k \in \mathbb{Z}$, where e_{rp} is a matrix unit. Clearly

$$A/I(H,S) = \bigoplus_{i\in X}\langle f_i(c_i)\rangle \subseteq \bigoplus_{i\in X} M_i.$$

Assume that (1) holds. If A is a graded ideal, then $A = I(H,S)$, by the remarks in Section 2.3, and we are done. Let us therefore suppose that A is a non-graded ideal, and hence $A = I(H,S) + \sum_{i\in X}\langle f_i(c_i)\rangle$, with X, f_i, and c_i as in the preceding paragraph. Also, as before,

$$A/I(H,S) = \bigoplus_{i\in X}\langle f_i(c_i)\rangle \subseteq \bigoplus_{i\in X} M_i = M.$$

As A is a semiprime ideal, $A/I(H,S)$ is semiprime in $L/I(H,S)$. Since M is a graded ideal, and hence a ring with local units (namely, of sums of distinct vertices of E^0 in M), it is easy to see that the intersection of any prime ideal of $L/I(H,S)$ with M is again a prime ideal of M. Hence $A/I(H,S)$ is also a semiprime ideal of M, which implies that $\langle f_i(c_i)\rangle$ is a semiprime ideal of M_i. Now

$$M_i \overset{\theta_i}{\cong} M_{\Lambda_i}(K[x,x^{-1}]),$$

and it is straightforward to show that the ideals of $M_{\Lambda_i}(K[x,x^{-1}])$ are of the form $M_{\Lambda_i}(\langle f(x)\rangle)$ for various $f(x) \in K[x,x^{-1}]$. Moreover, from the description of the isomorphism θ_i above, it follows that the ideal $\langle f_i(c_i)\rangle$ of M_i corresponds to the ideal $M_{\Lambda_i}(\langle f_i(x)\rangle)$. Since $M_{\Lambda_i}(K[x,x^{-1}])$ is Morita equivalent to the ring $K[x,x^{-1}]$, the ideal lattices of these rings are isomorphic, and under this isomorphism prime and semiprime ideals of $M_{\Lambda_i}(K[x,x^{-1}])$ correspond to prime and semiprime ideals, respectively, of $K[x,x^{-1}]$ (see e.g., [**6**, Propositions 3.3 and 3.5]). Since $\langle f_i(c_i)\rangle$ is a semiprime ideal of M_i, we see that $\langle f_i(x)\rangle$ is a semiprime ideal of $K[x,x^{-1}]$. By Lemma 3.2, we then conclude that $f_i(x)$ is a product of distinct irreducible polynomials in $K[x,x^{-1}]$, proving (2).

Conversely, assume that (2) holds. Thus $A = I(H,S) + \sum_{i\in X}\langle f_i(c_i)\rangle$, where we may assume that X is nonempty, the stated condition holds for the c_i, and each $f_i(x)$ is a product of distinct irreducible polynomials in $K[x,x^{-1}]$. As noted in the first paragraph of the proof, we have

$$A/I(H,S) = \bigoplus_{i\in X}\langle f_i(c_i)\rangle \subseteq \bigoplus_{i\in X} M_i = M,$$

where $M_i \cong M_{\Lambda_i}(K[x,x^{-1}])$ for each $i \in X$. Now, by Lemma 3.2, $\langle f_i(x)\rangle$ is a semiprime ideal of $K[x,x^{-1}]$ and, as noted earlier, the Morita equivalence of $K[x,x^{-1}]$ with $M_{\Lambda_i}(K[x,x^{-1}])$ then implies that $\langle f_i(c_i)\rangle$ is semiprime in M_i. Since this is true for all $i \in X$, and since we may view $\bigoplus_{i\in X} M_i$ as a direct sum of rings, we conclude that $A/I(H,S)$ is a semiprime ideal of $\bigoplus_{i\in X} M_i = M$. Now consider the exact sequence

$$0 \to M/\bar{A} \to \bar{L}/\bar{A} \to \bar{L}/M \to 0,$$

where $\bar{A} = A/I(H,S)$ and $\bar{L} = L/I(H,S)$. The graded ideal M is semiprime in $\bar{L}$, and hence $\bar{L}/M$ is a semiprime ring. Since $M/\bar{A}$ is also a semiprime ring, we conclude that $\bar{L}/\bar{A}$ is semiprime as well. Finally, since $L/A \cong \bar{L}/\bar{A}$, we see that A is a semiprime ideal of L, proving (1). □

Theorem 3.3 has the following consequence; in the context of results to be presented in the sequel, this property of semiprime ideals is somewhat unexpected.

PROPOSITION 3.4. *Let E be a graph, K a field, and $L := L_K(E)$. Then the semiprime ideals of L form a complete sublattice S of the lattice of ideals of L, that is, S is closed under taking arbitrary sums and intersections.*

In particular, the union of a chain of semiprime ideals of L is again a semiprime ideal.

PROOF. Suppose that $\{A_t : t \in T\}$ is a set of semiprime ideals of L. Since each A_t is an intersection of prime ideals, clearly $\bigcap_{t\in T} A_t$ is a semiprime ideal of L, showing that S is closed under taking intersections.

Setting $H_t = A_t \cap E^0$ for each $t \in T$, Theorem 3.3 implies that $A_t = \mathrm{gr}(A_t) + \sum_{i_t\in X_t} \langle f_{i_t}(c_{i_t})\rangle$, where each c_{i_t} is a cycle without exits in $E^0\backslash H_{i_t}$, based at a vertex v_{i_t}, and each $f_{i_t}(x)$ is a product of distinct irreducible polynomials in $K[x, x^{-1}]$. Let $A = \sum_{t\in T} A_t$ and $H = A \cap E^0$. Since a sum of graded ideals is again a graded ideal, $\sum_{t\in T} \mathrm{gr}(A_t) \subseteq \mathrm{gr}(A)$. It is possible that $\mathrm{gr}(A)$ contains some of the vertices v_{i_t} and hence the ideals $\langle f_{i_t}(c_{i_t})\rangle$. Removing such ideals from the sum, we can write $A = \mathrm{gr}(A) + \sum_{i\in I}\langle f_i(c_i)\rangle$, where each c_i is a cycle without exits in $E^0\backslash H$, and each $f_i(x)$ is a product of distinct irreducible polynomials in $K[x, x^{-1}]$. Therefore, by Theorem 3.3, A is a semiprime ideal of L, showing that S is closed under taking sums.

The final claim follows from the fact that the union of a chain of ideals is equal to its sum. □

We conclude this section by describing the graphs E for which every ideal of $L_K(E)$ is semiprime, and the ones for which every semiprime ideal of $L_K(E)$ is prime.

PROPOSITION 3.5. *Let E be a graph, K a field, and $L := L_K(E)$. Then the following are equivalent.*

(1) *Every ideal of L is semiprime.*
(2) *Every ideal of L is graded.*
(3) *The graph E satisfies Condition* (K).

PROOF. Assume that (1) holds. Suppose also that there is a non-graded ideal I of L, and let $H = I \cap E^0$. By Theorem 2.2, $I = \mathrm{gr}(I) + \sum_{i\in X}\langle f_i(c_i)\rangle$, where each c_i is a cycle without exits in $E^0\backslash H$, based at a vertex v_i, and each $f_i(x) \in K[x]$. Then for any $i \in X$, the ideal $A = \mathrm{gr}(I) + \langle (v_i + c_i)^2\rangle$ is not semiprime, since the ideal $B = \mathrm{gr}(I) + \langle v_i + c_i\rangle$ satisfies $B^2 \subseteq A$ and $B \not\subseteq A$. This contradicts (1), and hence every ideal of L must be graded, proving (2).

Now (2) implies (1), as noted after Lemma 3.2, and the equivalence of (2) and (3) is well-known (see [**18**, Theorem 6.16] or [**1**, Theorem 3.3.11]). □

PROPOSITION 3.6. *Let E be a graph, K a field, and $L := L_K(E)$. Then the following are equivalent.*

(1) *Every semiprime ideal of L is a prime ideal.*
(2) *The prime ideals of L form a chain under set inclusion.*

(3) *Every ideal of L is both graded and prime, and the ideals of L form a chain under set inclusion.*
(4) *The graph E satisfies Condition* (K), *and the admissible pairs of E form a chain under the partial order of the admissible pairs.*

PROOF. Assume that (1) holds. To prove (2), suppose, on the contrary, that there are two prime ideals A, B of L such that $A \not\subseteq B$ and $B \not\subseteq A$. Choose an element $a \in A \backslash B$ and an element $b \in B \backslash A$. Since aLb is contained in both A and B, we have $aLb \subseteq A \cap B$. Now $A \cap B$ is a semiprime ideal and is hence prime, by supposition. Thus one of a and b must belong to $A \cap B$. This means that either $a \in B$ or $b \in A$; a contradiction. Hence the prime ideals of L form a chain under set inclusion, proving (2).

Assume that (2) holds. We now claim that every ideal of L is graded. Suppose that there is a non-graded ideal A in L, and set $H = A \cap E^0$. Now $\mathrm{gr}(A)$ is a semiprime ideal and hence must be prime, since by hypothesis it is the intersection of a chain of prime ideals, so that Proposition 2.1 applies. This means that $E^0 \backslash H$ is downward directed, by Theorem 2.3, and hence $E^0 \backslash H$ can contain at most one cycle without exits. Thus, by Theorem 2.2, A must be of the form $A = \mathrm{gr}(A) + \langle f(c) \rangle$, where c is the unique cycle without exits in $E^0 \backslash H$ and $f(x) \in K[x]$. Now for any two distinct irreducible $p(x), q(x) \in K[x]$ with nonzero constant terms, the ideals $P = \mathrm{gr}(A) + \langle p(c) \rangle$ and $Q = \mathrm{gr}(A) + \langle q(c) \rangle$ are prime, by Theorem 2.3, with the property that $P \not\subseteq Q$ and $Q \not\subseteq P$, contradicting (2). Thus every ideal I of L must be graded, and hence also semiprime. Since the prime ideals of L form a chain, intersections of prime ideals of L are also prime (again using Proposition 2.1). Therefore every ideal of L is prime, proving (3).

Assume that (3) holds. Then E satisfies Condition (K), by Proposition 3.5, and the admissible pairs of E form a chain, by the comments in Section 2.3, proving (4).

Finally, if (4) holds, then every ideal of L is graded, by Proposition 3.5. Since admissible pairs correspond to generating sets for graded ideals, this means that the ideals of L form a chain. Since semiprime ideals are intersections of prime ideals, and the intersection of any chain of prime ideals is prime, (1) follows. □

4. The Kaplansky Conjecture and the Kaplansky Property

In this section the Kaplansky conjecture regarding Theorem (K-1), mentioned in the Introduction, is shown to be true for Leavitt path algebras L. We then show that the Kaplansky Property holds for semiprime ideals of L, and give sufficient conditions under which prime ideals of L satisfy this property. In particular, we use the Kaplansky Property to characterize the totally ordered sets that can be realized as prime spectra of Leavitt path algebras.

Our first result shows that Kaplansky's conjecture holds in Leavitt path algebras, and it is stated in a slightly stronger form, in the sense that we do not need the additional hypothesis that the ring is semiprime (as all Leavitt path algebras are semiprime), and that we only need to consider graded prime ideals.

Recall that a ring R is called *von Neumann regular* if for every $r \in R$ there exists a $p \in R$ such that $r = rpr$.

PROPOSITION 4.1. *Let E be a graph, K a field, and $L := L_K(E)$. Then L is von Neumann regular if and only if L/P is von Neumann regular for any graded prime ideal P of L.*

PROOF. Assume that L/P is von Neumann regular for all graded prime ideals P. We claim that E contains no cycles. Suppose, on the contrary, that there is a cycle c, based at a vertex v, in the graph E, and let $H = \{u \in E^0 : u \not\geq v\}$. Clearly H is a hereditary saturated set of vertices and $E^0 \backslash H$ is downward directed. Now $P = I(H, B_H)$ is a prime ideal, by Theorem 2.3, and $L/P \cong L_K(E\backslash(H, B_H))$, by the remarks in Section 2.3. By hypothesis, $L_K(E\backslash(H, B_H))$ is von Neumann regular, and by [**5**, Theorem 1], $E\backslash(H, B_H)$ must then be acyclic, contradicting c being a cycle in $E\backslash(H, B_H)$. Hence E must be acyclic and we appeal to [**5**, Theorem 1] to conclude that L is von Neumann regular.

The converse follows from the fact that any quotient of a von Neumann regular ring is again von Neumann regular. □

REMARK 4.2. As mentioned in the Introduction, Fisher and Snider showed in [**7**, Theorem 1.1] that a semiprime ring R with all its prime factor rings regular is von Neumann regular, provided the union of any ascending chain of semiprime ideals of R is again a semiprime ideal. Since a Leavitt path algebra L is always semiprime, and since Proposition 3.4 shows that the union of an ascending chain of semiprime ideals in L is semiprime, we can appeal to [**7**, Theorem 1.1] to get an alternate proof of Proposition 4.1.

Next we explore the Kaplansky Property (KAP). The following proposition gives conditions for (KAP) to hold for a given family of ideals in a ring R. The argument used here is fundamentally due to Kaplansky [**10**, Theorem 11]. For convenience, we say that a pair of ideals (P, Q) in a ring is a *Kaplansky pair*, if the ideals P, Q satisfy the covering property stated in Theorem (K-2).

PROPOSITION 4.3. *Let R be an arbitrary ring. Suppose that $\mathcal{F}$ is a nonempty family of ideals of R such that $\mathcal{F}$ is closed under taking unions and intersections of chains of ideals. Then* (KAP) *holds for ideals in $\mathcal{F}$.*

PROOF. Suppose that $A \subsetneq B$ are two elements of $\mathcal{F}$, and let $x \in B\backslash A$. Since the union of a chain of ideals in $\mathcal{F}$ is again a member of $\mathcal{F}$, we apply Zorn's Lemma to obtain an ideal $M \in \mathcal{F}$ maximal with respect to the property that $A \subseteq M \subsetneq B$ and $x \notin M$. Let $C = M + \langle x \rangle$. Since the intersection of a chain of ideals in $\mathcal{F}$ is a member of $\mathcal{F}$, we can similarly apply Zorn's Lemma to obtain an ideal $N \in \mathcal{F}$ minimal with respect to the property that $C \subseteq N \subseteq B$. (Note that if $C \in \mathcal{F}$, then $C = N$.) We claim that (M, N) is a Kaplansky pair. For, suppose that there is an ideal $P \in \mathcal{F}$ that satisfies $M \subsetneq P \subseteq N$. Then $x \in P$, by the maximality of M, and hence $C \subseteq P$. The minimality of N then implies that $P = N$. Thus (KAP) holds for ideals belonging to $\mathcal{F}$. □

COROLLARY 4.4. *Let E be a graph, K a field, and $L := L_K(E)$. Then the semiprime ideals of L satisfy* (KAP).

PROOF. By Proposition 3.4, the set of semiprime ideals of L is closed under taking unions and intersections of chains. We then appeal to Proposition 4.3 to conclude that (KAP) holds for semiprime ideals of L. □

Since the intersection of a chain of prime ideals in any ring is again a prime ideal (Proposition 2.1), the following corollary follows immediately from Proposition 4.3.

COROLLARY 4.5. *Let R be an arbitrary ring. If the union of any chain of prime ideals in R is again a prime ideal, then* (KAP) *holds for the prime ideals of R.*

The sufficient condition in Corollary 4.5 is not necessary, as the next example shows.

EXAMPLE 4.6. Consider the following row-finite graph E.

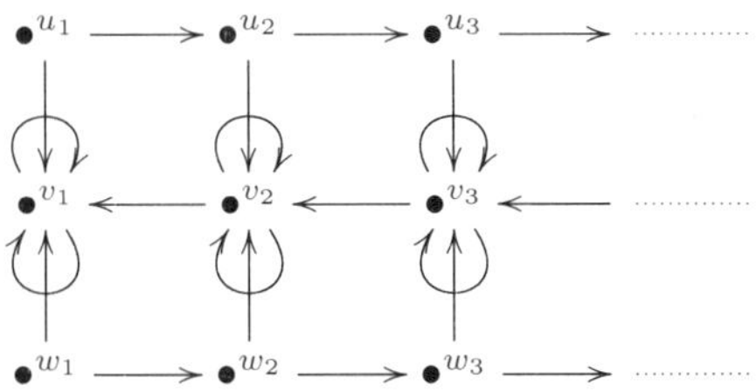

We claim that (KAP) holds for the prime ideals of $L := L_K(E)$. To see this, first observe that, since E satisfies Condition (K), all the ideals of L are necessarily graded, by Proposition 3.5. In addition, E is row-finite, and hence the ideals of L correspond to the hereditary saturated subsets of E^0, by the remarks in Section 2.3. Setting $U = \{u_1, u_2, u_3, \dots\}$, $V = \{v_1, v_2, v_3, \dots\}$, and $W = \{w_1, w_2, w_3, \dots\}$, we see that the proper hereditary saturated subsets of E^0 are $\emptyset$, $U \cup V$, $W \cup V$, V, and $H_n = \{v_1, \dots, v_n\}$, for each $n \geq 1$. It is easy to verify that V is the only hereditary saturated set for which $E^0 \backslash V$ is not downward directed. Thus, by Theorem 2.3, the prime ideals of L are 0, $\langle U \cup V \rangle$, $\langle W \cup V \rangle$, and $P_n = \langle H_n \rangle$, for each $n \geq 1$. It follows that (P_n, P_{n+1}) is a Kaplansky pair for each $n \geq 1$. Since $P_n, P_{n+1} \subseteq \langle U \cup V \rangle$ and $P_n, P_{n+1} \subseteq \langle W \cup V \rangle$, it then follows that $\mathrm{Spec}(L)$ satisfies (KAP). However, the union of the (well-ordered ascending) chain

$$P_1 \subseteq P_2 \subseteq P_3 \subseteq \cdots$$

of prime ideals of L is the ideal $\langle V \rangle$, which is not prime.

REMARK 4.7. In contrast to Example 4.6, we note that if E is row-finite and if an ideal P of $L_K(E)$ is the union of a well-ordered strictly ascending chain

$$P_1 \subsetneq \cdots \subsetneq P_\lambda \subsetneq P_{\lambda+1} \subsetneq \cdots \qquad (\lambda < \kappa)$$

of prime ideals in $L_K(E)$, where the ordinal κ has uncountable cofinality, then P is a prime ideal. To justify this, by Theorems 2.2 and 2.3, it suffices to show that $E^0 \backslash H$ is downward directed, where $H = P \cap E^0$. This is equivalent to showing that $T(u) \cap T(v) \nsubseteq H$ for all $u, v \in E^0 \backslash H$ (where, for any vertex u, $T(u) = \{w \in E^0 : u \geq w\}$ is the *tree* of u). Suppose, on the contrary, that $T(u) \cap T(v) \subseteq H$. Now the set $T(u) \cap T(v)$ is countable, as E is row-finite, and since κ has uncountable cofinality, there exists a $\lambda < \kappa$ such that $T(u) \cap T(v) \subseteq H_\lambda$, where $H_\lambda = P_\lambda \cap E^0$. This is a contradiction, since P_λ is a prime ideal and so $u, v \in E^0 \backslash H_\lambda$ implies that $T(u) \cap T(v) \nsubseteq H_\lambda$, again by Theorem 2.3. Hence P must be a prime ideal.

REMARK 4.8. The claim in Remark 4.7 remains true if we replace "row-finite" by "row-countable", as $T(u) \cap T(v)$ would then still be at most countable, and the same argument applies.

We next explore additional sufficient conditions for (KAP) to hold for prime ideals of $L_K(E)$. We begin with an interesting property of prime ideals.

PROPOSITION 4.9. *Let E be a graph, K a field, and $L := L_K(E)$. If $P \subseteq A$ are ideals of L, where P is prime, then either $P = A$ or $P \subseteq \mathrm{gr}(A)$.*

PROOF. If P is graded, then clearly the given condition holds. Let us therefore assume that P is a non-graded prime ideal, and suppose that there is an ideal A such that $P \subseteq A$ and $P \nsubseteq \mathrm{gr}(A)$. In particular, A is a non-graded ideal, and hence, by Theorem 2.2, $A = I(H', S') + \sum_{i \in X} \langle f_i(c_i) \rangle$, with $H' = A \cap E^0$ and $S' \subseteq B_{H'}$, where each c_i is a cycle without exits in $E^0 \backslash H'$, and each $f_i(x) \in K[x]$ is a polynomial of smallest degree such that $f_i(c_i) \in A$. By Theorem 2.3 and Remark 3.1, we have $P = I(H, B_H) + \langle p(c) \rangle$, where $H = P \cap E^0$, c is a cycle without exits, based at a vertex u, in $E \backslash (H, B_H)$, $v \geq u$ for all $v \in E^0 \backslash H$, and $p(x) \in K[x]$ is an irreducible polynomial with nonzero constant term. Clearly, c is the only cycle without exits in $E^0 \backslash H$. As $P \nsubseteq \mathrm{gr}(A) = I(H', S')$, we see that $p(c) \notin I(H', S')$ and c is also a cycle without exits in $E^0 \backslash H'$. Since $v \geq u$ for all $v \in E^0 \backslash H$ and hence also all $v \in E^0 \backslash H'$, we conclude that $c_i = c$ for all $i \in X$. Moreover $H' = H$, because if there were a $v \in H' \backslash H$, then $v \geq u$ would imply that $u \in H'$, contradicting $p(c) \notin I(H', S')$. It follows also that $S' = B_H$. Thus A is of the form $A = I(H, B_H) + \langle f(c) \rangle$, where $f(x) \in K[x]$ has nonzero constant term. Clearly, $\langle p(c) \rangle \subseteq \langle f(c) \rangle$ in $L/I(H, B_H)$. Since $f(x)$ was chosen to be a polynomial of smallest degree in $K[x]$ such that $f(c) \in A$, and since $p(x)$ is irreducible in $K[x]$, it follows that $\langle p(c) \rangle = \langle f(c) \rangle$ in $L/I(H, B_H)$. Therefore $A = I(H, B_H) + \langle p(c) \rangle = P$, as desired. □

PROPOSITION 4.10. *Let E be a graph, K a field, and $L := L_K(E)$. If* $\mathrm{Spec}(L)$ *is totally ordered by set inclusion, then* (KAP) *holds for prime ideals in L.*

PROOF. If the prime ideals of L form a chain, then every semiprime ideal of L is prime, by Proposition 3.6. Thus, by Proposition 3.4, $\mathrm{Spec}(L)$ is closed under taking unions of chains. Applying Corollary 4.5, we conclude that (KAP) holds for prime ideals in L. □

Proposition 4.10 and [**3**, Theorem 27] can be used to completely characterize totally ordered sets which can be realized as the prime spectra of Leavitt path algebras. To achieve this, let us recall some notation and a result from [**3**].

DEFINITION 4.11. Let $(P, \leq)$ be a partially ordered set. For any $S \subseteq P$ and $p \in P$, we write $p = \mathrm{glb}(S)$ in case S has a greatest lower bound in P, and it is equal to p. Also let

$$\mathcal{R}(P) = \{p \in P : p \neq \mathrm{glb}(S) \text{ for all } S \subseteq P \text{ downward directed without a least element}\}.$$

We view $\mathcal{R}(P)$ as a partially ordered subset of P.

NOTATION 4.12. We assign the indicated names in case the partially ordered set $(P, \leq)$ satisfies the following properties.

- (GLB) Every downward directed subset of P has a greatest lower bound in P.
- (DC) For every downward directed subset S of P and every $p \in \mathcal{R}(P)$ satisfying $p \geq \mathrm{glb}(S)$, we have $p \geq s$ for some $s \in S$.
- (DD) For every downward directed subset S of P such that $\mathrm{glb}(S) \in P$, there exists a downward directed subset T of $\mathcal{R}(P)$ satisfying $\mathrm{glb}(S) = \mathrm{glb}(T)$.
- (KAP) For all $p, q \in P$ such that $p < q$, there exist $p', q' \in P$ such that $p \leq p' < q' \leq q$ and there is no $t \in P$ satisfying $p' < t < q'$.

PROPOSITION 4.13 (Proposition 38 in [**3**]). *Let* $(P, \leq)$ *be a partially ordered set satisfying* (GLB), *and suppose that for every downward directed* $S \subseteq P$ *and every* $p \in P$ *satisfying* $p > \text{glb}(S)$, *we have* $p \geq s$ *for some* $s \in S$. *Then* P *satisfies* (KAP) *if and only if* P *satisfies* (DD).

THEOREM 4.14. *A totally ordered set* $(P, \leq)$ *can be realized as the prime spectrum of a Leavitt path algebra if and only if* P *satisfies* (GLB) *and* (KAP).

PROOF. [**3**, Theorem 27] states, among other things, that a partially ordered set that satisfies (GLB), (DC), and (DD) can be realized as the prime spectrum of a Leavitt path algebra. Now, since P is totally ordered, it necessarily satisfies (DC) and the condition in Proposition 4.13. Thus, if P satisfies (GLB) and (KAP), then it also satisfies (DC) and (DD), and hence it can be realized as the prime spectrum of a Leavitt path algebra, by [**3**, Theorem 27].

Conversely, if P is order-isomorphic to the prime spectrum of a Leavitt path algebra, then it satisfies (KAP), by Proposition 4.10. Moreover, the prime spectrum of any ring satisfies (GLB) by Proposition 2.1, from which the result follows. □

REMARK 4.15. Given a field K and a graph E, it can be shown that the ideal extension $\mathcal{E}(K, L)$ of K by $L := L_K(E)$ is a unital ring, such that there is a one-to-one inclusion-preserving correspondence between the prime ideals of L and the prime ideals of $\mathcal{E}(K, L)$, except for an extra maximal ideal in the latter. Thus Theorem 4.14 implies that any totally ordered set with a maximal element, that satisfies (GLB) and (KAP), can be realized as the prime spectrum of a unital ring.

5. Union-Prime Ideals of Leavitt Path Algebras

Following [**8**], we call a non-prime ideal in a ring R, which is the union of an ascending chain of prime ideals of R, a *union-prime ideal*. In this section we study the union-prime ideals in Leavitt path algebras, since the absence of such ideals helps (KAP) hold for prime ideals. To measure the extent to which union-prime ideals are prevalent in Leavitt path algebras, we investigate the $P^{\Uparrow}$-index, introduced in [**8**], for such rings, and show that $P^{\Uparrow}(L_K(E)) \leq |E^0|$ for any graph E. Examples are constructed of Leavitt path algebras having $P^{\Uparrow}$-index κ, for any prescribed infinite cardinal κ.

PROPOSITION 5.1. *Let* E *be a graph,* K *a field, and* $L := L_K(E)$. *Then the union of any strictly ascending chain of prime ideals of* L *is a graded ideal, and the intersection of any strictly descending chain of prime ideals of* L *is a graded ideal.*

In particular, every union-prime ideal in L *is graded, and hence semiprime.*

PROOF. Let $\{P_i : i \in X\}$ be a strictly ascending chain of prime ideals of L, where $(X, \leq)$ is a totally ordered set, and $P_i \subseteq P_j$ whenever $i \leq j$ $(i, j \in X)$. Since this chain of prime ideals is assumed to be strictly ascending, for each $i \in X$ we can find some $i' \in X$ such that $P_i \subsetneq P_{i'}$. Then $P_i \subseteq \text{gr}(P_{i'})$, by Proposition 4.9, and hence $\bigcup_{i \in X} P_i = \bigcup_{i' \in X} \text{gr}(P_{i'})$ is a graded ideal.

Proving that the intersection of a strictly descending chain of prime ideals is graded can be accomplished in an entirely analogous manner.

The final claim follows from the aforementioned fact that every graded ideal of a Leavitt path algebra is semiprime. □

DEFINITION 5.2. Let $\tilde{P} = \{P_i : i \in X\}$ be a chain of prime ideals in a ring R. The cardinality of the set of non-prime unions of subchains of $\tilde{P}$ is called the *index* of $\tilde{P}$, and is denoted by $\mathrm{Ind}(\tilde{P})$. The $P^{\Uparrow}$-*index* of R, denoted by $P^{\Uparrow}(R)$, is the supremum of the set of cardinals $\mathrm{Ind}(\tilde{P})$, taken over all chains of primes $\tilde{P}$ in R.

We require the following lemma in order to give a bound for the $P^{\Uparrow}$-index of an arbitrary Leavitt path algebra.

LEMMA 5.3. *Let E be a graph, K a field, and $L := L_K(E)$. Also let Λ be a chain of prime ideals of L, and let $\{P_i : i \in X\}$ be the set of the union-prime ideals that are unions of subchains of Λ, where $(X, \leq)$ is a totally ordered set such that $P_i \subseteq P_j$ whenever $i \leq j$ $(i, j \in X)$.*

Then for every $i \in X$ that is not a greatest element, there is a graded prime ideal Q_i of L such that $P_i \subsetneq Q_i \subsetneq P_j$ for all $j > i$ $(j \in X)$.

PROOF. Given $i \in X$ that is not a greatest element, there exists $j \in X$ such that $j > i$, and hence there is a prime ideal $I \in \Lambda$ such that $P_i \subseteq I \subseteq P_j$, since otherwise we would have $P_i = P_j$. Thus $Q_i' = \bigcap\{I \in \Lambda : P_i \subseteq I\}$, being the intersection of a nonempty set of prime ideals, is a prime ideal (Proposition 2.1). Then $Q_i = \mathrm{gr}(Q_i')$ is also a prime ideal, by Theorem 2.3, and it is graded, by definition. Clearly $Q_i \subseteq Q_i' \subseteq P_j$ for all $j > i$ $(i \in X)$. Moreover, for all $I \in \Lambda$ such that $I \subseteq P_i$, we have $I \subsetneq Q_i'$, and hence $I \subseteq Q_i$, by Proposition 4.9. It follows that $P_i \subseteq Q_i$. Finally, since for all $j > i$, the ideals P_i and P_j are not prime, we conclude that $P_i \subsetneq Q_i \subsetneq P_j$, as desired. □

PROPOSITION 5.4. *Let E be a graph, K a field, and $L := L_K(E)$. Then $P^{\Uparrow}(L) \leq |E^0|$.*

PROOF. If E^0 is finite, then by Theorem 2.3, there can be no infinite chains of prime ideals in L, and hence there can be no union-prime ideals in L, giving $P^{\Uparrow}(L) = 0 \leq |E^0|$. Thus we may assume that E^0 is infinite.

Let Λ be a chain of prime ideals of L, and let $\{P_i : i \in X\}$ be the set of the union-prime ideals that are unions of subchains of Λ, where $(X, \leq)$ is a totally ordered set such that $P_i \subseteq P_j$ whenever $i \leq j$ $(i, j \in X)$. Also let

$$X' = \begin{cases} X & \text{if } X \text{ has no greatest element} \\ X \setminus \{x\} & \text{if } x \in X \text{ is a greatest element} \end{cases}.$$

Our goal is to define an injective map $f : X' \to E^0$.

Given $i \in X'$, by Lemma 5.3, there is a graded prime ideal Q_i of L such that $P_i \subsetneq Q_i \subsetneq P_j$ for all $j > i$ $(j \in X)$. Also, by Proposition 5.1, each P_l $(l \in X)$ is a graded ideal. Hence, by the remarks in Section 2.3, we can write $P_i = I(H_i, S_i)$ and $Q_i = I(H_i', S_i')$, for some hereditary saturated $H_i, H_i' \subseteq E^0$ and $S_i \subseteq B_{H_i}, S_i' \subseteq B_{H_i'}$. Since $P_i \subsetneq Q_i$, we have $H_i \cup S_i \subsetneq H_i' \cup S_i'$, by the order-preserving correspondence between admissible pairs and graded ideals discussed in Section 2.3. Thus, we can define $f : X' \to E^0$ by choosing arbitrarily for each $i \in X'$ some $u_i \in (H_i' \cup S_i') \setminus (H_i \cup S_i)$ and setting $f(i) = u_i$.

To show that f is injective, let $P_i = I(H_i, S_i)$ and $Q_i = I(H_i', S_i')$ be as above, let $j > i$ $(j \in X)$, and set $P_j = I(H_j, S_j)$, where $H_j \subseteq E^0$ is hereditary saturated and $S_j \subseteq B_{H_j}$. By the aforementioned correspondence between admissible pairs and graded ideals, $H_i' \cup S_i' \subseteq H_j \cup S_j$, from which it follows that $u_j \neq u_i$, showing that $f : X' \to E^0$ is injective. Therefore

$$|X| \leq |X'| + 1 \leq |E^0| + 1 = |E^0|,$$

which implies that $P^{\Uparrow}(L) \leq |E^0|$. □

We conclude this note by constructing, for each cardinal κ, a directed graph E such that $P^{\Uparrow}(L_K(E)) = \kappa$. To facilitate the construction, we require the following notation from [**3**].

Definition 5.5. Given a partially ordered set $(P, \leq)$ we define a graph E_P by letting

$$E_P^0 = \{v_p : p \in P\} \quad \text{and} \quad E_P^1 = \{e_{p,q}^i : i \in \mathbb{N}, \text{ and } p, q \in P \text{ satisfy } p > q\},$$

where $s(e_{p,q}^i) = v_p$ and $r(e_{p,q}^i) = v_q$ for all $i \in \mathbb{N}$.

Example 5.6. For every ordinal γ define a partially ordered set $(P_\gamma, \leq_\gamma)$ by

$$P_\gamma = \{p_\lambda : \lambda < \gamma\} \cup \{r_\lambda : \lambda < \gamma \text{ is an infinite limit ordinal}\},$$

where for all ordinals $\lambda < \mu < \gamma$ we let $p_\lambda <_\gamma p_\mu$, $p_\lambda <_\gamma r_\mu$, $r_\lambda <_\gamma p_\mu$, and $r_\lambda <_\gamma r_\mu$, where μ is assumed to be an infinite limit ordinal wherever appropriate. For instance, $E_{P_{\omega+2}}$ can be visualized as follows.

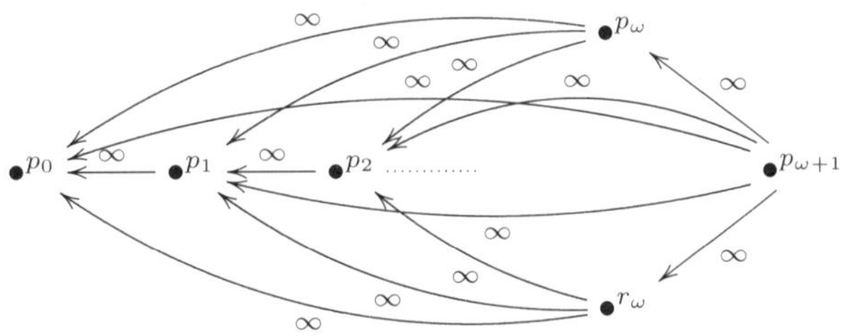

(The symbol ∞ appearing adjacent to an edge indicates that there are countably infinitely many edges from one vertex to another.) We note that given two ordinals $\lambda < \gamma$, we have $P_\lambda \subseteq P_\gamma$, and $\leq_\lambda$ is the restriction of $\leq_\gamma$ to P_λ. Thus we may view E_{P_λ} as a subgraph of E_{P_γ} whenever $\lambda \leq \gamma$. Finally, for each ordinal γ let $L_\gamma = L_K(E_{P_\gamma})$, for an arbitrary field K.

Let γ be an ordinal. Given an ordinal $\lambda < \gamma$, we note that $E_{P_\gamma}^0 \setminus E_{P_\lambda}^0$ is downward directed if and only if λ is not an infinite limit ordinal, since in the case where λ is an infinite limit ordinal $E_{P_\gamma}^0 \setminus E_{P_\lambda}^0$ contains the two sinks v_{p_λ} and v_{r_λ}. Noting that $E_{P_\lambda}^0$ is necessarily hereditary and saturated, we conclude, by Theorem 2.3, that $P_\lambda = \langle E_{P_\lambda}^0 \rangle \subseteq L_\gamma$ is a prime ideal if and only if λ is not an infinite limit ordinal. Moreover, if λ is an infinite limit ordinal, then $P_\lambda = \bigcup_{\alpha<\lambda} P_\alpha$, and hence P_λ is a union-prime ideal.

Now let κ be any infinite cardinal. Let γ be the smallest ordinal such that the set of all infinite limit ordinals less than γ has cardinality κ (such exists by a straightforward set-theoretic argument). Then $P^{\Uparrow}(L_\gamma) = \kappa$.

Acknowledgement

We thank the referee for a very thorough reading of the original version.

References

[1] G. Abrams, P. Ara, and M. Siles Molina, *Leavitt path algebras.* Lecture Notes in Mathematics, Vol. 2191, Springer-Verlag, London, 2017. MR3729290

[2] G. Abrams and G. Aranda Pino, *The Leavitt path algebras of arbitrary graphs*, Houston J. Math. **34** (2008), no. 2, 423–442. MR2417402

[3] G. Abrams, G. Aranda Pino, Z. Mesyan, and C. Smith, *Realizing posets as prime spectra of Leavitt path algebras*, J. Algebra **476** (2017), 267–296. MR3608152

[4] G. Abrams, G. Aranda Pino, F. Perera, and M. Siles Molina, *Chain conditions for Leavitt path algebras*, Forum Math. **22** (2010), no. 1, 95–114. MR2604365

[5] G. Abrams and K. M. Rangaswamy, *Regularity conditions for arbitrary Leavitt path algebras*, Algebr. Represent. Theory **13** (2010), no. 3, 319–334. MR2630124

[6] P. N. Ánh and L. Márki, *Morita equivalence for rings without identity*, Tsukuba J. Math. **11** (1987), no. 1, 1–16. MR899719

[7] J. W. Fisher and R. L. Snider, *On the von Neumann regularity of rings with regular prime factor rings*, Pacific J. Math. **54** (1974), 135–144. MR0360679

[8] B. Greenfeld, L. H. Rowen, and U. Vishne, *Unions of chains of primes*, J. Pure Appl. Algebra **220** (2016), no. 4, 1451–1461. MR3423457

[9] M. Hochster, *Prime ideal structure in commutative rings*, Trans. Amer. Math. Soc. **142** (1969), 43–60. MR0251026

[10] I. Kaplansky, *Commutative rings*, Revised edition, The University of Chicago Press, Chicago, Ill., and London, 1974. MR0345945

[11] I. Kaplansky, *Algebraic and analytic aspects of operator algebras*, American Mathematical Society, Providence, R.I., 1970. Conference Board of the Mathematical Sciences Regional Conference Series in Mathematics, No. 1. MR0312283

[12] W. J. Lewis, *The spectrum of a ring as a partially ordered set*, J. Algebra **25** (1973), 419–434. MR0314811

[13] W. J. Lewis and J. Ohm, *The ordering of* Spec R, Canad. J. Math. **28** (1976), no. 4, 820–835. MR0409428

[14] K. M. Rangaswamy, *The theory of prime ideals of Leavitt path algebras over arbitrary graphs*, J. Algebra **375** (2013), 73–96. MR2998948

[15] K. M. Rangaswamy, *On generators of two-sided ideals of Leavitt path algebras over arbitrary graphs*, Comm. Algebra **42** (2014), no. 7, 2859–2868. MR3178047

[16] S. Sarussi, *Totally ordered sets and the prime spectra of rings*, Comm. Algebra **45** (2017), no. 1, 411–419. MR3556583

[17] T. P. Speed, *On the order of prime ideals*, Algebra Universalis **2** (1972), 85–87. MR0306061

[18] M. Tomforde, *Uniqueness theorems and ideal structure for Leavitt path algebras*, J. Algebra **318** (2007), no. 1, 270–299. MR2363133

Department of Mathematics, University of Colorado, Colorado Springs, CO, 80918, USA

Email address: abrams@math.uccs.edu

Department of Mathematics, Bar Ilan University, Ramat Gan 5290002, Israel

Email address: beeri.greenfeld@gmail.com

Department of Mathematics, University of Colorado, Colorado Springs, CO, 80918, USA

Email address: zmesyan@uccs.edu

Department of Mathematics, University of Colorado, Colorado Springs, CO, 80918, USA

Email address: krangasw@uccs.edu

Contemporary Mathematics
Volume **715**, 2018
https://doi.org/10.1090/conm/715/14401

The conditions $(\mathrm{C_i}), \mathrm{i} = 1, 2, 3, 11, 12$, in rings, modules, categories, and lattices

Toma Albu

Dedicated to the memory of Bruno J. Müller (1934-2014)

ABSTRACT. In this survey paper we discuss the latticial counterparts of the conditions (C_i), $i = 1, 2, 3, 11, 12$, for modules. Applications are given to Grothendieck categories and module categories equipped with a hereditary torsion theory.

Introduction

In this survey paper we illustrate a general strategy which consists on putting a module-theoretical definition/result in a latticial frame, in order to translate that definition/result to Grothendieck categories and to module categories equipped with a hereditary torsion theory. More specifically, we present latticial counterparts of known definitions and results about modules satisfying the conditions (C_i), $i = 1, 2, 3, 11, 12$.

As it is well-known, Bruno Müller and his collaborators obtained important results in Module Theory concerning the conditions (C_i), $i = 1, 2, 3,$ and this is the main reason to present here their various extensions.

In Section 1 we list some definitions and results about lattices, especially from [**6**] and [**38**]. In Section 2 we define the conditions (C_i), $i = 1, 2, 3, 11, 12$, for lattices, and present some of their basic properties. Section 3 is devoted to linear modular lattices that are defined with the aid of the concept of a linear morphism of lattices. In Section 4 we present the latticial version of the well-known module-theoretical concept of a preradical. The last section illustrates our general strategy explained above by presenting applications of the latticial results from the previous sections to Grothendieck categories and to module categories equipped with a hereditary torsion theory.

2010 *Mathematics Subject Classification.* 06C05, 06C99, 06B35, 16D80, 16S90, 18E15.

Key words and phrases. Modular lattice, upper continuous lattice, complement, pseudo-complement, essential element, closed element, compact element, compactly generated lattice, (C_i) condition, $i = 1, 2, 3, 11, 12$, CS module, CC lattice, continuous lattice, quasi-continuous lattice, linear modular lattice, lattice preradical, The Osofsky-Smith Theorem, Grothendieck category, hereditary torsion theory.

1. Lattice background

This section presents some known definitions and results on lattices that will be freely used throughout the paper. For all other undefined notation and terminology on lattices, the reader is referred to [**5**], [**6**], [**20**], [**21**], [**27**], and/or [**38**].

Basic concepts

A *partially ordered set* (more briefly, a *poset*) is a pair $(P, \leqslant)$ consisting of a non-empty set P and a binary relation $\leqslant$ on P which is reflexive, anti-symmetric, and transitive. Very often, a poset $(P, \leqslant)$ will be denoted just by P.

By a *lattice* we mean a poset $(L, \leqslant)$ such that any pair of elements a, b in L has a *greatest lower bound* $a \wedge b$ (also called *meet*) and a *least upper bound* $a \vee b$ (also called *join*).

All posets and lattices considered in this paper are assumed to have a least element denoted by 0 and a last element denoted by 1, i.e., they are *bounded,* and L will always denote such a lattice. The notation $L \simeq L'$ means that the lattices L and L' are isomorphic. We denote by $\mathbb{N}$ the set $\{1, 2, \ldots\}$ of all positive integers.

For a poset P and elements $a \leqslant b$ in P we write

$$b/a := [a, b] = \{ \, x \in P \mid a \leqslant x \leqslant b \, \}.$$

A *subfactor* of P is any interval b/a of P with $a \leqslant b$. An *initial interval* of b/a is any interval c/a for some $c \in b/a$.

A lattice L is called *modular* provided

$$a \wedge (b \vee c) = b \vee (a \wedge c),$$

for all a, b, c in L with $b \leqslant a$. By $\mathcal{M}$ we shall denote the class of all (bounded) modular lattices.

Complements and pseudo-complements

An element $c \in L$ is a *complement in* L if there exists an element $a \in L$ such that $a \wedge c = 0$ and $a \vee c = 1$; we say in this case that c is a *complement* of a *in* L. One denotes by $D(L)$ the set of all complements of L. By a *complement interval* of L we mean any interval $d/0$ of L with $d \in D(L)$. The lattice L is called *indecomposable* if $L \neq \{0\}$ and $D(L) = \{0, 1\}$. The lattice L is said to be *complemented* if every element of L has a complement in L.

For a lattice L and a, b, $c \in L$, the notation $a = b \overset{\cdot}{\vee} c$ will mean that $a = b \vee c$ and $b \wedge c = 0$, and we say then that a is a *direct join* of b and c. Also, for a non-empty subset S of L, we use the *direct join* notation $a = \overset{\cdot}{\bigvee}_{b \in S} b$ if S is an independent subset of L and $a = \bigvee_{b \in S} b$. Recall that a non-empty subset S of L is called *independent* if $0 \notin S$, and for every $x \in S$, $n \in \mathbb{N}$, and subset $T = \{t_1, \ldots, t_n\}$ of S with $x \notin T$, $x \wedge (t_1 \vee \cdots \vee t_n) = 0$. Alternatively, we say that a finite family $(x_i)_{i \in I}$ of elements of a lattice L is *independent* if $x_i \neq 0$ and $x_i \wedge (\bigvee_{j \in I \setminus \{i\}} x_j) = 0$ for every $i \in I$, and in that case, necessarily $x_p \neq x_q$ for each $p \neq q$ in I.

An element $b \in L$ is a *pseudo-complement in* L if there exists an element $a \in L$ such that $a \wedge b = 0$ and b is maximal with this property; we say in this case that b is a *pseudo-complement* of a, and $P(a)$ will denote the set, possibly empty, of all pseudo-complements of a in L. One denotes by $P(L)$ the set of all pseudo-complement elements of L.

As in [**5**], [**6**], L is called *pseudo-complemented* if every element of L has a pseudo-complement, and *strongly pseudo-complemented* if for all $a, b \in L$ with $a \wedge b = 0$, there exists a pseudo-complement p of a in L such that $b \leqslant p$. Every upper continuous modular lattice L is strongly pseudo-complemented.

Complete and upper continuous lattices

A poset in which every subset S has a least upper bound $\bigvee_{s \in S} s$ (also denoted by $\bigvee S$) and a greatest lower bound $\bigwedge_{x \in S} x$ (also denoted by $\bigwedge S$) is called a *complete lattice.*

A lattice L is said to be *upper continuous* if L is complete and

$$a \wedge \left(\bigvee S\right) = \bigvee \{\, a \wedge s \mid s \in S \,\},$$

for every $a \in L$ and every chain (or, equivalently, upward directed subset) $S \subseteq L$. Recall that a subset S of L is said to be *upward directed* if for every $x, y \in S$ there exists $z \in S$ with $x \leqslant z$ and $y \leqslant z$.

Essential elements

An element $e \in L$ is *essential in* L if $e \wedge x \neq 0$ for every $x \neq 0$ in L. One denotes by $E(L)$ the set of all essential elements of L. The lattice L is called *uniform* if $L \neq \{0\}$ and $x \wedge y \neq 0$ for every non-zero elements $x, y \in L$. An element u of L is called *uniform* if the interval $u/0$ of L is a uniform lattice. As in [**5**], [**6**], L is called *E-complemented* ("E" for essential) if for each $a \in L$ there exists $b \in L$ such that $a \wedge b = 0$ and $a \vee b \in E(L)$.

An element $c \in L$ is said to be *closed* if $c \notin E(a/0)$ for all $a \in L$ with $c < a$. One denotes by $C(L)$ the set of all closed elements of L. For an element $a \in L$, we say that $c \in L$ is a *closure* of a in L if $a \in E(c/0)$ and $c \in C(L)$. A lattice L is called *essentially closed* if for each element $a \in L$ there exists a closure of a in L. Every strongly pseudo-complemented lattice (hence every upper continuous) modular lattice is essentially closed by Theorem [**6**, Theorem 1.2.24]. We say that $a \in L$ has a *unique closure* in L if a has exactly one closure in L.

Compact elements

Like in the *Heine-Borel Theorem* from the topology of metric spaces, an element $c \in L$ is called *compact in* L if whenever $c \leqslant \bigvee_{x \in A} x$ for a subset A of L, there is a finite subset F of A such that $c \leqslant \bigvee_{x \in F} x$. The lattice L is said to be *compact* if 1 is a compact element in L, and *compactly generated* if it is complete and every element of L is a join of compact elements.

Simple lattices, atoms, and socle

A lattice L is said to be *simple* if it has exactly two elements, so, L is simple if $L = \{0, 1\}$ and $0 \neq 1$. An element $a \in L$ is said to be an *atom* if $a \neq 0$ and $a/0 = \{0, a\}$, i.e., $a/0$ is a simple lattice. We denote by $A(L)$ the set, possibly empty, of all atoms of L. The *socle* $\mathrm{Soc}(L)$ of a complete lattice L is the join of all atoms of L, i.e., $\mathrm{Soc}\,(L) := \bigvee A(L)$; if L has no atoms, then $\mathrm{Soc}\,(L) = 0$. As in [**32**], a lattice L is called *semi-atomic* if 1 is a join of atoms of L.

2. The conditions $(\mathrm{C_i})$, $\mathrm{i} = 1, 2, 3, 11, 12$, in lattices

In this section we present after [**14**] the definitions and some of the basic properties of the conditions (C_i), $i = 1,\ 2,\ 3,\ 11,\ 12,$ that are the latticial counterparts of the known corresponding conditions on modules (see [**31**], [**23**], [**34**], [**35**], [**36**]).

The conditions $(\mathrm{C_i})$: a brief (incomplete) chronology

1936 *John Von Neumann* [**39**] introduces *regular* rings.
1965 *Yuzo Utumi* [**37**] introduces the conditions (C_i), $i = 1, 2, 3$, for rings.
1971 *Louis Jeremy* [**29**] defines *continuous* and *quasi-continuous* modules.
1977 *A.V. Chatters & Charudatta R. Hajarnavis* [**19**] introduce *CS* rings.
1978 *V.K. Goel & Surender Kumar Jain* [**25**] introduce and investigate *quasi-continuous* modules under the name *π-injective.*
1981 *Manabu Harada & Kiyochi Oshiro* [**28**] use the term of *extending* for modules satisfying the condition (C_1) (i.e., *CS* modules).
1990 *Saad H. Mohamed & Bruno J. Müller* publish their seminal monograph "*Continuous and Discrete Modules*" [**31**].
1993 *Adnan Tercan & Patrick F. Smith* [**34**], [**35**] introduce the conditions (C_{11}) and (C_{12}) for modules.
2016 *Toma Albu & Mihai Iosif & Adnan Tercan* [**14**] introduce and investigate the conditions (C_i), $i = 1,\ 2,\ 3,\ 11,\ 12,$ for lattices, with applications to Grothendieck categories and torsion theories.
2016 *Toma Albu & Mihai Iosif* [**12**] establish new results on C_{11} and C_{12} lattices.
2017 *Toma Albu & Mihai Iosif* [**13**] investigate the behavior of the (C_{11}) condition under *lattice preadicals,* that have been introduced earlier by them in [**11**].
2017 *Toma Albu* [**7**] introduces *fully-invariant extending* modular lattices, the latticial counterpart of fully-invariant extending modules [**18**].

Basic definitions

Recall that for a lattice L we use throughout this paper the following notation:

$P(L) =$ the set of all *pseudo-complement* elements of L (P for "*Pseudo*"),
$E(L) =$ the set of all *essential* elements of L (E for "*Essential*"),
$C(L) =$ the set of all *closed* elements of L (C for "*Closed*"),
$D(L) =$ the set of all *complement* elements of L (D for "*Direct summand*").

Also, recall that for any $a \in L$, we have denoted by $P(a)$ the set, possibly empty, of all pseudo-complements of a in L, so $P(L) = \bigcup_{a \in L} P(a)$.

Definitions 2.1. ([**14**, Definitions 1.1]). For a lattice L one may consider the following conditions:

(C_1) For every $x \in L$ there exists $d \in D(L)$ such that $x \in E(d/0)$.
(C_2) For every $x \in L$ such that $x/0 \simeq d/0$ for some $d \in D(L)$, one has $x \in D(L)$.
(C_3) For every $d_1,\ d_2 \in D(L)$ with $d_1 \wedge d_2 = 0$, one has $d_1 \vee d_2 \in D(L)$.
(C_{11}) For every $x \in L$ there exists a pseudo-complement p of x with $p \in D(L)$, in other words, $D(L) \cap P(x) \neq \varnothing$.
(C_{12}) For every $x \in L$ there exist $d \in D(L)$, $e \in E(d/0)$, and a lattice isomorphism $x/0 \simeq e/0$.

DEFINITIONS 2.2. ([**14**, Definitions 1.2] and [**12**, Definition 0.1]). A lattice L is called *CC* or *extending* if it satisfies (C_1), *continuous* if it satisfies (C_1) and (C_2), and *quasi-continuous* if it satisfies (C_1) and (C_3).

L is said to be a C_{11} (respectively, C_{12}) lattice if it satisfies the condition (C_{11}) (respectively, (C_{12})), and a C_{11}^{+} (respectively, C_{12}^{+}) lattice if $d/0$ is a C_{11} (respectively, C_{12}) lattice for every $d \in D(L)$.

Connections between the conditions $(\mathbf{C_i})$, $\mathbf{i = 1, 2, 3, 11, 12}$

In this subsection we present the connections between the conditions (C_i), $i = 1, 2, 3, 11, 12$, and characterizes essentially closed CC lattices in terms of closeness; in particular, it explains the term of CC, acronym for *C*losed elements are *C*omplements.

PROPOSITION 2.3. ([**14**, Proposition 1.10]). *The following statements hold for a lattice* $L \in \mathcal{M}$.

(1) L *is uniform* $\Longrightarrow$ L *is quasi-continuous* $\Longrightarrow$ L *is CC.*
(2) *If* L *is indecomposable, then* L *is CC* $\Longleftrightarrow$ L *is uniform.*
(3) *If additionally* L *is essentially closed, then*

$$L \text{ is } CC \Longleftrightarrow C(L) \subseteq D(L) \Longleftrightarrow C(L) = D(L).$$

(4) *If additionally* L *is strongly pseudo-complemented, then*

$$L \text{ is } CC \Longleftrightarrow C(L) \subseteq D(L) \Longleftrightarrow C(L) = D(L) \Longleftrightarrow P(L) = D(L).$$

(5) L *satisfies* $(C_2) \Longrightarrow L$ *satisfies* (C_3).
(6) L *satisfies* $(C_1) \Longrightarrow L$ *satisfies* (C_{11}).
(7) L *satisfies* $(C_{11}) \Longrightarrow L$ *satisfies* (C_{12}). □

Characterizations of conditions $(\mathbf{C_{11}})$ and $(\mathbf{C_{12}})$

The next two results characterize modular C_{11} lattices and essentially closed modular C_{12} lattices.

PROPOSITION 2.4. ([**14**, Proposition 1.8]). *A modular lattice* L *is* $C_{11} \Longleftrightarrow \forall\, x \in L,\ \exists\, d \in D(L)$ *with* $d \wedge x = 0$ *and* $d \vee x \in E(L)$. □

PROPOSITION 2.5. ([**14**, Proposition 1.9]). *An essentially closed modular lattice* L *is* $C_{12} \Longleftrightarrow \forall c \in C(L),\ \exists\, d \in D(L),\ e \in E(d/0)$ *such that* $c/0 \simeq e/0$. □

Characterizations of quasi-continuous lattices

Strongly pseudo-complemented (in particular, upper continuous) modular quasi-continuous modular lattices enjoy the following characterizations.

PROPOSITION 2.6. ([**14**, Proposition 1.18]). *The following statements are equivalent for a strongly pseudo-complemented modular lattice* L.

(1) L *is quasi-continuous.*
(2) $1 = p_1 \dot{\vee} p_2$ *for every* $p_1, p_2 \in P(L)$ *which are pseudo-complements of each other.*
(3) *For every* $x_1, x_2 \in L$ *with* $x_1 \wedge x_2 = 0$, *there exist* $d_1, d_2 \in L$ *with* $1 = d_1 \dot{\vee} d_2$ *and* $x_1 \leqslant d_1$, $x_2 \leqslant d_2$. □

Inheritance properties of conditions $(\mathbf{C_i})$, $\mathbf{i = 1, 2, 3, 11, 12}$

In this subsection we discuss the inheritance properties of conditions (C_i), $i = 1, 2, 3, 11, 12$ under complement intervals and direct joins.

PROPOSITION 2.7. ([**14**, Proposition 1.15]). *Let $L \in \mathcal{M}$ be a strongly pseudo-complemented lattice (in particular, an upper continuous lattice), and let $d \in D(L)$. If L satisfies (C_i), $i = 1, 2, 3,$ then $d/0$ also satisfies (C_i), $i = 1, 2, 3,$ in other words, the conditions (C_i), $i = 1, 2, 3,$ are inherited by complement intervals.* □

The condition (C_1) is, in general, not inherited by direct joins, as this is well-known for modules (see, e.g., [**23**]), in contrast with the condition (C_{11}). As we may see below, the condition (C_{11}) is inherited by finite direct joins in any modular lattice L.

PROPOSITION 2.8. ([**14**, Proposition 2.5]). *Let $L \in \mathcal{M}$, and let $(a_i)_{1 \leqslant i \leqslant n}$ be a finite independent family of elements of L such that $1 = \dot{\bigvee}_{1 \leqslant i \leqslant n} a_i$ and $a_i/0$ satisfies (C_{11}) for all $1 \leqslant i \leqslant n$. Then L satisfies (C_{11}).* □

For infinite direct joins, in order to apply Zorn's Lemma, we have to impose in the next result the additional assumption that L is upper continuous.

THEOREM 2.9. ([**14**, Theorem 2.6]). *Let L be an upper continuous modular lattice, and let $(a_i)_{i \in I}$ be an independent family of elements of L such that $1 = \dot{\bigvee}_{i \in I} a_i$ and $a_i/0$ satisfies (C_{11}) for all $i \in I$. Then L satisfies (C_{11}).* □

Using now Proposition 2.3 (6), we deduce at once from Theorem 2.9 the following consequence.

COROLLARY 2.10. *Let L be an upper continuous modular lattice, and let $(a_i)_{i \in I}$ be an independent family of elements of L such that $a_i/0$ satisfies (C_1) for all $i \in I$. Then $(\bigvee_{i \in I} a_i)/0$ satisfies (C_{11}).* □

Observe that any uniform lattice is C_{11}, so, again by Theorem 2.9, we obtain the next result.

COROLLARY 2.11. *Let L be an upper continuous modular lattice, and let $(a_i)_{i \in I}$ be an independent family of uniform elements of L. Then $(\bigvee_{i \in I} a_i)/0$ satisfies (C_{11}).* □

In contrast to conditions (C_i), $i = 1, 2, 3,$ the condition (C_{11}) is not inherited by complement intervals, as this follows in module case from [**35**, Example 4]. The next three results provide some positive results in this trend.

PROPOSITION 2.12. ([**14**, Proposition 2.10]). *Let $L \in \mathcal{M}$ be a lattice which satisfies (C_{11}) and (C_3). Then any complement interval of L satisfies (C_{11}).* □

PROPOSITION 2.13. ([**14**, Proposition 2.11]). *The following statements are equivalent for a lattice $L \in \mathcal{M}$ and a direct join decomposition $1 = m_1 \dot{\vee} m_2$ in L.*

(1) *$m_1/0$ satisfies (C_{11}).*
(2) *$\forall x \in m_1/0$, $\exists d \in D(L)$ with $m_2 < d$, $d \wedge x = 0$, and $d \vee x \in E(L)$.* □

PROPOSITION 2.14. ([**14**, Proposition 2.12]). *Let $L \in \mathcal{M}$ be a lattice satisfying (C_{11}) and having a direct join decomposition $1 = m_1 \dot{\vee} m_2$. Suppose that $k \vee m_2 \in D(L)$ for every $k \in D(L)$ with $k \wedge m_2 = 0$. Then $m_1/0$ satisfies (C_{11}).* □

For the definition of the Goldie dimension of a modular lattice that intervene in the next results, see, e.g., [**6**, Section 2.2] or [**32**, Section 1.5].

PROPOSITION 2.15. ([**12**, Proposition 1.8]). *The following assertions are equivalent for a lattice $L \in \mathcal{M}$ with finite Goldie dimension.*

(1) *Any $d \in D(L)$ is a join of an independent (finite) set of uniform elements of L.*
(2) *L is a C_{11}^+ lattice.*
(3) *L is a C_{12}^+ lattice.* □

PROPOSITION 2.16. ([**12**, Proposition 1.10]). *The following assertions are equivalent for an indecomposable modular lattice L.*

(1) *L is uniform.*
(2) *L is C_{11}.*
(3) *L is C_{12} and L has at least a uniform element.* □

THEOREM 2.17. ([**12**, Theorem 1.14]). *Let L be a compactly generated modular C_{11}^+ lattice such that the lattice $1/\operatorname{Soc}(L)$ has finite Goldie dimension. Then there exist $d_1, d_2 \in L$ with $d_1 \dot{\vee} d_2 = 1$ such that the lattice $d_1/0$ is semi-atomic and the lattice $d_2/0$ has finite Goldie dimension. In particular, 1 is a join of an independent family of uniform elements of L.* □

The next result is the latticial counterpart of a known result saying that a module M_R satisfies the ACC on essential submodules if and only if the module $M/\operatorname{Soc}(M)$ is Noetherian.

Recall that a poset $(P, \leqslant)$ is called *Noetherian* (respectively, *Artinian*) if it satisfies the ACC (respectively, DCC), i.e., if there is no strictly ascending (respectively, descending) chain

$$x_1 < x_2 < \ldots \text{ (respectively, } x_1 > x_2 > \ldots)$$

in P.

The switch from $\operatorname{Soc}(L)$ to $E(L)$ in the next result is assured by the equality $\operatorname{Soc}(L) = \bigwedge E(L)$ holding in any compactly generated modular lattice L (see [**38**, Chapter III, Proposition 6.7]).

PROPOSITION 2.18. ([**12**, Proposition 1.15]). *The following statements are equivalent for a compactly generated modular lattice L.*

(1) *$1/\operatorname{Soc}(L)$ is a Noetherian lattice.*
(2) *$E(L)$ is a Noetherian poset.*
(3) *$1/e$ is a Noetherian lattice for every $e \in E(L)$.* □

COROLLARY 2.19. ([**12**, Corollary 1.16]). *Let L be a compactly generated modular C_{11}^+ lattice. If $E(L)$ is a Noetherian poset, then there exist $d_1, d_2 \in L$ with $d_1 \dot{\vee} d_2 = 1$ such that the lattice $d_1/0$ is semi-atomic and the lattice $d_2/0$ is Noetherian.* □

It is not clear whether the results in Proposition 2.18 and Corollary 2.19 are true or not when we replace "Noetherian" by "Artinian".

The Latticial Osofsky-Smith Theorem

We end this section by stating a latticial counterpart involving CC lattices of the following renown result of Module Theory.

THE OSOFSKY-SMITH THEOREM [**33**]. *A finitely generated (respectively, cyclic) right R-module such that all of its finitely generated (respectively, cyclic) subfactors are CS modules is a finite direct sum of uniform submodules.* □

Recall that a module M is said to be *CS* (or *extending*) if every submodule of M is essential in a direct summand of M, or, equivalently, if any complement submodule of M is a direct summand of M. The name CS is an acronym for *C*omplements submodules are direct *S*ummands. Recall that in Module Theory one says that a submodule N of M is a *complement* if there exists a submodule L of M such that $N \cap L = 0$ and N is maximal in the set of all submodules P of M such that $P \cap L = 0$, i.e., the element N of the lattice $\mathcal{L}(M)$ of all submodules of M is a pseudo-complement element in this lattice. Consequently, a module M is CS if and only if the lattice $\mathcal{L}(M)$ is CC.

The following latticial version of the Osofsky-Smith Theorem was established in [**3**], and applications of it to Grothendieck categories and module categories equipped with a hereditary torsion theory were given in [**4**].

THEOREM 2.20. (THE LATTICIAL OSOFSKY-SMITH THEOREM [**3**]). *Let L be a compact, compactly generated, modular lattice. Assume that all compact subfactors of L are CC. Then 1 is a finite direct join of uniform elements of L.* □

A right R-module M is said to be *CF* if every closed submodule of M is finitely generated. More generally, we say that a lattice L is *CK* (acronym for *C*losed are *K*ompact) if every closed element of L is compact, i.e., $C(L) \subseteq K(L)$, where $K(L)$ denotes the set of all compact elements of L (K for "*Kompakt*").

An immediate consequence of Theorem 2.20 is the next result.

PROPOSITION 2.21. ([**6**, Proposition 5.2.5]). *Let L be a compact, compactly generated, modular lattice. Assume that all compact subfactors of L are CK. Then $D(L)$ is a Noetherian poset, in particular, 1 is a finite direct join of indecomposable elements of L.* □

3. Linear modular lattices

In this section we present and list some of its basic properties of the concept of a *linear morphism* of lattices, introduced in [**8**]. It captures the property of a linear mapping $\varphi : M \longrightarrow N$ between modules M_R and N_R to have a kernel $\operatorname{Ker}(\varphi)$ and to verify the Fundamental Theorem of Isomorphism $M/\operatorname{Ker}(\varphi) \simeq \operatorname{Im}(\varphi)$. The linear morphisms of lattices are closely related to *quasi-frame morphisms* introduced in [**40**], where a *quasi-frame* is nothing else than an upper continuous modular lattice, and a *quasi-frame morphism* is by definition any mapping between two quasi-frames that commutes with arbitrary joins and preserves intervals.

Linear morphisms of lattices are the basic tools in defining the concepts of *lattice preradicals* that will be discussed in Section 4, *fully invariant-extending lattices* (see [**7**]), *injective lattices* (see [**8**]), etc.

DEFINITION 3.1. ([**8**, Definitions 1.1]). A mapping $f : L \longrightarrow L'$ between the lattices L and L' is called a *linear morphism* if there exist $k \in L$, called a *kernel* of f, and $a' \in L'$ such that the following two conditions are satisfied.

(1) $f(x) = f(x \vee k)$, $\forall\, x \in L$.

(2) f induces a lattice isomorphism $\bar{f} : 1/k \xrightarrow{\sim} a'/0'$, $\bar{f}(x) = f(x)$, $\forall\, x \in 1/k$, where $0'$ is the least element of L'.

For any morphism of right R-modules $\varphi : M_R \longrightarrow M'_R$, the mapping

$$f : \mathcal{L}(M_R) \longrightarrow \mathcal{L}(M'_R),\ f(N) := \varphi(N),\ \forall\, N \in \mathcal{L}(M_R),$$

is clearly a linear morphism with kernel $\mathrm{Ker}\,(\varphi)$ of their submodule lattices.

The main properties of linear morphisms of lattices are collected in the next result.

PROPOSITION 3.2. ([**8**, Proposition 1.3]). *The following assertions are true for a linear morphism $f : L \longrightarrow L'$ with a kernel k.*

(1) *For $x, y \in L$, $f(x) = f(y) \iff x \vee k = y \vee k$.*
(2) *$f(k) = 0'$ and k is the greatest element of L having this property, so, the kernel of a linear morphism is uniquely determined.*
(3) *f commutes with arbitrary joins, i.e., $f\left(\bigvee_{i\in I} x_i\right) = \bigvee_{i\in I} f(x_i)$ for any family $(x_i)_{i\in I}$ of elements of L, provided both joins exist.*
(4) *f is an increasing mapping.*
(5) *f preserves intervals, i.e., $f(v/u) = f(v)/f(u)$ for any $u \leqslant v$ in L.*
(6) *If $L \in \mathcal{M}$, then for any $a \in L$, the restriction*
$$f_a : a/0 \longrightarrow L',\ f_a(x) = f(x),\ \forall\, x \in a/0,$$
of f to $a/0$ is a linear morphism with kernel $a \wedge k$. □

The class $\mathcal{M}$ of all (bounded) modular lattices can be considered in a natural way a category with morphisms the usual lattice morphisms, i.e., mappings commuting with finite meets and joins.

With the aid of linear morphisms of lattices, the same class $\mathcal{M}$ can be organized as a new category, denoted by $\mathcal{LM}$ ($\mathcal{L}$ for "*Linear*"). In contrast with the category $\mathcal{M}$, the subobjects in $\mathcal{LM}$ of a modular lattice L are its initial intervals, and this is essential in our subsequent results.

PROPOSITION 3.3. ([**8**, Proposition 2.2]). *The following statements hold.*

(1) *The class $\mathcal{M}$ of all (bounded) modular lattices becomes a category, denoted by $\mathcal{LM}$, if for any $L, L' \in \mathcal{M}$ one takes as morphisms from L to L' all the linear morphisms from L to L'.*
(2) *The isomorphisms in the category $\mathcal{LM}$ are exactly the isomorphisms in the category $\mathcal{M}$.*
(3) *The monomorphisms in the category $\mathcal{LM}$ are exactly the injective linear morphisms.*
(4) *The epimorphisms in the category $\mathcal{LM}$ are exactly the surjective linear morphisms.*
(5) *The subobjects of $L \in \mathcal{LM}$ can be viewed as the intervals $a/0$, $a \in L$.* □

DEFINITIONS 3.4. ([**16**, Definitions 2.1]). Let $\varnothing \neq \mathcal{C} \subseteq \mathcal{L}$. We say that:

(1) $\mathcal{C}$ is an *abstract class* if it is closed under lattice isomorphisms, i.e., if $L, K \in \mathcal{L}$, $K \simeq L$, and $L \in \mathcal{C}$, then $K \in \mathcal{C}$.
(2) $\mathcal{C}$ is *hereditary* if it is an abstract class and for any $L \in \mathcal{L}$ and any $a \leqslant b \leqslant c$ in L such that $c/a \in \mathcal{C}$, it follows that $b/a \in \mathcal{C}$.

For any non-empty subclass $\mathcal{C}$ of $\mathcal{M}$ we shall denote by $\mathcal{LC}$ the full subcategory of $\mathcal{LM}$ having $\mathcal{C}$ as the class of its objects.

PROPOSITION 3.5. ([**11**, Proposition 2.3]). *The following assertions are equivalent for an abstract subclass* $\mathcal{C}$ *of* $\mathcal{M}$.

(1) $\mathcal{C}$ *is hereditary.*

(2) *For any* $L \in \mathcal{C}$, *the subobjects of* L *in the category* $\mathcal{LC}$ *can be viewed as the initial intervals* $a/0$ *of* $L = 1/0$, $a \in L$.

In this case, the monomorphisms in the category $\mathcal{LC}$ *are precisely the injective linear morphisms.* □

4. Lattice preradicals

The aim of this section is to present after [**11**] the latticial counterpart of the module-theoretical concept of preradical.

By a *module preradical* we understand a preradical on the category Mod-R of all unital right R-modules over an associative ring R with identity element, i.e., a subfunctor p of the identity functor $1_{\text{Mod-}R}$ of Mod-R; this means that p assigns to each right R-module M a submodule $p(M)$ of M such that each morphism $f : M \longrightarrow N$ in Mod-R induces by restriction a morphism $p(f) : p(M) \longrightarrow p(N)$, i.e., $f(p(M)) \subseteq p(N)$. The most known examples of module preradicals are the *Jacobson radical* "Jac", the *socle* "Soc", the *radical* "Rad", and the *torsion radical* t_τ associated with any hereditary torsion theory τ on Mod-R.

The main ingredient in defining the concept of a *lattice preradical* is that of a *linear morphism* of lattices, introduced and investigated in [**8**], and discussed in the previous section.

DEFINITION 4.1. ([**11**, Definitions 2.1]). Let $\mathcal{C}$ be a hereditary subclass of $\mathcal{M}$. A *lattice preradical* on $\mathcal{C}$ is any functor $r : \mathcal{LC} \longrightarrow \mathcal{LC}$ satisfying the following two conditions.

(1) $r(L) \leqslant L$, i.e., $r(L)$ is a subobject of L, for any $L \in \mathcal{LC}$.

(2) For any morphism $f : L \longrightarrow L'$ in $\mathcal{LC}$, $r(f) : r(L) \longrightarrow r(L')$ is the restriction of f, i.e., $f(r(L)) \subseteq r(L')$.

In other words, a lattice preradical is nothing else than a subfunctor of the identity functor $1_{\mathcal{LC}}$ of the category $\mathcal{LC}$.

Examples of lattice preradicals, including the counterparts of the well-known module preradicals "Jac", "Soc", and "Rad" may be found in [**9**] and [**11**]. The connections between lattice preradicals and module preradicals has been investigated in [**10**], where it is shown that to any lattice preradical one may associate in a canonical way a module preradical, but not conversely.

Let $\mathcal{C}$ be a hereditary subclass of $\mathcal{M}$, and let $r : \mathcal{LC} \longrightarrow \mathcal{LC}$ be a lattice preradical on $\mathcal{C}$. By Proposition 3.5, for every $L \in \mathcal{C}$ and $a \in L$, the subobject $r(a/0)$ of L in $\mathcal{LC}$ is necessarily an initial interval of $a/0$. We denote

$$r(a/0) := a^r/0.$$

If $a \leqslant b$ in L, then $a/0$, $b/0$ are in $\mathcal{C}$ because $\mathcal{C}$ is hereditary. The inclusion mapping $\iota : a/0 \hookrightarrow b/0$ is clearly a linear morphism, thus it is a morphism in $\mathcal{LC}$. Applying now r we obtain $r(\iota) : a^r/0 \longrightarrow b^r/0$ as a restriction of ι, and so

$$a \leqslant b \Longrightarrow a^r \leqslant b^r.$$

For any lattice $L \in \mathcal{C}$ and any a, b in L such that $a \dot{\vee} b = 1$, the mapping

$$q : L \longrightarrow a/0, \; q(x) := (x \vee b) \wedge a,$$

is a surjective linear morphism with kernel b (see [**11**, Example 0.2(3)]), so q is a morphism in $\mathcal{LC}$. This is the latticial counterpart of the canonical projection $M \oplus M' \longrightarrow M$ for two modules M_R and M'_R.

PROPOSITION 4.2. ([**11**, Proposition 1.3]). *For any modular lattice L and any finite independent family $(a_i)_{1 \leqslant i \leqslant n}$ of L, with n a positive integer, one has*

$$\Big(\dot{\bigvee_{1\leqslant i\leqslant n}} a_i\Big)^r = \dot{\bigvee_{1\leqslant i\leqslant n}} a_i^r.$$

□

The property of a lattice preradical to commute with arbitrary direct joins, as this happens in the module case, needs the additional assumption that the considered lattice be upper continuous.

PROPOSITION 4.3. ([**11**, Proposition 1.4]). *For any upper continuous modular lattice L and any independent family $(a_i)_{i \in I}$ of elements of L one has*

$$\Big(\dot{\bigvee_{i\in I}} a_i\Big)^r = \dot{\bigvee_{i\in I}} a_i^r.$$

□

DEFINITION 4.4. ([**11**, Definition 1.2]). Let $r : \mathcal{LM} \longrightarrow \mathcal{LM}$ be a lattice preradical. We say that r is *hereditary* or *left exact* if for all $L \in \mathcal{LM}$, $a^r = a \wedge 1^r$ for every $a \in L$.

For instance, by [**11**, Example 3.6], the assignment $L \mapsto \mathrm{Soc}\,(L)/0$ defines a hereditary preradical on the full subcategory $\mathcal{LM}_u$ of $\mathcal{LM}$ consisting of all upper continuous modular lattices.

Condition $(\mathbf{C_{11}})$ under lattice preradicals

The next result provides the behavior of the condition (C_{11}) under hereditary preradicals on hereditary subclasses of $\mathcal{M}$.

THEOREM 4.5. ([**13**, Theorem 2.4]). *Let $\mathcal{C}$ be a hereditary subclass of $\mathcal{M}$, and let r be a hereditary preradical on $\mathcal{C}$. Then, the following assertions are equivalent for an essentially closed lattice $L \in \mathcal{C}$ such that 1^r has a unique closure in L.*

(1) *L is a C_{11} lattice.*

(2) *There exist $a_1, a_2 \in L$ such that $1 = a_1 \dot{\vee} a_2$, $a_1^r \in E(a_1/0)$, $a_2^r = 0$, and the lattices $a_1/0$ and $a_2/0$ are both C_{11}.* □

5. Applications to Grothendieck categories and torsion theories

The aim of this section is to provide applications of the latticial results of the previous sections to Grothendieck categories and module categories equipped with hereditary torsion theories.

A latticial strategy

We first present a ***general strategy*** which consists on putting a *module-theoretical* definition/result into a *latticial frame* (we call it *latticization*), in order to translate that definition/result to *Grothendieck categories* (we call it *absolutization*) and *module categories* equipped with *hereditary torsion theories* (we call it *relativization*).

More precisely, if $\mathbb{P}$ is a problem, involving subobjects or submodules, to be investigated in Grothendieck categories or in module categories with respect to hereditary torsion theories, our strategy consists of the following three steps:

I. *Translate/formulate,* if possible, the problem $\mathbb{P}$ into a *latticial setting.*
II. *Investigate* the obtained problem $\mathbb{P}$ in this *latticial frame.*
III. *Back* to Grothendieck categories and module categories equipped with hereditary torsion theories.

This approach is very natural because we ignore the specific context of Grothendieck categories and module categories equipped with hereditary torsion theories, focusing only on those latticial properties which are relevant to our given specific categorical or relative module-theoretical problem $\mathbb{P}$. The renowned *Hopkins-Levitzki Theorem* and *Osofsky-Smith Theorem* from Ring and Module Theory are among the most relevant illustrations of the power of this strategy (see [**2**] and [**6**]).

Grothendieck categories

In this section $\mathcal{G}$ will denote a *Grothendieck category,* i.e., an Abelian category with exact direct limits (or satisfying the axiom AB5 of Grothendieck) and with a generator. Recall that an object $G \in \mathcal{G}$ is said to be generator of $\mathcal{G}$ if every object of $\mathcal{G}$ is an epimorphic image of a direct sum sum of copies of G. For any object X of $\mathcal{G}$, $\mathcal{L}(X)$ will denote the lattice of all subobjects of X. It is well-known that $\mathcal{L}(X)$ is an upper continuous modular lattice (see, e.g., [**38**, Chapter 4, Proposition 5.3, and Chapter 5, Section 1].

Recall that an object X of $\mathcal{G}$ is said to be *Noetherian* (respectively, *Artinian*) if the lattice $\mathcal{L}(X)$ is Noetherian (respectively, Artinian). More generally, if $\mathbb{P}$ is any property on lattices, we say that an object $X \in \mathcal{G}$ is/has $\mathbb{P}$ if the lattice $\mathcal{L}(X)$ is/has $\mathbb{P}$. Similarly, a subobject Y of an object $X \in \mathcal{G}$ is/has $\mathbb{P}$ if the element Y of the lattice $\mathcal{L}(X)$ is/has $\mathbb{P}$. Thus, we obtain the concepts of an *uniform* object, *compact* object, (C_i), $i = 1, 2, 3, 11, 12$, condition for an object, *CC* object, C_{11} object, C_{12} object, *quasi-continuous* object, *continuous* object, *pseudo-complement* subobject of an object, *essential* subobject of an object, *closed* subobject of an object, *complement* subobject of an object, etc. For a complement (respectively, compact) subobject of an object $X \in \mathcal{G}$ one uses the well-established term of a *direct summand* (respectively, *finitely generated* subobject) of X, and for this reason, instead of saying that X is a CC object we shall say that X is a CS object (acronym for *C*losed subobjects are direct *S*ummands).

For all undefined notation and terminology on Abelian categories and Grothendieck categories the reader is referred to [**1**], [**15**], and/or [**38**].

Categorical results

Of course, all the notions and results of the previous sections have categorical versions obtained by specializing them from an arbitrary modular lattice L to the upper continuous modular lattice $\mathcal{L}(X)$ of any object X of a Grothendieck category $\mathcal{G}$. No further proofs are required. We shall present below some of them.

The next result is an immediate application of Theorem 2.20 to the upper continuous modular lattice $L = \mathcal{L}(X)$, $X \in \mathcal{G}$.

THEOREM 5.1. (CATEGORICAL O-ST [**4**]). *Let $\mathcal{G}$ be a Grothendieck category, and let $X \in \mathcal{G}$ be a finitely generated, locally finitely generated object such that every finitely generated subfactor object of X is CS. Then X is a finite direct sum of uniform objects.* □

Recall that the category $\mathcal{G}$ is called *locally finitely generated* if it has a family of finitely generated generators, or equivalently, if the lattices $\mathcal{L}(X)$ are compactly generated for all objects X of $\mathcal{G}$. We say that an object $X \in \mathcal{G}$ is *locally finitely generated* if the lattice $\mathcal{L}(X)$ is compactly generated.

PROPOSITION 5.2. ([**14**, Proposition 3.1]). *An object X of a Grothendieck category $\mathcal{G}$ is quasi-continuous if and only if for any subobjects X_1 and X_2 of X with $X_1 \cap X_2 = 0$, there exist subobjects D_1 and D_2 of X with $X = D_1 \oplus D_2$ and $X_1 \subseteq D_1$, $X_2 \subseteq D_2$.* □

THEOREM 5.3. ([**14**, Theorem 3.2]). *Any direct sum of C_{11} objects of a Grothendieck category is also a C_{11} object.* □

PROPOSITION 5.4. ([**12**, Proposition 2.1]). *Let $\mathcal{G}$ be a Grothendieck category, and let $X \in \mathcal{G}$ be such that $X = U \oplus V$ for two uniform objects U, V of $\mathcal{G}$. Then any closed subobject of X is C_{11}. In particular, X is a C_{11}^{+} object.* □

PROPOSITION 5.5. ([**12**, Proposition 2.2]). *The following assertions are equivalent for an object X with finite Goldie dimension of a Grothendieck category $\mathcal{G}$.*

(1) *Any direct summand of X is a (finite) direct sum of uniform objects of $\mathcal{G}$.*
(2) *X is C_{11}^{+}.*
(3) *X is C_{12}^{+}.* □

The next result is difficult to be proved directly without passing through its latticial counterpart, where the condition that the considered lattice is compactly generated is essential.

THEOREM 5.6. ([**12**, Theorem 2.3]). *Let X be a locally finitely generated object of a Grothendieck category $\mathcal{G}$. Assume that X is a C_{11}^{+} object and $X/\operatorname{Soc}(X)$ has finite Goldie dimension. Then, there exist $X_1, X_2 \in \mathcal{G}$ such that $X = X_1 \oplus X_2$, X_1 is semisimple object, and X_2 has finite Goldie dimension. In particular, X is a direct sum of uniform objects of $\mathcal{G}$.* □

COROLLARY 5.7. ([**12**, Corollary 2.4]). *Let X be a locally finitely generated C_{11}^{+} object of a Grothendieck category $\mathcal{G}$ satisfying the ascending chain condition on essential subobjects. Then, there exist $X_1, X_2 \in \mathcal{G}$ such that $X = X_1 \oplus X_2$, X_1 is semisimple, and X_2 is Noetherian.* □

Torsion theories

Throughout the remainder of this paper R will be an associative ring with unit element $1 \neq 0$ and Mod-R will denote the category of all unital right R-modules. We often write M_R to emphasize that M is a right R-module; $\mathcal{L}(M_R)$, or just $\mathcal{L}(M)$, stands for the lattice of all submodules of M. The notation $N \leqslant M$ means that N is a submodule of M. Basic references on modules may be found by the reader in [**17**], [**30**], and or [**41**].

The concept of a *torsion theory* for Abelian categories has been introduced by S.E. Dickson [**22**] in 1966. For our purposes, we present it only for module categories in one of the many equivalent ways that can be done.

A *hereditary torsion theory* on Mod-R is a pair $\tau = (\mathcal{T}, \mathcal{F})$ of nonempty subclasses $\mathcal{T}$ and $\mathcal{F}$ of Mod-R such that $\mathcal{T}$ is a *localizing subcategory* of Mod-R in the Gabriel's sense [**24**] (this means that $\mathcal{T}$ is a Serre class of Mod-R which is closed under direct sums) and $\mathcal{F} = \{ F_R \,|\, \mathrm{Hom}_R(T, F) = 0, \forall T \in \mathcal{T} \}$. So, any hereditary torsion theory $\tau = (\mathcal{T}, \mathcal{F})$ is uniquely determined by its first component $\mathcal{T}$.

Recall that a nonempty subclass $\mathcal{T}$ of Mod-R is a *Serre class* if for any short exact sequence

$$0 \longrightarrow X' \longrightarrow X \longrightarrow X'' \longrightarrow 0$$

in Mod-R, one has

$$X \in \mathcal{T} \Longleftrightarrow X' \in \mathcal{T} \;\;\&\;\; X'' \in \mathcal{T}.$$

One says that $\mathcal{T}$ is *closed under direct sums* if for any family $(X_i)_{i \in I}$, I arbitrary set, with $X_i \in \mathcal{T}$, $\forall\, i \in I$, it follows that $\bigoplus_{i \in I} X_i \in \mathcal{T}$.

The prototype of a hereditary torsion theory is the pair $(\mathcal{A}, \mathcal{B})$ in Mod-$\mathbb{Z}$, where $\mathcal{A}$ is the class of all torsion Abelian groups, and $\mathcal{B}$ is the class of all torsion-free Abelian groups.

Throughout the remainder of this paper $\tau = (\mathcal{T}, \mathcal{F})$ will be a fixed hereditary torsion theory on Mod-R. For any module M_R we denote

$$\tau(M) := \sum_{N \leqslant M,\, N \in \mathcal{T}} N.$$

Since $\mathcal{T}$ is a localizing subcategory of Mod-R, it follows that $\tau(M) \in \mathcal{T}$, and we call it the *τ-torsion submodule* of M. The assignment $M \mapsto \tau(M)$ defines a module preradical, in fact a radical, usually denoted by t_τ.

Note that, as for Abelian groups, we have

$$M \in \mathcal{T} \Longleftrightarrow \tau(M) = M \quad \text{and} \quad M \in \mathcal{F} \Longleftrightarrow \tau(M) = 0.$$

The members of $\mathcal{T}$ are called *τ-torsion modules* and the members of $\mathcal{F}$ are called *τ-torsion-free modules*.

For any $N \leqslant M$ we denote by $\overline{N}$ the submodule of M such that $\overline{N}/N = \tau(M/N)$, called the *$\tau$-saturation* of N (in M). One says that N is *τ-saturated* if $\overline{N} = N$, or equivalently, if $M/N \in \mathcal{F}$, and the set of all τ-saturated submodules of M is denoted by $\mathrm{Sat}_{\,\tau}(M)$, i.e.,

$$\mathrm{Sat}_\tau(M) \;:=\; \{\, N \mid N \leqslant M \text{ and } M/N \in \mathcal{F} \,\}.$$

It is well-known that $\mathrm{Sat}_\tau(M)$ is an upper continuous modular lattice (see [**38**, Chapter 9, Proposition 4.1]).

For all undefined notation and terminology on torsion theories the reader is referred to [**15**], [**26**], and/or [**38**].

Relative concepts

A module M_R is said to be τ-CC if the lattice $\mathrm{Sat}_\tau(M)$ is CC. More generally if $\mathbb{P}$ is any property on lattices, we say that a module M_R is/has τ-$\mathbb{P}$ if the lattice $\mathrm{Sat}_\tau(M)$ is/has $\mathbb{P}$.

Thus, we obtain the concepts of a τ-*Artinian* module, τ-*Noetherian* module, τ-*uniform* module, τ-*compact* module, τ-*compactly generated* module, condition τ-(C_i), τ-C_{11} module, τ-C_{12} module, τ-*quasi-continuous* module, τ-*continuous* module, etc. We say that a submodule N of M_R is/has τ-$\mathbb{P}$ if its τ-saturation $\overline{N}$, which is an element of $\mathrm{Sat}_\tau(M)$, is/has $\mathbb{P}$. Thus, we obtain the concepts of a τ-*pseudo-complement* submodule of a module, τ-*complement* submodule of a module, τ-*essential* submodule of a module, τ-*closed* submodule of a module, τ-*independent* set/family of submodules of a module, etc.

In the sequel we shall use the well-established term of a τ-*direct summand* of a module instead of that of a τ-*complement* submodule of a module and of a τ-*CS* module instead of that of a τ-*CC* module.

Relative results

In this subsection, we present relative versions with respect to a hereditary torsion theory on Mod-R of some module-theoretical results related to conditions (C_i). Their proofs are immediate applications of the lattice-theoretical results obtained in the previous sections.

All the notions and results from the previous sections for an arbitrary modular lattice L can now be easily specialized for the particular case when $L = \mathrm{Sat}_\tau(M_R)$. by using the description from [**4**, Proposition 5.3] of the relative concepts that intervene in their statements. We present below some of them.

PROPOSITION 5.8. ([**14**, Proposition 4.4]). *If M_R is a module satisfying the condition τ-(C_i), $i = 1, 2, 3$, then any τ-direct summand of M also satisfies the condition τ-(C_i). In particular, any τ-direct summand of a τ-CS module is also τ-CS.* □

THEOREM 5.9. (RELATIVE O-ST [**4**]) *Let M_R be a τ-compact, τ-compactly generated module. Assume that all τ-compact subfactors of M are τ-CS. Then there exists a finite τ-independent family $(U_i)_{1\leqslant i\leqslant n}$ of τ-uniform submodules of M such that $M/(\sum_{1\leqslant i\leqslant n} U_i) \in \mathcal{T}$.* □

PROPOSITION 5.10. ([**14**, Proposition 4.2]). *A module M_R satisfies the condition τ-(C_{11}) if and only if for every $N \leqslant M$ there exist $D \leqslant M$ and $K \leqslant M$ such that $M/(D+K) \in \mathcal{T}$, $D \cap K \in \mathcal{T}$, $N \cap D \in \mathcal{T}$, and $(N+D) \cap X \notin \mathcal{T}$ for every $X \leqslant M$ with $X \notin \mathcal{T}$.* □

THEOREM 5.11. ([**14**, Theorem 4.3]). *Let $(N_i)_{i\in I}$ be a τ-independent family of submodules of a module M_R such that all N_i are τ-C_{11}. Then $\sum_{i\in I} N_i$ is also τ-C_{11}.* □

THEOREM 5.12. ([**12**, Theorem 3.2]). *Let M_R be a τ-compactly generated module. Suppose that any τ-direct summand of M is τ-C_{11} and that $M/\operatorname{Soc}_\tau(M)$ has finite τ-Goldie dimension. Then there exist τ-independent submodules N_1, N_2 of M such that $M/(N_1+N_2) \in \mathcal{T}$, N_1 is τ-semisimple, and N_2 has finite τ-Goldie dimension.* □

COROLLARY 5.13. ([**12**, Corollary 3.5]). *Let M_R be a τ-compactly generated module. Suppose that any τ-direct summand of M is τ-C_{11} and for any ascending chain $E_1 \subseteq E_2 \subseteq \ldots$ of τ-essential submodules of M, there exists an $n \in \mathbb{N}$ such that $E_{m+1}/E_m \in \mathcal{T}$ for all $m \geqslant n$. Then, there exist τ-independent submodules N_1, N_2 of M such that $M/(N_1 + N_2) \in \mathcal{T}$, N_1 is τ-semisimple, and N_2 is τ-Noetherian.* □

According to our definition above of a τ-$\mathbb{P}$ submodule of a module, we say that a non-empty class $\mathcal{H}$ of right R-modules is *τ-hereditary* if for any $M \in \mathcal{H}$ and $N \in \mathrm{Sat}_\tau(M)$ one has $N \in \mathcal{H}$, or equivalently $\overline{N} \in \mathcal{H}$ for any $N \leqslant M$ and any $M \in \mathcal{H}$.

By a *preradical* on a τ-hereditary subclass $\mathcal{H}$ of Mod-R we mean a subfunctor of the identity functor $1_{\mathcal{H}}$ of the full subcategory $\mathcal{H}$ of Mod-R in Mod-R.

THEOREM 5.14. ([**13**, Theorem 5.3]). *Let $\mathcal{H}$ be a τ-hereditary class of right R-modules and let r be a left exact preradical on* Mod-R. *Then, the following assertions are equivalent for a module $M_R \in \mathcal{H} \cap \mathcal{F}$ such that $r(M)$ has a unique τ-closure in M.*

(1) *M is a τ-C_{11} module.*
(2) *There exist submodules N and P of M such that $M/(N + P) \in \mathcal{T}$, $N \cap P \in \mathcal{T}$, $r(N)$ is τ-essential in N, $r(P) \in \mathcal{T}$, and both N and P are τ-C_{11} modules.* □

COROLLARY 5.15. ([**13**, Corollary 5.4]). *The following statements are equivalent for a module $M_R \in \mathcal{F}$ such that $\mathrm{Soc}_\tau(M)$ has a unique τ-closure in M.*

(1) *M is a τ-C_{11} module.*
(2) *There exist submodules N and P of X such that $M/(N + P) \in \mathcal{T}$, $N \cap P \in \mathcal{T}$, $\mathrm{Soc}_\tau(N)$ is τ-essential in N, $\mathrm{Soc}_\tau(P) \in \mathcal{T}$, and N and P are both τ-C_{11} modules.* □

COROLLARY 5.16. ([**13**, Corollary 5.5]). *Let $\mathcal{H}$ be a τ-hereditary class of right R-modules and let r be a left exact preradical on* Mod-R. *Then, the following assertions are equivalent for a module $M_R \in \mathcal{H} \cap \mathcal{F}$ such that $r(M)$ is a τ-closed submodule of M.*

(1) *M is a τ-C_{11} module.*
(2) *There exists a submodule N of M such that such that $M/(r(M)+N) \in \mathcal{T}$, $r(M) \cap N \in \mathcal{T}$, and $r(M)$ and N are both τ-C_{11} modules.* □

References

[1] T. Albu, *"Lessons on the Grothendieck Category $\sigma[M]$"*, Editura Universităţii din Bucureşti, Bucharest, 2004, 122 pages. MR2882883

[2] T. Albu, *A seventy years jubilee: The Hopkins-Levitzki Theorem*, in "Ring and Module Theory", Trends Math., Edited by T. Albu, G. F. Birkenmeier, A. Erdoğan, and A. Tercan, Birkhäuser/Springer Basel AG, Basel, 2010, pp. 1–26. MR2744040

[3] T. Albu, *The Osofsky-Smith Theorem for modular lattices, and applications (I)*, Comm. Algebra **39** (2011), no. 12, 4488–4506. MR2863447

[4] T. Albu, *The Osofsky-Smith Theorem for modular lattices, and applications (II)*, Comm. Algebra **42** (2014), no. 6, 2663–2683. MR3169733

[5] T. Albu, *"Topics in Lattice Theory with Applications to Rings, Modules, and Categories"*, Lecture Notes, XXIII Brazilian Algebra Meeting, Maringá, Paraná, Brasil, 2014, 80 pages.

[6] T. Albu, *"Chain Conditions in Modular Lattices with Applications to Grothendieck Categories and Torsion Theories"*, Monograph Series of the Parana's Mathematical Society, No. 1, Sociedade Paranaense de Matemática, Maringá, 2015, 134 pages. MR3362808

[7] T. Albu, *Fully invariant-extending modular lattices, and applications*, in preparation.
[8] T. Albu and M. Iosif, *The category of linear modular lattices*, Bull. Math. Soc. Sci. Math. Roumanie (N.S.) **56 (104)** (2013), no. 1, 33–46. MR3076689
[9] T. Albu and M. Iosif, *On socle and radical of modular lattices*, Ann. Univ. Buchar. Math. Ser. **5 (LXIII)** (2014), no. 2, 187–194. MR3296458
[10] T. Albu and M. Iosif, *Lattice preradicals versus module preradicals*, Ann. Univ. Buchar. Math. Ser. **6 (LXIV)** (2015), no. 1, 19–34. MR3376327
[11] T. Albu and M. Iosif, *Lattice preradicals with applications to Grothendieck categories and torsion theories*, J. Algebra **444** (2015), 339–366. MR3406182
[12] T. Albu and M. Iosif, *New results on C_{11} and C_{12} lattices with applications to Grothendieck categories and torsion theories*, Front. Math. China **11** (2016), no. 4, 815–828. MR3531032
[13] T. Albu and M. Iosif, *Modular C_{11} lattices and lattice preradicals*, J. Algebra Appl. **16** (2017), no. 6, 1750116, 19 pages. MR3635146
[14] T. Albu, M. Iosif, and A. Tercan, *The conditions (C_i) in modular lattices, and applications*, J. Algebra Appl. **15** (2016), no. 1, 1650001, 19 pages. MR3393930
[15] T. Albu and C. Năstăsescu, *"Relative Finiteness in Module Theory"*, Monographs and Textbooks in Pure and Applied Mathematics, vol. 84, Marcel Dekker, Inc., New York and Basel, 1984. MR749933
[16] T. Albu and P. F. Smith, *Localization of modular lattices, Krull dimension, and the Hopkins-Levitzki theorem (I)*, Math. Proc. Cambridge Philos. Soc. **120** (1996), no. 1, 87–101. MR1373350
[17] F. W. Anderson and K. R. Fuller, *"Rings and Categories of Modules"*, 2nd ed., Graduate Texts in Mathematics, vol. 13, Springer-Verlag, New York, 1992. MR1245487
[18] G. F. Birkenmeier, B. J. Müller, and S. T. Rizvi, *Modules in which every fully invariant submodule is essential in a direct summand*, Comm. Algebra **30** (2002), no. 3, 1395–1415. MR1892606
[19] A. W. Chatters and C. R. Hajarnavis, *Rings in which every complement right ideal is a direct summand*, Quart. J. Math. Oxford Ser. (2) **28** (1977), no. 109, 61–80. MR0437595
[20] G. Birkhoff, *"Lattice Theory"*, Third edition. Colloquium Publications, Vol. XXV, American Mathematical Society, Providence, R.I., 1967. MR0227053
[21] P. Crawley, R.P. Dilworth, *"Algebraic Theory of Lattices"*, Prentice-Hall, Englewood Cliffs, New Jersey, 1973.
[22] S. E. Dickson, *A torsion theory for Abelian categories*, Trans. Amer. Math. Soc. **121** (1966), 223–235. MR0191935
[23] N. V. Dung, D. V. Huynh, P. F. Smith, and R. Wisbauer, *"Extending Modules"*, Pitman Research Notes in Mathematics Series, vol. 313, Longman Scientific & Technical, Harlow, 1994. MR1312366
[24] P. Gabriel, *Des catégories abéliennes*, Bull. Soc. Math. France **90** (1962), 323–448. MR0232821
[25] V. K. Goel and S. K. Jain, *π-injective modules and rings whose cyclics are π-injective*, Comm. Algebra **6** (1978), no. 1, 59–73. MR0491819
[26] J. S. Golan, *"Torsion Theories"*, Pitman Monographs and Surveys in Pure and Applied Mathematics, vol. 29, Longman Scientific & Technical, Harlow; John Wiley & Sons, Inc., New York, 1986. MR880019
[27] G. Grätzer, *"General Lattice Theory"*, Second Edition, Birkhäuser Verlag, Basel Boston Berlin, 2003.
[28] M. Harada and K. Oshiro, *On extending property on direct sums of uniform modules*, Osaka J. Math. **18** (1981), no. 3, 767–785. MR635733
[29] L. Jeremy, *Sur les modules et anneaux quasi continus*, C. R. Acad. Sci. Paris Sér. A-B **273** (1971), A80–A83. MR0289561
[30] T.Y. Lam, *"Lectures on Modules and Rings"*, Springer-Verlag, Berlin Heidelberg New York, 1999.
[31] S. H. Mohamed and B. J. Müller, *"Continuous and Discrete Modules"*, London Mathematical Society Lecture Note Series, vol. 147, Cambridge University Press, Cambridge, 1990. MR1084376
[32] C. Năstăsescu and F. Van Oystaeyen, *"Dimensions of Ring Theory"*, Mathematics and its Applications, vol. 36, D. Reidel Publishing Co., Dordrecht, 1987. MR894033

[33] B. L. Osofsky and P. F. Smith, *Cyclic modules whose quotients have all complement submodules direct summands*, J. Algebra **139** (1991), no. 2, 342–354. MR1113780

[34] P. F. Smith and A. Tercan, *Generalizations of CS-modules*, Comm. Algebra **21** (1993), no. 6, 1809–1847. MR1215548

[35] P. F. Smith and A. Tercan, *Direct summands of modules which satisfy* (C_{11}), Algebra Colloq. **11** (2004), no. 2, 231–237. MR2058772

[36] A. Tercan and C. C. Yücel, *"Module Theory, Extending Modules and Generalizations"*, Frontiers in Mathematics, Birkhäuser/Springer, 2016. MR3468915

[37] Y. Utumi, *On continuous rings and self injective rings*, Trans. Amer. Math. Soc. **118** (1965), 158–173. MR0174592

[38] B. Stenström, *"Rings of Quotients"*, Springer-Verlag, Berlin Heidelberg New York, 1975. Die Grundlehren der Mathematischen Wissenschaften, Band 217; An Introduction to Methods of Ring Theory. MR0389953

[39] J. Von Neumann, *On regular rings*, Proc. Nat. Acad. Sci. **22** (1936), 707-713.

[40] S. Virili, *A point-free approach to L-surjunctivity and stable finiteness*, arXiv: 1410.1643v3 [math.RA], 13 Jan 2015, 34 pages.

[41] R. Wisbauer, *"Foundations of Module and Ring Theory"*, A Handbook for Study and Research, Revised and translated from the 1988 German edition, Algebra, Logic and Applications, vol. 3, Gordon and Breach Science Publishers, Philadelphia Reading Paris Tokyo Melbourne, 1991. MR1144522

Simion Stoilow Institute of Mathematics of the Romanian Academy, P.O. Box 1-764, RO-010145 Bucharest 1, Romania

Email address: `Toma.Albu@imar.ro`

Contemporary Mathematics
Volume **715**, 2018
https://doi.org/10.1090/conm/715/14402

On K-semi discrete modules

Ahmed Y. Ammar, Atif S. El-Araby, Mahmoud A. Kamal, and Nawal S. Mahmoud

Dedicated to the memory of Bruno J. Müller

ABSTRACT. Discrete (Quasi-discrete) modules form an important class in module theory, they are studied extensively by many authors. The decomposition theory for quasi-discrete modules plays an important role for better understanding of such modules. Here we make use of the K-smallness to introduce two new notions, namely the notions of K-semi discrete modules (or modules with the condition (KD_1)), and of K-supplements. When $M \cap K$ is a small in M, the classes of modules with (KD_1), and of semi discrete modules (or modules with the condition (D_1)) coincide. The K-supplements are supplements while the converse does not hold true. We study the properties of K-supplement submodules in relation with the known supplement submodules. We also study the properties of K-semi discrete submodules which are in analogy with that of semi discrete modules. We give a decomposition theorem for such modules. In fact, we prove that, a module M has the condition (KD_1), yields $M \cap K = M_1 \oplus S$, where M_1 is a direct summand of M and satisfies the condition (D_1), and S is a small submodule of M.

1. Introduction

Throughout all modules are unital right R-modules over an arbitrary associative ring R with unity. We use $L \leq M$ and $N \leq^{\oplus} M$ to denote that L is a submodule of M and N is a direct summand of M, respectively.

Semi-discrete modules were defined in Mohamed [2], and in Singh [3], under the name semi-dual-continuous modules, and in [1] semi-discrete modules were given under the name lifting modules. It was shown in [2] that every quasi-discrete module is a direct sum of hollow submodules, unique up to isomorphism, and is fully relatively projective. On the other hand, a direct sum $H(I) := \oplus_{i \in I} H_i$ of hollows is a quasi-discrete module if and only if it complements direct summands by [2, Theorem 4.48], every local summand is a summand, and the H_i is $H(I \setminus \{i\})$-projective for every $i \in I$, where $H(I \setminus \{i\}) = \oplus_{j \in I \setminus \{i\}} H_j$.

A submodule A of a module M is called small in M (denoted by $A \leq^s M$) if $A + B \neq M$ for any proper submodule B of M. A module H is called hollow if every proper submodule of H is small. A submodule P of a module M is called a supplement of A in M if P is minimal with respect to $M = A + P$, or equivalently,

---

2010 *Mathematics Subject Classification.* Primary 16D40, 16D70, 16D90.
Key words and phrases. projective modules, discrete modules.

if $M = A+P$ and $A \cap P \leq^s P$. A module M is called weakly supplemented if every submodule has a supplement.

A module M is called semi discrete (or has (D_1)) if for every submodule A of M, there is a decomposition $M = M_1 \oplus M_2$ such that $M_1 \leq A$ and $M_2 \cap A \leq^s M$. A module M is called a discrete module if it is semi discrete, and satisfies (D_2): if for all $A \leq M$ such that $M/A \cong B$, and $B \leq^{\oplus} M$, then $A \leq^{\oplus} M$. A module M is called a quasi discrete module if it is semi discrete, and satisfies (D_3): if for all $M_1 \leq^{\oplus} M$, $M_2 \leq^{\oplus} M$, with $M_1 + M_2 = M$, then $M_1 \cap M_2 \leq^{\oplus} M$.

Let M be a submodule of a module G, and let K be a fixed submodule of G. A submodule S of G is called a K-small submodule in G in case of, $K + Q = G$; whenever Q is a submodule of G with $S + Q = G$. If S is K-small in G, then A is K-small in G for all $A \leq S + K$.

Let M be a module, and B be a submodule of M. B is called a coclosed submodule of M if B is not C-small in M, for all proper submodule C of B.

We observe that:

(1) In any module M, the trivial submodules 0 and M are coclosed submodules of M.

(2) There are modules M that contains no non-trivial coclosed submodules. For instance, hollow modules contain no non-trivial coclosed submodules.

(3) If B is a non-zero coclosed submodule of M, then B is not small in M.

(4) Every supplement submodule of a module M is coclosed in M.

2. K-Supplement Submodules

In this section, we consider K to be a fixed submodule of a module G, M is a submodule of G, and all modules in question are submodules of M.

DEFINITION 2.1. A submodule A of a module M is called a K-supplement of B in M if A is minimal in the set $\{X : M = X + B$, and $X \cap (B + K)$ is $(X \cap K)$-small in $X\}$.

DEFINITION 2.2. A submodule A of a module M is called a K-weak supplement of B in M when $M = A + B$, and $A \cap (B + K)$ is $(A \cap K)$-small in A.

Observe that K-supplement implies K-weak supplement.

LEMMA 2.3. If $S \leq N$, N is a K-weak supplement of L in M, and S is $(N \cap K)$-small in M, then S is $(N \cap K)$-small in N.

PROOF. Let $N = S+Y$. Then $M = N+L = S+L+Y$. As S is $(N\cap K)$-small in M, we have $M = (N \cap K)+L+Y$; and hence $N = Y+N\cap(L+(N\cap K))$. Since N is a K -weak supplement of L in M, we have $N = (N \cap K) + Y$. Therefore, S is $(N \cap K)$-small in N. □

LEMMA 2.4. If N is a K-weak supplement of L in M, and S is a submodule of N such that $M = S + L$, then S is a K-weak supplement of $N \cap K$ in N.

PROOF. Let N be a submodule of M containing a submodule S, and that $S + L = M$. Then $N = S + (N \cap L) = S + (N \cap (L + K))$. As $N \cap (L + K)$ is $(N \cap K)$-small in N, we have that $N = S + (N \cap K)$.

Now $S \cap [(N \cap K) + K] = (S \cap K)$ is $(S \cap K)$-small in S; and hence S is a K-weak supplement of $N \cap K$ in N. □

LEMMA 2.5. If N is a K-weak supplement of L in M, and N is a $(N\cap K)$-small submodule in M, then $N=N\cap K$.

PROOF. Since $M=N+L$, and N is $(N\cap K)$-small in M, then $M=(N\cap K)+L$. Hence $N=(N\cap K)+(N\cap L)$. Since N is a K-weak supplement of L in M, we have $N=N\cap(L+(N\cap K))$ is $(N\cap K)$-small in N; and hence $(L\cap N)$ is $(N\cap K)$-small in N. It follows that $N=N\cap K$. □

LEMMA 2.6. Let A and B submodules of a module M. Then the following are equivalent:

1) A is a K-supplement of B in M;
2) A is a $(K\cap M)$-supplement of B in M.

PROOF. Let $\Lambda=\{X : M=X+B$, and $X\cap(B+K)$ is $(X\cap K)$-small in $X\}$, and let $\Gamma=\{X : M=X+B$, and $X\cap(B+(K\cap M))$ is $(X\cap(K\cap M))$-small in $X\}$. Since $X\cap(B+K)=X\cap M\cap(B+K)=X\cap(B+(K\cap M))$, and $X\cap K=X\cap(K\cap M)$, therefore $\Lambda=\Gamma$, and therefore A is a K-supplement of B in M if and only if A is a $K\cap M$-supplement of B in M. □

PROPOSITION 1. Let A and B be submodules of a module M. Then the following are equivalent:

1) A is a K-supplement of B in M;
2) A is a supplement of B in M, and A is a K-weak supplement of B in M.

PROOF. 1)⇒2): Since A is a K-supplement of B in M, we have that A is minimal in the set $\{X : M=X+B$, and $X\cap(B+K)$ is $(X\cap K)$-small in $X\}$. Now let $X\leq A$ with $M=X+B$, and let $X\cap(B+K)+Q=X$, it follows that $X\cap(B+K)+Q+B=M$, and $B\cap(X+K)+K\cap(B+X)+Q+B=M$. This implies that

$$(K\cap M)+Q+B=M;\ \text{and that}\ A\cap(B+(K\cap M))+Q=A,$$

since $A\cap(B+K)$ is $(A\cap K)$-small in A, then $A\cap K+Q=A$; and hence, by the modular law, $X\cap K+Q=X$. Thus $X\cap(B+K)$ is $(X\cap K)$-small in X.

By the minimality of A, we have that $X=A$. Therefore, A is a minimal with respect to $M=A+B$; i.e. A is a supplement of B in M, and A is a K-weak supplement of B in M.

2)⇒1): Since A is a supplement of B in M, and A is a K-weak supplement of B in M, we have that $M=A+B$, and A is minimal with this property and $A\in\{X : M=X+B$, and $X\cap(B+K)$ is $(X\cap K)$-small in $X\}$. Now if $X\leq A$ with $M=X+B$, and $X\cap(B+K)$ is $(X\cap K)$-small in X, then $X=A$ (due to A minimal). □

Observe that, supplements need not be $K-$supplement. For example, let R be a ring, and let $M=(1,0)R\oplus(0,1)R$, $A=(1,0)R, B=(0,1)R$, and $K=(1,1)R$. A is a supplement of B in M, but A is not a K-supplement of B in M.

COROLLARY 1. Let $A\leq K$ or $B\leq K$, where A, and B are submodules of a module M. Then the following are equivalent:

1) A is a K-supplement of B in M;
2) A is a supplement of B in M.

LEMMA 2.7. Let N be a supplement of L in M. If K is $(N\cap K)-$small in M, then N is a K-supplement of L in M.

PROOF. If N is a supplement of L in M, then $M = N + L$, and $(N \cap L) \leq^s N$. It remains to show that $N \cap (L + K)$ is $(N \cap K)$-small in N. To this end, let $N = N \cap (L + K) + Q$. Hence $M = N \cap (L + K) + Q + L = K + Q + L$. Since K is $(N \cap K)$-small in M, we have $M = (N \cap K) + Q + L$. It follows that $N = (N \cap K) + Q + (N \cap L)$. Since $(N \cap L) \leq^s N$, we have $N = (N \cap K) + Q$. Therefore, N is a K-weak supplement of L in M, and N is a K-supplement of L in M. □

LEMMA 2.8. Let $A \leq N \leq M$. If A is a K-weak supplement of B in M, then A is a K-weak supplement of $B \cap N$ in N.

PROOF. Let A be a K-weak supplement of B in M. Then it follows that $M = A + B$, and $A \cap (B + K)$ is $(A \cap K)$-small in A.

As $A \leq N$ and $A \cap ((B \cap N) + K) \leq A \cap (B + K)$, we have that $N = A + (B \cap N)$, and that $A \cap (B + K)$ is $(A \cap K)$-small in A. Therefore, A is a K-weak supplement of $B \cap N$ in N. □

COROLLARY 2. Let $A \leq N \leq M$. If A is a K-supplement of B in M, then A is a K-supplement of $B \cap N$ in N.

LEMMA 2.9. Let $A \leq N \leq M$. If A is a K-weak supplement of B in N, and N is a K-weak supplement of H in M, then A is a K-weak supplement of $B + H$ in M.

PROOF. Since $A + B = N$, and $N + H = M$, we have that $A + (B + H) = M$. To prove that $A \cap ((B + H) + K)$ is $(A \cap K)-$small in A, let $A \cap ((B + H) + K) + Q = A$, then $A \cap ((B + H) + K) + Q + (B + H) = M$, thus

$$(B + H) \cap (A + K) + K \cap ((B + H) + A) + Q + (B + H) = M;$$

and hence $(B + H) + (K \cap M) + Q = M$, and $N = N \cap ((B + H) + (K \cap M)) + Q$. Since $N \cap ((B + H) + (K \cap M))$ is a $(N \cap K)$-small in N, we have that $N = (N \cap K) + Q$, and that $A = (A \cap K) + Q$. Therefore $A \cap ((B + H) + K)$ is $(A \cap K)$-small in A, and A is a K-weak supplement of $B + H$ in M. □

COROLLARY 3. Let $A \leq N \leq M$. If A is a K-supplement of B in N and N is a K-supplement of H in M, then A is a K-supplement of $B + H$ in M.

DEFINITION 2.10. A submodule B of a module M is called a K-coclosed submodule of M if the following condition holds: $[A + K \leq B + K$ and B is $A + K$-small in $M] \Rightarrow A + K = B + K$.

LEMMA 2.11. Every K-weak supplement in M is a K-coclosed in M.

PROOF. Let B be a K-weak supplement of C in M. To show that B is K-coclosed in M, let $A + K \leq B + K$ and B be a $(A + K)$-small in M. We have that $M = B + C$, and $B \cap (C + K)$ is $(B \cap K)$-small in B. Since B is $(A + K)$-small in M, it follows that $M = A + K + C$. By the modular law $B + K = A + K + C \cap (B + K)$. It follows that $B + K = A + K + B \cap (C + K)$.

Since $B \cap (C + K)$ is $(B \cap K)$-small in B; and hence in $B + K$, we have that $B + K = A + K$. Therefore, B is K-coclosed in M. □

COROLLARY 4. Every K-supplement in M is K-coclosed in M.

3. K-Semi Discrete Modules

DEFINITION 3.1. A module M is said to have the condition (KD_1) if, whenever $A \leq L \leq^{\oplus} M$, there exists a decomposition of L, $L = L_1 \oplus L_2$, where $L_1 \leq A$, $L_2 \cap (A + (L \cap K))$ is $[L_2 \cap (L_1 + (L \cap K))]$-small in L, and $A \cap K$ is $(L_1 \cap K)$-small in L. M is called a K-semi discrete module if it satisfies the condition (KD_1).

Observe that, the condition (D_1) is automatically obtained from the condition (KD_1) when $K = 0$.

PROPOSITION 2. If a module M has the condition (KD_1), then every direct summand of M has condition (KD_1).

PROOF. It is clear. □

LEMMA 3.2. Let $M = \sum_{i \in F} H_i$, where H_i is a hollow module for all $i \in F$, and F is a finite set. Then for any submodule A of M, there is a subset $K \subseteq F$ such that $\sum_{i \in K} H_i$ is a supplement of A in M.

PROOF. It is clear. □

PROPOSITION 3. If a module $M = H \oplus K$, where H is a hollow module with finite length and K is a simple module, then M has (KD_1).

PROOF. To avoid the trivial cases, let A be an arbitrary non-zero proper submodule of M. Observe that M has a decomposition of indecomposable submodules with local endomorphism rings (by the Krull-Schmidt Theorem), hence complements direct summands. Thus, each summand L of M is either hollow or simple, which allows us to consider only a summand L of M containing A. If $K \leq A$, then the proof done.

Now let $A \cap K = 0$ we have the following cases:

Case 1): If $A \oplus K = M$, then this decomposition satisfies the condition for the given A.

Case 2): If $A \oplus K \neq M$, then in this case we have that $A \oplus K = K \oplus (H \cap (A \oplus K))$, where $H \cap (A \oplus K) \neq H$. It follows that

$$(A + K)/K = [K \oplus (H \cap (A \oplus K))]/K \leq^{s} M/K.$$

Hence A is K-small in M. Therefore, the trivial decomposition $M = 0 \oplus M$ satisfies the condition (KD_1) for the given A. □

REMARK 3.3. It is easy to show that if a module $M = H \oplus K$, where H is a hollow module, and K is a simple module, then M does not have (D_1). So, Proposition 3 gives us an example of a module with (KD_1), and not having (D_1).

In the following, we provide a number of additional conditions on modules with (KD_1) to satisfy the condition (D_1).

LEMMA 3.4. Let M be a module with (KD_1). If $(K \cap M)$ is small in M, then M has (D_1).

PROOF. Assume that $A \leq M \leq^{\oplus} M$. By (KD_1) for M, there exists a decomposition $M = M_1 \oplus M_2$, where $M_1 \leq A$, $M_2 \cap (A + (M \cap K))$ is $[M_2 \cap (M_1 + (M \cap K))]$-small in M, and $A \cap K$ is $(M_1 \cap K)$-small in M. Since $M \cap K \leq^{s} M$, it follows that $M_2 \cap (M_1 + (M \cap K)) \leq^{s} M$.

As $M_2 \cap (A + (M \cap K))$ is $[M_2 \cap (M_1 + (M \cap K))]$-small in M, it follows that $M_2 \cap (A + (M \cap K)) \leq^s M$; and thus, $M_2 \cap A \leq^s M$, and $M_2 \cap A \leq^s M_2$. Therefore, M has (D_1). □

LEMMA 3.5. If $M = M_1 \oplus M_2$, and S is M_1-small in M, then $S \cap M_2 \leq^s M_2$.

PROOF. Let $(S \cap M_2) + Q = M_2$, then $(S \cap M_2) + Q + M_1 = M$. Since S is M_1-small in M, then $Q + M_1 = M$, and $Q = M_2$. Therefore $S \cap M_2 \leq^s M_2$. □

LEMMA 3.6. If M has (KD_1), and $M \leq K$, then M has (D_1).

PROOF. Let $A \leq M \leq^\oplus M$. By using the condition (KD_1) for M, there exists a decomposition $M = M_1 \oplus M_2$, where $M_1 \leq A$, and $A \cap K$ is $(M_1 \cap K)$-small in M. It follows that A is M_1-small in M; and thus, by Lemma 3.5, $A \cap M_2 \leq^s M_2$. Therefore, M has (D_1). □

PROPOSITION 4. Let M be a module which has the condition (KD_1). Then every K-supplement submodule is a direct summand of M.

PROOF. Let M be a module with (KD_1), and A be a K-supplement submodule of a submodule B in M. By using the condition (KD_1) for M, there exists a decomposition

$$M = M_1 \oplus M_2,$$

where $M_1 \leq A$, $M_2 \cap (A + (M \cap K))$ is $[M_2 \cap (M_1 + (M \cap K))]$-small in M, and $(A \cap K)$ is $(M_1 \cap K)$-small in M. Since A is a K-supplement of B in M, we have that $M = A + B$, and $A \cap (B + K)$ is $(A \cap K)$-small in A. By the modular law, we have that $A = M_1 \oplus (A \cap M_2)$; and hence $M = M_1 + (A \cap M_2) + B$.

Since $M_2 \cap A$ is $[M_2 \cap (M_1 + (M \cap K))]$-small in M (due to $M_2 \cap (A + (M \cap K))$ is an $[M_2 \cap (M_1 + (M \cap K))]$-small in M), $M = M_1 + (M_2 \cap (M_1 + (M \cap K))) + B$. Thus $M = M_1 + (M_1 \cap (M_2 + (M \cap K))) + ((M \cap K) \cap (M_1 + M_2)) + B = M_1 + (M \cap K) + B$.

Again, by the modular law, $A = M_1 + (A \cap ((M \cap K) + B))$. Since $A \cap (B + K)$ is $(A \cap K)$-small in A, we have $A = M_1 + (A \cap K)$. Hence $M = M_1 + (A \cap K) + B$. Since $A = M_1 + (A \cap K)$, we have $M = M_1 + (M_1 \cap K) + B = M_1 + B$.

Because A is a K-supplement of B in M, we have that A is a K-supplement of B in M, $M = M_1 + B$, and $M_1 \leq A$. Therefore $A = M_1 \leq^\oplus M$. □

LEMMA 3.7. Let M be a module which has (KD_1). If A is a coclosed submodule of M, which is contained in $M \cap K$, then A has (D_1).

PROOF. Let M be a module with (KD_1) and A be a coclosed submodule of M, $A \leq M \cap K$, and let $B \leq A$. Then by (KD_1) for M, we have that there exists a decomposition $M = M_1 \oplus M_2$, where $M_1 \leq B$, and B is M_1-small in M. By the modular law, we have that $B = M_1 \oplus (M_2 \cap B)$.

Since B is M_1-small in M, we have $M_2 \cap B \leq^s M_2$, and $M_2 \cap B \leq^s M$. Since A is coclosed in M, we have $M_2 \cap B \leq^s A$. It follows that $B = M_1 \oplus (M_2 \cap B)$, where $M_1 \leq^\oplus A$, and $M_2 \cap B \leq^s A$. Therefore, A has (D_1). □

PROPOSITION 5. Let M be a module with (KD_1). Then $M \cap K = M_1 \oplus S$, where M_1 is a direct summand of M with (D_1), and S is a small submodule of M.

PROOF. By (KD_1) for M, and as $M \cap K \leq M$, there exists a decomposition $M = M_1 \oplus M_2$, where $M_1 \leq M \cap K$ and $M \cap K$ is M_1-small in M. We have that $M \cap K = M_1 \oplus (M_2 \cap K)$. Since $M_2 \cap K$ is M_1-small in M, it follows that $M_2 \cap K \leq^s M_2$, and $M_2 \cap K \leq^s M$.

As M_1 is a coclosed submodule of M (due to $M_1 \leq^{\oplus} M$) and $M_1 \leq M \cap K$, we have that M_1 has (D_1). Therefore $M \cap K = M_1 \oplus S$, where $M_1 \leq^{\oplus} M$, M_1 has (D_1), and $S := M_2 \cap K \leq^s M$. □

LEMMA 3.8. If M has (KD_1), and $M \cap K$ is a coclosed submodule of M, then $M \cap K \leq^{\oplus} M$ and $M \cap K$ has (D_1).

PROOF. By Proposition 5, and as M has (KD_1), we have that $M \cap K = M_1 \oplus S$, where $M_1 \leq^{\oplus} M$ with (D_1), and $S \leq^s M$. Since $M \cap K$ is a coclosed submodule of M, we have that $S \leq^s M \cap K$; and hence $M \cap K = M_1$ is a direct summand of M, and has (D_1), by Lemma 3.7. □

DEFINITION 3.9. Let M be a nonzero module, and $(K \cap M)$ be a proper submodule of M. M is called a K-hollow module if every proper submodule is $(K \cap M)$-small in M.

LEMMA 3.10. Every K-hollow module M must be indecomposable.

PROOF. Let $M = M_1 \oplus M_2$, where $M_1 \neq 0$, and $M_2 \neq 0$. As M_1 and M_2 are $(K \cap M)$-small in M, we have that $M = (K \cap M) + M_2 = (K \cap M)$. But $M \neq (K \cap M)$. Therefore, M must be indecomposable. □

PROPOSITION 6. A module M is a K-hollow module if and only if M is indecomposable and has (KD_1).

PROOF. Let M be a K-hollow module, then, by Lemma 3.10, M is indecomposable. Now each non-zero proper submodule A of M can be written as $M = 0 + M$, such that $0 \leq A$, and that $M \cap (A + (M \cap K)) = A + (M \cap K)$ is $(M \cap K)-$small in M. Now let $(A \cap K) + Q = M$. If $Q \neq M$, then Q is a $(K \cap M)$-small in M, hence $(A \cap K) + (M \cap K) = (M \cap K) = M$. But $M \neq (K \cap M)$. Thus $Q = M$, and $(A \cap K) \leq^s M$. Therefore, M has (KD_1).

For the converse, let A be a proper submodule of M. Since M is indecomposable and has (KD_1), then $M = 0 + M$, where $0 \leq A$, $M \cap (A + (M \cap K)) = A + (M \cap K)$ is a $(M \cap K)$-small in M, and $(A \cap K) \leq^s M$, and hence A is a $(M \cap K)$-small in M. Therefore, M is a K-hollow module. □

References

[1] J. Clark, C. Lomp, N. Vanaja, and R. Wisbauer, *Lifting modules; Supplements and projectivity in module theory*, Frontiers in Mathematics, Birkhäuser Verlag, Basel, 2006. MR2253001

[2] S. H. Mohamed and B. J. Müller, *Continuous and discrete modules*, London Mathematical Society Lecture Note Series, vol. 147, Cambridge University Press, Cambridge, 1990. MR1084376

[3] S. Singh, *Semidual continuous modules over Dedekind domains* (English, with Arabic summary), J. Univ. Kuwait Sci. **11** (1984), no. 1, 33–40. MR776699

Math Dept., Faculty of Education, Ain Shams Univ., Roxy, Cairo. Egypt
Email address: ahmedammar500@hotmail.com

Math. Dept., Faculty of Education, Ain Shams Univ., Roxy Cairo, Egypt
Email address: atif3331@hotmail.com

Math. Dept., Faculty of Education, Ain Shams Univ., Roxy Cairo, Egypt
Email address: mahmoudkamal333@hotmail.com

Math. Dept., Faculty of Education, Ain Shams univ., Roxy Cairo, Egypt
Email address: sayednawal@yahoo.com

Contemporary Mathematics
Volume **715**, 2018
https://doi.org/10.1090/conm/715/14403

Nonlinear Lie triple higher derivation on triangular algebras

Mohammad Ashraf and Aisha Jabeen

Abstract. Let $\mathcal{R}$ be a commutative ring with identity. A triangular algebra is an algebra of the form $\mathfrak{A} = \begin{bmatrix} \mathcal{A} & \mathcal{M} \\ 0 & \mathcal{B} \end{bmatrix}$ where $\mathcal{A}$ and $\mathcal{B}$ are unital algebras over $\mathcal{R}$ and $\mathcal{M}$ is a $(\mathcal{A}, \mathcal{B})$-bimodule which is faithful as a left $\mathcal{A}$-module as well as a right $\mathcal{B}$-module. In this paper, we study nonlinear Lie triple higher derivation on $\mathfrak{A}$ and show that under certain assumptions on $\mathfrak{A}$, every nonlinear Lie triple higher derivation on $\mathfrak{A}$ is of standard form, i.e., a nonlinear Lie triple higher derivation can be expressed as the sum of additive higher derivation and the nonlinear functionals vanishing on all Lie triple product on $\mathfrak{A}$. Finally, application of the main theorem to some classical examples of triangular algebra has been given.

1. Introduction

Let $\mathcal{R}$ be a commutative ring with identity element and $\mathcal{A}$ be an algebra over $\mathcal{R}$. For any $x, y \in \mathcal{A}$, $[x, y]$ will denote the commutator $xy - yx$, while $x \circ y$ will represent the anticommutator $xy + yx$. An $\mathcal{R}$-linear (resp. nonlinear) map $d : \mathcal{A} \to \mathcal{A}$ is called derivation (resp. nonlinear derivation) if $d(xy) = d(x)y + xd(y)$ for all $x, y \in \mathcal{A}$. An $\mathcal{R}$-linear (resp. nonlinear) map $d : \mathcal{A} \to \mathcal{A}$ is called a Jordan derivation (resp. nonlinear Jordan derivation) if $d(x \circ y) = d(x) \circ y + x \circ d(y)$ holds for all $x, y \in \mathcal{A}$. More generally, an $\mathcal{R}$-linear (resp. nonlinear) map $L : \mathcal{A} \to \mathcal{A}$ is called a Lie derivation (resp. nonlinear Lie derivation) if $L([x, y]) = [L(x), y] + [x, L(y)]$ for all $x, y \in \mathcal{A}$. An $\mathcal{R}$-linear (resp. nonlinear) map $L : \mathcal{A} \to \mathcal{A}$ is called a Lie triple derivation (resp. nonlinear Lie triple derivation) if $L([[x, y], z]) = [[L(x), y], z] + [[x, L(y)], z] + [[x, y], L(z)]$ for all $x, y, z \in \mathcal{A}$. It is not difficult to observe that any derivation is a Lie derivation and any Lie derivation is a Lie triple derivation. However, the converse statements are in general not true. In fact if $d : \mathcal{A} \to \mathcal{A}$ is a derivation on an algebra $\mathcal{A}$ and $g : \mathcal{A} \to Z(\mathcal{A})$ be an additive mapping with $g([\mathcal{A}, \mathcal{A}]) = 0$, then it can be seen that $d + g$ is a Lie derivation, which is not necessarily a derivation. If $h : \mathcal{A} \to Z(\mathcal{A})$ is an additive mapping such that $h([[\mathcal{A}, \mathcal{A}], \mathcal{A}]) = 0$, then $d + h$ is a Lie triple derivation on $\mathcal{A}$, which is not necessarily a Lie derivation on $\mathcal{A}$. Clearly, every derivation is also a Jordan derivation. Also, every Lie derivation is a Lie triple derivation. The well-known formula $[[x, y], z] = x \circ (y \circ z) - y \circ (x \circ z)$ for all $x, y, z \in \mathcal{A}$ implies that every Jordan derivation is also a Lie triple derivation. Hence, studying Lie triple derivations enables us to treat both important classes of Jordan and Lie derivations simultaneously. In [**C**] Cheung initiated the study of

2010 *Mathematics Subject Classification.* 16W25, 15A78.

Lie derivations of triangular algebras $\mathfrak{A}$ and gave a sufficient condition under which every Lie derivation on $\mathfrak{A}$ is a sum of derivation on $\mathfrak{A}$ and a mapping from $\mathfrak{A}$ to its center $Z(\mathfrak{A})$. Further, Lie derivations on triangular algebras were studied in [**H**, **P**], whereas the study of Lie triple derivations on triangular algebras can be found in [**A**, **I**, **O**].

Recently many authors have studied different kinds of higher derivations in rings and algebras. Let $\mathbb{N}$ be the set of nonnegative integers and $\mathfrak{L} = \{L_n\}_{n\in\mathbb{N}}$ be a sequence of linear maps $L_n : \mathcal{A} \to \mathcal{A}$ such that $d_0 = id_{\mathcal{A}}$. Then $\mathfrak{L}$ is called

(*i*) a *higher derivation* on $\mathfrak{L}$ if $L_n(xy) = \sum\limits_{p+q=n} L_p(x)L_q(y)$ for all $x, y \in \mathcal{A}$ and for each $n \in \mathbb{N}$.

(*ii*) a *Jordan higher derivation* on $\mathfrak{L}$ if $L_n(x^2) = \sum\limits_{p+q=n} L_p(x)L_q(x)$ for all $x, y \in \mathcal{A}$ and for each $n \in \mathbb{N}$.

(*iii*) a *Jordan triple higher derivation* on $\mathfrak{L}$ if $L_n(xyx) = \sum\limits_{p+q+r=n} L_p(x)L_q(y)L_r(x)$ for all $x, y \in \mathcal{A}$ and for each $n \in \mathbb{N}$.

(*iv*) a *Lie higher derivation* on $\mathfrak{L}$ if $L_n[x, y] = \sum\limits_{p+q=n} [L_p(x), L_q(y)]$ for all $x, y \in \mathcal{A}$ and for each $n \in \mathbb{N}$.

(*v*) a *Lie triple higher derivation* on $\mathfrak{L}$ if $L_n([[x, y], z]) = \sum\limits_{p+q+r=n} [[L_p(x), L_q(y)], L_r(z)]$ for all $x, y, z \in \mathcal{A}$ and for each $n \in \mathbb{N}$.

Note that for $n = 1$ the above higher derivation L_1 is a derivation, a Jordan derivation, a Jordan triple derivation, a Lie derivation and a Lie triple derivation on an algebra $\mathcal{A}$. If the linear assumption in the above definitions is dropped, that is, if $\mathfrak{L} = \{L_n\}_{n\in\mathbb{N}}$ is a sequence of mappings $L_n : \mathcal{A} \to \mathcal{A}$, (not necessarily linear), then the corresponding higher derivation, Jordan higher derivation, Jordan triple higher derivation, Lie higher derivation and Lie triple higher derivation is said to be nonlinear higher derivation, nonlinear Jordan higher derivation, nonlinear Jordan triple higher derivation, nonlinear Lie higher derivation and nonlinear Lie triple higher derivation respectively. Further note that if $\mathfrak{L} = \{L_n\}_{n\in\mathbb{N}}$ is a sequence of additive mappings $L_n : \mathcal{A} \to \mathcal{A}$, then the corresponding higher derivation, Jordan higher derivation, Jordan triple higher derivation, Lie higher derivation and Lie triple higher derivation is said to be additive higher derivation, additive Jordan higher derivation, additive Jordan triple higher derivation, additive Lie higher derivation and additive Lie triple higher derivation respectively.

It can be easily seen that every higher derivation on $\mathcal{A}$ is a Jordan higher derivation on $\mathcal{A}$ but the converse need not be true in general. In the year 2002, Ferrero and Haetinger [**E**] proved the converse part and obtained that every Jordan higher derivation of a 2-torsion free semiprime ring is a higher derivation on semiprime ring. Further, Wei and Xiao [**M**] studied higher derivations on triangular algebras and obtained several interesting results. Also, every higher derivation is a Lie higher derivation and every Lie higher derivation is a Lie triple higher derivation but the converse need not be true in general. Note that if $D = \{d_n\}_{n\in\mathbb{N}}$ is a higher derivation on $\mathcal{A}$ for any $n \in \mathbb{N}$ and $L_n = d_n + f_n$ where $f_n : \mathcal{A} \to Z(\mathcal{A})$ where $Z(\mathcal{A})$ is the center of $\mathcal{A}$, is a linear (resp. nonlinear) mapping, then it is easy to see that $\{L_n\}_{n\in\mathbb{N}}$ is a Lie higher derivation (resp. nonlinear Lie higher derivation) if and only if $f_n([x,y]) = 0$ for each $x, y \in \mathcal{A}$, and $\{L_n\}_{n\in\mathbb{N}}$ is a Lie triple higher derivation (resp. nonlinear Lie triple higher derivation) if and only if $f_n([[x,y],z]) = 0$ for each $x, y, z \in \mathcal{A}$. But it is not a higher derivation (resp. nonlinear higher derivation) if $f_n \neq 0$ for some n. The function f_n, $n \in \mathbb{N}$ is called a linear functional (resp. nonlinear functional). Lie triple higher derivation (resp. nonlinear Lie triple higher derivation) of the above kind are called *standard*. The natural problem that one considers in this context is whether or not every Lie triple higher derivation (resp. nonlinear Lie triple higher derivation) is standard. Several authors have made important contributions to the related topics see for reference [**F**, **P**, **I**, **M**, **N**, **O**] where further references can be found. Cheung [**B**] studied the mappings on triangular algebras and obtained many important results. He described derivations, Lie derivations, commuting maps and automorphisms of triangular algebras. Further Qi and Hou [**K**] described the structure of Lie higher derivations on nest algebras. Motivated by these works Xiao and Wei [**N**] studied nonlinear Lie higher derivations on a triangular algebra and showed that under certain restrictions every nonlinear Lie higher derivation on triangular algebra has standard form. In the present paper, we have investigated the structure of nonlinear Lie triple higher derivations on triangular algebras and proved that under certain suitable restrictions on $\mathfrak{A}$ every nonlinear Lie triple higher derivation has the standard form.

2. Triangular Algebras

Throughout this paper, $\mathcal{R}$ will always denote a commutative ring with identity element. Let $\mathcal{A}$ and $\mathcal{B}$ be unital algebras over $\mathcal{R}$ and let $\mathcal{M}$ be $(\mathcal{A},\mathcal{B})$-bimodule which is faithful as a left $\mathcal{A}$-module and also as a right $\mathcal{B}$-module. The $\mathcal{R}$-algebra

$$\mathfrak{A} = Tri(\mathcal{A},\mathcal{M},\mathcal{B}) = \left\{ \begin{bmatrix} a & m \\ 0 & b \end{bmatrix} \middle|\, a \in \mathcal{A}, m \in \mathcal{M}, b \in \mathcal{B} \right\}$$

under the usual matrix operations is called a triangular algebra. The center of $\mathfrak{A}$ is

$$Z(\mathfrak{A}) = \left\{ \begin{bmatrix} a & 0 \\ 0 & b \end{bmatrix} \middle|\, am = mb \ \forall\ m \in \mathcal{M} \right\}.$$

Define two natural projections $\pi_{\mathcal{A}} : \mathfrak{A} \to \mathcal{A}$ and $\pi_{\mathcal{B}} : \mathfrak{A} \to \mathcal{B}$ by

$$\pi_{\mathcal{A}} \begin{bmatrix} a & m \\ 0 & b \end{bmatrix} = a \text{ and } \pi_{\mathcal{B}} \begin{bmatrix} a & m \\ 0 & b \end{bmatrix} = b.$$

Moreover, $\pi_{\mathcal{A}}(Z(\mathfrak{A})) \subseteq Z(\mathcal{A})$ and $\pi_{\mathcal{B}}(Z(\mathfrak{A})) \subseteq Z(\mathcal{B})$ and there exists a unique algebraic isomorphism $\tau : \pi_{\mathcal{A}}(Z(\mathfrak{A})) \to \pi_{\mathcal{B}}(Z(\mathfrak{A}))$ such that $am = m\tau(a)$ for all $a \in \pi_{\mathcal{A}}(Z(\mathfrak{A})), m \in \mathcal{M}$.

Let $1_{\mathcal{A}}$ (resp.$1_{\mathcal{B}}$) be the identity of the algebra $\mathcal{A}$ (resp.$\mathcal{B}$) and let I be the identity of triangular algebra $\mathfrak{A}$. Throughout, this paper we shall use the following notions: $P = \begin{bmatrix} 1_{\mathcal{A}} & 0 \\ 0 & 0 \end{bmatrix}$, $Q = I - P = \begin{bmatrix} 0 & 0 \\ 0 & 1_{\mathcal{B}} \end{bmatrix}$ and $\mathfrak{A}_{11} = P\mathfrak{A}P$, $\mathfrak{A}_{12} = P\mathfrak{A}Q$, $\mathfrak{A}_{22} = Q\mathfrak{A}Q$. Thus

$$\mathfrak{A} = P\mathfrak{A}P + P\mathfrak{A}Q + Q\mathfrak{A}Q = \mathfrak{A}_{11} + \mathfrak{A}_{12} + \mathfrak{A}_{22},$$

where $\mathfrak{A}_{11}$ is a subalgebra of $\mathfrak{A}$ isomorphic to $\mathcal{A}$, $\mathfrak{A}_{22}$ is a subalgebra of $\mathfrak{A}$ isomorphic to $\mathcal{B}$ and $\mathfrak{A}_{12}$ is $(\mathfrak{A}_{11}, \mathfrak{A}_{22})$-bimodule isomorphic to $\mathcal{M}$. Also, $\pi_{\mathcal{A}}(Z(\mathfrak{A}))$ and $\pi_{\mathcal{B}}(Z(\mathfrak{A}))$ are isomorphic to $PZ(\mathfrak{A})P$ and $QZ(\mathfrak{A})Q$ respectively. Then there is an algebra isomorphism $\tau : PZ(\mathfrak{A})P \to QZ(\mathfrak{A})Q$ such that $am = m\tau(a)$ for all

$m \in P\mathfrak{A}Q$. From now, we always assume that triangular algebra $\mathfrak{A} = Tri(\mathcal{A}, \mathcal{M}, \mathcal{B})$ is 2-torsion free.

3. Main Result

Qi and Hou [**K**] obtained the structure of a Lie higher derivation defined on a nest algebra and proved that if $\mathcal{N}$ is a nest on a Banach space X over the real or complex field $\mathbb{F}$ and if there exists a nontrivial idempotent element in $\mathcal{N}$ which is complemented in X, then $\{L_n\}_{n\in\mathbb{N}}$ is a Lie higher derivation of the nest algebra $\tau(\mathcal{N})$ if and only if each L_n has the standard form $L_n(T) = d_n(T) + f_n(T)$ for all $T \in \tau(\mathcal{N})$, where $\{d_n\}_{n\in\mathbb{N}}$ is a higher derivation and $\{f_n\}_{n\in\mathbb{N}}$ is a sequence of linear functionals satisfying the condition $f_n([U, W]) = 0$ for all $U, W \in \tau(\mathcal{N})$ for each $n \in \mathbb{N}$. Motivated by this work Xiao and Wei [**N**] studied nonlinear Lie higher derivation on triangular algebras and proved that any nonlinear Lie higher derivation on a triangular algebra has the standard form. In the present paper, we shall extend the result by Xiao and Wei [**N**] for a nonlinear Lie triple higher derivation on a triangular algebra and show that it has also standard form. In fact we prove the following theorem:

THEOREM 3.1. *Let* $\mathfrak{A} = \begin{bmatrix} \mathcal{A} & \mathcal{M} \\ 0 & \mathcal{B} \end{bmatrix}$ *be a triangular algebra and* $\mathfrak{L} = \{L_n\}_{n\in\mathbb{N}}$ *be nonlinear Lie triple higher derivation on* $\mathfrak{A}$. *If* $\mathfrak{A}$ *satisfies the following conditions:*

(1) $\pi_{\mathcal{A}}(Z(\mathfrak{A})) = Z(\mathcal{A})$ *and* $\pi_{\mathcal{B}}(Z(\mathfrak{A})) = Z(\mathcal{B})$,
(2) $Z(\mathcal{A}) = \{a \mid [[a, x], y] = 0 \text{ for all } x, y \in \mathcal{A}\}$ *or*

$$Z(\mathcal{B}) = \{b \mid [[b, x], y] = 0 \text{ for all } x, y \in \mathcal{B}\}.$$

Then $\mathfrak{L}$ *is of standard form. More precisely, there exist an additive higher derivations* $\{d_n\}_{n\in\mathbb{N}}$ *on* $\mathfrak{A}$ *and a sequence of nonlinear functionals* $\{h_n\}_{n\in\mathbb{N}}$ *which annihilates all Lie triple product* $[[X, Y], Z]$ *for all* $X, Y, Z \in \mathfrak{A}$ *such that* $L_n(T) = d_n(T) + h_n(T)$ *for all* $T \in \mathfrak{A}$ *and* $n \in \mathbb{N}$.

In order to prove our main result we will use the following result:

THEOREM 3.2. [**I**, Theorem 2.1] *Let* $\begin{bmatrix} \mathcal{A} & \mathcal{M} \\ 0 & \mathcal{B} \end{bmatrix}$ *be triangular algebra with* $\pi_{\mathcal{A}}(Z(\mathfrak{A})) = Z(\mathcal{A})$ *and* $\pi_{\mathcal{B}}(Z(\mathfrak{A})) = Z(\mathcal{B})$. *Then every nonlinear Lie triple derivation on* $\mathfrak{A}$ *is of standard form.*

PROOF OF THEOREM 3.1. In order to prove the theorem, we will use the method of induction on n. For $n = 1$, L_1 is nonlinear Lie triple derivation on $\mathfrak{A}$. By [**I**, Theorem 2.1] it follows that there exist an additive derivation d_1 and a nonlinear mapping h_1 satisfying $h_1([[X, Y], Z]) = 0$ for all $X, Y, Z \in \mathfrak{A}$ such that $L_1(T) = d_1(T) + h_1(T)$ for all $T \in \mathfrak{A}$. Moreover, L_1 and d_1 satisfy the following properties:

$$\mathbf{C}_1 : \begin{cases} L_1(0) = 0, L_1(P), L_1(Q) \in \mathfrak{A}_{12} + Z(\mathfrak{A}); & L_1(\mathfrak{A}_{12}) \subseteq \mathfrak{A}_{12}, \\ L_1(\mathfrak{A}_{11}) \subseteq \mathfrak{A}_{11} + \mathfrak{A}_{12} + Z(\mathfrak{A}); & L_1(\mathfrak{A}_{22}) \subseteq \mathfrak{A}_{22} + \mathfrak{A}_{12} + Z(\mathfrak{A}), \\ d_1(P), d_1(Q) \in \mathfrak{A}_{12}; & d_1(\mathfrak{A}_{12}) \subseteq \mathfrak{A}_{12}, \\ d_1(\mathfrak{A}_{11}) \subseteq \mathfrak{A}_{11} + \mathfrak{A}_{12}; & d_1(\mathfrak{A}_{22}) \subseteq \mathfrak{A}_{22} + \mathfrak{A}_{12}. \end{cases}$$

We assume that the result holds for all $1 < s < n, n \in \mathbb{N}$. Then there exist $\{d_i\}_{i=0}^{i=s}$ an additive higher derivation and non zero central mapping h_s satisfying

$h_s([[X,Y],Z]) = 0$ for all $X,Y,Z \in \mathfrak{A}$ such that $L_s(T) = d_s(T) + h_s(T)$ for all $T \in \mathfrak{A}$. Thus L_s and d_s satisfy the following properties:

$$\mathbf{C}_s : \begin{cases} L_s(0) = 0, L_s(P), L_s(Q) \in \mathfrak{A}_{12} + Z(\mathfrak{A}); & L_s(\mathfrak{A}_{12}) \subseteq \mathfrak{A}_{12}, \\ L_s(\mathfrak{A}_{11}) \subseteq \mathfrak{A}_{11} + \mathfrak{A}_{12} + Z(\mathfrak{A}); & L_s(\mathfrak{A}_{22}) \subseteq \mathfrak{A}_{22} + \mathfrak{A}_{12} + Z(\mathfrak{A}), \\ d_s(P), d_s(Q) \in \mathfrak{A}_{12}; & d_s(\mathfrak{A}_{12}) \subseteq \mathfrak{A}_{12}, \\ d_s(\mathfrak{A}_{11}) \subseteq \mathfrak{A}_{11} + \mathfrak{A}_{12}; & d_s(\mathfrak{A}_{22}) \subseteq \mathfrak{A}_{22} + \mathfrak{A}_{12}. \end{cases}$$

Our aim to show that above conditions also hold for n.

Claim 1. $L_n(0) = 0, L_n(P), L_n(Q) \in \mathfrak{A}_{12} + Z(\mathfrak{A})$ and $L_n(\mathfrak{A}_{12}) \subseteq \mathfrak{A}_{12}$.
Using condition $\mathbf{C}_s$, we find that

$$\begin{aligned} L_n(0) &= L_n[[0,0],0] \\ &= \sum_{p+q+r=n} [[L_p(0), L_q(0)], L_r(0)] \\ &= [[L_n(0),0],0] + [[0, L_n(0)],0] + [[0,0], L_n(0)] \\ &\quad + \sum_{\substack{p+q+r=n \\ p,q,r>0}} [[L_p(0), L_q(0)], L_r(0)] \\ &= 0. \end{aligned}$$

As we know that $PM_{12}Q = [[M_{12}, P], P]$ for all $M_{12} \in \mathfrak{A}_{12}$, we have

$$\begin{aligned} L_n(M_{12}) &= L_n(PM_{12}Q) \\ &= L_n([[M_{12}, P], P]) \\ &= [[L_n(M_{12}), P], P] + [[M_{12}, L_n(P)], P] + [[M_{12}, P], L_n(P)] \\ &\quad + \sum_{\substack{p+q+r=n \\ p,q,r>0}} [[L_p(M_{12}), L_q(P)], L_r(P)]. \end{aligned}$$

Applying condition $\mathbf{C}_s$, it follows that

$$L_n(M_{12}) = [[L_n(M_{12}), P], P] + [[M_{12}, L_n(P)], P] + [[M_{12}, P], L_n(P)]. \tag{3.1}$$

Let us assume $L_n(M_{12}) = R_{11} + R_{12} + R_{22}$ and $L_n(P) = P_{11} + P_{12} + P_{22}$. Then (3.1) gives $R_{11} = 0, R_{22} = 0$ and $P_{11}M_{12} = M_{12}P_{22}$ for all $M_{12} \in \mathfrak{A}_{12}$ which leads to $P_{11} + P_{22} \in Z(\mathfrak{A})$. Therefore, $L_n(P) \in \mathfrak{A}_{12} + Z(\mathfrak{A})$ and $L_n(\mathfrak{A}_{12}) \subseteq \mathfrak{A}_{12}$. Similarly, we can show that $L_n(Q) \in \mathfrak{A}_{12} + Z(\mathfrak{A})$.

Claim 2. $L_n(\mathfrak{A}_{ii}) \subseteq \mathfrak{A}_{ii} + \mathfrak{A}_{12} + Z(\mathfrak{A})$ for $i = 1, 2$.
Since

$$\begin{aligned} 0 &= [[A_{11}, B_{22}], T] \\ &= L_n[[A_{11}, B_{22}], T] \\ &= [[L_n(A_{11}), B_{22}], T] + [[A_{11}, L_n(B_{22})], T] + [[A_{11}, B_{22}], L_n(T)] \\ &\quad + \sum_{\substack{p+q+r=n \\ p,q,r>0}} [[L_p(A_{11}), L_q(B_{22})], L_r(T)]. \end{aligned}$$

On using condition $\mathbf{C}_s$, we get

$$[[L_n(A_{11}), B_{22}], T] + [[A_{11}, L_n(B_{22})], T] + \sum_{\substack{p+q+r=n \\ p,q,r>0}} [[L_p(A_{11}), L_q(B_{22})], L_r(T)] = 0.$$

Multiplying from right as well as left side by P and Q respectively, we get

$$PL_n(B_{22})P \in Z(\mathcal{A}) \text{ and } QL_n(A_{11})Q \in Z(\mathcal{B})$$

which implies that $\tau(PL_n(B_{22})P) = QL_n(A_{11})Q$. Now we obtain

$$\begin{aligned} L_n(A_{11}) &= (PL_n(A_{11})P - \tau^{-1}(QL_n(A_{11})Q)) + PL_n(A_{11})Q \\ &\quad +(\tau^{-1}(QL_n(A_{11})Q) + QL_n(A_{11})Q) \end{aligned}$$

and

$$\begin{aligned} L_n(B_{22}) &= (PL_n(B_{22})P + \tau(PL_n(B_{22})P)) + PL_n(B_{22})Q \\ &\quad +(QL_n(B_{22})Q - \tau(PL_n(B_{22})P) \end{aligned}$$

which gives $L_n(\mathfrak{A}_{ii}) \subseteq \mathfrak{A}_{ii} + \mathfrak{A}_{12} + Z(\mathfrak{A})$ for $i = 1, 2$.

REMARK 3.3. Now we define $f_{n1}(A_{11}) = QL_n(A_{11})Q$ and $f_{n2}(B_{22}) = PL_n(B_{22})P$ for any $A_{11} \in \mathfrak{A}_{11}$ and $B_{22} \in \mathfrak{A}_{22}$. From Claim 3 it follows that $f_{n1} : \mathfrak{A}_{11} \to QZ(\mathfrak{A})Q$ is a mapping such that $f_{n1}([[X_{11}, Y_{11}], Z_{11}]) = 0$ for all $X_{11}, Y_{11}, Z_{11} \in \mathfrak{A}_{11}$ and $f_{n2} : \mathfrak{A}_{22} \to PZ(\mathfrak{A})P$ is a nonlinear mapping such that $f_{n2}([[X_{22}, Y_{22}], Z_{22}]) = 0$ for all $X_{22}, Y_{22}, Z_{22} \in \mathfrak{A}_{22}$. Now set

$$f_n(T) = f_{n1}(PTP) + \tau^{-1}(f_{n1}(PTP)) + f_{n2}(QTQ) + \tau(f_{n2}(QTQ)).$$

Obviously,$f_n(T) \in Z(\mathfrak{A})$ and $f_n([[U, V], W]) = 0$ for all $U, V, W \in \mathfrak{A}$. Now define a new mapping as follows

$$\delta_n(T) = L_n(T) - f_n(T) \text{ for all } T \in \mathfrak{A}.$$

From Claim 1 and Claim 2, it is clear that

Claim 3. $\delta_n(0) = 0, \delta_n(\mathfrak{A}_{12}) \subseteq \mathfrak{A}_{12}, \delta_n(P), \delta_n(Q) \subseteq \mathfrak{A}_{12}$ and $\delta_n(\mathfrak{A}_{ii}) \subseteq \mathfrak{A}_{ii} + \mathfrak{A}_{12}$ for $i = 1, 2$.

Claim 4. For any $A_{11} \in \mathfrak{A}_{11}, M_{12} \in \mathfrak{A}_{12}$ and $B_{22} \in \mathfrak{A}_{22}$, we have

(*i*) $\delta_n(A_{11}M_{12}) = \delta_n(A_{11})M_{12} + A_{11}\delta_n(M_{12}) + \sum\limits_{\substack{p+q=n \\ 0<p,q<n}} d_p(A_{11})d_q(M_{12})$,

(*ii*) $\delta_n(M_{12}B_{22}) = \delta_n(M_{12})B_{22} + M_{12}\delta_n(B_{22}) + \sum\limits_{\substack{p+q=n \\ 0<p,q<n}} d_p(M_{12})d_q(B_{22})$.

Since $A_{11}M_{12} = [[A_{11}, M_{12}], Q]$ this implies that

$$\begin{aligned} \delta_n(A_{11}M_{12}) &= L_n([[A_{11}, M_{12}], Q]) \\ &= \sum_{p+q+r=n} [[L_p(A_{11}), L_q(M_{12})], L_r(Q)] \\ &= [[\delta_n(A_{11}), M_{12}], Q] + [[A_{11}, \delta_n(M_{12})], Q] + [[A_{11}, M_{12}], \delta_n(Q)] \\ &\quad + \sum_{\substack{p+q+r=n \\ p,q,r>0}} [[d_p(A_{11}), d_q(M_{12})], d_r(Q)]. \end{aligned}$$

As $\delta_n(\mathfrak{A}_{11}) \subseteq \mathfrak{A}_{11}+\mathfrak{A}_{12}, \delta_n(M_{12}), \delta_n(Q) \in \mathfrak{A}_{12}$ and using condition $\mathbf{C}_s$, the equation becomes

$$\delta_n(A_{11}M_{12}) = \delta_n(A_{11})M_{12} + A_{11}\delta_n(M_{12}) + \sum_{\substack{p+q=n \\ 0<p,q<n}} d_p(A_{11})d_q(M_{12}).$$

Also, we know that $M_{12}B_{22} = [[M_{12}, B_{22}], Q]$ for $M_{12} \in \mathfrak{A}_{12}$ and $B_{22} \in \mathfrak{A}_{22}$. Similarly, we can show that

$$\delta_n(M_{12}B_{22}) = \delta_n(M_{12})B_{22} + M_{12}\delta_n(B_{22}) + \sum_{\substack{p+q=n \\ 0<p,q<n}} d_p(M_{12})d_q(B_{22}).$$

Claim 5. For any $A_{ii}, B_{ii} \in \mathfrak{A}_{ii}$ $(i = 1, 2)$, we have

$$\delta_n(A_{ii}B_{ii}) = \delta_n(A_{ii})B_{ii} + A_{ii}\delta_n(B_{ii}) + \sum_{\substack{p+q=n \\ 0<p,q<n}} d_p(A_{ii})d_q(B_{ii}).$$

For any $A_{11}, B_{11} \in \mathfrak{A}_{11}$ and $M_{12} \in \mathfrak{A}_{12}$.

$$\begin{aligned}
&\delta_n(A_{11}B_{11}M_{12}) \\
&= \delta_n((A_{11}B_{11})M_{12}) \\
&= \delta_n(A_{11}B_{11})M_{12} + A_{11}B_{11}\delta_n(M_{12}) + \sum_{\substack{p+q=n \\ 0<p,q<n}} d_p(A_{11}B_{11})d_q(M_{12}) \\
&= \delta_n(A_{11}B_{11})M_{12} + A_{11}B_{11}\delta_n(M_{12}) + \sum_{\substack{p+q+r=n \\ p,q,r>0}} d_p(A_{11})d_q(B_{11})d_r(M_{12})
\end{aligned}$$

and

$$\begin{aligned}
&\delta_n(A_{11}B_{11}M_{12}) \\
&= \delta_n(A_{11}(B_{11}M_{12})) \\
&= \delta_n(A_{11})B_{11}M_{12} + A_{11}\delta_n(B_{11}M_{12}) + \sum_{\substack{p+q=n \\ 0<p,q<n}} d_p(A_{11})d_q(B_{11}M_{12}) \\
&= \delta_n(A_{11})B_{11}M_{12} + A_{11}\delta_n(B_{11})M_{12} + A_{11}B_{11}\delta_n(M_{12}) \\
&\quad + \sum_{\substack{p+q=n \\ 0<p,q<n}} A_{11}d_p(B_{11})d_q(M_{12}) + \sum_{\substack{p+q+r=n \\ p,q,r>0}} d_p(A_{11})d_q(B_{11})d_r(M_{12}) \\
&= \delta_n(A_{11})B_{11}M_{12} + A_{11}\delta_n(B_{11})M_{12} + A_{11}B_{11}\delta_n(M_{12}) \\
&\quad + \sum_{\substack{p+q=n \\ 0<p,q<n}} d_p(A_{11})d_q(B_{11})M_{12} + \sum_{\substack{p+q+r=n \\ p,q,r>0}} d_p(A_{11})d_q(B_{11})d_r(M_{12}).
\end{aligned}$$

By the condition $\mathbf{C}_s$, the above expressions become

$$\delta_n(A_{11}B_{11})M_{12} = \{\delta_n(A_{11})B_{11} + A_{11}\delta_n(B_{11}) + \sum_{\substack{p+q=n \\ 0<p,q<n}} d_p(A_{11})d_q(B_{11})\}M_{12}.$$

Since $\delta_n(\mathfrak{A}_{11}) \subseteq \mathfrak{A}_{11} + \mathfrak{A}_{12}$ and $P\mathfrak{A}Q$ is faithful as left $P\mathfrak{A}P$-module, the above relation implies that

$$\delta_n(A_{11}B_{11})P = \{\delta_n(A_{11})B_{11} + A_{11}\delta_n(B_{11}) + \sum_{\substack{p+q=n \\ 0<p,q<n}} d_p(A_{11})d_q(B_{11})\}P. \tag{3.2}$$

Also, $[[A_{11}, Q], Q] = 0$ for all $A_{11} \in \mathfrak{A}_{11}$

$$\begin{aligned}
0 &= L_n([[A_{11}, Q], Q]) \\
&= [[L_n(A_{11}), Q], Q] + [[A_{11}, L_n(Q)], Q] + [[A_{11}, Q], L_n(Q)] \\
&\quad + \sum_{\substack{p+q+r=n \\ p,q,r>0}} [[L_p(A_{11}), L_q(Q)], L_r(Q)] \\
&= [[\delta_n(A_{11}), Q], Q] + [[A_{11}, \delta_n(Q)], Q] + [[A_{11}, Q], \delta_n(Q)] \\
&\quad + \sum_{\substack{p+q+r=n \\ p,q,r>0}} [[d_p(A_{11}), d_q(Q)], d_r(Q)].
\end{aligned}$$

Since $\delta_n(\mathfrak{A}_{11}) \subseteq \mathfrak{A}_{11} + \mathfrak{A}_{12}, \delta_n(Q) \in \mathfrak{A}_{12}$. The above equation implies that

$$0 = \delta_n(A_{11})Q + A_{11}\delta_n(Q) + \sum_{\substack{p+q=n \\ 0<p,q<n}} d_p(A_{11})d_q(Q) \text{ for all } A_{11} \in \mathfrak{A}_{11}. \tag{3.3}$$

On substituting A_{11} by B_{11} and $A_{11}B_{11}$ in (3.3) respectively, we get

$$0 = \delta_n(B_{11})Q + B_{11}\delta_n(Q) + \sum_{\substack{p+q=n \\ 0<p,q<n}} d_p(B_{11})d_q(Q) \tag{3.4}$$

and

$$0 = \delta_n(A_{11}B_{11})Q + A_{11}B_{11}\delta_n(Q) + \sum_{\substack{p+q=n \\ 0<p,q<n}} d_p(A_{11}B_{11})d_q(Q). \tag{3.5}$$

Now left multiplying A_{11} in (3.4) and combining it with (3.5) gives

$$\delta_n(A_{11}B_{11})Q + \sum_{\substack{p+q+r=n \\ 0<r,0<p}} d_p(A_{11})d_q(B_{11})d_r(Q) = A_{11}\delta_n(B_{11})Q$$

which implies that

$$\delta_n(A_{11}B_{11})Q + \sum_{p=1}^{n-1} d_p(A_{11}) \sum_{\substack{q+r=n-p \\ 0<r}} d_q(B_{11})d_r(Q) = A_{11}\delta_n(B_{11})Q.$$

Now using the condition $\mathbf{C}_s$, we find that

$$\delta_n(A_{11}B_{11})Q - \sum_{p=1}^{n-1} d_p(A_{11})d_{n-p}(B_{11})Q = A_{11}\delta_n(B_{11})Q$$

which gives

$$\delta_n(A_{11}B_{11})Q = A_{11}\delta_n(B_{11})Q + \sum_{p=1}^{n-1} d_p(A_{11})d_{n-p}(B_{11})Q.$$

Hence,

$$\delta_n(A_{11}B_{11})Q = \{\delta_n(A_{11})B_{11} + A_{11}\delta_n(B_{11}) + \sum_{\substack{p+q=n \\ 0<p,q<n}} d_p(A_{11})d_q(B_{11})\}Q. \tag{3.6}$$

Now adding the (3.2) and (3.6), we have

$$\delta_n(A_{11}B_{11}) = \delta_n(A_{11})B_{11} + A_{11}\delta_n(B_{11}) + \sum_{\substack{p+q=n \\ 0<p,q<n}} d_p(A_{11})d_q(B_{11}).$$

Similarly, we can obtain that

$$\delta_n(A_{22}B_{22}) = \delta_n(A_{22})B_{22} + A_{22}\delta_n(B_{22}) + \sum_{\substack{p+q=n \\ 0<p,q<n}} d_p(A_{22})d_q(B_{22})$$

for all $A_{22}, B_{22} \in \mathfrak{A}_{22}$.

Claim 6. For any $A_{11} \in \mathfrak{A}_{11}, B_{22} \in \mathfrak{A}_{22}$ and $M_{12}, N_{12} \in \mathfrak{A}_{12}$, we have

(*i*) $\delta_n(A_{11} + M_{12}) - \delta_n(A_{11}) - \delta_n(M_{12}) \in Z(\mathfrak{A})$,

(*ii*) $\delta_n(B_{22} + N_{12}) - \delta_n(B_{22}) - \delta_n(N_{12}) \in Z(\mathfrak{A})$.

Since $A_{11}X_{12} = [[A_{11} + M_{12}, X_{12}], Q]$ for all $A_{11} \in \mathfrak{A}_{11}$ and $M_{12}, N_{12} \in \mathfrak{A}_{12}$. Now

$$\begin{aligned}
&\delta_n(A_{11}X_{12})\\
&= L_n[[A_{11} + M_{12}, X_{12}], Q]\\
&= \sum_{p+q+r=n} [[L_p(A_{11} + M_{12}), L_q(X_{12})], L_r(Q)]\\
&= [[\delta_n(A_{11} + M_{12}), X_{12}], Q] + [[A_{11} + M_{12}, \delta_n(X_{12})], Q]\\
&\quad +[[A_{11} + M_{12}, X_{12}], \delta_n(Q)] + \sum_{\substack{p+q+r=n\\ p,q,r>0}} [[d_p(A_{11} + M_{12}), d_q(X_{12})], d_r(Q)].
\end{aligned}$$

Using the condition $\mathbf{C}_s$ and $\delta_n(X_{12}), \delta_n(Q) \in \mathfrak{A}_{12}, \delta_n(\mathfrak{A}_{11}) \subseteq \mathfrak{A}_{11} + \mathfrak{A}_{12}$, we get

$$\delta_n(A_{11}X_{12}) = [[\delta_n(A_{11} + M_{12}), X_{12}], Q] + A_{11}\delta_n(X_{12}).$$

Now using Claim $4(i)$, we have

$$[[\delta_n(A_{11} + M_{12}) - \delta_n(A_{11}), X_{12}], Q] = 0. \tag{3.7}$$

Assume that $\delta_n(A_{11} + M_{12}) - \delta_n(A_{11}) = \begin{bmatrix} a & m \\ 0 & b \end{bmatrix}$ and by (3.7), we have

$$\delta_n(A_{11} + M_{12}) - \delta_n(A_{11}) - P(\delta_n(A_{11} + M_{12}) - \delta_n(A_{11}))Q \in Z(\mathfrak{A}). \tag{3.8}$$

It is clear that $PTQ = [[T, P], P]$ for all $T \in \mathfrak{A}$. Therefore,

$$\begin{aligned}
&P(\delta_n(A_{11} + M_{12}) - \delta_n(A_{11}))Q\\
&= [[\delta_n(A_{11} + M_{12}) - \delta_n(A_{11}), P], P]\\
&= [[L_n(A_{11} + M_{12}), P], P] - [[L_n(A_{11}), P], P]\\
&= L_n[[(A_{11} + M_{12}), P], P] - [[M_{12}, L_n(P)], P]\\
&\quad -[[M_{12}, P], L_n(P)] - \sum_{\substack{p+q+r=n\\ p,q,r>0}} [[L_p(M_{12}), L_q(P)], L_r(P)].
\end{aligned}$$

Applying condition $\mathbf{C}_s$ and using Claim 1, we have

$$P(\delta_n(A_{11} + M_{12}) - \delta_n(A_{11}))Q = \delta_n(M_{12}).$$

This implies that

$$\delta_n(A_{11} + M_{12}) - \delta_n(A_{11}) - \delta_n(M_{12}) \in Z(\mathfrak{A}).$$

Similarly, we can show that

$$\delta_n(B_{22} + N_{12}) - \delta_n(B_{22}) - \delta_n(N_{12}) \in Z(\mathfrak{A}).$$

Claim 7. δ_n is additive on $\mathfrak{A}_{11}, \mathfrak{A}_{12}$ and $\mathfrak{A}_{22}$ respectively.
For any $M_{12}, N_{12} \in \mathfrak{A}_{12}$

$$\begin{aligned}
&\delta_n(M_{12} + N_{12})\\
&= L_n[P, [P + M_{12}, Q + N_{12}]]\\
&= \sum_{p+q+r=n} [L_p(P), [L_q(P + M_{12}), L_r(Q + N_{12})]]\\
&= [\delta_n(P), [P + M_{12}, Q + N_{12}]] + [P, [\delta_n(P + M_{12}), Q + N_{12}]]\\
&\quad +[P, [P + M_{12}, \delta_n(Q + N_{12})]]\\
&\quad + \sum_{\substack{p+q+r=n\\ p,q,r>0}} [d_p(P), [d_q(P + M_{12}), d_r(Q + N_{12})]].
\end{aligned}$$

Using condition $\mathbf{C}_s$, Claim 3 and Claim 6, we get

$$\delta_n(M_{12} + N_{12}) = \delta_n(M_{12}) + \delta_n(N_{12}).$$

Now,

$$\begin{aligned} \delta_n((A_{11}+B_{11})M_{12}) &= \delta_n(A_{11}M_{12})+\delta_n(B_{11}M_{12}) \\ &= \delta_n(A_{11})M_{12}+A_{11}\delta_n(M_{12})+\sum_{\substack{p+q=n\\0<p,q<n}} d_p(A_{11})d_q(M_{12}) \\ &\quad +\delta_n(B_{11})M_{12}+B_{11}\delta_n(M_{12})+\sum_{\substack{p+q=n\\0<p,q<n}} d_p(B_{11})d_q(M_{12}) \end{aligned} \tag{3.9}$$

for all $A_{11}, B_{11} \in \mathfrak{A}_{11}$ and $M_{12} \in \mathfrak{A}_{12}$. On the other hand,

$$\begin{aligned} \delta_n((A_{11}+B_{11})M_{12}) &= \delta_n(A_{11}+B_{11})M_{12}+(A_{11}+B_{11})\delta_n(M_{12}) \\ &\quad +\sum_{\substack{p+q=n\\0<p,q<n}} d_p(A_{11}+B_{11})d_q(M_{12}). \end{aligned} \tag{3.10}$$

Combining (3.9), (3.10) and applying condition $\mathbf{C}_s$, we have

$$\delta_n(A_{11}+B_{11})M_{12} = \delta_n(A_{11})M_{12}+\delta_n(B_{11})M_{12}. \tag{3.11}$$

Since $\delta_n(\mathfrak{A}_{11}) \subseteq \mathfrak{A}_{11}+\mathfrak{A}_{12}$ and $P\mathfrak{A}Q$ is faithful as a left $P\mathfrak{A}P$. Then (3.11) implies that

$$\delta_n(A_{11}+B_{11})P = \delta_n(A_{11})P+\delta_n(B_{11})P. \tag{3.12}$$

Replace A_{11} for $A_{11}+B_{11}$ in (3.3), we get

$$0 = \delta_n(A_{11}+B_{11})Q+(A_{11}+B_{11})\delta_n(Q)+\sum_{\substack{p+q=n\\0<p,q<n}} d_p(A_{11}+B_{11})d_q(Q) \tag{3.13}$$

for all $A_{11} \in \mathfrak{A}_{11}$. Combining (3.13) with (3.3) and (3.4), we obtain

$$\delta_n(A_{11}+B_{11})Q = \delta_n(A_{11})Q+\delta_n(B_{11})Q. \tag{3.14}$$

Addition of (3.12) and (3.14) implies that

$$\delta_n(A_{11}+B_{11}) = \delta_n(A_{11})+\delta_n(B_{11}).$$

Similarly, we can deduce that

$$\delta_n(A_{22}+B_{22}) = \delta_n(A_{22})+\delta_n(B_{22}).$$

Claim 8. For any $A_{ij} \in \mathfrak{A}_{ij}$ with $1 \leq i \leq j \leq 2$, we have

$$\delta_n(A_{11}+A_{12}+A_{22})-\delta_n(A_{11})-\delta_n(A_{12})-\delta_n(A_{22}) \in Z(\mathfrak{A}).$$

For $M_{12} \in \mathfrak{A}_{12}$ and $A_{22} \in \mathfrak{A}_{22}$ by Claim 3 and Claim 7, we have

$$\begin{aligned} 0 &= \delta_n(0) \\ &= \delta_n(M_{12}A_{22}-M_{12}A_{22}) \\ &= \delta_n(M_{12}A_{22})+\delta_n(-M_{12}A_{22}) \end{aligned}$$

which implies that $\delta_n(-M_{12}A_{22}) = -\delta_n(M_{12}A_{22})$. By Claim 4 and Claim 7, we get

$$\begin{aligned}
&\delta_n(A_{11}M_{12} - M_{12}A_{22}) \\
&= \delta_n(A_{11}M_{12}) - \delta_n(M_{12}A_{22}) \\
&= \delta_n(A_{11})M_{12} + A_{11}\delta_n(M_{12}) + \sum_{\substack{p+q=n \\ 0<p,q<n}} d_p(A_{11})d_q(M_{12}) \\
&\quad -\delta_n(M_{12})A_{22} - M_{12}\delta_n(A_{22}) - \sum_{\substack{p+q=n \\ 0<p,q<n}} d_p(M_{12})d_q(A_{22}).
\end{aligned} \tag{3.15}$$

Also, it is clear that $A_{11}M_{12}-M_{12}A_{22} = [[A_{11}+A_{12}+A_{22}, M_{12}], Q]$ for all $A_{ij}, M_{ij} \in \mathfrak{A}_{ij}$ with $i \leq i \leq j \leq 2$. Using Claim 3 and by condition $\mathbf{C}_s$, we obtain

$$\begin{aligned}
&\delta_n(A_{11}M_{12} - M_{12}A_{22}) \\
&= L_n[[A_{11} + A_{12} + A_{22}, M_{12}], Q] \qquad (3.16) \\
&= \sum_{p+q+r=n} [[L_p(A_{11} + A_{12} + A_{22}), L_q(M_{12})], L_r(Q)] \\
&= [[\delta_n(A_{11} + A_{12} + A_{22}), M_{12}], Q] + [[A_{11} + A_{12} + A_{22}, \delta_n(M_{12})], Q] \\
&\quad +[[A_{11} + A_{12} + A_{22}, M_{12}], \delta_n(Q)] \\
&\quad + \sum_{\substack{p+q+r=n \\ p,q,r>0}} [[d_p(A_{11} + A_{12} + A_{22}), d_q(M_{12})], d_r(Q)] \\
&= [[\delta_n(A_{11} + A_{12} + A_{22}), M_{12}], Q] + A_{11}\delta_n(M_{12}) - \delta_n(M_{12})A_{22} \\
&\quad + \sum_{\substack{p+q=n \\ 0<p,q<n}} (d_p(A_{11})d_q(M_{12}) - d_p(M_{12})d_q(A_{22})). \qquad (3.17)
\end{aligned}$$

Combining (3.15) and (3.17) implies that

$$\delta_n(A_{11})M_{12} - M_{12}\delta_n(A_{22}) = [[\delta_n(A_{11} + A_{12} + A_{22}), M_{12}], Q]$$

which is also written as

$$[[\delta_n(A_{11} + A_{12} + A_{22}) - \delta_n(A_{11}) - \delta_n(A_{22}), M_{12}], Q] = 0.$$

Let us assume $\delta_n(A_{11} + A_{12} + A_{22}) - \delta_n(A_{11}) - \delta_n(A_{22}) = \begin{bmatrix} a & m \\ 0 & b \end{bmatrix}$. It can be easily seen that

$$\begin{aligned}
&\delta_n(A_{11} + A_{12} + A_{22}) - \delta_n(A_{11}) - \delta_n(A_{22}) \\
&\quad -P(\delta_n(A_{11} + A_{12} + A_{22}) - \delta_n(A_{11}) - \delta_n(A_{22}))Q \in Z(\mathfrak{A}).
\end{aligned}$$

To determine the term $P(\delta_n(A_{11}+A_{12}+A_{22})-\delta_n(A_{11})-\delta_n(A_{22}))Q$, we use $P\mathfrak{A}Q=[[T,P],P]$. Therefore,

$$\begin{aligned}
&P(\delta_n(A_{11}+A_{12}+A_{22})-\delta_n(A_{11})-\delta_n(A_{22}))Q\\
&= [[\delta_n(A_{11}+A_{12}+A_{22})-\delta_n(A_{11})-\delta_n(A_{22}),P],P]\\
&= [[L_n(A_{11}+A_{12}+A_{22}),P],P]-[[L_n(A_{11}),P],P]-[[L_n(A_{22}),P],P]\\
&= L_n[[A_{11}+A_{12}+A_{22},P],P]-[[A_{11}+A_{12}+A_{22},L_n(P)],P]\\
&\quad -L_n[[A_{11},P],P]-[[A_{11}+A_{12}+A_{22},P],L_n(P)]+[[A_{11},L_n(P)],P]\\
&\quad +[[A_{22},P],L_n(P)]-\sum_{\substack{p+q+r=n\\ p,q,r>0}}[[L_p(A_{11}+A_{12}+A_{22}),L_q(P)],L_r(P)]\\
&\quad +\sum_{\substack{p+q+r=n\\ p,q,r>0}}[[L_p(A_{11}),L_q(P)],L_r(P)]+[[A_{11},P],L_n(P)]+[[A_{22},L_n(P)],P]\\
&\quad +\sum_{\substack{p+q+r=n\\ p,q,r>0}}[[L_p(A_{22}),L_q(P)],L_r(P)]\\
&= \delta_n[[A_{11}+A_{12}+A_{22},P],P]-[[A_{12}],\delta_n(P)],P]-[[A_{12},P],\delta_n(P)]\\
&\quad -\sum_{\substack{p+q+r=n\\ p,q,r>0}}[[d_p(A_{11}+A_{12}+A_{22}),d_q(P)],d_r(P)]-\delta_n[[A_{11},P],P]\\
&\quad +\sum_{\substack{p+q+r=n\\ p,q,r>0}}[[d_p(A_{11}),d_q(P)],d_r(P)]-\delta_n[[A_{22},P],P]\\
&\quad +\sum_{\substack{p+q+r=n\\ p,q,r>0}}[[d_p(A_{22}),d_q(P)],d_r(P)]\\
&= \delta_n(A_{12}).
\end{aligned}$$

Therefore, $\delta_n(A_{11}+A_{12}+A_{22})-\delta_n(A_{11})-\delta_n(A_{12})-\delta_n(A_{22})\in Z(\mathfrak{A})$ as required.

REMARK 3.4. Now we establish a mapping $g_n:\mathfrak{A}\to Z(\mathfrak{A})$ by

$$g_n(T)=\delta_n(T)-\delta_n(PTP)-\delta_n(PTQ)-\delta_n(QTQ) \text{ for all } T\in\mathfrak{A}.$$

Then define a mapping $d_n(T)=\delta_n(T)-g_n(T)$ for all $T\in\mathfrak{A}$. It is easy to verify d_n satisfy the following property

$$d_n(A_{11}+A_{12}+A_{22})=d_n(A_{11})+d_n(A_{12})+d_n(A_{22}).$$

From the definition of d_n and g_n that

$$L_n=\delta_n+f_n=d_n+g_n+f_n=d_n+h_n$$

where $h_n=g_n+f_n$.

Claim 9. $\{d_n\}_{i=0}^{i=n}$ is a higher derivation on triangular algebra $\mathfrak{A}$.
Suppose $X,Y\in\mathfrak{A}$ such that $X=X_{11}+X_{12}+X_{22}$ and $Y=Y_{11}+Y_{12}+Y_{22}$ where $X_{ij},Y_{ij}\in\mathfrak{A}_{ij}$ with $1\le i\le j\le 2$. Then

$$\begin{aligned}
d_n(X+Y) &= d_n\big((X_{11}+X_{12}+X_{22})+(Y_{11}+Y_{12}+Y_{22})\big)\\
&= d_n\big((X_{11}+Y_{11})+(X_{12}+Y_{12})+(X_{22}+Y_{22})\big)\\
&= \delta_n(X_{11}+Y_{11})+\delta_n(X_{12}+Y_{12})+\delta_n(X_{22}+Y_{22})\\
&= \delta_n(X_{11})+\delta_n(Y_{11})+\delta_n(X_{12})+\delta_n(Y_{12})+\delta_n(X_{22})+\delta_n(Y_{22})\\
&= \delta_n(X_{11})+\delta_n(X_{12})+\delta_n(X_{22})+\delta_n(Y_{11})+\delta_n(Y_{12})+\delta_n(Y_{22})\\
&= d_n(X_{11}+X_{12}+X_{22})+d_n(Y_{11}+Y_{12}+Y_{22})\\
&= d_n(X)+d_n(Y).
\end{aligned}$$

By Claim 4 and Claim 5, we have

$$\begin{aligned} d_n(XY) &= d_n\big((X_{11}+X_{12}+X_{22})(Y_{11}+Y_{12}+Y_{22})\big) \\ &= \delta_n(X_{11}Y_{11}+X_{11}Y_{12}+X_{12}Y_{22}+X_{22}Y_{22}) \\ &= \delta_n(X_{11})Y_{11}+X_{11}\delta_n(Y_{11})+\sum_{\substack{p+q=n\\0<p<n}} d_p(X_{11})d_q(Y_{11}) \\ &\quad +\delta_n(X_{11})Y_{12}+X_{11}\delta_n(Y_{12})+\sum_{\substack{p+q=n\\0<p<n}} d_p(X_{11})d_q(Y_{12}) \\ &\quad +\delta_n(X_{12})Y_{22}+X_{12}\delta_n(Y_{22})+\sum_{\substack{p+q=n\\0<p<n}} d_p(X_{12})d_q(Y_{22}) \\ &\quad +\delta_n(X_{22})Y_{22}+X_{22}\delta_n(Y_{22})+\sum_{\substack{p+q=n\\0<p<n}} d_p(X_{22})d_q(Y_{22}). \end{aligned} \tag{3.18}$$

On the other hand, we have

$$\begin{aligned} & d_n(X)Y+Xd_n(Y)+\sum_{\substack{p+q=n\\0<p<n}} d_p(X)d_q(Y) \\ = \;& (\delta_n(X_{11})+\delta_n(X_{12})+\delta_n(X_{22}))Y+\sum_{\substack{p+q=n\\0<p<n}} d_p(X_{11})d_q(Y_{11}) \\ & +X(\delta_n(Y_{11})+\delta_n(Y_{12})+\delta_n(Y_{22}))+\sum_{\substack{p+q=n\\0<p<n}} d_p(X_{11})d_q(Y_{12}) \\ & +\sum_{\substack{p+q=n\\0<p<n}} d_p(X_{11})d_q(Y_{22})+\sum_{\substack{p+q=n\\0<p<n}} d_p(X_{12})d_q(Y_{11}) \\ & +\sum_{\substack{p+q=n\\0<p<n}} d_p(X_{12})d_q(Y_{12})+\sum_{\substack{p+q=n\\0<p<n}} d_p(X_{12})d_q(Y_{22})) \\ & +\sum_{\substack{p+q=n\\0<p<n}} d_p(X_{22})d_q(Y_{11})+\sum_{\substack{p+q=n\\0<p<n}} d_p(X_{22})d_q(Y_{12}) \\ & +\sum_{\substack{p+q=n\\0<p<n}} d_p(X_{22})d_q(Y_{22}). \end{aligned}$$

By using condition $\mathbf{C}_s$, and from Claim 3, we have

$$\begin{aligned} & d_n(X)Y+Xd_n(Y)+\sum_{\substack{p+q=n\\0<p<n}} d_p(X)d_q(Y) \\ = \;& \delta_n(X_{11})Y_{11}+X_{11}\delta_n(Y_{11})+\sum_{\substack{p+q=n\\0<p<n}} d_p(X_{11})d_q(Y_{11})+\delta_n(X_{11})Y_{12} \\ & +X_{11}\delta_n(Y_{12})+\sum_{\substack{p+q=n\\0<p<n}} d_p(X_{11})d_q(Y_{12})+\delta_n(X_{12})Y_{22}+X_{12}\delta_n(Y_{22}) \\ & +\sum_{\substack{p+q=n\\0<p<n}} d_p(X_{12})d_q(Y_{22})+\delta_n(X_{22})Y_{22}+X_{22}\delta_n(Y_{22}) \\ & +\sum_{\substack{p+q=n\\0<p<n}} d_p(X_{22})d_q(Y_{22}). \end{aligned} \tag{3.19}$$

Combining (3.18) and (3.19) ,we get

$$d_n(XY) = d_n(X)Y + Xd_n(Y) + \sum_{\substack{p+q=n \\ 0<p<n}} d_p(X)d_q(Y).$$

Therefore, $\{d_n\}_{i=0}^{i=n}$ is an additive higher derivation on $\mathfrak{A}$. This implies that, $\{d_n\}_{n\in\mathbb{N}}$ is an additive higher derivation on $\mathfrak{A}$. Finally, there exist a map $h_n : \mathfrak{A} \to Z(\mathfrak{A})$ such that $h_n[[X,Y],Z] = L_n[[X,Y],Z] - d_n[[X,Y],Z] = 0$ for all $X, Y, Z \in \mathfrak{A}$. This completes the proof of the theorem. □

4. Applications

In this section we study the application of Theorem 3.1 to some classical examples of a triangular algebra such as a upper triangular matrix algebra and a block upper triangular matrix algebra. For the definitions of a upper triangular matrix algebra and a block upper triangular matrix algebra see [**B**] and references therein.

COROLLARY 4.1. *Every nonlinear Lie triple higher derivation has standard form on a upper triangular matrix algebra $T_n(\mathcal{R})$.*

PROOF. Consider $n = 2$, then $T_2(\mathcal{R}) = \begin{pmatrix} \mathcal{R} & \mathcal{R} \\ 0 & \mathcal{R} \end{pmatrix}$. It can be easily seen that $T_2(\mathcal{R})$ satisfies conditions (1) and (2) of the Theorem 3.1. Therefore, there exist an additive derivation d and a nonlinear functional f vanishing all Lie triple product of $T_2(\mathcal{R})$ such that $L(T) = d(T) + f(T)$ for all $T \in T_2(\mathcal{R})$.
If $n \geq 3$, then $T_n(\mathcal{R}) = \begin{pmatrix} \mathcal{A} & \mathcal{M} \\ 0 & \mathcal{B} \end{pmatrix}$, where $\mathcal{A} = \mathcal{R}, \mathcal{M} = M_{1\times(n-1)}(\mathcal{R})$ and $\mathcal{B} = T_{n-1}(\mathcal{R})$. Therefore, the result follows from the Theorem 3.1. □

COROLLARY 4.2. *Every nonlinear Lie triple higher derivation has standard form on the block upper triangular matrix algebra $B_n^{\bar{k}}(\mathcal{R})$.*

References

[A] D. Benkovič and D. Eremita, *Multiplicative Lie n-derivations of triangular rings*, Linear Algebra Appl. **436** (2012), no. 11, 4223–4240. MR2915278

[B] W. S. Cheung, *Maps on Triangular Algebras*, Ph.D. Dissertation, University of Victoria, 2000.

[C] W.-S. Cheung, *Commuting maps of triangular algebras*, J. London Math. Soc. (2) **63** (2001), no. 1, 117–127. MR1802761

[D] W.-S. Cheung, *Lie derivations of triangular algebras*, Linear Multilinear Algebra **51** (2003), no. 3, 299–310. MR1995661

[E] M. Ferrero and C. Haetinger, *Higher derivations and a theorem by Herstein*, Quaest. Math. **25** (2002), no. 2, 249–257. MR1916335

[F] D. Han, *Lie-type higher derivations on operator algebras*, Bull. Iranian Math. Soc. **40** (2014), no. 5, 1169–1194. MR3273831

[H] P. Ji and W. Qi, *Characterizations of Lie derivations of triangular algebras*, Linear Algebra Appl. **435** (2011), no. 5, 1137–1146. MR2807225

[I] P. Ji, R. Liu, and Y. Zhao, *Nonlinear Lie triple derivations of triangular algebras*, Linear Multilinear Algebra **60** (2012), no. 10, 1155–1164. MR2983757

[J] J. Li and Q. Shen, *Characterizations of Lie higher and Lie triple derivations on triangular algebras*, J. Korean Math. Soc. **49** (2012), no. 2, 419–433. MR2933607

[K] X.-f. Qi and J.-c. Hou, *Lie higher derivations on nest algebras*, Commun. Math. Res. **26** (2010), no. 2, 131–143. MR2662638

[L] X. F. Qi, *Characterization of Lie higher derivations on triangular algebras*, Acta Math. Sin. (Engl. Ser.) **29** (2013), no. 5, 1007–1018. MR3040390

[M] F. Wei and Z. Xiao, *Higher derivations of triangular algebras and its generalizations*, Linear Algebra Appl. **435** (2011), no. 5, 1034–1054. MR2807218

[N] Z. Xiao and F. Wei, *Nonlinear Lie higher derivations on triangular algebras*, Linear Multilinear Algebra **60** (2012), no. 8, 979–994. MR2955278

[O] Z. Xiao and F. Wei, *Lie triple derivations of triangular algebras*, Linear Algebra Appl. **437** (2012), no. 5, 1234–1249. MR2942345

[P] W. Yu and J. Zhang, *Nonlinear Lie derivations of triangular algebras*, Linear Algebra Appl. **432** (2010), no. 11, 2953–2960. MR2639258

Department of Mathematics, Aligarh Muslim University, Aligarh 202002, India
Email address: mashraf80@hotmail.com

Department of Mathematics, Aligarh Muslim University, Aligarh 202002, India
Email address: ajabeen329@gmail.com

Contemporary Mathematics
Volume **715**, 2018
https://doi.org/10.1090/conm/715/14404

On universal localization of Noetherian rings

John A. Beachy and Christine M. Leroux

Dedicated to the memory of Bruno J. Mueller

ABSTRACT. We determine the universal localization $R_{\Gamma(S)}$ at certain semiprime ideals S of a Noetherian ring R that is a finite extension of its center. For this class of semiprime ideals, $R_{\Gamma(S)}$ is Noetherian, which need not be true in the general case.

Throughout this note, R will denote an associative ring with identity, and S will denote a semiprime right Goldie ideal of R. (That is, R/S is a semiprime right Goldie ring.) The study of the universal localization $R_{\Gamma(S)}$ of R at S was initiated by P. M. Cohn in [**6**] in the Noetherian case, and continued by the first author (in [**1**], [**2**], and [**3**]) and by others. We note that Goldie's localization, defined and studied in [**9**] and [**10**] for Noetherian rings, is a homomorphic image of the universal localization (see [**2**]). The major results in this paper are based on results in [**13**], the second author's dissertation. The description of $R_{\Gamma(S)}$ when R is a finite extension of its center depends on the seminal work of B. J. Mueller in [**15**].

For any ideal I of R, the set of elements $c \in R$ that are regular modulo I will be denoted by $\mathcal{C}(I)$. For any positive integer n, let $\Gamma_n(S)$ denote the set of all matrices C such that C belongs to the $n \times n$ matrix ring $M_n(R)$ and the image of C in $M_n(R/S)$ is a regular element. This will be abbreviated by saying that C is regular modulo S. The union over all $n > 0$ of $\Gamma_n(S)$ will be denoted by $\Gamma(S)$.

Following Cohn, the universal localization $R_{\Gamma(S)}$ of R at S is defined to be the universal $\Gamma(S)$-inverting ring. That is, $R_{\Gamma(S)}$ satisfies the following conditions: there is a ring homomorphism $\lambda : R \to R_{\Gamma(S)}$ under which each element of $\Gamma(S)$ is invertible, and for any $\Gamma(S)$-inverting ring homomorphism $\phi : R \to T$ there exists a unique ring homomorphism $\theta : R_{\Gamma(S)} \to T$ with $\phi = \theta\lambda$, as in the following diagram.

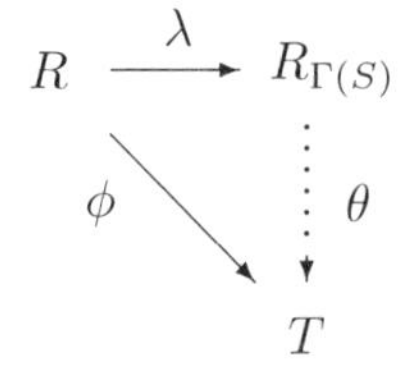

2010 *Mathematics Subject Classification.* Primary 16P20 16S10 16S85.

The ring $R_{\Gamma(S)}$ can be constructed as follows (see [**6**] and [**7**] for details). For each n and each $n \times n$ matrix $[c_{ij}]$ in $\Gamma(S)$, take a set of n^2 symbols $[d_{ij}]$, and take a ring presentation of $R_{\Gamma(S)}$ consisting of all of the elements of R, as well as all of the elements d_{ij} as generators; as defining relations take all of the relations holding in R, together with all of the relations $[c_{ij}][d_{ij}] = I$ and $[d_{ij}][c_{ij}] = I$ which define all of the inverses of the matrices in $\Gamma(S)$.

The universal localization was shown by Cohn in [**6**] to have a number of desirable properties: $\lambda : R \to R_{\Gamma(S)}$ is an epimorphism in the category of rings, and if S is right localizable then $R_{\Gamma(S)} \cong R_S$, where R_S is the right ring of fractions defined by $\mathcal{C}(S)$. The ring $R_{\Gamma(S)}$ modulo its Jacobson radical is naturally isomorphic to $Q_{cl}(R/S)$, the classical ring of right quotients of R/S, and Theorem 1.1 of [**1**] shows that $R_{\Gamma(S)}$ can be defined as the ring universal with respect to this property.

On the other hand, the universal localization also presents some difficulties. In general, it is quite hard to determine $\ker(\lambda)$. If $\mathcal{C}(S)$ is not a right denominator set, then $R_{\Gamma(S)}$ is not flat as a left R-module (Corollary 3.2 of [**1**]). Even when R is finitely generated as a module over its Noetherian center, $R_{\Gamma(S)}$ may not be Noetherian, as shown by the first of the following examples.

1. Examples

In Example 4 of [**1**], R_1 is the ring of lower triangular 2×2 matrices over the ring of integers $\mathbb{Z}$ and S is the semiprime ideal $\begin{bmatrix} p\mathbb{Z} & 0 \\ \mathbb{Z} & q\mathbb{Z} \end{bmatrix}$, where p and q are distinct prime numbers. Then $(R_1)_{\Gamma(S)} = \begin{bmatrix} \mathbb{Z}_{(p)} & 0 \\ \mathbb{Q} & \mathbb{Z}_{(q)} \end{bmatrix}$ is neither right nor left Noetherian, where $\mathbb{Q}$ is the field of rational numbers and $\mathbb{Z}_{(p)}$ and $\mathbb{Z}_{(q)}$ are the standard localizations of $\mathbb{Z}$.

To construct a similar example in case S is a prime ideal proved to be more difficult. We note such an example from [**4**]. A well-known example of Nagata (details can be found in [**17**] and [**18**]) provides two commutative Noetherian domains $A \subset B$ such that A is a local ring with maximal ideal M, and B contains two maximal ideals, J_1 and J_2, such that $J_1 \cap J_2 = M$. Let

$$R_2 = \left\{ \begin{bmatrix} b & 0 & 0 \\ x & a & 0 \\ y & z & c \end{bmatrix} \middle| \, b, c \in B; a \in A; x, y, z \in J_1; b - a, c - a \in J_1 \right\}$$

and let

$$P = \left\{ \begin{bmatrix} b & 0 & 0 \\ x & a & 0 \\ y & z & c \end{bmatrix} \middle| \, b, x, y, z, c \in J_1; a \in M \right\}.$$

Then it can be shown that the universal localization $(R_2)_{\Gamma(P)}$ of R_2 at the maximal ideal P is not Noetherian.

We note that $P \subset M_3(J_1)$, which is an ideal of the ring $M_3(B)$. Since B is a commutative Noetherian domain, $\cap_{n=1}^{\infty}(J_1)^n = (0)$ by Theorem 77 of [**12**], so $\cap_{n=1}^{\infty} P^n = (0)$ and thus $\cap_{n=1}^{\infty} P^n \subseteq Q$ for all prime ideals Q of R_2.

The relevant properties of these examples illustrate some of the obstructions to having a Noetherian universal localization. To provide an explanation, we need to utilize the notion of a link between two prime ideals of R. Following [**11**], we say that there is a link from a prime ideal $P \subset R$ to a prime ideal $Q \subset R$, denoted by $P \rightsquigarrow Q$, if there is an ideal A with $PQ \subseteq A \subset P \cap Q$ such that $(P \cap Q)/A$ is nonzero and torsionfree both as a right (R/Q)-module and as a left (R/P)-module. For a prime ideal $P \subset R$, we say that the prime ideal Q belongs to the clique of P, denoted by $\mathrm{Clq}(P)$, if there exist prime ideals $\{P = P_1, \ldots, P_m\}$ such that for $i = 1, \ldots, m-1$, either $P_i \rightsquigarrow P_{i+1}$ or $P_{i+1} \rightsquigarrow P_i$. When R is a Noetherian ring that is finitely generated as a module over its center Z, it has been shown by Mueller in [**14**] that

$$\mathrm{Clq}(P) = \{Q \in \mathrm{Spec}(R) \mid Q \cap Z = P \cap Z\}\ .$$

In the ring R_1 of the first example given above, we note that the semiprime ideal S is the intersection of two prime ideals from different cliques. In the second example, in the ring R_2 the prime ideal P belongs to a clique that contains at least one other prime, since P is not localizable. Therefore the intersection of the powers of P is contained in another prime ideal in $\mathrm{Clq}(P)$.

In Theorem 2.3, the main result of this paper, we show that if R is finitely generated as a module over its Noetherian center, and the semiprime ideal S is the intersection of prime ideals in the same clique of R, then $R_{\Gamma(S)}$ is a homomorphic image of a central localization, provided that the intersection of the symbolic powers of S is not contained in any other prime ideal of the given clique. We conclude that under these hypotheses $R_{\Gamma(S)}$ is Noetherian.

To illustrate this procedure, consider the ring $R_1 = \begin{bmatrix} \mathbb{Z} & 0 \\ \mathbb{Z} & \mathbb{Z} \end{bmatrix}$ and the prime ideal $P = \begin{bmatrix} p\mathbb{Z} & 0 \\ \mathbb{Z} & \mathbb{Z} \end{bmatrix}$, where p is a prime number. Then P is not right localizable, but $N = \begin{bmatrix} p\mathbb{Z} & 0 \\ \mathbb{Z} & p\mathbb{Z} \end{bmatrix}$, the intersection of the prime ideals in the clique of P, *is* right localizable, yielding $\begin{bmatrix} \mathbb{Z}_{(p)} & 0 \\ \mathbb{Z}_{(p)} & \mathbb{Z}_{(p)} \end{bmatrix}$ as the localization at N. Then $(R_1)_{\Gamma(P)}$ is obtained by factoring out the intersection of powers of the extension $\begin{bmatrix} p\mathbb{Z}_{(p)} & 0 \\ \mathbb{Z}_{(p)} & \mathbb{Z}_{(p)} \end{bmatrix}$ of P, and so $(R_1)_{\Gamma(P)} \cong \mathbb{Z}_{(p)}$. (In this case, the construction can be confirmed by noting that P is left localizable.)

2. Main Theorems

In commutative algebra, the Krull Intersection Theorem has been an extremely useful result and there are several interesting results that follow from it. See Kaplansky [**12**] or Eisenbud [**8**] for more information. We begin by developing a noncommutative extension of the Krull Intersection Theorem that leads to information regarding the structure of the universal localization in a particular case.

THEOREM 2.1. *Let R be a right fully bounded Noetherian ring, and let S be a semimaximal ideal of R. Suppose that $S = \cap_{i=1}^m P_i$ for maximal ideals $\{P_i\}_{i=1}^m$ in $\mathrm{Clq}(P)$, where $P = P_1$, and let $K = \cap_{n=1}^\infty S^n$. If $K \not\subseteq Q$ for any $Q \in \mathrm{Clq}(P)$ such that $Q \neq P_i$ for $i = 1, \ldots, m$, then $KS = K$.*

PROOF. Let I be a right ideal of R that is maximal with respect to the property that $I \cap K = PK$. (Note that $KS \cap K = KS$, so the given set is nonempty, and then I exists by Zorn's lemma.) It is well-known that $(K+I)/I$ is an essential submodule of R/I. Define $f : K \to (K+I)/I$ by $f(a) = a + I$ for every $a \in K$. We note that

f is an onto R-homomorphism, and that $\ker(f) = \{a \in K : a \in I\} = I \cap K = KS$, and thus $(K/KS)_R \cong ((K+I)/I)_R$.

By assumption, R/S is a semisimple Artinian ring, and since $(K+I)/I$ is a finitely generated right R/S-module, it is isomorphic to a direct sum $\oplus_{j=1}^t X_j$ of simple modules X_j. Since R/I is an essential extension of $(K+I)/I$, we can identify R/I with a submodule of the injective envelope of $(K+I)/I$, say $R/I \subseteq \oplus_{j=1}^t E(X_j)$. The projection of R/I into each component is a cyclic submodule Rx_j, so in fact we have $R/I \subseteq \oplus_{j=1}^t Rx_j$.

By assumption R is a right fully bounded Noetherian ring, and so Theorem 9.12 of [**11**] implies that each submodule Rx_j is Artinian. Furthermore, Corollary 12.14 of [**11**] implies that the annihilator of each composition factor of Rx_j belongs to the clique of the annihilator of X_j. It follows that each annihilator of a composition factor of R/I belongs to the clique of P, and so the same is true for the factor module $R/(K+I)$.

Thus there exists a set of maximal ideals $Q_1, \ldots Q_k$ whose product annihilates $R/(K+I)$, and so $\prod_{i=1}^k Q_i \subseteq K+I$. Because K annihilates each composition factor of $R/(K+I)$, we have $K \subseteq Q_i$ for $i = 1, \ldots, k$. By assumption this implies that each ideal Q_i is one of the original set $\{P_1, \ldots, P_m\}$ of maximal ideals. We conclude that $S^n \subseteq K + I$ for some positive integer n. Since S annihilates $(K+I)/I \cong K/KS$, we have $S^{n+1} \subseteq I$, and therefore $K \subseteq I$, so $KS = I \cap K = K$. □

In commutative ring theory, one approach to determining the kernel of a localization map is to utilize symbolic powers. In [**3**] this notion was extended to the setting of universal localization. Let R be a left Noetherian ring, with semiprime ideal S, and let $\lambda : R \to R_{\Gamma(S)}$ be a universal localization of R at S. The nth *symbolic power of* S, denoted by $S^{(n)}$, is defined as $S^{(n)} = \lambda^{-1}(R_{\Gamma(S)}\lambda(S^n)R_{\Gamma(S)})$. We note that Proposition 1.3 of [**2**] shows that this definition is equivalent to the one given by B. J. Mueller in [**14**], where $S^{(n)}$ is defined to be the intersection of all ideals I containing S^n for which $\mathcal{C}(S) \subseteq \mathcal{C}(I)$.

For our purposes, the importance of the symbolic powers is related to their connection to the kernel of the universal localization map. In [**3**] it is shown that if R is a right Noetherian ring, with semiprime ideal S, and $\lambda : R \to R_{\Gamma(S)}$ is the universal localization of R at S, then $S^{(n)} = \lambda^{-1}(J(R_{\Gamma(S)})^n)$.

COROLLARY 2.2. *Let R be a right fully bounded Noetherian ring, and let S be a semimaximal ideal of R. Suppose that $S = \cap_{i=1}^m P_i$ for maximal ideals $\{P_i\}_{i=1}^m$ in* $\mathrm{Clq}(P)$, *where $P = P_1$, and let $K = \cap_{n=1}^\infty S^n$. Let $\lambda : R \to R_{\Gamma(S)}$ be the canonical mapping of R into the universal localization $R_{\Gamma(S)}$. Then* $\ker(\lambda) = K$.

PROOF. Since R is a right fully bounded Noetherian ring and S is a semimaximal ideal, it follows that R/S is semisimple Artinian by Proposition 9.4 of [**11**]. Therefore each factor S^i/S^{i+1} is torsionfree over R/S since it is a module over a semisimple Artinian ring, and so $S^n = S^{(n)}$ for each n. It follows from Proposition 1.9 of [**3**] that $\ker(\lambda) \subseteq \cap_{i=1}^\infty S^{(n)} = K$.

Since K is finitely generated as a right ideal and $KS = K$ by Theorem 2.1, it follows directly from Lemma 1.5 of [**1**] that $K \subseteq \ker(\lambda)$. □

When R is a Noetherian ring that is finitely generated as a right module over its center we will construct the universal localization in two steps, first using an Ore localization, and then following it by a universal localization. This procedure uses an extension of Proposition 1.7 of [**1**], which shows that if P is a prime ideal minimal over the semiprime Goldie ideal N, then $R_{\Gamma(P)}$ is the universal localization of $R_{\Gamma(N)}$ at a maximal ideal. For the ideal I of R, we denote the ideal of $R_{\Gamma(S)}$ generated by $\lambda(I)$ by I^e.

THEOREM 2.3. *Let R be a ring that is finitely generated as a module over its Noetherian center, and let S be a semiprime ideal of R, with $S = \cap_{i=1}^{m} P_i$ for prime ideals $\{P_i\}_{i=1}^{m}$ minimal over S. Suppose that each prime ideal P_i belongs to* Clq(P), *where $P = P_1$, and let $N = \cap Q_i$ be the intersection of all prime ideals Q_i in* Clq(P).

(a) *The following conditions are equivalent:*

(1) *$R_{\Gamma(S)}$ is a homomorphic image of R_N;*

(2) *$\cap_{n=1}^{\infty} S^{(n)} \not\subseteq Q$ for any $Q \in$ Clq(P) such that $Q \neq P_i$ for $i = 1, \ldots, m$.*

(b) *If the conditions given in part (a) hold, then $R_{\Gamma(S)} = R_N / \cap_{n=1}^{\infty} (S^e)^n$.*

PROOF. Since R is assumed to be module-finite over its Noetherian center, and N is the intersection of a clique of prime ideals, it follows that N satisfies the hypotheses of Theorem 14.21 of [**11**], and so N is classically right localizable. Since R is a right fully bounded Noetherian ring, the same is true of R_N, by Lemma 2.1 of [**5**]. Furthermore, it follows from Section 4.3 of [**14**] that S^e is a semi-maximal ideal of R_N. Then Lemma 14.18 of [**11**] gives us that $NR_N = J(R_N)$, and that $R_N/J(R_N) \cong Q_{cl}(R/N)$, so $R_N/(NR_N) \cong Q_{cl}(R/N)$.

Let $\eta : R \to R_N$ and $\lambda : R_N \to (R_N)_{\Gamma(S^e)}$ be the standard ring homomorphisms. A modification of the proof of Proposition 1.7 of [**1**] shows that $\lambda\eta$ is the canonical ring homomorphism from R into $R_{\Gamma(S)}$. (That $(R_N)_{\Gamma(S^e)}$ corresponds to $R_{\Gamma(S)}$ also follows from our Proposition 3.2 given below.) Since R_N is a right fully bounded Noetherian ring and S^e is a semi-maximal ideal, it follows that R_N/S^e is semisimple Artinian by Proposition 9.4 of [**11**]. Therefore each factor $(S^e)^i/(S^e)^{i+1}$ is torsionfree over R_N/S^e since it is a module over a semisimple Artinian ring, and so $(S^e)^n = (S^e)^{(n)}$ for each n. Thus by Theorem 1.10 of [**3**], we have $\ker(\lambda) \subseteq \cap(S^e)^n$. Letting $K = \cap_{i=1}^{\infty}(S^e)^n$, it follows that $\ker(\lambda) \subseteq K$.

(1) $\Rightarrow$ (2) Suppose that $R_{\Gamma(S)}$ is a homomorphic image of R_N. We know that $R_{\Gamma(S)}$ modulo its Jacobson radical is isomorphic to $Q_{cl}(R/S)$. so its Jacobson radical must be S^e. Since R_N is a right fully bounded Noetherian ring, by Theorem 9.13 of [**11**] we have $\cap_{i=1}^{\infty}(S^e)^n \subseteq \ker(\lambda)$, since $R_{\Gamma(S)}$ is also a right fully bounded Noetherian ring. Thus $\ker(\lambda) = K$. Because S^e is the Jacobson radical of $R_{\Gamma(S)}$, the maximal ideals of $R_{\Gamma(S)}$ correspond to the prime ideals of R minimal over S, and so we cannot have $K \subseteq Q^e$ for any other prime ideal Q minimal over N. It follows that $K \not\subseteq Q$ for any $Q \in$ Clq(P) such that $Q \neq P_i$ for $i = 1, \ldots, m$.

(2) $\Rightarrow$ (1) Suppose that the given condition holds. Then it follows from Corollary 2.2 that $K = \ker(\lambda)$. Hence if we verify that $(R_N)_{\Gamma(S^e)} = R_N/K$, we will obtain the desired result.

Since $N = \cap Q_i$ where the prime ideals Q_i are the elements of the clique of P, all maximal ideals of R_N are of the form Q^e, where $Q \in$ Clq(P). However, since $K \not\subseteq Q^e$ for any $P \neq Q \in$ Clp(P), it follows that S^e/K is the intersection of the maximal ideals of R_N/K and so $S^e/K = J(R_N/K)$. It is well-known that two prime ideals P^e and Q^e of R_N are linked in R_N if and only if P and Q are linked in R. (See the remarks on page 12 of [**5**].) It follows that the prime ideals of R_N minimal over S^e form a clique in R_N/K, and hence S^e/K is a localizable semiprime ideal. But then since $S^e/K = J(R_N/K)$ and the classical localization of R_N at S^e coincides with $(R_N)_{\Gamma(S^e)} = R_N/K$, we have $R_{\Gamma(S)} = (R_N)_{\Gamma(S^e)} = R_N/K$. □

COROLLARY 2.4. *If the ring R and semiprime ideal S satisfy the hypotheses of Theorem 2.3, then $R_{\Gamma(S)}$ is a Noetherian ring.*

3. Some general results

We conclude this note with several propositions that shed some light on the general case of universal localization at a semiprime Goldie ideal. If Δ is any set of matrices over R, then there exists a universal Δ-inverting ring R_Δ, which can be constructed in a fashion similar that of $R_{\Gamma(S)}$ (see Theorem 7.2.4 of [**7**]).

LEMMA 3.1. *Let Γ and Δ be sets of matrices over the ring R, with $\Delta \subseteq \Gamma$. Let $\lambda : R \to R_\Gamma$ be the universal Γ-inverting map and let $\eta : R \to R_\Delta$ be the universal Δ-inverting map.*

If $\Sigma = \eta(\Gamma)$, and $\epsilon : R_\Delta \to (R_\Delta)_\Sigma$ is the universal Σ-inverting map, then there exists a natural isomorphism $\gamma : (R_\Delta)_\Sigma \to R_\Gamma$ with $\gamma\epsilon\eta = \lambda$.

PROOF. Since $\Delta \subseteq \Gamma$, it follows that $\lambda : R \to R_\Gamma$ is a Δ-inverting map. Then since $\eta : R \to R_\Delta$ is a universal Δ-inverting map, there exists a unique homomorphism $\alpha : R_\Delta \to R_\Gamma$ with $\alpha\eta = \lambda$.

Since $\epsilon : R_\Delta \to (R_\Delta)_\Sigma$ inverts $\Sigma = \eta(\Gamma)$, it follows that $\epsilon\eta$ is Γ-inverting. Since $\lambda : R \to R_\Gamma$ is a universal Γ-inverting map, there exists a unique homomorphism $\beta : R_\Gamma \to (R_\Delta)_\Sigma$ with $\beta\lambda = \epsilon\eta$.

Since $\alpha\eta = \lambda$ and λ is Γ-inverting, it follows that $\alpha : R_\Delta \to R_\Gamma$ inverts $\Sigma = \eta(\Gamma)$. Since $\epsilon : R_\Delta \to (R_\Delta)_\Sigma$ is a universal Σ-inverting map, there exists a unique homomorphism $\gamma : (R_\Delta)_\Sigma \to R_\Gamma$ with $\gamma\epsilon = \alpha$.

This leads to the following diagram.

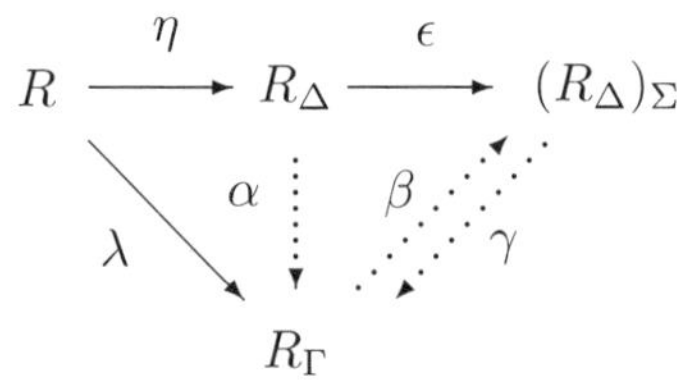

We claim that $\beta = \gamma^{-1}$, which will show that γ is the required isomorphism. We have

$$(\gamma\beta)\lambda = \gamma(\beta\lambda) = \gamma(\epsilon\eta) = (\gamma\epsilon)\eta = \alpha\eta = \lambda$$

and then $\gamma\beta = 1$ since λ is an epimorphism in the category of rings. On the other hand,

$$(\beta\gamma)(\epsilon\eta) = \beta(\gamma\epsilon)\eta = \beta(\alpha\eta) = \beta\lambda = \epsilon\eta$$

and so $\beta\gamma = 1$ since both η and ϵ are epimorphisms in the category of rings. □

PROPOSITION 3.2. *Let S be a semiprime right Goldie ideal of the ring R, and suppose that $\mathcal{C} \subseteq \mathcal{C}(S)$ is a right denominator set. If $R\mathcal{C}^{-1}/SR\mathcal{C}^{-1}$ is the classical ring of right quotients of R/S via the embedding induced by the canonical mapping $\eta : R \to R\mathcal{C}^{-1}$, then the universal localization $R_{\Gamma(S)}$ of R at S can be constructed as the universal localization of $R\mathcal{C}^{-1}$ at the semiprime ideal $SR\mathcal{C}^{-1}$.*

PROOF. In order to use Lemma 3.1, let $\overline{\eta} : R/S \to R\mathcal{C}^{-1}/SR\mathcal{C}^{-1}$ be the mapping induced by η, and let $\pi : R \to R/S$ and $\widehat{\pi} : R\mathcal{C}^{-1} \to R\mathcal{C}^{-1}/SR\mathcal{C}^{-1}$ be the canonical projections, as in the following diagram.

$$\begin{array}{ccc} R & \xrightarrow{\eta} & RC^{-1} \\ {\scriptstyle\pi}\downarrow & & \downarrow{\scriptstyle\widehat{\pi}} \\ R/S & \xrightarrow{\overline{\eta}} & RC^{-1}/SRC^{-1} \end{array}$$

The role of Δ in Lemma 3.1 will be played by the set $\mathcal{C}$, while the role of Σ will be played by $\eta(\Gamma(S))$. It follows from Lemma 3.1 that the universal localization $R_{\Gamma(S)}$ is naturally isomorphic to the universal inverting ring of the set $\eta(\Gamma(S))$ (defined on $R\mathcal{C}^{-1}$). We claim that any ring homomorphism out of $R\mathcal{C}^{-1}$ inverts $\eta(\Gamma(S))$ if and only if it inverts $\Gamma(SR\mathcal{C}^{-1})$, which will complete the proof.

Since $\overline{\eta}$ is the embedding of R/S in its classical ring of left quotients, a matrix C is regular modulo S if and only if $\overline{\eta}\pi(C)$ is invertible, which occurs if and only if $\widehat{\pi}\eta(C)$ is invertible. This shows that $\eta(\Gamma(S)) \subseteq \Gamma(SR\mathcal{C}^{-1})$.

To prove the reverse inclusion, let C be a matrix over $R\mathcal{C}^{-1}$ that is regular modulo $SR\mathcal{C}^{-1}$. By assumption, $R\mathcal{C}^{-1}/SR\mathcal{C}^{-1}$ is a simple Artinian ring, so a matrix is regular modulo $SR\mathcal{C}^{-1}$ if and only if it is invertible modulo $SR\mathcal{C}^{-1}$. In $R\mathcal{C}^{-1}$ we can find a common denominator $d \in \mathcal{C}$ for the entries of C, and hence we can write $C = C^*(dI)^{-1}$, where C^* has entries in $\eta(R)$, say $C^* = \eta(C')$, and I is the identity matrix of the appropriate size. Because dI is invertible over $R\mathcal{C}^{-1}$, it follows that $\widehat{\pi}(C)$ is invertible if and only if $\widehat{\pi}(C^*)$ is invertible. Furthermore, any ring homomorphism out of $R\mathcal{C}^{-1}$ will invert C if and only if it inverts C^*. In particular, since C is regular modulo $SR\mathcal{C}^{-1}$, this implies that $\widehat{\pi}$ inverts C, so $\overline{\eta}\pi(C') = \widehat{\pi}\eta C'$ is invertible, and therefore $C' \in \Gamma(S)$. This shows that any ring homomorphism out of $R\mathcal{C}^{-1}$ will invert $\eta(\Gamma(S))$ if and only if it inverts $\Gamma(SR\mathcal{C}^{-1})$, and allows us to conclude that the universal $\Gamma(SR\mathcal{C}^{-1})$-inverting ring is naturally isomorphic to the universal $\eta(\Gamma(S))$-inverting ring. □

PROPOSITION 3.3. *Let S be a semiprime right Goldie ideal of R, and let Z be the center of R. If $S \cap Z = N$, then $R_{\Gamma(S)}$ is an algebra over Z_N.*

PROOF. Let $\lambda : R \to R_{\Gamma(S)}$ be the natural mapping. Each element $c \in Z \setminus N$ is regular modulo S, so it is inverted by λ. The universal property defining the localization Z_N yields a homomorphism $\mu : Z_N \to R_{\Gamma(S)}$ such that the following diagram is commutative.

$$\begin{array}{ccc} Z & \subseteq & R \\ \downarrow & & \downarrow{\scriptstyle\lambda} \\ Z_N & \xrightarrow{\mu} & R_{\Gamma(S)} \end{array}$$

To show that μ defines an algebra structure on $R_{\Gamma(S)}$ it is only necessary to show that the image of μ is in the center of $R_{\Gamma(S)}$. For $b, d \in Z$ with $d \in Z \setminus N$, we have $\mu(d^{-1}b) = \lambda(d)^{-1}\lambda(b)$, and it is routine to show that such elements lie in the center of $R_{\Gamma(S)}$, using the fact that each element of $R_{\Gamma(S)}$ has the form $\lambda(a)\lambda(C)^{-1}\lambda(r^t)$, for some $a, r \in R^n$ and $C \in \Gamma_n(S)$. □

The final two results are well-known in case the semiprime ideal S is localizable.

PROPOSITION 3.4. *Let S be a semiprime ideal of R. If $R_{\Gamma(S)}$ is right Noetherian and I is any ideal of R, then $I^e = \lambda(I)R_{\Gamma(S)}$.*

PROOF. We need to show that $\lambda(I)R_{\Gamma(S)}$ is a left ideal of $R_{\Gamma(S)}$, where $\lambda : R \to R_{\Gamma(S)}$ is the canonical ring homomorphism. Let $x \in \lambda(I)R_{\Gamma(S)}$, with $x = \sum_{i=1}^{m} \lambda(a_i)q_i$, where $q_i \in R_{\Gamma(S)}$ and $a_i \in I$, for each i. If q is any element of $R_{\Gamma(S)}$, then since $\Gamma(S)$ contains all permutation matrices and is closed under products, we may assume without loss of generality that q is the $(1,1)$–entry of $\lambda(C)^{-1}$ for some $C \in \Gamma_n(S)$. In the matrix ring $M_n(R_{\Gamma(S)})$, we have

$$\lambda(C)M_n(\lambda(I)) \subseteq M_n(\lambda(I))$$

since I is an ideal of R. Thus

$$M_n(\lambda(I)) \subseteq \lambda(C)^{-1}M_n(\lambda(I)) \,.$$

Since $R_{\Gamma(S)}$ is right Noetherian, so is $M_n(R_{\Gamma(S)})$. The ascending chain of right ideals

$$\lambda(C)^{-1}M_n(\lambda(I))M_n(R_{\Gamma(S)}) \subseteq \lambda(C)^{-2}M_n(\lambda(I))M_n(R_{\Gamma(S)}) \subseteq \cdots$$

must terminate at some point, say

$$\lambda(C)^{-k}M_n(\lambda(I))M_n(R_{\Gamma(S)}) = \lambda(C)^{-k-1}M_n(\lambda(I))M_n(R_{\Gamma(S)}) \,.$$

This implies that

$$M_n(\lambda(I))M_n(R_{\Gamma(S)}) = \lambda(C)^{-1}M_n(\lambda(I))M_n(R_{\Gamma(S)}) \,.$$

The element $q\left(\sum_{i-1}^{m} \lambda(a_i)q_i\right)$ is the $(1,1)$–entry of the matrix

$$\lambda(C)^{-1}\left(\sum_{i=1}^{m} \begin{bmatrix} \lambda(a_i) & 0 \\ 0 & 0 \end{bmatrix}\begin{bmatrix} q_i & 0 \\ 0 & 0 \end{bmatrix}\right)$$

which belongs to $M_n(\lambda(I))M_n(R_{\Gamma(S)})$, and so it must have the form required to be an element of $\lambda(I)R_{\Gamma(S)}$. □

PROPOSITION 3.5. *Let S be a semiprime ideal of R. If $R_{\Gamma(S)}$ is Noetherian as a right module over R, then $R_{\Gamma(S)} = \lambda(R)$, where $\lambda : R \to R_{\Gamma(S)}$ is the canonical mapping.*

PROOF. Let $q \in R_{\Gamma(S)}$ and suppose that q is an entry in the matrix $\lambda(C)^{-1}$, for $C \in \Gamma_n(S)$. It is sufficient to show that $\lambda(C)^{-1}$ has entries in $\lambda(R)$. By assumption, the matrix ring $T = M_n(R_{\Gamma(S)})$ is Noetherian as a module over R, so we may consider the ascending chain of submodules $\lambda(C)^{-1}T \subseteq \lambda(C)^{-2}T \subseteq \ldots$. By assumption this chain must terminate after finitely many steps, say $\lambda(C)^{-k}T = \lambda(C)^{-k-l}T$. Therefore $\lambda(C)^{-1}T = M_n(\lambda(R))$, and so the entries of $\lambda(C)^{-1}$ belong to $\lambda(R)$. □

References

[1] J. A. Beachy, *Inversive localization at semiprime Goldie ideals*, Manuscripta Math. **34** (1981), no. 2-3, 211–239. MR620449

[2] J. A. Beachy, *On noncommutative localization*, J. Algebra **87** (1984), no. 1, 213–221. MR736776

[3] J. A. Beachy, *On universal localization at semiprime Goldie ideals*, Ring theory (Granville, OH, 1992), World Sci. Publ., River Edge, NJ, 1993, pp. 41–57. MR1344222

[4] J. A. Beachy and W. D. Blair, *Examples in noncommutative localization*, J. Algebra **99** (1986), no. 1, 108–113. MR836636

[5] A. D. Bell. Notes on Localization in Noncommutative Noetherian Rings. *Cuadernos de Algebra* **9**, University of Granada, 1989, 77 pp.

[6] P. M. Cohn, *Inversive localization in Noetherian rings*, Comm. Pure Appl. Math. **26** (1973), 679–691. Collection of articles dedicated to Wilhelm Magnus. MR0364324

[7] P. M. Cohn, *Free ideal rings and localization in general rings*, New Mathematical Monographs, vol. 3, Cambridge University Press, Cambridge, 2006. MR2246388

[8] D. Eisenbud, *Commutative algebra with a view toward algebraic geometry*, Graduate Texts in Mathematics, vol. 150, Springer-Verlag, New York, 1995. MR1322960

[9] A. W. Goldie, *Localization in non-commutative noetherian rings*, J. Algebra **5** (1967), 89–105. MR0207747

[10] A. W. Goldie, *A note on noncummutative localization*, J. Algebra **8** (1968), 41–44. MR0223394

[11] K. R. Goodearl and R. B. Warfield Jr., *An introduction to noncommutative Noetherian rings*, 2nd ed., London Mathematical Society Student Texts, vol. 61, Cambridge University Press, Cambridge, 2004. MR2080008

[12] I. Kaplansky, *Commutative rings*, Revised edition, The University of Chicago Press, Chicago, Ill.-London, 1974. MR0345945

[13] C. M. Leroux, *On universal localization of noncommutative Noetherian rings*, ProQuest LLC, Ann Arbor, MI, 2013. Thesis (Ph.D.)–Northern Illinois University. MR3167327

[14] B. J. Mueller. *Localization of Non-commutative Noetherian Rings at Semiprime Ideals*. Mathematisches Institut Der Universität München, 1974.

[15] B. J. Müller, *Localization in non-commutative Noetherian rings*, Canad. J. Math. **28** (1976), no. 3, 600–610. MR0409536

[16] B. J. Müller, *Localization in fully bounded Noetherian rings*, Pacific J. Math. **67** (1976), no. 1, 233–245. MR0427351

[17] L. W. Small and J. T. Stafford, *Localisation and completions of Noetherian PI algebras*, J. Algebra **70** (1981), no. 1, 156–161. MR618384

[18] O. Zariski and P. Samuel. *Commutative Algebra Volume II*, volume 29 of *Graduate Texts in Mathematics*. Springer-Verlag, New-York, 1960.

Department of Mathematical Sciences, Northern Illinois University, DeKalb, IL 60115, USA

Email address: jbeachy@niu.edu

Mathematics Department, SUNY Orange, Middletown, NY 10940, USA

Email address: christine.leroux@sunyorange.edu

Contemporary Mathematics
Volume **715**, 2018
https://doi.org/10.1090/conm/715/14405

A survey of intrinsic extensions of rings

Gary F. Birkenmeier and Enoch K. S. Lee

Dedicated to the memory of Professor Bruno J. Müller.

ABSTRACT. Let R be a subring of T. Then T is called a right intrinsic extension of R if $X \cap R \neq 0$ for each nonzero right ideal X of T. In this paper, we survey the mathematical literature on intrinsic extensions and their various generalizations and specializations. Examples, results, applications and open problems are provided.

The following problems and questions lead to the motivation and development of the "intrinsic" concept.

(I) [**FU64**, p. 505] Characterizing the condition that a ring extension R of S is a left quotient ring of S wholly in terms of the relative left ideal structures of R and S.

(II) If R is a subring of T and $\{e_1, \dots, e_n\}$ is a complete set of orthogonal idempotents of R with some property (e.g., primitive, centrally primitive, etc.), when does $\{e_1, \dots, e_n\}$ have this property in T?

(III) Let $R[G]$ be the group ring of the ring R over the group G. If H is a subgroup of G, how is the ideal structure of $R[H]$ related to that of $R[G]$?

(IV) Find conditions on R that ensure an effective transfer of information between R and its center.

As we will discuss in this paper, the notion of an intrinsic extension and its generalizations provides a framework in which to consider these problems and questions.

In this paper, all rings are associative with a unity and subrings have the same unity as their overring unless indicated otherwise. From [**FU64**], a ring extension T of R is *left (right) intrinsic over* R in case $X \cap R \neq 0$ for each nonzero left (right) ideal of T (equivalently, for each $0 \neq t \in T$ there exists $t_1 \in T$ such that $0 \neq t_1 t \in R$ ($0 \neq tt_1 \in R$)), denoted $R \leq_l^{int} T$ ($R \leq_r^{int} T$). Observe that if T is a ring extension of R and $R_R \leq^{ess} T_R$ (i.e., T_R is an essential extension of R_R) then $R \leq_r^{int} T$. In general, the converse is not true because $\mathbb{Z} \leq_r^{int} \mathbb{H}$, but $\mathbb{Z}_\mathbb{Z} \not\leq^{ess} \mathbb{H}_\mathbb{Z}$ where $\mathbb{Z}$ and $\mathbb{H}$ denote the rings of integers and real quaternions, respectively. In [**BL14a**], the intrinsic notion is generalized to encompass several similar notions. A ring R is called *right* $\mathcal{C}$ *intrinsic* in an extension ring T (or T is a *right* $\mathcal{C}$ *intrinsic* extension

2010 *Mathematics Subject Classification.* Primary 16D50; 16D60; 16P60; Secondary 16S85.
Key words and phrases. Essential extensions, intrinsic extensions, rings of quotients, Goldie rings.

of R) if and only if for each $0 \neq X \in \mathcal{C}$, $X \cap R \neq 0$ where $\mathcal{C}$ is a distinguished subset of the set of right ideals of T. Some particular cases of this definition are: R is *ideal intrinsic* in T if $\mathcal{C} = \{\text{ideals of } T\}$, denoted $R \trianglelefteq^{int} T$ [**BPR09**, p. 330]; T is *ideal intrinsic over its center*, IIC, if R is the center of T and $\mathcal{C} = \{\text{ideals of } T\}$ [**ABP09**]; R is a *right McCoy* ring, then $T = R[x]$ and $\mathcal{C} = \{r.ann_T(f(x)) \mid f(x) \in T\}$, where $r.ann_T(-)$ denotes the right annihilator of $-$ in T ($l.ann_T(-)$ denotes the left annihilator of $-$ in T) [**Nie06**]; R is *right annihilator ideal intrinsic* in T if $\mathcal{C} = \{A \trianglelefteq T \mid A = r.ann_T(X), \text{ where } X \trianglelefteq T\}$, see [**Pas11**, pp. 143–144] for examples. R is *right direct summand intrinsic* in T if $\mathcal{C} = \{D \mid D \text{ is a direct summand of } T \text{ as a right ideal}\}$, denoted by $R \leq_r^{int\oplus} T$. Similarly, R is defined to be left direct summand intrinsic in T, denoted $R \leq_l^{int\oplus} T$ [**BL14a**].

In Section 1, examples, basic results and properties which transfer from a base ring R to its right or left intrinsic extensions are provided. Also the intrinsic extension concept is defined for modules. Moreover, we give a historical development of solutions to Problem I in the papers [**FU64**], [**Hut69**], [**O'M74**] and [**Utu63**].

In Section 2, we show that if T is a ring extension of R and $\{e_1, \ldots, e_n\}$ is a complete set of (centrally) primitive orthogonal idempotents of R, then $\{e_1, \ldots, e_n\}$, in general, is not such a set in T (even when $T = Q(R)$). However, by using the right direct summand intrinsic condition and various other properties (e.g., extending and FI-extending), $\{e_1, \ldots, e_n\}$ or a complete set of orthogonal idempotents constructed from it is shown to be (centrally) primitive in T, thus providing answers to Question II.

Section 3 shows that the notion of a dense (or rational) extension of a ring or module can be generalized to an intrinsic type notion. From [**BL14b**], a ring extension T of R is a *left (right) dense intrinsic extension* of R if for each $x, y \in T$ with $x \neq 0$, there exists $t \in T$ such that $0 \neq tx$ and $ty \in R$ ($0 \neq xt$ and $yt \in R$), denoted $R \leq_l^{dint} T$ ($R \leq_r^{dint} T$). Observe that if $R_R \leq^{den} T_R$ (i.e., T_R is a dense extension of R_R), then $R \leq_r^{dint} T$. In general, the converse is not true because $\mathbb{Z} \leq_r^{dint} \mathbb{H}$, but $\mathbb{Z}_{\mathbb{Z}} \not\leq^{den} \mathbb{H}_{\mathbb{Z}}$. Examples and basic results are given. Various conditions (e.g., right extending, right Kasch, etc.) are shown to transfer from a base ring R to its right dense intrinsic extensions.

In Section 4, examples and basic results on ideal intrinsic extensions are provided. Moreover for a large subclass of semiprime rings the FI-extending property transfers to ideal intrinsic extensions. Question III is considered via examples of ideal intrinsic extensions of group rings and intersection theorems. Solutions to Problem IV are obtained by considering IIC rings.

Standard terminology and notation are adhered to as much as possible. The fields of rational, real and complex numbers are denoted by $\mathbb{Q}$, $\mathbb{R}$ and $\mathbb{C}$, respectively. $X \leq Y$ and $X \trianglelefteq R$ denote X is a subobject of Y and X is an ideal of R, respectively. We use $Q(R)$, $Q^l(R)$ and $Q_s(R)$ to symbolize the maximal right, maximal left and symmetric rings of quotients. Likewise $Mat_n(R)$, $Z(R_R)$, $Z({}_RR)$ and $\mathbf{B}(R)$ symbolize the ring of n-by-n matrices over R, the right singular ideal, the left singular ideal and the set of central idempotents of R. For the trivial (also called split-null) extension of an (R, R)-bimodule M by a ring R we use $\mathscr{S}(R; M)$. A ring R is *right cononsingular* if for each $X_R \leq R_R$, $l.ann_R(X) = 0$ implies that $X_R \leq^{ess} R_R$. R is a *right Kasch* ring if the only nonzero dense right ideal of R is R. Recall a right R-module is: *extending* [**Har82**] if each submodule is essential in a direct summand; *FI-extending* [**BMR02**] if each fully invariant submodule is essential in a direct summand; *PI-extending* [**BTY14**] (also denoted π-extending [**BTY16**]) if each projection invariant submodule is essential in a direct summand; C_{11} [**ST93**] if each submodule has at least one relative complement which is a direct summand; $\mathcal{G}$-extending [**ABT09**] if for each submodule X there is a direct summand D such that $(X \cap D)_R \leq^{ess} X_R$ and $(X \cap D)_R \leq^{ess} D_R$.

1. Intrinsic extensions

We begin this section with examples and constructions most of which are taken from [**BL14b**] and [**BL14a**], except for (iii) and (viii).

EXAMPLE 1.1. *Let T be an overring (or extension) of a ring R.*

(i) *Assume $R_R \leq^{ess} T_R$. Then $R \leq_r^{int} T$.*

(ii) *Let T be a division ring, and let R be any subring of T. Then $R \leq_r^{int} T$ and $R \leq_l^{int} T$. Furthermore if V is an extension of T such that $T \leq_r^{int} V$, then V is a division ring such that $T \leq_l^{int} V$. In particular, let $T = \mathbb{R}$, $R = \mathbb{Z}$ and $V = \mathbb{H}$. Then $R \leq_r^{int} T \leq_r^{int} V$, but $R_R \not\leq^{ess} T_R$.*

(iii) *Let R be a commutative reduced (i.e., R has no nonzero nilpotent elements) ring such that $a^2 + b^2 \neq 0$ for all nonzero $a, b \in R$. Let $T = R[x]/\langle x^2+1\rangle$ where $\langle x^2+1\rangle$ denotes the ideal generated by x^2+1, and identify R with its image in T. Then $R \leq_r^{int} T$; but in general, $R_R \not\leq^{ess} T_R$. In particular, for $R = \mathbb{Z}$ then $R \leq_r^{int} T$, but $R_R \not\leq^{ess} T_R$.*

(iv) *If $X_T \leq^{ess} T_T$ and $X \subseteq R$, then $R \leq_r^{int} T$. In particular, let A be a ring and $T = A[x]/\langle x^4\rangle$, $I = \langle x^3\rangle/\langle x^4\rangle$, and let R be the subring of T generated by $A \cup I$. It is straightforward to check that $I_T \leq^{ess} T_T$. Hence $R \leq_r^{int} T$. Observe that $R_R \not\leq^{ess} T_R$ because $(x + \langle x^4\rangle)R \cap R = 0$. Let S be the subring of T generated by $A \cup \{x^2 + \langle x^4\rangle, x^3 + \langle x^4\rangle\}$. Since $I \subseteq S$, $S \leq_r^{int} T$. Note that $R \leq S \leq T$, but $R \not\leq_r^{int} S$ and $I_S \not\leq^{ess} S_S$.*

Recall [**Goo76**, pp. 114–120], *R is a* right essential product *if R is a subdirect product of a family of rings $\{R_i\}_{i\in I}$, and R contains an essential right ideal of $T = \prod R_i$. Therefore right essential products provide intrinsic extensions which, in general, are not essential extensions* [**Goo76**, p. 120, Exercise 6].

(v) *If $R \leq_r^{int} T$ and S is an intermediate ring between R and T, then $S \leq_r^{int} T$. However R may not be intrinsic in S. See the rings R, S, and T in the first part (iv) or take $R = \mathbb{Q}$, $S = \mathbb{Q}[e]$, and $T = \mathbb{R}$ for illustrations.*

(vi) *Let $R_1 \leq_r^{int} R_2 \leq_r^{int} \cdots \leq_r^{int} R_k \leq_r^{int} R_{k+1} \leq_r^{int} \cdots$. If $T = \cup R_i$, then $R_j \leq_r^{int} T$ for every $j \in \mathbb{Z}^+$. In particular, let $R = \left(\begin{smallmatrix} \mathbb{Z}_4 & 2\mathbb{Z}_4 \\ 0 & \mathbb{Z}_4 \end{smallmatrix}\right)$, $S = \left(\begin{smallmatrix} \mathbb{Z}_4 & \mathbb{Z}_4 \\ 0 & \mathbb{Z}_4 \end{smallmatrix}\right)$, and $T = \left(\begin{smallmatrix} \mathbb{Z}_4 & \mathbb{Z}_4 \\ \mathbb{Z}_4 & \mathbb{Z}_4 \end{smallmatrix}\right)$. Then $R_R \leq^{ess} S_R$ and $S_S \leq^{den} T_S$. Hence $R \leq_r^{int} S \leq_r^{int} T$ so $R \leq_r^{int} T$, but $R_R \not\leq^{ess} T_R$.*

(vii) *Recall, R is a* right Utumi ring *if R is right nonsingular and right cononsingular* [**Ste75**, pp. 251–252]. *From* [**Utu63**, Theorem 2.2], *assume $Z(R_R) = 0$. Then R is a right Utumi ring if and only if $R \leq_l^{int} Q(R)$. In particular, there exists a right and left nonsingular ring such that $R \leq_l^{int} Q(R)$ but ${}_RR \not\leq^{ess} {}_RQ(R)$. Let R be a right but not left Ore domain (see* [**Lam99**, p. 308] *for a discussion of domains satisfying a one-sided (but not the two-sided) Ore condition). Then $R \leq_l^{int} Q(R)$ because $Q(R)$ is a division ring. Also, R is left and right FI-extending, but*

${}_R R \not\leq^{ess} {}_R Q(R)$. To see this, assume for contradiction that ${}_R R \leq^{ess} {}_R Q(R)$. Since $Z({}_R R) = 0$, then ${}_R R \leq^{den} {}_R Q(R)$. So $R \hookrightarrow Q(R) \hookrightarrow Q^l(R)$. From [**Lam99**, p. 377], *$Q^l(R)$ is a simple, regular right selfinjective ring which is not a division ring. Let $0 \neq x \in Q^l(R)$ be a nilpotent element. There exists $r \in R$ such that $0 \neq rx \in Q(R)$. However, $r^{-1}rx = x \in Q(R)$, a contradiction.*

(viii) *By a Zorn's lemma argument, there exists an intermediate ring S between R and T such that S is maximum among right intrinsic extensions of R which are subrings of T.*

In the next part of this section (results 1.2 – 1.14), we survey the papers [**Utu63**], [**FU64**], [**Hut69**], and [**O'M74**] in which the intrinsic concept is introduced to provide a solution to Problem I. Note that in these papers the rings are not assumed to have unity elements, and a ring A is called a *right (left) quotient ring* of a ring B if $B_B \leq^{ess} A_B$ (${}_B B \leq^{ess} {}_B A$). Thus, if A is a right (left) quotient ring of B, then $B \leq_r^{int} A$ ($B \leq_l^{int} A$), by Example 1.1(i); but, by Example 1.1(ii), the converse is not true. However, with suitable conditions on A and B the converse is true leading to a solution of Problem I. Our first result seems to be a precursor for the intrinsic notion and the formulation of Problem I, and it motivates the definition of a right Utumi ring.

THEOREM 1.2. [**Utu63**, Theorem 2.2] *and* [**Ste75**, pp. 251–252] *Let R be a right nonsingular ring. Then the following conditions are equivalent.*

(1) *If $A \cap B = 0$, $A \neq 0$ for two right ideals A and B, then $l.ann_R(B) \neq 0$.*
(2) *Every closed right ideal is an annihilator.*
(3) *Every nonzero left ideal of $Q(R)$ has nonzero intersection with R (i.e., $R \leq_l^{int} Q(R)$).*
(4) *R is a right cononsingular.*

From [**FU64**, pp. 506–507] an extension T of R is called *right (left) strongly intrinsic*, denoted $R \leq_r^{sint} T$ ($R \leq_l^{sint} T$), if $R \leq_r^{int} T$ ($R \leq_l^{int} T$) and for each closed right (left) ideal A of R there is a right (left) ideal B of T such that $B \cap R = A$.

LEMMA 1.3. [**FU64**, Lemma 1.2] *Assume R is a right Utumi ring. Then $R \leq_r^{int} T$ if and only if $R \leq_r^{sint} T$.*

THEOREM 1.4. [**FU64**, Theorem 3.1] *Assume R is a right Utumi ring which does not contain any nonzero strongly regular ideals. Then any right intrinsic extension of R is a right quotient ring of R.*

In 1969, Hutchinson [**Hut69**] carried on the investigation started by Faith and Utumi and discovered a relation between right strongly intrinsic extensions and the classical right quotient ring of a semiprime right Goldie ring.

LEMMA 1.5. [**Hut69**, Lemma 4.4] *Assume R is a right selfinjective ring and $Z(R_R) = 0$. Then $R \leq_r^{int} T$ if and only if $R \leq_r^{sint} T$.*

THEOREM 1.6. [**Hut69**, Theorem 4.5] *Assume R is a semiprime right Goldie ring and T is an extension of R with $Z(T_T) = 0$. Then $R \leq_r^{int} T$ if and only if $R \leq_r^{sint} T$.*

Let $T = \prod R_\alpha$ and $R \subseteq T$ be a subdirect product of $\{R_\alpha\}$. Following Hutchinson [**Hut69**], the ring R is called a *right essential subdirect sum* of $\{R_\alpha\}$ if R is an essential right R-submodule of T, i.e. T is an essential ring extension of R. Note that a right essential subdirect sum is a right essential product (see Example 1.1(iv)), but not vice-versa. Then we have the following result.

THEOREM 1.7. [**Hut69**, Theorem 3.4] *Suppose that R is an extension of a semiprime right Goldie ring S. Then, R is right strongly intrinsic over S if and only if one of the following is true:*

(i) *$Q(S) = Q(R)$ is semisimple Artinian;*
(ii) *S and R are right essential subdirect sums of right Ore domains $\{S_\alpha\}_{\alpha=1}^n$ and $\{R_\alpha\}_{\alpha=1}^n$, respectively, where $S_\alpha \leq_r^{int} R_\alpha$ for each α;*
(iii) *S and R are right essential subdirect sums of nonzero rings $\{S_\alpha\}_{\alpha=1}^2$ and $\{R_\alpha\}_{\alpha=1}^2$, respectively, where $S_\alpha \leq_r^{int} R_\alpha$ for $\alpha = 1, 2$. Furthermore, condition (i) holds for S_1 and R_1, and condition (ii) holds for S_2 and R_2.*

The following corollary provides a characterization of right strongly intrinsic extensions of prime right Goldie rings.

COROLLARY 1.8. [**Hut69**, Corollary 3.5] *T is a right strongly intrinsic extension of a prime right Goldie ring R if and only if either T is a right quotient of R or T and R are right Ore domains such that $R \leq_r^{int} T$.*

In addition to continuing the studies along the direction established by Faith *et al.*, O'Meara [**O'M74**] also considered the case when right intrinsic extension rings with some special properties are right rings of quotients. In fact, suppose $\mathscr{C}$ is the class of right nonsingular prime rings that contain uniform right ideals. Two questions were attempted:

(A) If $R \in \mathscr{C}$ and R is right intrinsic over a prime ring S, is R a right ring of quotients of S?
(B) If $S \in \mathscr{C}$ and R is right intrinsic over the prime ring S, is R a right ring of quotients of S?

For question (A), the answer is yes (see Theorem 1.9) if the ring R is not an integral domain. Question (B) can be considered an extension on the previous results.

THEOREM 1.9. [**O'M74**, Theorem 2.1] *Let $R \in \mathscr{C}$ and assume R is not an integral domain. If R is a right intrinsic extension of a prime ring S, then R is a right quotient ring of S.*

(Note that this result was further extended to Theorem 4.2 of [**O'M74**].) As a consequence, we have:

THEOREM 1.10. [**O'M74**, Theorem 3.1] *Let S be a prime right nonsingular ring (not an integral domain), and suppose S contains uniform right ideals. Then, the maximal right quotient ring of S is a left quotient ring if and only if the maximal right quotient ring of S is left intrinsic if and only if S is right cononsingular if and only if S is a right Utumi ring.*

Note that a right Goldie ring must contain a uniform right ideal [**Goo76**, Lemma 3.18]. Theorem 1.10 can be considered a one-sided version of Theorem 3.2 of [**FU64**]. O'Meara also noticed the following interesting one-sided version of [**Utu63**, Theorem 3.3].

THEOREM 1.11. [**O'M74**, Thorem 3.2] *Let S be a prime right nonsingular ring (not an integral domain) and suppose S contains uniform right ideals. Then every right ring of quotients of S is also a left ring of quotients of S if and only if the closed right ideals of S are its right annihilator ideals.*

As a consequence, a criterion for a prime right Goldie ring to be a left Goldie ring is obtained. Namely

THEOREM 1.12. [**O'M74**, Theorem 3.3] *Suppose R is a prime right Goldie ring that is not a right Ore domain. Then R is a left Goldie ring if and only the closed right ideals of R are the right annihilator ideals.*

O'Meara further showed the following results that partially address Question (B).

THEOREM 1.13. [**O'M74**, Theorem 3.7] *If S is a prime nonsingular ring (not an integral domain), assuming that S contains a uniform right ideal and a nonzero finite dimensional right annihilator ideal B, then right intrinsic extensions of S are right rings of quotients of S.*

COROLLARY 1.14. [**O'M74**, Corollary 3.8] *Suppose S is a prime right Goldie ring that is not a right Ore domain. If R is a right intrinsic extension of S, then R is a right ring of quotients of S.*

For the remainder of this section, we assume rings have unity element.

The next four results discuss the transfer of various properties from a base ring R to its intrinsic extensions.

PROPOSITION 1.15. [**BL14a**, Proposition 1.14] *Assume $R \leq_r^{int} T$.*

(i) *$Z(R_R) \subseteq Z(T_T)$ (see* [**Hut69**, Lemma 4.1]*).*
(ii) *If $u.dim(R_R) < \infty$, then $u.dim(T_T) \leq u.dim(R_R)$.*
(iii) *If R is simple, prime, or semiprime, then T is simple, prime, or semiprime, respectively.*
(iv) *If R is a domain and V is some ring extension of T such that $R \leq_l^{int} V$, then T is a domain.*
(v) *If T has involution, $*$, and $R = R^*$, then $R \leq_l^{int} T$.*

We note that every right Utumi ring is left nonsingular. To see this, observe that $Q(R)$ is regular and $R \leq_l^{int} Q(R)$. By Proposition 1.15(i), $Z({}_RR) = 0$.

PROPOSITION 1.16.

(i) [**BL14b**, Proposition 3.21] *Let R be a right primitive ring and $R \leq_r^{int} T$. Then either T is right primitive, or for each maximal right ideal M of R such that R/M is a faithful simple right R-module, $MT = T$.*
(ii) [**BL14a**, Proposition 1.14 (viii)] *Let $R = F[H]$ and $T = F[G]$, where G is a group and $H \leq G$, F is a field and $R \leq_r^{int} T$ (in fact $R \trianglelefteq^{int} T$ is enough). If R is primitive, then so is T.*

THEOREM 1.17. [**BL14b**, Theorem 3.9] *Suppose $R \leq_r^{int} T$. If R_R satisfies any of the following conditions, then so must T_T:*

(i) *$\mathcal{G}$-extending;*
(ii) *C_{11};*

(iii) *PI-extending;*
(iv) *FI-extending.*

THEOREM 1.18. [**BL14b**, Theorem 3.10(i)] *Suppose S is an intermediate ring between R and T. If R is quasi-Baer, $R \leq_r^{int} T$ and $R \leq_l^{int} S$, then S is quasi-Baer.*

Observe that Theorem 1.18 generalizes [**BPR06**, Theorem 1.9(i)]. The next definition allows us to generalize the intrinsic concept from rings to modules.

DEFINITION 1.19. [**BL14b**, Definition 2.4] *Let R be a subring of T. Suppose M is a right R-module and N is a right T-module such that $M_R \leq N_R$. We say M is* (right) intrinsic *in N (denoted $M_R \leq^{int} N_T$) if for any $0 \neq n \in N$, there exists $t \in T$, such that $0 \neq nt \in M$ (i.e., each nonzero T-submodule of N has nonzero intersection with M).*

From the above definition, it is clear that if we take $R = T$, then we have the usual notion of essentiality for modules. We apply this definition in the following result.

PROPOSITION 1.20. [**BL14b**, Proposition 4.2] *Let R be a subring of T, suppose N is a (T,T)-bimodule, and M an (R,R)-bimodule such that $M_R \leq^{int} N_T$. If either of the following conditions holds, then $\mathscr{S}(R;M) \leq_r^{int} \mathscr{S}(T;N)$.*

(i) *$R \leq_r^{int} T$ and $_RM$ is faithful;*
(ii) *$_TN$ is faithful.*

For the next proposition, let us recall that if R is a commutative ring and G is a group, then the group ring $R[G]$ is endowed with a natural involution. We denote this involution by $*$, and it is defined by $(\sum a_g g)^* = \sum a_g g^{-1}$.

PROPOSITION 1.21. [**BL14a**, Proposition 2.15] *Let R be a commutative domain, F its field of fractions, and G a group with $H \leq G$.*

(i) *Assume $R[H] \leq_r^{int} R[G]$. If $R[H]$ is a domain, then $F[G]$ is a domain.*
(ii) *Assume G is torsionfree. There exists $M \leq G$ that is maximal with respect to $F[M]$ being a domain. M can be chosen to contain a maximal elementary amenable subgroup of G which contains $Cen(G)$. If $M \neq G$, then there exists $0 \neq X_{F[G]} \leq F[G]_{F[G]}$ such that $X \cap F[M] = 0 = X^* \cap F[M]$ and $F[M]$ contains no nonzero ideals of $F[G]$.*

Observe that from Proposition 1.21(i), whenever $F[H]$ is a solution to the Zero Divisor Problem [**Pas11**] and $H \leq G$ such that $F[H] \leq_r^{int} F[G]$, then $F[G]$ is a solution to the Zero Divisor Problem (i.e., $F[H]$ is a domain implies $F[G]$ is a domain). The next proposition further illustrates Proposition 1.21.

PROPOSITION 1.22. [**BL14a**, Proposition 2.16] *Let R be a right Ore domain with a classical right ring of quotients D and G a torsionfree group with $H \leq G$ such that H is Abelian and $[G : H] = 2$. Then $R[H]$ and $D[G]$ are right Ore domains and $R[H] \leq_r^{int} D[G]$. If R is commutative then additionally $R[H] \leq_l^{int} D[G]$.*

2. Direct summand intrinsic extensions

The following questions motivate this section and arise naturally when one considers a base ring R and a ring extension T of R (see Question II in the Introduction):

(1) If R has a complete set of (centrally) orthogonal primitive idempotents or a complete set of triangulating idempotents, when does T have such a set?

(2) If T does have such a set, can it be constructed from the corresponding set in R by addition and/or multiplication?

In general these questions have a negative answer. Observe that, if R is a right Ore domain, then $Q(R)$ is a division ring. So $\{1\}$ is a complete set of orthogonal primitive idempotents in both R and $Q(R)$. However, if R is a domain that is not right Ore, then $\{1\}$ is a complete set of orthogonal primitive idempotents in R; but $Q(R)$ has no complete set of orthogonal primitive idempotents [**Lam99**, pp. 377–379]. Even when both R and $Q(R)$ each have a complete set of orthogonal primitive idempotents, it may not be true that the complete set of orthogonal primitive idempotents of $Q(R)$ can be constructed (via addition and/or multiplication) from a complete set of orthogonal primitive idempotents of R. For example [**BPR08**, Example 6(ii)], let $R = \mathbb{Z}[G]$ be the group ring of the cyclic group $G = \{1, g\}$ of order 2 over $\mathbb{Z}$. Then $\{1\}$ is the unique complete set of orthogonal primitive idempotents of R. However, in $Q(R) = \mathbb{Q}[G]$, $\{\frac{1}{2}(1+g), \frac{1}{2}(1-g)\}$ is a complete set of orthogonal primitive idempotents in $Q(R)$. Some work related to Questions (1) and (2) appears in [**BP03**] and [**BPR08**].

In this section these questions are answered affirmatively by using the right direct summand intrinsic condition with various other properties (e.g., uniform and FI-extending).

EXAMPLE 2.1. *Let T be an extension of a ring R.*

(i) [**BL14a**] *If $R \leq_r^{int} T$, then $R \leq_r^{int\oplus} T$. If R is von Neumann regular, then $R \leq_r^{int} T$ if and only if $R \leq_r^{int\oplus} T$.*

(ii) [**BL14a**] *If R contains all idempotents of T, then $R \leq_r^{int\oplus} T$ and $R \leq_l^{int\oplus} T$. In particular, if R is an Abelian ring (i.e., all idempotents are central) and $T = R[x]$, or T_T is indecomposable and R is any subring, then $R \leq_r^{int\oplus} T$ and $R \leq_l^{int\oplus} T$. Concretely, if $R = \mathbb{Z}$ and $T = \mathbb{Z}[x]$ then $R \leq_r^{int\oplus} T$ and $R \leq_l^{int\oplus} T$, but $R \not\leq_r^{int} T$ and $R \not\leq_l^{int} T$.*

(iii) *Let A and B be rings, M an (A, B)-bimodule, $R = \left(\begin{smallmatrix} A & 0 \\ 0 & B \end{smallmatrix}\right)$ and $T = \left(\begin{smallmatrix} A & M \\ 0 & B \end{smallmatrix}\right)$. Assume for each nonzero idempotent $\alpha = \left(\begin{smallmatrix} c & m \\ 0 & e \end{smallmatrix}\right) \in T$, $r.\mathrm{ann}_B(m) \cap eB \neq 0$ when $e \neq 0$. Then $R \leq_r^{int\oplus} T$. To see this, consider the following cases:*

(a) *Assume $m = 0$. Then we are done.*

(b) *Assume $m \neq 0$ and $e \neq 0$. Then there exists $0 \neq eb \in eB$ such that $0 \neq \alpha\left(\begin{smallmatrix} 0 & 0 \\ 0 & eb \end{smallmatrix}\right) = \left(\begin{smallmatrix} 0 & 0 \\ 0 & eb \end{smallmatrix}\right) \in R \cap \alpha T$.*

(c) *Assume $e = 0$. Then $0 \neq \alpha\left(\begin{smallmatrix} 1 & 0 \\ 0 & 0 \end{smallmatrix}\right) = \left(\begin{smallmatrix} c & 0 \\ 0 & 0 \end{smallmatrix}\right) \in R \cap \alpha T$. Therefore $R \leq_r^{int\oplus} T$, but $R \not\leq_r^{int} T$ and R does not contain all the idempotents of T. Concretely, let $A = B = \mathbb{Z}$ and $M = \mathbb{Z}_4$.*

(iv) *There exists an intermediate ring S between R and T such that S is maximal among right direct summand intrinsic extensions of R which are subrings of T.*

THEOREM 2.2. [**BL14a**, Theorem 2.2]

(i) *Assume* $R \leq_r^{int \oplus} T$ $(R \leq_r^{int} T)$ *and* $e = e^2 \in R$. *If* eR_R *is uniform (e.g., when* e *is primitive and* R_R *is extending), then* e *is a primitive idempotent in* T $(eT_T$ *is uniform).*

(ii) *Assume* $R \leq_r^{int \oplus} T$ $(R \leq_r^{int} T)$ *and* $\{e_1, \ldots, e_n\}$ *is a complete set of orthogonal idempotents in* R. *If each* e_iR_R *is uniform, then* $\{e_1, \ldots, e_n\}$ *is a complete set of orthogonal primitive idempotents in* T *(each* e_iT_T *is uniform).*

(iii) *Let* $\{e_1, \ldots, e_n\}$ *be a set of idempotents in* R *such that* $R = \bigoplus_{i=1}^n e_iR$, *where each* e_iR_R *is uniform. If* $R \leq_r^{int \oplus} T$ $(R \leq_r^{int} T)$, *then* $T = \bigoplus_{i=1}^n e_iT$, *where each* e_iT_T *is indecomposable (uniform).*

COROLLARY 2.3. *Assume* R_R *is extending,* $R \leq_r^{int \oplus} T$, *and* $\{e_i\}_{i=1}^n$ *is a set of nonzero idempotents of* R. *Then* $R = \bigoplus_{i=1}^n e_iR$ *is an indecomposable decomposition of* R *if and only if* $T = \bigoplus_{i=1}^n e_iT$ *is an indecomposable decomposition of* T.

Note that the hypothesis for Theorem 2.2(ii), that each e_iR_R is uniform, is more general than simply assuming R_R is extending. Let D be a commutative domain, not a field. If R is the ring of upper triangular n-by-n matrices over D for $(n > 1)$, and $T = Q(R) = Mat_n(F)$ where F is the field of fractions of D, then R_R is not extending. However, $\{e_1, \ldots, e_n\}$, where each e_i is the $n \times n$ matrix with 1 in the (i, i)-position and zeroes elsewhere, is a complete set of orthogonal primitive idempotents for R and T where each e_iR_R and e_iT_T are uniform.

COROLLARY 2.4. [**BL14a**, Corollary 2.4] *Let* R *be a right Utumi ring. If* R *is left or right extending and* $\{e_1, \ldots, e_n\}$ *is a complete set of primitive idempotents for* R, *then* $\{e_1, \ldots, e_n\}$ *is a complete set of primitive idempotents for* $Q(R)$.

THEOREM 2.5. [**BL14a**, Theorem 2.5] *Suppose* R *is a subring of* T. *Assume* $R \leq_r^{int \oplus} T$ *and* R_R *is FI-extending.*

(i) $\mathbf{B}(T) \subseteq \mathbf{B}(R)$. *In particular, if* $Cen(R) \subseteq Cen(T)$ *(e.g.,* $T = Q(R)$*), then* $\mathbf{B}(T) = \mathbf{B}(R)$.

(ii) *Assume* $\{c_1, \ldots, c_n\}$ *is a complete set of centrally primitive idempotents for* R. *Then* T *has a complete set of centrally primitive idempotents* $\{e_1, \ldots, e_m\}$, *where* $m \leq n$ *and each* $e_i = \sum_{j \in J_i} c_j$ *and* $\{J_1, \ldots, J_m\}$ *is a partition of* $\{1, \ldots, n\}$.

Note that in Theorem 2.5, when $\mathbf{B}(T) = \mathbf{B}(R)$, then in (ii) $\{c_1, \ldots, c_n\} = \{e_1, \ldots, e_m\}$. Also, observe that for $R = \mathbb{C}$ and $T = \mathbb{H}$, $R \leq_r^{int} T$ and $\mathbf{B}(R) = \mathbf{B}(T)$ but $Cen(R) = \mathbb{C} \not\subseteq \mathbb{R} = Cen(T)$.

COROLLARY 2.6. [**BL14a**, Corollary 2.6]

(i) *If* R *is a prime ring and* $R \leq_r^{int \oplus} T$, *then* T *is an indecomposable ring.*

(ii) *Let* R *be a right Utumi ring. If* R *is left or right FI-extending and if* R *has a complete set of centrally primitive idempotents* $\{c_1, \ldots, c_n\}$, *then* $Q(R)$ *has a complete set of centrally primitive idempotents.*

Recall an idempotent $e \in R$ is *left (right) semicentral* if $Re = eRe$ $(eR = eRe)$. $S_l(R)$ $(S_r(R))$ denote the set of left (right) semicentral idempotents of R. Taking e to be an idempotent of R, observe that $S_l(eRe) = \{0, e\}$ if and only if $S_r(eRe) =$

$\{0, e\}$; when this occurs we say e is *semicentral reduced.* If 1 is semicentral reduced, then we say R is *semicentral reduced.*

As in [**BHKP00**], an ordered set $\{b_1, \ldots, b_n\}$ of nonzero distinct idempotents in R is called a set of *left triangulating idempotents* of R if all the following statements hold:

(i) $1 = b_1 + \ldots + b_n$;
(ii) $b_1 \in S_l(R)$; and
(iii) $b_{k+1} \in S_l(c_kRc_k)$, where $c_k = 1 - (b_1 + \ldots + b_k)$, for $1 \leq k \leq n-1$.

Similarly, we define a set of *right triangulating idempotents* of R using condition (i), $b_1 \in S_r(R)$, and $b_{k+1} \in S_r(c_kRc_k)$. From part (iii) of the above definition, a set of left (right) triangulating idempotents is a set of pairwise orthogonal idempotents. A set $\{b_1, \ldots, b_n\}$ of left (right) triangulating idempotents is said to be *complete* if each b_i is also semicentral reduced.

Note that from [**BHKP00**, Corollary 1.7 and Theorem 2.10] the number of elements in a complete set of left triangulating idempotents is unique for a given ring R (which has such a set). This is also the number of elements in any complete set of right triangulating idempotents of R. Following [**BPR08**, Defintion 2], we say a ring R has *triangulating dimension* n, written $Tdim(R) = n$, if R has a complete set of left (or right) triangulating idempotents with exactly n elements. Note that R is semicentral reduced if and only if $Tdim(R) = 1$. If R has no complete set of left triangulating idempotents, then we say R has *infinite triangulating dimension,* denoted $Tdim(R) = \infty$. A generalized triangular matrix representation is called *complete* if each R_i is semicentral reduced. See Example 4.1(viii) for more details on generalized triangular matrix representations. By [**BHKP00**, Proposition 1.3], R has a (complete) set of left triangulating idempotents if and only if R has a (complete) generalized triangular matrix representation.

From [**BHKP00**, Proposition 2.14] one can see that most of the standard finiteness conditions imply finite triangulating dimension. Thus, our next result is applicable to a large class of rings.

THEOREM 2.7. [**BL14a**, Theorem 2.8] *Assume $R \leq_r^{int\oplus} T$ where R_R is FI-extending and $\{b_1, \ldots, b_n\}$ is a complete set of left triangulating idempotents of R. Then T has a complete set of left triangulating idempotents $\{e_1, \ldots, e_m\}$ where $m \leq n$, each $e_i = \sum_{j \in J_i} b_j$ and $\{J_1, \ldots, J_m\}$ is a partition of $\{1, \ldots, n\}$.*

COROLLARY 2.8. [**BL14a**, Corollary 2.9] *Let R be a right Utumi ring. If R is right or left FI-extending and if R has a complete set of left triangulating idempotents, then $Q(R)$ has a complete set of left triangulating idempotents and $Tdim(Q(R)) \leq Tdim(R)$.*

COROLLARY 2.9. [**BL14a**, Corollary 2.10] *Suppose $R \leq_r^{int} T$.*

(i) *If R_R is FI-extending, $Tdim(R) = n$ and T is semiprime, then $T = \bigoplus_{i=1}^m T_i$ where each T_i is a prime ring and $m \leq n$.*
(ii) *If $R = \bigoplus_{i=1}^n R_i$ where each R_i is a prime ring, then $T = \bigoplus_{i=1}^m T_i$ where each T_i is a prime ring and $m \leq n$.*
(iii) *If R_R is FI-extending, $Z(T_T) = 0$ and R has n minimal prime ideals, then T has m minimal prime ideals, where $m \leq n$.*

The following example illustrates Theorems 2.5 and 2.7 and shows that the previous results are applicable to algebras that occur in functional analysis.

EXAMPLE 2.10. [**BL14a**, Example 2.11] *Let* $R = \left(\begin{smallmatrix} B(H) & 0 \\ 0 & B(H) \end{smallmatrix}\right)$ *and* $T = \left(\begin{smallmatrix} B(H) & A \\ 0 & B(H) \end{smallmatrix}\right)$, *and let* $A = B(H)/K(H)$ *where* H *denotes a Hilbert space,* $B(H)$ *is the ring of bounded linear operators on* H *and* $K(H)$ *is the set of compact operators on* H. *Since* $K(H)$ *is a norm closed ideal of* $B(H)$, A *is a Banach algebra with the quotient norm. Clearly,* R *and* T *are Banach algebras. The norm on* T *is given by* $\left\|\left(\begin{smallmatrix} f & h \\ 0 & g \end{smallmatrix}\right)\right\| = \|f\|_{B(H)} + \|h\|_A + \|g\|_{B(H)}$ *and* R *is endowed with the subspace norm. Because* $B(H)$ *is a prime ring,* R_R *is FI-extending. A routine calculation shows that* R *is not right intrinsic in* T. *However,* $R \leq_r^{int\oplus} T$. *To see this, let* $0 \neq \left(\begin{smallmatrix} c & a \\ 0 & e \end{smallmatrix}\right) = \left(\begin{smallmatrix} c & a \\ 0 & e \end{smallmatrix}\right)^2 \in T$. *Observe that* c *and* e *cannot both be zero. First, assume* $c \neq 0$. *Then* $0 \neq \left(\begin{smallmatrix} c & a \\ 0 & e \end{smallmatrix}\right)\left(\begin{smallmatrix} 1 & 0 \\ 0 & 0 \end{smallmatrix}\right) = \left(\begin{smallmatrix} c & 0 \\ 0 & 0 \end{smallmatrix}\right) \in R$. *Now suppose that* $e \neq 0$. *Since* $K(H)_{B(H)} \leq^{ess} B(H)_{B(H)}$, *there exists* $k \in B(H)$ *such that* $0 \neq ek \in K(H)$. *Then* $0 \neq \left(\begin{smallmatrix} c & a \\ 0 & e \end{smallmatrix}\right)\left(\begin{smallmatrix} 0 & 0 \\ 0 & ek \end{smallmatrix}\right) = \left(\begin{smallmatrix} 0 & 0 \\ 0 & ek \end{smallmatrix}\right) \in R$. *Therefore,* $R \leq_r^{int\oplus} T$. *Observe that* $\left(\begin{smallmatrix} 1 & 0 \\ 0 & 0 \end{smallmatrix}\right)$ *and* $\left(\begin{smallmatrix} 0 & 0 \\ 0 & 1 \end{smallmatrix}\right)$ *form a complete set of left triangulating idempotents for both* R *and* T *and a complete set of centrally primitive idempotents for* T *(in fact,* $\mathbf{B}(T) \subsetneq \mathbf{B}(R)$*). Since* R_R *is FI-extending, this example illustrates both Theorems 2.5 and 2.7. Observe that from* [**BL14a**, Theorem 1.16], T_T *is not FI-extending since* $l.ann_{B(H)}(B(H)/K(H)) = K(H)$ *is not generated by an idempotent. Therefore, in Corollary 2.9, we cannot replace* $R \leq_r^{int} T$ *with* $R \leq_r^{int\oplus} T$.

Recall from the introduction of this section that for $R = \mathbb{Z}[C_2]$ and $T = \mathbb{Q}[C_2]$, then $R_R \leq^{den} T_R$, $\{1_R\}$ is a complete set of (centrally) primitive idempotents and a complete set of left triangulating idempotents in R, but it is not such a set in T (in fact, $Tdim(R) = 1$, but $Tdim(T) = 2$). Thus, the condition that if $\{e_1, \ldots, e_n\}$ is a complete set of orthogonal idempotents where each e_iR_R is uniform and the FI-extending condition are not superfluous in Theorems 2.2(ii), 2.5 and 2.7, respectively. Furthermore, this example shows that in Corollary 2.3, the extending condition is necessary.

3. Dense intrinsic extensions

In this section, we consider an intrinsic type of condition which generalizes the notion of dense (or rational) extension. Recall [**BL14b**] a ring T is a *right (left) dense intrinsic extension* of a ring R if for each $x, y \in T$ with $0 \neq x$, there exists $t \in T$ such that $0 \neq xt$, $yt \in R$ ($0 \neq tx$ and $ty \in R$), denoted $R \leq_r^{dint} T$ ($R \leq_l^{dint} T$).

EXAMPLE 3.1. *Let* T *be a ring extension of a ring* R.

(i) $R_R \leq^{den} T_R \implies R \leq_r^{dint} T \implies R \leq_r^{int} T$.

(ii) *To see that* $R \leq_r^{int} T \not\Rightarrow R \leq_r^{dint} T$, *let* A *be a ring and assume* $X \trianglelefteq A$ *which is right essential but not right dense. Write* $R = \left(\begin{smallmatrix} A & X \\ 0 & A \end{smallmatrix}\right)$ *and* $T = \left(\begin{smallmatrix} A & A \\ 0 & A \end{smallmatrix}\right)$. *To see* $R_R \leq^{ess} T_R$ *let* $0 \neq \left(\begin{smallmatrix} a & b \\ 0 & c \end{smallmatrix}\right) \in T$ *with* $b \notin X$. *Since* $X_A \leq^{ess} A_A$, *let* $\alpha \in A$ *be such that* $0 \neq b\alpha \in X$. *So* $0 \neq \left(\begin{smallmatrix} a & b \\ 0 & c \end{smallmatrix}\right)\left(\begin{smallmatrix} 0 & 0 \\ 0 & \alpha \end{smallmatrix}\right) = \left(\begin{smallmatrix} 0 & b\alpha \\ 0 & c\alpha \end{smallmatrix}\right) \in R$. *Now assume for contradiction that* $R \leq_r^{dint} T$. *Since* $X_A \not\leq^{den} A_A$, *there exists* $r, s \in A$ *with* $r \neq 0$, *but there cannot exist* $y \in A$ *such that* $0 \neq ry$ *and* $sy \in X$. *By assumption* $R \leq_r^{dint} T$, *so there exists* $\left(\begin{smallmatrix} a & b \\ 0 & c \end{smallmatrix}\right) \in T$ *such that* $0 \neq \left(\begin{smallmatrix} 0 & 0 \\ 0 & r \end{smallmatrix}\right)\left(\begin{smallmatrix} a & b \\ 0 & c \end{smallmatrix}\right) = \left(\begin{smallmatrix} 0 & 0 \\ 0 & rc \end{smallmatrix}\right)$ *and* $\left(\begin{smallmatrix} 0 & s \\ 0 & 0 \end{smallmatrix}\right)\left(\begin{smallmatrix} a & b \\ 0 & c \end{smallmatrix}\right) = \left(\begin{smallmatrix} 0 & sc \\ 0 & 0 \end{smallmatrix}\right) \in R$, *a contradiction. For a concrete example, let* $A = \mathbb{Z}_4$ *and* $X = 2\mathbb{Z}_4$.

(iii) [**BL14b**] *Let* T *be a division ring, and let* R *be any subring of* T. *Then* $R \leq_r^{dint} T$ *and* $R \leq_l^{dint} T$.

(iv) [**BL14b**, Proposition 2.10] *Suppose A is a subring of B. Let γ denote matrices of the form:* $\gamma = \begin{pmatrix} a & x & 0 & 0 \\ 0 & b & 0 & 0 \\ 0 & 0 & b & y \\ 0 & 0 & 0 & a \end{pmatrix}$. *Let $R = \{\gamma \mid a \in A; b, x, y \in B\}$ and $T = \{\gamma \mid a, b, x, y \in B\}$. If $A \leq_r^{int} B$ ($A \leq_r^{dint} B$), then $R \leq_r^{int} T$ ($R \leq_r^{dint} T$).*

(v) [**BL14b**, Lemma 3.1] *If $R \leq_r^{int} T$ and $Z(T_T) = 0$, then $R \leq_r^{dint} T$. This result yields many examples from the intrinsic examples in Section 1. For instance from Example 1.1(iii), take R to be a reduced commutative ring such that $a^2 + b^2 \neq 0$ for all nonzero $a, b \in R$. Then $R \leq_r^{int} T$ and $Z(T_T) = 0$ where $T = R[x]/\langle x^2+1 \rangle$. So $R \leq_r^{dint} T$. Also using Example 1.1(vii), we have that R is a right Utumi ring if and only if $R \leq_l^{dint} Q(R)$. Moreover Example 1.1(vii) provides concrete examples of rings R such that $R \leq_l^{dint} Q(R)$ but ${}_RR \not\leq^{ess} {}_RQ(R)$.*

(vi) *There exists an intermediate ring S between R and T such that S is maximal among right dense intrinsic extensions of R which are subrings of T.*

PROPOSITION 3.2. [**BL14b**, Lemma 2.3] *Suppose A, B, C are rings such that A is a subring of B and B is a subring of C:*

(i) *If $A \leq_r^{dint} B$, $B \leq_r^{dint} C$, and $B_B \leq^{ess} C_B$ then $A \leq_r^{dint} C$.*
(ii) *If $A \leq_r^{dint} C$ then $B \leq_r^{dint} C$.*

EXAMPLE 3.3. [**BL14b**, Example 2.8] *Suppose A and B are rings with $A \leq_r^{dint} B$. Let $R = \left(\begin{smallmatrix} A & B \\ 0 & A \end{smallmatrix}\right)$, $S = \left(\begin{smallmatrix} A & B \\ 0 & B \end{smallmatrix}\right)$ and $T = \left(\begin{smallmatrix} B & B \\ 0 & B \end{smallmatrix}\right)$. It is routine to check that $R \leq_r^{dint} S \leq_r^{dint} T$. Since $S_S \leq^{den} T_S$, Proposition 3.2(i) yields that $R \leq_r^{dint} T$. If $A_A \not\leq^{ess} B_A$, then $R_R \not\leq^{ess} S_R$. In particular, take $A = \mathbb{Z}$ and $B = \mathbb{H}$.*

Observe that Proposition 1.15(i) motivates the question: If $R \leq_r^{int} T$, under what conditions does $Z(R_R) = 0$ imply $Z(T_T) = 0$?

PROPOSITION 3.4. *Let T be a ring extension of R.*

(i) [**FU64**, Lemma 1.7] *Assume $R \leq_r^{sint} T$. Then $Z(R_R) = 0$ implies $Z(T_T) = 0$.*
(ii) [**BL14b**, Proposition 3.4] *Assume $R \leq_r^{dint} T$ and $R \leq_l^{int} T$. Then $Z(R_R) = 0$ implies $Z(T_T) = 0$.*

The next three theorems indicate the transfer of various properties from a base ring to its dense intrinsic extensions.

THEOREM 3.5. [**BL14b**, Theorem 3.7] *Suppose $R \leq_r^{dint} T$. If R_R satisfies any of the following conditions then so must T_T:*

(i) *extending;*
(ii) *quasi-continuous.*

COROLLARY 3.6. [**BL14b**, Propositions 3.12 and 3.13] *Assume R_R is extending and $R \leq_r^{dint} T$.*

(i) *$R \leq_r^{sint} T$.*
(ii) *If $Z(R_R) = 0$, then T is Baer and right cononsingular.*

Following [**BPR06**], a ring R is *right essential Baer (resp., quasi-Baer)* if the right annihilator of any nonempty subset (resp., ideal) of R is essential in a right ideal generated by an idempotent.

THEOREM 3.7. [**BL14b**, Proposition 3.11] *Suppose* $R \leq_r^{int} T$ *and* $R \leq_l^{dint} T$.

(i) *If* R *is right essentially quasi-Baer, then* T *is right essentially quasi-Baer.*
(ii) *If* R *is right essentially Baer, then* T *is right essentially Baer.*

THEOREM 3.8. [**BL14b**, Theorem 3.17] *Let* $R \leq_r^{dint} T$. *If* R *is a right Kasch ring then* T *is a right Kasch ring.*

Note that the above theorem shows one of the advantages of studying the dense intrinsic property. If we restrict ourselves to the notion of dense or rational extensions, the above theorem would be trivial. Assume $R_R \leq^{den} T_R$, and R is a right Kasch ring. Then $R = Q(R) = T$ (see [**Lam99**, Corollary 13.24]) and we have learned nothing from moving to the maximal ring of quotients. Observe that in the very simple example, $\mathbb{Q} \leq_r^{dint} \mathbb{R}$, both rings are right Kasch. So a dense intrinsic extension of a right Kasch ring need not be the same ring as is the case with dense extensions. The next proposition shows a way to construct a variety of Kasch rings with nontrivial dense intrinsic extensions that remain Kasch thereby illustrating Theorem 3.8. This proposition generalizes and sharpens [**BL14a**, Proposition 2.13]. Since any field is a QF-ring by [**Lam99**, Exercise 15.14], $F[C_2]$, $K[C_2]$ are also QF-rings where F is a subfield of K. Therefore, $F[C_2]$ and $K[C_2]$ are both Kasch. Note that because $F[C_2]$ is a QF-ring, it has no proper essential extensions. However, the next proposition demonstrates that there are many proper intrinsic and dense intrinsic extensions of $F[C_2]$.

PROPOSITION 3.9. [**BL14b**, Proposition 3.18] *Let* $F \subseteq K$ *be fields of characteristic not* 2 *and let* C_p *denote the cyclic group of order* p. *Then the following are equivalent:*

(i) $p = 2$;
(ii) $F[C_p] \leq_r^{int} K[C_p]$;
(iii) $F[C_p] \leq_r^{dint} K[C_p]$.

EXAMPLE 3.10. [**BL14b**, Example 3.19] *Let* γ *denote the matrix given in Example* 3.1*(iv). Let* $F \subseteq K$ *be a field extension of degree* $n > 1$. *Suppose* $R = \{\gamma \mid a \in F; b, x, y \in K\}$ *and* $T = \{\gamma \mid a, b, x, y \in K\}$. *Since* $F \leq_r^{dint} K$, *Example* 3.1*(iv) shows that* $R \leq_r^{dint} T$. *In* [**Lam99**, Exercise 16.16], *it is shown that* R *and* T *are both Kasch rings. Note that* $R_R \not\leq {}^{ess} T_R$, *and* R *is not QF but* T *is QF.*

The following definition generalizes the dense intrinsic notion from rings to modules.

DEFINITION 3.11. [**BL14b**, Defintion 2.4] *Let* R *be a subring of* T. *Suppose* M *is a right* R*-module and* N *is a right* T*-module such that* $M_R \leq N_R$. *We say* M *is* (right) dense intrinsic *in* N *(denoted* $M_R \leq^{dint} N_T$*) if for any* $n_1, n_2 \in N$ *with* $0 \neq n_1$, *there exists* $t \in T$ *such that* $0 \neq n_1 t$ *and* $n_2 t \in M$.

From the above definition, it is clear that if we take $R = T$, then we have the usual notion of density for modules. Recall from [**Goo76**, p. 55], $M_R \leq^{den} N_R$ if and only if $Hom_R({}^{P}\!/_{M}, N) = 0$ for all $M_R \leq P_R \leq N_R$. Our next proposition provides a dense intrinsic partial analogue for the aforementioned result on the density of R-submodules of M.

PROPOSITION 3.12. [**BL14b**, Proposition 2.5] *Let M be a right R-module and N a right T-module where R is a subring of T. If $M_R \leq^{dint} N_T$, then $Hom_T({}^{P}\!/_{MT}, N) = 0$ for all $MT_T \leq P_T \leq N_T$.*

The next result is a revision of [**BL14b**, Proposition 4.3].

PROPOSITION 3.13. *Let $R \leq_r^{int} T$, M a (R, R)-bimodule, N a (T, T)-bimodule such that ${}_TN$ is faithful, and $Z(N_T) = 0$. If $M_R \leq^{dint} N_T$, then $\mathscr{S}(R; M) \leq_r^{dint} \mathscr{S}(T; N)$.*

Observe that Proposition 3.13 is illustrated by taking $R = \mathbb{Z}$, $M = 2\mathbb{Z}$, $T = \mathbb{C}$ and $N = \mathbb{H}$.

4. Ideal intrinsic extensions

In the first part of this section, examples and basic results are provided on ideal intrinsic extensions including a theorem showing that the FI-extending condition transfers from a semiprime base ring R to an ideal intrinsic extension T when $\mathbf{B}(R) \subseteq \mathbf{B}(T)$. We note that examples of ideal intrinsic extensions appear in the study of group rings and related rings (e.g., crossed products) under the heading of "intersection theorems" (see [**Pas11**, p. 336]) and [**Pas13**]). Example 4.1(vii) exhibits an intersection theorem.

In the remainder of this section, we focus on the important case when R is an ideal intrinsic extension of its center, $Cen(R) \trianglelefteq^{int} R$. These rings are called IIC rings in [**ABP09**] from whence most of the IIC results mentioned in this section are taken. Rings with this property have also been called "rings with large center" [**Wis96**, p. 198]. One motivation for studying this class of rings is the well known result of Rowen [**Row73**] which shows that if R is a semiprime PI-ring (i.e., polynomial identity ring), then R is IIC.

EXAMPLE 4.1. *Let T be a ring extension of R.*

(i) *Assume $R \leq_r^{int} T$. Then $R \trianglelefteq^{int} T$.*
(ii) *Let T be a simple ring and let R be any subring of T. Then $R \trianglelefteq^{int} T$. If T is not a division ring, then $Cen(T) \not\leq_r^{int} T$ and $Cen(T) \not\leq_l^{int} T$.*
(iii) *Recall, T is* strongly right bounded [**FP84**] *if every nonzero right ideal contains a nonzero ideal. Then for T strongly right bounded, $R \leq_r^{int} T$ if and only if $R \trianglelefteq^{int} T$.*
(iv) *Let $\bar{R} = RI_n$, where I_n is the unity element of $Mat_n(R)$. Then $\bar{R} \trianglelefteq^{int} Mat_n(R)$.*
(v) [**Pas11**, Lemma 9.1.4] *Let F be an algebraic field extension of K, and let A be a torsionfree Abelian group. If $T = F[A]$ and $R = K[A]$, then $R \trianglelefteq^{int} T$.*
(vi) [**BPR13**, Lemma 9.2.3(i)] *Let R be a semiprime ring, G a group of ring automorphisms of R, and $T = R * G$ (i.e., the skew group ring). If G is X-outer, then $R \trianglelefteq^{int} T$.*
(vii) [**Pas11**, p. 296, Lemma 4.9] *Let G be a group with a normal abelian subgroup A, and let $H = \{x \in G \mid [A : C_A(x)] < \infty\}$. Then H is a normal subgroup of G containing A. Let K be a field, and let I be a nonzero ideal of $K[G]$. Then $I \cap K[H] \neq 0$ (i.e., $K[H] \trianglelefteq^{int} K[G]$).*

(viii) *Recall, an n-by-n generalized (or formal) matrix ring R is a square array*

$$R = \begin{pmatrix} R_1 & M_{12} & \cdots & M_{1n} \\ M_{21} & R_2 & \ddots & \vdots \\ \vdots & \ddots & \ddots & M_{n-1,n} \\ M_{n1} & \cdots & M_{n,n-1} & R_n \end{pmatrix}$$

where each R_i is a ring, each M_{ij} is an (R_i, R_j)-bimodule and there exist (R_i, R_k)-bimodule homomorphisms $\theta_{ijk} : M_{ij} \otimes_{R_j} M_{jk} \to M_{ik}$ for all $i, j, k = 1, \ldots, n$ (with $M_{ii} = R_i$). For $m_{ij} \in M_{ij}$ and $m_{jk} \in M_{jk}$, $m_{ij}m_{jk}$ denotes $\theta_{ijk}(m_{ij} \otimes m_{jk})$. The homomorphisms θ_{ijk} must satisfy the associativity relation: $(m_{ij}m_{jk})m_{kl} = m_{ij}(m_{jk}m_{kl})$ for all $m_{ij} \in M_{ij}$, $m_{jk} \in M_{jk}$, $m_{kl} \in M_{kl}$ and all $i, j, k, l = 1, \ldots, n$. Observe that θ_{iii} is determined by the ring multiplication in R_i, while θ_{ijj} and θ_{jjk} are determined by the bimodule scalar multiplications.

With these conditions, addition on R is componentwise and multiplication on R is row-column matrix multiplication. A Morita context is a 2-by-2 generalized matrix ring. An n-by-n generalized upper (lower) triangular matrix ring *is a generalized matrix ring with $M_{ij} = 0$ for $j < i$ ($M_{ij} = 0$ for $i < j$). Note that $\{E_{ii} \in R \mid E_{ii}$ is the matrix with $1 \in R_i$ in the (i,i)-position and 0 elsewhere, $i = 1, \ldots, n\}$ is a complete set $\{E_{ii}\}_{i=1}^n$ of orthogonal idempotents in the above constructed generalized matrix R. Note that if R has a complete set of orthogonal idempotents, $\{e_1, \ldots, e_n\}$, then R is isomorphic to a n-by-n generalized matrix ring.*

In [**ABvW16**] *the class $\mathcal{T}_n$ consists of those n-by-n generalized matrix rings for which $\theta_{iji} = 0$ for all $i \neq j$. Thus $\mathcal{T}_n$ is exactly the class of n-by-n generalized matrix rings in which the diagonal entries of the product of two matrices is the product of the corresponding entries of the diagonals of the factor matrices. So all n-by-n generalized upper (or lower) triangular matrices are in $\mathcal{T}_n$.*

In [**ABvW16**, Theorem 4.12(3)], *it is shown that every n-by-n generalized matrix ring R has a subring $S \in \mathcal{T}_n$ such that $S \trianglelefteq^{int} R$.*

(ix) *There exists an intermediate ring S between R and T such that S is maximal among ideal intrinsic extensions of R which are subrings of T.*

PROPOSITION 4.2. *If $I \trianglelefteq^{ess} R$ (i.e., I is an ideal of R such that I has nonzero intersection with every nonzero ideal of R), then the subring generated by 1 and I is ideal intrinsic in R. In particular, if R is a prime ring and $0 \neq I \trianglelefteq R$, then the subring generated by 1 and I is ideal intrinsic in R.*

PROPOSITION 4.3. *Assume that S is an intermediate ring between rings R and T.*

(i) *If $R \trianglelefteq^{int} S$ and $S \trianglelefteq^{int} T$, then $R \trianglelefteq^{int} T$.*
(ii) *If $R \trianglelefteq^{int} T$, then $S \trianglelefteq^{int} T$.*

PROPOSITION 4.4. [**BL14a**, pp.5139–5140] *Assume $R \trianglelefteq^{int} T$. If R is simple, prime, or semiprime, then T is simple, prime, or semiprime, respectively.*

THEOREM 4.5. [**BPR09**, Proposition 1.5(ii)] *Assume $R \trianglelefteq^{int} T$ and $\mathbf{B}(R) \subseteq \mathbf{B}(T)$ (e.g., $T = Q(R)$). If R is semiprime and (right) FI-extending, then T is semiprime and (right) FI-extending.*

COROLLARY 4.6. [**BKP00**, Theorem 2.4] *and* [**BMR02**, Theorem 4.7] *Assume R is IIC.*

(i) *R is semiprime if and only if $Cen(R)$ is semiprime.*
(ii) *Assume R is semiprime. Then the following conditions are equivalent:*
   (a) *$Cen(R)$ is extending;*
   (b) *R is right FI-extending;*
   (c) *R is FI-extending;*
   (d) *R is quasi-Baer;*
   (e) *$Cen(R)$ is Baer.*

EXAMPLE 4.7. *The following are some examples of IIC rings:*

(i) *If R is a biregular ring (i.e., R is* biregular *if for each $x \in R$ there is a central idempotent $e \in R$ such that $RxR = eR$), then $Cen(R) \trianglelefteq^{int} R$. In particular (c.f.* [**ABP02**, Example 1.5]*), let F be a field and let $R = \{(a_n)_{n=1}^{\infty} \in \prod_{n=1}^{\infty} Mat_n(F) \mid a_n = kI_n$ eventually for some $k \in F\}$, which is an F-subalgebra of $\prod_{n=1}^{\infty} Mat_n(F)$, where I_n is the unity matrix in $Mat_n(F)$. Then as in* [**ABP02**, Example 1.5], *R is an almost PI-ring (hence IIC), but R is neither a PI-ring nor is $Cen(R) \leq_r^{int} R$.*
(ii) *Regular rings with primitive factor rings Artinian (*[**Goo91**, Theorem 6.2, Theorem 6.6, and Lemma 6.20]*) are IIC. In particular, regular rings of bounded index (i.e., there is a positive integer $n \geq 1$ such that $x^n = 0$ whenever x is a nilpotent element of R)* [**Goo91**, Corollary 7.10].
(iii) *Semiprime rings in which every nonzero one-sided ideal has nonzero intersection with the center of the ring. Constructions of such rings which are neither left nor right duo are provided in* [**AB96**] *(e.g., $\mathbb{H}[x]$). We note that there are semiprime IIC rings whose nonzero proper one-sided ideals have no nonzero intersection with the center. For example, the n-by-n matrix ring over a field for $n > 1$ is such a ring.*
(iv) *Reduced rings with essential socle are IIC.*
(v) [**Goo91**, Theorem 9.25] *Regular right selfinjective directly finite rings are IIC.*
(vi) *If R is simple, then $R[x]$ is IIC.*
(vii) [**DI71**, Corollary 3.7 and Theorem 3.9] *Separable algebras are IIC.*
(viii) [**AH89**, Theorem 2] *Let R be an algebraic extension of $Cen(R)$ which satisfies ACC on right annihilators. Then $Cen(R) \trianglelefteq^{int} R$.*

Henceforth we do not assume that R must have a unity.

The next five results exhibit basic properties of IIC rings.

PROPOSITION 4.8. [**ABP09**, Proposition 1.2] *Let R be an IIC ring. Then the following are equivalent.*

(i) *R is semiprime.*
(ii) *R is right nonsingular.*
(iii) *R is left nonsingular.*

PROPOSITION 4.9. *Let $X \trianglelefteq R$ such that X is semiprime as a ring.*

(i) [**ABP09**, Proposition 1.3(iii)] *If R is IIC, then X is IIC.*
(ii) [**ABP09**, Proposition 1.4(i)] *If $X \trianglelefteq^{ess} R$ and X is IIC, then R is semiprime IIC.*
(iii) [**ABP09**, Proposition 1.6(iii)] *If X and R/X are IIC, then R is IIC.*

PROPOSITION 4.10. [**ABP09**, Proposition 1.11]

(i) *If a ring R is IIC and $I \trianglelefteq R$ is ideal essentially closed in R (i.e., I has no ideal essential extension in R), then R/I is IIC.*
(ii) *Let a ring R be IIC such that $\mathbf{P}(R)$ is not ideal essential in R. Then there exists a proper ideal K of R such that:*
  (a) *$\mathbf{P}(R)$ is ideal essential in K.*
  (b) *${}^{R}\!/_{K}$ is a semiprime IIC ring.*

PROPOSITION 4.11. [**ABP09**, Proposition 4.12] *Let R be a direct sum or a direct product of rings $\{R_i \mid i \in \Gamma\}$. Then R is IIC if and only if each R_i is IIC.*

THEOREM 4.12. [**ABP09**, Theorem 4.9] *Any semiprime IIC ring is a subdirect product of prime IIC rings.*

In the following two results we consider connections between R and $Cen(R)$ when R is IIC.

PROPOSITION 4.13. [**ABP09**, Proposition 2.1] *Let R be an IIC ring.*

(i) *If $Cen(R)$ has unity $1_{Cen(R)}$, then R has unity $1_{Cen(R)}$.*
(ii) *$Cen(R)$ is a field if and only if R is simple with $R^2 \neq 0$.*
(iii) *$Cen(R)$ is a domain if and only if R is a prime ring.*
(iv) *$Cen(R)$ is a reduced ring if and only if R is a semiprime ring.*

PROPOSITION 4.14. [**ABP09**, Proposition 2.4] *Let R be a semiprime IIC ring. Then the following conditions are equivalent.*

(i) *$Cen(R)$ has DCC on ideals.*
(ii) *R has DCC on ideals.*
(iii) *R is a finite direct sum of simple rings with unity.*

The semiprime IIC condition on a ring R imposes structural conditions on $Q(R)$.

THEOREM 4.15. [**ABP09**, Theorem 3.4] *Let R be a semiprime IIC ring. Then we have the following.*

(i) *$Q(Cen(R)) = Cen(Q(R))$.*
(ii) *$u.dim(Cen(R)) = n$ if and only if $Q(R) = \oplus_{i=1}^{n} Q_i$, where Q_i is a simple ring with unity.*
(iii) *$Q(R)$ is a subdirect product of simple regular rings with unity.*
(iv) *The Martindale right (left) ring of quotients of R is $Q_s(R)$.*

The next four results indicate that the IIC condition transfers to various ring extensions.

Let R be an algebra over a commutative ring C with unity. We say that R is *algebraic IIC* if every nonzero algebra ideal has nonzero intersection with $Cen(R)$. Thus if R is IIC, then it is algebraic IIC.

THEOREM 4.16. [**ABP09**, Theorem 4.12] *Let C be a commutative ring with unity and R an algebra over C. Then R is algebra IIC if and only if the Dorroh extension $\mathcal{D}(R; C)$ is IIC.*

THEOREM 4.17. [**ABP09**, Theorem 4.3] *Assume R has a unity. Then the IIC condition is a Morita invariant.*

THEOREM 4.18. [**ABP09**, Theorem 4.4] *Let R be a ring with $l.ann_R(R) = r.ann_R(R) = 0$ (e.g., R is semiprime or R has unity) and S an overring of R with a free basis $\{s_i \in Cen(S) \mid i \in \Gamma\}$ over R. Then R is IIC if and only if S is IIC.*

COROLLARY 4.19. [**ABP09**, Corollary 4.5] *Let R be an IIC ring with unity and $R[M]$ the monoid ring, where M is a commutative monoid. Then R is IIC if and only if $R[M]$ is IIC. In particular, R is IIC if and only if the polynomial ring $R[X]$ is IIC, where X is a set of commuting indeterminates.*

In the remaining results of this section, we show that every semiprime ring R has a largest IIC ideal, $\mathfrak{I}_C(R)$. Moreover, it is shown that $\mathfrak{I}_C(R)$ has several radical-like properties. We investigate this ideal and show that if R is semiprime, then $\mathfrak{I}_C(Q(R))$ is a ring direct summand of $Q(R)$, and $\mathfrak{I}_C(R) \subseteq \mathfrak{I}_C(Q(R))$. Applications are made to C^*-algebras.

THEOREM 4.20. [**ABP09**, Theorem 6.1] *Let R be a semiprime ring,* **IIC** $:= \{I \trianglelefteq R \mid I$ *is an IIC ring* $\}$, *and* $\mathfrak{I}_C(R) := \sum_{I \in \mathbf{IIC}} I$. *Then we have the following.*

(i) **IIC** *is closed under arbitrary sums and intersections.*
(ii) *$\mathfrak{I}_C(R)$ is an IIC ring.*
(iii) *$\mathfrak{I}_C(R)_R$ is essentially closed in R_R.*
(iv) $\mathfrak{I}_C({}^{R}\!/_{\mathfrak{I}_C(R)}) = 0$.

PROPOSITION 4.21. [**ABP09**, Proposition 6.3] *Let R be a semiprime ring.*

(i) *If $K \trianglelefteq R$ such that $K \cap Cen(R) = 0$, then $K \subseteq ann_R(\mathfrak{I}_C(R))$.*
(ii) *$Z(R_R) \subseteq ann_R(\mathfrak{I}_C(R))$ and $Z({}_RR) \subseteq ann_R(\mathfrak{I}_C(R))$.*
(iii) *${}^{R}\!/_{ann_R(\mathfrak{I}_C(R))}$ is a semiprime IIC ring.*

THEOREM 4.22. [**ABP09**, Theorem 6.4] *Let R be a semiprime ring. Then we have the following.*

(i) *There exists an ideal $M \trianglelefteq R$ maximal with respect to the property that $M \cap Cen(R) = 0$.*
(ii) *${}^{R}\!/_{M}$ is a semiprime IIC ring.*
(iii) *$M_R \leq^{ess} ann_R(\mathfrak{I}_C(R))_R$.*
(iv) *$\mathfrak{I}_C(R) = ann_R(M)$.*

COROLLARY 4.23. [**ABP09**, Corollary 6.5] *Let R be a semiprime ring. Then either R is IIC or there is $0 \neq M \trianglelefteq R$ such that ${}^{R}\!/_{M}$ is semiprime IIC, $M \cap Cen(R) = 0$, and $\mathfrak{I}_C(R) \oplus M$ is essential in R_R.*

PROPOSITION 4.24. [**ABP09**, Proposition 6.7] *Let R be a semiprime ring. Then we have the following.*

(i) *$Cen(\mathfrak{I}_C(R)) = \mathfrak{C}$, where $\mathfrak{C} = \{c \in Cen(R) \mid$ for each $a \in R$ with $ca \neq 0$ there exists $y \in RaR$ such that $0 \neq cy \in Cen(R)\}$.*
(ii) *$[Cen(\mathfrak{I}_C(R)) \oplus Cen(ann_R(\mathfrak{I}_C(R)))]_{Cen(R)} \leq^{ess} Cen(R)_{Cen(R)}$.*
(iii) *If T is an intermediate ring between R and $Q(R)$ such that T contains all central idempotents of $Q(R)$ (e.g., $T = Q(R)$), then $T = \mathfrak{I}_C(T) \oplus ann_T(\mathfrak{I}_C(T))$.*
(iv) *$\mathfrak{I}_C(R) \subseteq \mathfrak{I}_C(T)$ for any intermediate ring T between R and $Q(R)$.*

LEMMA 4.25. [**ABP09**, Lemma 6.8] *Let A be a semiprime Banach algebra.*

(i) *If $V \trianglelefteq A$, then V_A is essential in its norm closure $\bar{V}_A$.*

(ii) *$\mathfrak{I}_C(A)$ and $ann_A(\mathfrak{I}_C(A))$ are norm closed in A. In particular, if A is a C^*-algebra, then $\mathfrak{I}_C(A)$ and $ann_A(\mathfrak{I}_C(A))$ are C^*-subalgebras of A.*

Recall from [**AM13**, pp. 55–65], the local multiplier algebra of a C^*-algebra A, $M_{loc}(A)$, is the topological completion of the bounded part of $Q_s(A)$, $Q_b(A)$. The final result is an application of our previous results to C^*-algebras.

PROPOSITION 4.26. [**ABP09**, Proposition 6.9(ii)] *Let A be a C^*-algebra. Then $M_{loc}(A) = \mathfrak{I}_C(M_{loc}(A)) \oplus ann_{M_{loc}(A)}(\mathfrak{I}_C(M_{loc}(A))$ (C^*-direct sum).*

OPEN PROBLEMS.

1. Determine classes of rings closed with respect to the various types of intrinsic extensions.
2. Determine classes of rings which are closed with respect to the various types of intrinsic subrings.

Acknowledgement

The authors wish to thank the referee for his comments, which improved our paper.

References

[AB96] H. E. Abulkheir and G. F. Birkenmeier, *Right complement bounded semiprime rings*, Acta Math. Hungar. **70** (1996), no. 3, 227–235. MR1374389

[ABP02] E. P. Armendariz, G. F. Birkenmeier, and J. K. Park, *Rings containing ideals with bounded index*, Comm. Algebra **30** (2002), no. 2, 787–801. MR1883025

[ABP09] E. P. Armendariz, G. F. Birkenmeier, and J. K. Park, *Ideal intrinsic extensions with connections to PI-rings*, J. Pure Appl. Algebra **213** (2009), no. 9, 1756–1776. MR2518175. Corrigendum to "Ideal intrinsic extensions with connections to PI-rings", J. Pure Appl. Algebra **215** (2011), 99–100. MR2678703

[ABT09] E. Akalan, G.F. Birkenmeier, and A. Tercan, *Goldie extending modules*, Comm. Algebra **37** (2009), 663–683; Corrigendum **38**(2010), 4747-4748. MR2805141; Corrigendum **41**(2013), 2005. MR3062842

[ABvW16] P. N. Ánh, G. F. Birkenmeier, and L. van Wyk, *Idempotents and structures of rings*, Linear Multilinear Algebra **64** (2016), no. 10, 2002–2029. MR3521154

[AH89] E. P. Armendariz and C. R. Hajarnavis, *A note on semiprime rings algebraic over their centres*, Comm. Algebra **17** (1989), no. 7, 1627–1631. MR1006517

[AM13] P. Ara and M. Mathieu, *Local Multipliers of C^*-Algebras*, Springer Monographs in Mathematics, Springer-Verlag London, Ltd., London, 2003. MR1940428

[BHKP00] G. F. Birkenmeier, H. E. Heatherly, J. Y. Kim, and J. K. Park, *Triangular matrix representations*, J. Algebra **230** (2000), no. 2, 558–595. MR1775804

[BKP00] G. F. Birkenmeier, J. Y. Kim, and J. K. Park, *Quasi-Baer ring extensions and biregular rings*, Bull. Austral. Math. Soc. **61** (2000), no. 1, 39–52. MR1819314

[BL14a] G. F. Birkenmeier and M. J. Lennon, *Extending sets of idempotents to ring extensions*, Comm. Algebra **42** (2014), no. 12, 5134–5151. MR3223631

[BL14b] G. F. Birkenmeier and M. J. Lennon, *Dense intrinsic extensions*, Houston J. Math. **40** (2014), no. 1, 21–42. MR3210552

[BMR02] G. F. Birkenmeier, B. J. Müller, and S. T. Rizvi, *Modules in which every fully invariant submodule is essential in a direct summand*, Comm. Algebra **30** (2002), no. 3, 1395–1415. MR1892606

[BP03] G. F. Birkenmeier and J. K. Park, *Triangular matrix representations of ring extensions*, J. Algebra **265** (2003), no. 2, 457–477. MR1987012

[BPR06] G. F. Birkenmeier, J. K. Park, and S. T. Rizvi, *Ring hulls and applications*, J. Algebra **304** (2006), no. 2, 633–665. MR2264273

[BPR08] G. F. Birkenmeier, J. K. Park, and S. T. Rizvi, *On triangulating dimension of rings*, Comm. Algebra **36** (2008), no. 4, 1520–1526. MR2406606

[BPR09] G. F. Birkenmeier, J. K. Park, and S. T. Rizvi, *Hulls of semiprime rings with applications to C^*-algebras*, J. Algebra **322** (2009), no. 2, 327–352. MR2529092

[BPR13] G. F. Birkenmeier, J. K. Park, and S. T. Rizvi, *Extensions of Rings and Modules*, Birkhäuser/Springer, New York, 2013. MR3099829

[BTY14] G. F. Birkenmeier, A. Tercan, and C. C. Yücel, *The extending condition relative to sets of submodules*, Comm. Algebra **42** (2014), no. 2, 764–778. MR3169601

[BTY16] G. F. Birkenmeier, A. Tercan, and C. C. Yücel, *Projection invariant extending rings*, J. Algebra and Its Appl. **15** (2016), 1650121.

[DI71] F. DeMeyer and E. Ingraham, *Separable Algebras Over Commutative Rings*, Lecture Notes in Mathematics, Vol. 181, Springer-Verlag, Berlin-New York, 1971. MR0280479

[FP84] C. Faith and S. Page, *FPF Ring Theory: Faithful Modules and Generators of Mod-R*, London Mathematical Society Lecture Note Series, vol. 88, Cambridge University Press, Cambridge, 1984. MR754181

[FU64] C. Faith and Y. Utumi, *Intrinsic extensions of rings*, Pacific J. Math. **14** (1964), 505–512. MR0161873

[Goo76] K. R. Goodearl, *Ring Theory: Nonsingular Rings and Modules*, Pure and Applied Mathematics, No. 33, Marcel Dekker, Inc., New York-Basel, 1976. MR0429962

[Goo91] K. R. Goodearl, *Von Neumann Regular Rings*, 2nd ed., Robert E. Krieger Publishing Co., Inc., Malabar, FL, 1991. MR1150975

[Har82] M. Harada, *On modules with extending properties*, Osaka J. Math. **19** (1982), no. 1, 203–215. MR656240

[Hut69] J. J. Hutchinson, *Intrinsic extensions of rings*, Pacific J. Math. **30** (1969), 669–677. MR0257152

[Lam99] T. Y. Lam, *Lectures on Modules and Rings*, Graduate Texts in Mathematics, vol. 189, Springer-Verlag, New York, 1999. MR1653294

[Nie06] P. P. Nielsen, *Semi-commutativity and the McCoy condition*, J. Algebra **298** (2006), no. 1, 134–141. MR2215121

[O'M74] K. C. O'Meara, *Intrinsic extensions of prime rings*, Pacific J. Math. **51** (1974), 257–269. MR0480608

[Pas11] D. Passman, *The Algebraic Structure of Group Rings*, Dover, Mineola, 2011.

[Pas13] D. Passman, *Infinite Crossed Products*, Dover, Mineola, 2013.

[Row73] L. Rowen, *Some results on the center of a ring with polynomial identity*, Bull. Amer. Math. Soc. **79** (1973), 219–223. MR0309996

[ST93] P. F. Smith and A. Tercan, *Generalizations of CS-modules*, Comm. Algebra **21** (1993), no. 6, 1809–1847. MR1215548

[Ste75] B. Stenström, *Rings of Quotients: An Introduction to Methods of Ring Theory*, Springer-Verlag, New York-Heidelberg, 1975. Die Grundlehren der Mathematischen Wissenschaften, Band 217. MR0389953

[Utu56] Y. Utumi, *On quotient rings*, Osaka Math. J. **8** (1956), 1–18. MR0078966

[Utu63] Y. Utumi, *On rings of which any one-sided quotient rings are two-sided*, Proc. Amer. Math. Soc. **14** (1963), 141–147. MR0142568

[Wis96] R. Wisbauer, *Modules and Algebras: Bimodule Structure and Group Actions on Algebras*, Pitman Monographs and Surveys in Pure and Applied Mathematics, vol. 81, Longman, Harlow, 1996. MR1396313

Department of Mathematics, University of Louisiana, Lafayette, Louisiana 70504
Email address: gfb1127@louisiana.edu

Department of Mathematics and Computer Science, Auburn University at Montgomery, Montgomery, AL 36117
Email address: elee4@aum.edu

Contemporary Mathematics
Volume **715**, 2018
https://doi.org/10.1090/conm/715/14406

The reduced ring order and lower semi-lattices

W. D. Burgess and R. Raphael

To the memory of Bruno Müller and in recognition of his many contributions to ring theory.

ABSTRACT. Every reduced ring R has a natural partial order defined by $a \leq b$ if $a^2 = ab$; it generalizes the natural order on a boolean ring. The article examines when R is a lower semi-lattice in this order with examples drawn from weakly Baer rings (pp-rings) and rings of continuous functions. Locally connected spaces and basically disconnected spaces give rings $C(X)$ which are such lower semi-lattices. Liftings of countable orthogonal (in this order) sets over a surjective ring homomorphisms are studied.

1. Introduction. The subject of this article is an aspect of the *reduced ring order* (here called the **rr**-order). This partial order is defined in all reduced rings (rings with no non-trivial nilpotent elements). The global assumption in the sequel will be that *the rings considered are unitary and reduced* and, for some parts, also commutative. If R is a reduced ring the set of idempotents of R, which are here all central, is denoted $\mathbf{B}(R)$. A *regular ring* will always mean *von Neumann regular.*

DEFINITION 1.1. Let R be a reduced ring then for $a, b \in R$, $a \leq_{\text{rr}} b$ if $ab = a^2$.

This is readily seen to be a partial order which has been called variously the *Abian order* and the *Conrad order* (see, for example, [**6**] which dealt with orthogonal completions in this order). In a reduced ring R, if $a^2 = ab$ then, also, $a^2 = ba$. This order generalizes to all reduced rings the natural order on boolean rings.

In a reduced ring R and $r \in R$ the left and right annihilators of r coincide and are denoted $\operatorname{ann} r$. The following well-known property of reduced rings, obvious in the commutative case, will be used often. It was first shown in [**2**] and [**16**] and is found in e.g. [**13**, Lemma 12.6].

LEMMA 1.2. *If R is a reduced ring and $\mathfrak{p}$ is a minimal prime ideal of R then $\mathfrak{p}$ is completely prime, i.e., $R/\mathfrak{p}$ is a domain.*

The **rr**-order can be a very coarse one but there are important classes of rings where it plays an interesting role. To illustrate one extreme, if R is a domain, then 0 is the unique minimal element in the order and no other pairs of elements are related. In fact in any reduced ring 0 is the unique minimal element while non zero-divisors are maximal. At the other extreme, a reduced ring R is a lattice

---

2010 *Mathematics Subject Classification.* 06F25 16W80 46E25.
The authors would like to thank the referee for a thorough reading and for some helpful suggestions.

in the **rr**-order only when R is boolean (Proposition 2.8) but the main question asked below is when a reduced ring is a lower semi-lattice in this order (**rr**-lower semi-lattice). Here there are many rings where the answer is positive. Moreover, some classes of rings are characterized by this property, for example weakly Baer rings within almost weakly Baer rings (Theorem 2.6). It is shown that if R is a commutative reduced ring it can be embedded, as an essential extension, in a weakly Baer ring; this extension is proper if R is not already weakly Baer (Proposition 2.14 and Corollary 2.20).

If $R \to S$ is a surjective homomorphism of reduced rings when can a countable orthogonal family (in the **rr**-order) in S be lifted to an orthogonal family in R? This is answered in two cases, Theorem 2.10 and Proposition 3.12.

Among rings of continuous functions, two quite different classes of topological spaces yield **rr**-lower semi-lattices, locally connected spaces (Theorem 3.5) and basically disconnected spaces ([**14**, Example 3.2]). A characterization of these spaces is available (Theorem 3.2) but requires looking at each pair of functions. However, even if X is a compact, metric and connected space it does not follow that $\mathrm{C}(X)$ is an **rr**-lower semi-lattice (Theorem 3.5(2)).

Consider the case of $R = \prod_{\alpha \in A} D_\alpha$ where each D_α is a domain. Here, any pair of elements has an infimum, noted $\wedge_{\mathrm{rr}}$. Let (a_α) and (b_α) be elements of R. Define $(c_\alpha) \in R$ by $c_\alpha = a_\alpha$ when $a_\alpha = b_\alpha$ and $c_\alpha = 0$ when $a_\alpha \neq b_\alpha$. It is readily seen that $(c_\alpha) = (a_\alpha) \wedge_{\mathrm{rr}} (b_\alpha)$.

Such computations are not readily available in a general reduced ring R. The question can be posed in the following way. Embed R in $S = \prod_{\mathfrak{p} \in \mathrm{Min}\, R} R/\mathfrak{p}$, where $\mathrm{Min}\, R$ is the minimal spectrum. Given $a, b \in R$ then $c = a \wedge_{\mathrm{rr}} b$ can be found in S but that element need not be in R. What is the "best approximation" to c which can be found in R? This question will be considered in the sequel, sometimes directly, sometimes indirectly. See Corollary 2.2, below for an easy case. It is also shown that if $|\,\mathrm{Min}\, R| \leq 3$ then R is an **rr**-lower semi-lattice but not necessarily for more than three minimal prime ideals (Proposition 2.13).

2. The reduced ring order and lower semi-lattices. Since the **rr**-order on a domain R is very easy to understand and makes R into a lower semi-lattice, the first place to look for examples is where rings are built out of domains in a straightforward way.

Recall ([**11**, Chapter V2]) that a ring R can be represented as the ring of sections of a sheaf, called the *Pierce sheaf* over the boolean space $\mathrm{Spec}\,\mathbf{B}(R)$, where $\mathbf{B}(R)$ is the boolean algebra of central idempotents of R. (In a reduced ring, all the idempotents are central.) The stalks, for $x \in \mathrm{Spec}\,\mathbf{B}(R)$, are the rings $R_x = R/Rx$. For $a \in R$, put $a_x = a + Rx$ and define $\mathrm{z}(a) = \{x \in \mathrm{Spec}\,\mathbf{B}(R) \mid a_x = 0\}$, an open set in $\mathrm{Spec}\,\mathbf{B}(R)$. Its complement is $\mathrm{coz}\, a$.

An interesting case in this study is where the stalks of this representation are all domains. For a reduced ring R, what is wanted is that the Rx, as above, are precisely the minimal prime ideals of R. In the commutative case these rings were characterized in [**14**, Theorem 2.2] as the rings R in which the annihilator of an element is generated by idempotents; these rings are called *almost weakly Baer rings*. The name will here be abbreviated to *awB rings*. For the purposes of this article, the definition will be extended to *all reduced rings*. The awB rings include commutative (Von Neumann) regular rings and, more generally, strongly regular rings, and Baer rings, among others. The characterization of rings whose Pierce stalks are domains extends readily to general reduced rings as shown in Proposition 2.5.

The next proposition gives some information about the **rr**-order. If a reduced ring R is viewed as a subring of $\prod_{\mathfrak{p}\in \operatorname{Min} R} R/\mathfrak{p}$, a product of domains, and $r, s \in R$, to ask about $r \wedge_{\mathrm{rr}} s$ is to ask to when there is an element $t \in R$ which is zero when $r+\mathfrak{p}$ and $s+\mathfrak{p}$ differ, coincides with them when $t+\mathfrak{p} \neq 0$ and is the unique element **rr**-maximal among all such elements.

PROPOSITION 2.1. *Let R and S be reduced rings. (i) Let $a, b, c \in R$ where c commutes with a: if $a \leq_{rr} b$ then $ca \leq_{rr} cb$ and, in particular, if $e = e^2$ then $ae \leq_{rr} be$.*

(ii) If $a, b \in R$, $c = a \wedge_{rr} b$ and u is a central unit in R, then $uc = ua \wedge_{rr} ub$.

*(iii) If $\phi\colon R \to S$ is a ring homomorphism then ϕ preserves the **rr**-order on R.*

*(iv) If R and S are **rr**-lower semi-lattices and $\phi\colon R \to S$ a surjective homomorphism with kernel K, where K is generated by idempotents, then ϕ preserves the lower semi-lattice structure.*

PROOF. Notice that if $a^2 = ab$ then, also, $a^2 = ba$ since $(ab - ba)^2 = 0$.

(i) Clearly $a^2 = ab$ implies $(ca)^2 = cacb$. Moreover, idempotents are central in R.

(ii) By (i), $uc \leq_{\mathrm{rr}} ua$ and $uc \leq_{\mathrm{rr}} ub$. Suppose $d \leq_{\mathrm{rr}} ua$ and $d \leq_{\mathrm{rr}} ub$, then $u^{-1}d \leq_{\mathrm{rr}} a$ and $u^{-1}d \leq_{\mathrm{rr}} b$. Hence, $u^{-1}d \leq_{\mathrm{rr}} c$ giving $d \leq_{\mathrm{rr}} uc$ and $uc = ua \wedge_{\mathrm{rr}} ub$.

(iii) Obvious.

(iv) If $a, b \in R$ and $c = a \wedge_{\mathrm{rr}} b$ it must be shown that $\phi(c) = \phi(a) \wedge_{\mathrm{rr}} \phi(b)$. Put $s = \phi(a) \wedge_{\mathrm{rr}} \phi(b)$. Clearly $\phi(c) \leq_{\mathrm{rr}} s$, using (iii). Pick $t \in \phi^{-1}(s)$. There exist idempotents $e, f \in K$ and $r, r' \in R$ so that $t^2 - ta = er$ and $t^2 - tb = fr'$. Let $g = (1-e)(1-f)$. It follows that $(t^2-ta)g = 0 = (t^2-tb)g$ which implies $(tg)^2 = tga$ and $tg \leq_{\mathrm{rr}} a$ and, similarly, $tg \leq_{\mathrm{rr}} b$. Thus, $tg \leq_{\mathrm{rr}} c$. Note that $\phi(g) = 1$. Applying ϕ gives $\phi(tg) = \phi(t) = s \leq_{\mathrm{rr}} \phi(c)$. This shows that $\phi(c) = s$. □

COROLLARY 2.2. *Let R be a reduced ring and $\mathfrak{P}(R) = \prod_{\mathfrak{p}\in \mathrm{Min}} R/\mathfrak{p}$. If for $a, b \in R$, $a \wedge_{rr} b = 0$ in $\mathfrak{P}(R)$ then $a \wedge_{rr} b = 0$ in R. In particular, if $ab = 0$ then $a \wedge_{rr} b = 0$ in both rings.*

PROOF. This follows immediately from Proposition 2.1 (iii) and how infs are calculated in $\mathfrak{P}(R)$. □

In Proposition 2.1 (iv) the hypotheses on ϕ are required. If, for example, ϕ is a surjective homomorphism from a domain R onto a domain S with non-zero kernel and if $a \neq b$ are not in $\ker \phi$ with $\phi(a) = \phi(b)$, then $a \wedge_{\mathrm{rr}} b = 0$ but $\{a, b\}$ is sent to $\{\phi(a) = \phi(b)\}$ and $\phi(a) \wedge_{\mathrm{rr}} \phi(b) = \phi(a) \neq 0$.

In what follows the case of a not necessarily commutative reduced ring R which is a Pierce sheaf of domains will be examined. To simplify notation, for $a \in R$, $\operatorname{ann}_{\mathbf{B}} a = \operatorname{ann} a \cap \mathbf{B}(R)$. The awB rings are those for which $\operatorname{ann} a = (\operatorname{ann}_{\mathbf{B}} a)R$. A *weakly Baer ring* (wB ring) is a reduced ring such that for each $r \in R$ there is $e \in \operatorname{ann}_{\mathbf{B}} r$ such that $\operatorname{ann} r = Re$. This means that for $x \in \operatorname{Spec} \mathbf{B}(R)$, $r_x = 0_x$ if and only if $e \notin x$. (Such rings are also called *pp-rings*, for example in [**3**], or *Rickart rings* in [**4**].)

To simplify notation the following definition is included.

DEFINITION 2.3. A reduced ring R which is an **rr**-lower semi-lattice is called ***rr**-good.* If, for $r, s \in R$, $r \wedge_{\mathrm{rr}} s = 0$, the elements r and s are called **rr**-orthogonal.

The property of being **rr**-orthogonal is what is called a *Pierce property.*

LEMMA 2.4. *Let R be a reduced ring. Then, for $r, s \in R$, $r \wedge_{rr} s = 0$ if and only if for all $x \in \operatorname{Spec} \mathbf{B}(R)$, $r_x \wedge_{rr} s_x = 0_x$.*

PROOF. Suppose first that $r \wedge_{\mathrm{rr}} s = 0$. If, for some $x \in \operatorname{Spec} \mathbf{B}(R)$, there is $a \in R$ with $0_x \neq a_x \leq_{\mathrm{rr}} r_x$ and $a_x \leq_{\mathrm{rr}} s_x$ in R_x. Then there is $e \in \operatorname{Spec} \mathbf{B}(R) \setminus x$ such that $a^2e = aer = aes$ and $ae \neq 0$, a contradiction. Suppose now that r_x and s_x are **rr**-orthogonal in each stalk. If there is $0 \neq a$ with $a^2 = ar = as$ then, for some $x \in \operatorname{Spec} \mathbf{B}(R)$, $a_x \neq 0_x$ leading to a contradiction. □

Before going on to characterize the **rr**-good rings among the awB rings it will be shown that the Niefield-Rosenthal characterization of commutative rings whose Pierce stalks are domains extends readily to all reduced rings.

PROPOSITION 2.5. *Let R be a reduced ring. Then each of the stalks of the Pierce sheaf of R is a domain if and only if R is an awB ring.*

PROOF. Recall the notation for the Pierce sheaf: the base space is $\operatorname{Spec} \mathbf{B}(R)$ and the stalks are the rings R/Rx, $x \in \operatorname{Spec} \mathbf{B}(R)$. Suppose first that R is awB. If for $r, s \in R$ and $x \in \operatorname{Spec} \mathbf{B}(R)$, $r_x \neq 0_x$ and $(rs)_x = 0_x$ then there is $e \in \mathbf{B}(R) \setminus x$ such that $rse = 0$. Hence, by the awB property, there are idempotents $e_1, \dots, e_k \in \operatorname{ann} r$ and $t_1, \dots, t_k \in R$ such that $se = \sum_{i=1}^k t_i e_i$. However, the $e_i \in x$ showing that $se \in Rx$ and, thus, $(se)_x = s_x = 0_x$ because $e_x = 1_x$. Hence, R_x is a domain.
In the other direction, if each R_x is a domain then, for $r \in R$, if $s \in \operatorname{ann} r$ then $s_x = 0_x$ if $r_x \neq 0_x$. Hence, $\operatorname{coz} s \subseteq \mathrm{z}(r)$, where $\mathrm{z}(r)$ is an open set. Since $\mathrm{z}(r)$ is a union of clopen (closed and open) sets of the form $\operatorname{coz} e$, $e \in \mathbf{B}(R)$ and $er = 0$. The compactness of the closed set $\operatorname{coz} s$ implies that $\operatorname{coz} s$ is covered by a finite set of the $\operatorname{coz} e$. The union of this finite collection is also clopen, say $\operatorname{coz} f$, $f \in \mathbf{B}(R)$ and $f \in \operatorname{ann} r$. From this, $s = sf$. Hence, $\operatorname{ann} r = (\operatorname{ann}_{\mathbf{B}} r)R$. □

THEOREM 2.6. *Let R be an awB ring. Then R is **rr**-good if and only if R is a wB ring.*

PROOF. Suppose first that R is a wB ring. Given non-zero $a, b \in R$ with $a \neq b$, let $eR = \operatorname{ann}(a-b)$, $e = e^2$. The claim is that $ae = be = a \wedge_{\mathrm{rr}} b$. First, $ae \leq_{\mathrm{rr}} a$ and $ae \leq_{\mathrm{rr}} b$ since $ae \cdot a = a^2e = (ae)^2$ and $ae \cdot b = be \cdot b = (be)^2 = (ae)^2$. If $c \leq_{\mathrm{rr}} a$ and $c \leq_{\mathrm{rr}} b$ then $c(a-b) = 0$ so that $c = ce$. Then, $cae = c^2e = c^2$ showing that $c \leq_{\mathrm{rr}} ae$ and, similarly, $c \leq_{\mathrm{rr}} be$.
Now suppose that R is an awB ring which is **rr**-good. Consider $r \in R$. It must be shown that $\operatorname{ann} r = eR$ for some $e \in \mathbf{B}(R)$. Set $a = r + 1$ and $b = 1$. By assumption there is $c = a \wedge_{\mathrm{rr}} b$. Since $c^2 = ca = cb$, $c \in \operatorname{ann}(a-b)$. Notice that since $c^2 = cb = c1 = c$, $c \in \operatorname{ann}_{\mathbf{B}}(a-b)$. Let $f \in \operatorname{ann}_{\mathbf{B}}(a-b)$. Set $g = f - cf$. Now $ga = g(r+1) = g$ so that $g \leq_{\mathrm{rr}} a$ and $g \leq_{\mathrm{rr}} b$, showing that $g \leq_{\mathrm{rr}} c$. Hence, $g = gc$, but $gc = 0$ giving $g = 0$ and thus $f \leq_{\mathrm{rr}} c$. Hence, c is the unique maximal element of $\operatorname{ann}_{\mathbf{B}}(a-b)$ showing that $\operatorname{ann} r = cR$. □

The calculations above show the following in an awB ring R: r and s in R have a non-zero **rr**-lower bound if and only if there is $0 \neq e = e^2$ with $e(r-s) = 0$ and $er \neq 0$.

Constructions of wB rings will be discussed below. Among rings of continuous functions, $C(X)$ is wB if and only if X is basically disconnected (see the remark before Example 3.2 in [**14**]). According to [**14**, Example 3.2], the ring of continuous real valued functions $C(\beta\mathbf{N}\setminus\mathbf{N})$ is awB but not wB. Rings of continuous functions $C(X,\mathbf{Z})$, where X is a topological space and $\mathbf{Z}$ has the discrete topology, are easily seen all to be wB. Notice that the property **rr**-good is not a Pierce property. If R is **rr**-good then each Pierce stalk is **rr**-good but the example $C(\beta\mathbf{N}\setminus\mathbf{N})$ has stalks domains and, hence, **rr**-good but the ring is not, by Theorem 2.6.

Before pursuing consequences of Theorem 2.6 it is worthwhile to observe how rarely pairs of elements, even in a wB ring, can have an **rr**-supremum (denoted $\vee_{\mathrm{rr}}$). Moreover, it is shown below that R is an **rr**-upper semi-lattice if and only if it is boolean.

Assume for now that R is an awB ring and that $r,s\in R$. Consider the following pairwise disjoint subsets of $\operatorname{Spec}\mathbf{B}(R)$:

$$\begin{array}{rclcrcl} A_{r,s} & = & \operatorname{coz} r\cap\operatorname{coz} s\cap \mathrm{z}(r-s)\,, & & B_{r,s} & = & \operatorname{coz} r\cap \mathrm{z}(s)\\ C_{r,s} & = & \mathrm{z}(r)\cap\operatorname{coz} s\,, & & D_{r,s} & = & \mathrm{z}(r)\cap\mathrm{z}(s)\end{array}$$

If, in addition R is wB, each $\operatorname{coz} t$ is the cozero set of an idempotent. For the given elements r and s the following idempotents, e_1,e_2,e_3,e_4 are defined by: $\operatorname{coz} e_1 = A_{r,s}$, $\operatorname{coz} e_2 = B_{r,s}$, $\operatorname{coz} e_3 = C_{r,s}$ and $\operatorname{coz} e_4 = D_{r,s}$. Note that these idempotents are orthogonal. This notation will be used in the following proposition.

PROPOSITION 2.7. *Let R be an awB ring. (i) Given $r,s\in R$ then $r\vee_{rr} s$ exists only if* $(*)$ $\operatorname{coz} r\cup\operatorname{coz} s\subseteq A_{r,s}\cup B_{r,s}\cup C_{r,s}$. *(ii) If, in addition, R is wB then $(*)$ implies that $r\vee_{rr} s$ exists.*

PROOF. Notice that condition $(*)$ implies that $A_{r,s}\cup B_{r,s}\cup C_{r,s}\cup D_{r,s} = \operatorname{Spec}\mathbf{B}(R)$.

(i) Suppose $(*)$ fails but that r and s have an **rr**-upper bound c. If, for example, $x\in\operatorname{coz} r$ and $x\notin A_{r,s}\cup B_{r,s}\cup C_{r,s}$ then $r_x\neq 0$, $s_x\neq 0$ and $r_x\neq s_x$. But then $r_x^2 = r_xc_x$ shows $r_x = c_x$ but $s_x^2 = s_xc_x$ gives $s_x = c_x$, a contradiction. Hence, r and s do not have an upper bound. Similarly if $x\in\operatorname{coz} s$ and $x\notin A_{r,s}\cup B_{r,s}\cup C_{r,s}$.

(ii) Suppose that $\operatorname{coz} r\cup\operatorname{coz} s\subseteq\operatorname{coz}(e_1+e_2+e_3)$. Define $c=(e_1+e_2)r+e_3s$. Then, $r^2-rc = r^2-r(e_1+e_2)r-re_3s = 0$ since $re_3=0$ and $r(e_1+e_2)=r$, by $(*)$. Similarly $sc=s^2$ since c is also $(e_1+e_3)s+e_2r$. Thus, c is an upper bound of r and s. For any $t\in R$, $c+te_4$ is also an upper bound but $t=0$ gives a smallest one of the form $c+te_4$. It will be shown that any upper bound d has this form. If d is an upper bound and $x\in\operatorname{coz} e_1$, $r_x^2=r_xd_x$ giving $d_x=r_x$ and, similarly $d_x=s_x$. Thus c and d are equal over $\operatorname{coz} e_1$. Similarly for $\operatorname{coz} e_2$ and $\operatorname{coz} e_3$ showing that c and d can only differ over $D_{r,s}$. Hence, $c=r\vee_{\mathrm{rr}} s$. □

PROPOSITION 2.8. *A ring R is an **rr**-upper semi-lattice if and only if it is a boolean ring.*

PROOF. Each boolean ring is an upper semi-lattice in its natural order, which coincides with the **rr**-order. On the other hand, suppose R is any reduced ring which is an **rr**-upper semi-lattice. If $a\in R$, $a\neq 1$, put $c=a\vee_{\mathrm{rr}} 1$. Then, $1^2=1c$ so that $c=1$. Then $a^2=a$. □

The next result looks at surjective homomorphisms from a wB ring. The following lemma will be useful.

LEMMA 2.9. *Let R be a wB ring and $a, b \in R$. There is a largest $e \in \mathbf{B}(R)$ such that $a \wedge_{rr} eb = 0$.*

PROOF. By Theorem 2.6, R is **rr**-good. Let $c = a \wedge_{\mathrm{rr}} b$ and $e \in \mathbf{B}(R)$ be such that $\operatorname{ann} c = eR$. Then for $x \in \operatorname{Spec}\mathbf{B}(R)$, $c_x = 0_x$ if and only if $e_x = 1_x$. In other words, $c_x = a_x = b_x \neq 0_x$ if and only if $e_x = 0_x$. Then there is no x so that $a_x = (eb)_x \neq 0_x$. This shows that $a \wedge_{\mathrm{rr}} eb = 0$. Now suppose that $f \in \mathbf{B}(R)$ is such that $e < f$. Thus, there is $x \in \operatorname{Spec}\mathbf{B}(R)$ such that $f_x = 1_x$ while $e_x = 0_x$. Then, for each such x it follows that $c_x = a_x = b_x = (fb)_x \neq 0_x$ which shows that $a \wedge_{\mathrm{rr}} fb \neq 0$ by Lemma 2.4. □

The following result about lifting **rr**-orthogonal sets also gives another family of examples in the study of lifting orthogonal sets in [**15**]. See also Proposition 3.12, below.

THEOREM 2.10. *Let R and S be reduced rings where R is wB and $\phi\colon R \to S$ is a surjective ring homomorphism such that $\phi(\mathbf{B}(R)) = \mathbf{B}(S)$. If $\{s_n\}_{n\in\mathbf{N}}$ is an **rr**-orthogonal set in S then it lifts, via ϕ, to an **rr**-orthogonal set in R.*

PROOF. Let $r_1 \in R$ be any lifting of s_1. Suppose for $n \geq 1$ that $s_1, \dots, s_n$ have been lifted to $r_1, \dots, r_n$, respectively, where $\{r_1, \dots, r_n\}$ is **rr**-orthogonal in R. Let u be any lifting of s_{n+1}. Notice that for any pair $1 \leq i < j \leq n$, r_i and r_j never coincide and are non-zero in a Pierce stalk. Assume $\{x \in \operatorname{Spec}\mathbf{B}(R) \mid \text{for some } 1 \leq i \leq n, (r_i)_x = u_x \neq 0_x\} \neq \emptyset$ for otherwise $r_{n+1} = u$ can be used. As in Lemma 2.9 there are maximal idempotents $e_1, \dots, e_n$ such that $r_i \wedge_{\mathrm{rr}} e_i u = 0$, and, by what has been observed, $\{1 - e_1, \dots, 1 - e_n\}$ are orthogonal, in the sense of idempotents. This is because it is not possible to have, for $i \neq j$, $(r_i)_x = u_x = (r_j)_x \neq 0_x$, which would happen if $(e_i)_x = (e_j)_x = 0_x$, i.e., if $((1 - e_i)(1 - e_j))_x \neq 0_x$. Put $\varepsilon = 1 - ((1 - e_1) + \cdots + (1 - e_n))$. Then, for $x \in \operatorname{Spec}\mathbf{B}(R)$, $\varepsilon_x = 0_x$ exactly where one of the $(e_i)_x = 0_x$. Now put $r_{n+1} = \varepsilon u$. It first needs to be shown that $r_i \wedge_{\mathrm{rr}} \varepsilon u = 0$ for $i = 1, \dots, n$ or, by Lemma 2.4, that $(r_i \wedge_{\mathrm{rr}} \varepsilon u)_x = 0_x$ for all $x \in \operatorname{Spec}\mathbf{B}(R)$. If $(\varepsilon u)_x \neq 0_x$ then $\varepsilon_x = 1_x$ and then $(e_i)_x = 1_x$ for all i. This means that $(r_i \wedge_{\mathrm{rr}} \varepsilon u)_x = (r_i \wedge_{\mathrm{rr}} e_i u)_x = 0_x$. If $(\varepsilon u)_x = 0_x$ then, of course, $(r_i \wedge_{\mathrm{rr}} \varepsilon u)_x = 0_x$. By Lemma 2.4, each $r_i \wedge_{\mathrm{rr}} \varepsilon u = 0$.

It must now be shown that $\phi(r_{n+1}) = s_{n+1}$. The restriction of ϕ to $\mathbf{B}(R)$ is, by hypothesis, onto $\mathbf{B}(S)$; call the restriction β. The maximal ideals of $\mathbf{B}(S)$ are in one-to-one correspondence with those of $\mathbf{B}(R)$ containing $\ker\beta$. It is only necessary to look at those $x \in \operatorname{Spec}\mathbf{B}(R)$ where $\ker\beta \subseteq x$. Consider now $x \in \operatorname{Spec}\mathbf{B}(R)$ with $\ker\beta \subseteq x$ and $\beta(x) = y \in \operatorname{Spec}\mathbf{B}(S)$. (i) If $\varepsilon_x = 0_x$ then, for some i, $1 \leq i \leq n$, $(e_i)_x = 0_x$. This means that $(r_i \wedge_{\mathrm{rr}} u)_x \neq 0$ and then $(r_i)_x = u_x \neq 0_x$. Then, $\phi(r_i)_y = (s_i)_y = \phi(u)_y = (s_{n+1})_y = 0_y$, since s_i and s_{n+1} are orthogonal. However, $\phi(\varepsilon u)_y = 0_y$ as well. (ii) If $\varepsilon_x = 1_x$ then for all i, $1 \leq i \leq n$, $(r_i)_x \neq u_x$ or $(r_i)_x = u_x = 0_x$. In any case $\phi(r_i)_y = (s_i)_y \neq \phi(u)_i = (s_{n+1})_y$ or $\phi(r_i)_y = \phi(u)_y = 0_y$. Here, $\phi(u)_y = \phi(\varepsilon u)_y$. Hence, in all cases, $\phi(u)_y = \phi(\varepsilon u)_y = (s_{n+1})_y$. In other words, $\phi(r_{n+1}) = s_{n+1}$. □

Notice that in Theorem 2.10 that the hypotheses do not guarantee that ϕ preserves the **rr**-order infima but that, in any case, countable **rr**-orthogonal sets in S lift to **rr**-orthogonal sets in R. No example is known to date of a surjective homomorphism of reduced rings where the lifting of countable **rr**-orthogonal sets fails.

A strongly regular ring (see e.g., [**13**, page 199]) R is a wB ring where the stalks are division rings. In this case Proposition 2.1 (iii) can be strengthened. If $r \in R$ then there is $r' \in R$ with $r^2 r' = r$ and $(r')^2 r = r'$. The idempotent rr' will be denoted $\mathbf{e}(r)$. If $\phi\colon R \to S$ is a surjective ring homomorphism with kernel K then S is also strongly regular, $\phi(\mathbf{e}(r)) = \mathbf{e}(\phi(r))$ and K is generated by idempotents. Moreover, if $r \in K$ then $\mathbf{e}(r) \in K$ and *vice versa*.

COROLLARY 2.11. *Let R and S be strongly regular rings equipped with the **rr**-order, and $\phi\colon R \to S$ a surjective ring homomorphism. Then, ϕ is an **rr**-lower semi-lattice homomorphism. Moreover, a countable **rr**-order orthogonal set in S can be lifted to an **rr**-order orthogonal set in R.*

PROOF. Since the Pierce stalks of strongly regular rings are division rings, these rings are awB but, in addition, each principal ideal is generated by an idempotent so that they are, in fact, wB rings. Thus, R and S are **rr**-good (Theorem 2.6) and ϕ preserves the lower semi-lattice structure since all ideals are generated by idempotents (Proposition 2.1 (iv)). The lifting follows from Theorem 2.10. □

Notice that when $\phi\colon R \to S$ is a ring surjection of wB rings, ϕ need not be a **rr**-lower semi-lattice morphism. Consider $\phi\colon \mathbf{Z} \to \mathbf{Z}/6\mathbf{Z} \cong \mathbf{Z}/2\mathbf{Z} \times \mathbf{Z}/3\mathbf{Z}$. The **rr**-orthogonal set $\{7, 13\}$ goes to the pair $((1,1),(1,1))$, which is not **rr**-orthogonal.

Theorem 2.6 gives a class of reduced rings which are **rr**-good and shows that there are reduced rings which are not. The following example is a reduced ring which is not **rr**-good but which has a wB subring S with $\mathbf{B}(R) = \mathbf{B}(S)$. Section 3 presents examples of rings which are **rr**-good but which are not awB, for example $\mathrm{C}(\mathbf{R})$, which has no non-trivial idempotents. Example 2.12 will show up again in Example 2.16

EXAMPLE 2.12. *Fix a prime p and let R be the ring of sequences of integers which are eventually constant modulo p. Then, R lies between the **rr**-good rings S of eventually constant integer sequences and $P = \prod_{\mathbf{N}} \mathbf{Z}$ with $\mathbf{B}(R) = \mathbf{B}(S)$. However, R is not **rr**-good.*

PROOF. The rings S and P are even wB. Notice that an idempotent in R has finite or cofinite support. Consider r which is constantly 1 and $s = (1, p+1, 1, p+1, 1, \ldots)$. Thsn, r and s have non-zero **rr**-lower bounds, for example $(1,0,0,0,\ldots)$ or $(1,0,1,0,0,\ldots)$ but no largest one. □

In [**14**, Corollary 1.6] it is pointed out that in a commutative noetherian ring, the conditions awB and wB coincide. A commutative noetherian reduced ring R is an order in a finite product of fields and it is natural to ask if such a ring is **rr**-good. When R has three or fewer minimal prime ideals the answer is "yes" (neither commutativity nor the noetherian property is needed) but not in general. The proof for three minimal primes will show what can go wrong with four or more.

PROPOSITION 2.13. *Let R be a reduced ring.*
*(i) Suppose R has three or fewer minimal primes. Then R is **rr**-good.*
*(ii) If R has four minimal prime ideals it need not be **rr**-good.*

PROOF. (i) The case of three minimal primes, $\mathfrak{p}_1, \mathfrak{p}_2, \mathfrak{p}_3$, will be considered. Embed R in $R/\mathfrak{p}_1 \times R/\mathfrak{p}_2 \times R/\mathfrak{p}_3 = S$ via $\phi\colon R \to S$. The ring R will be viewed as a subring of S. Consider $a, b \in R$ and their images in S, say (a_1, a_2, a_3) and (b_1, b_2, b_3). If $a_i \neq b_i$ for $i = 1, 2, 3$ then a and b have no non-zero lower bounds in S and so, $a \wedge_{\mathrm{rr}} b = 0$ in R. If a and b differ in precisely two places, say, $a_2 \neq b_2$ and $a_3 \neq b_3$ then the inf in S is $(a_1, 0, 0)$. If $(a_1, 0, 0,) \in R$ then that is the inf in R otherwise $a \wedge_{\mathrm{rr}} b = 0$. Finally suppose a and b coincide in the first two places but not in the third. The inf in S is $(a_1, a_2, 0)$. If it is in R then it is $a \wedge_{\mathrm{rr}} b$. If not, there are three possibilities: $(a_1, 0, 0) \in R$, $(0, a_2, 0) \in R$ and neither is in R. In each case $a \wedge_{\mathrm{rr}} b$ exists in R.

(ii) Let K be a field. Let Q be a product of four copies of the field $K(x, y, z)$. The ring R is generated as a $K[x, y, z]$-subalgebra of Q by the following elements:

$$\begin{array}{rclcrcl} r & = & (x^2 + x, y, y, y) & \quad & b_1 & = & (0, z, 0, 0) \\ s & = & (x, y, y, y) & & b_2 & = & (0, 0, z, 0) \\ a & = & (0, y, y, 0) & & b_3 & = & (0, 0, 0, z) \\ b & = & (0, 0, y, y) & & 1 & = & (1, 1, 1, 1) \end{array}$$

The ring R is noetherian with four minimal prime ideals and will be seen not to be **rr**-good.

Note first that $Q_{\mathrm{cl}}(R) = Q$ since $r - s, b_1, b_2, b_3$ give the four idempotents of Q in $Q_{\mathrm{cl}}(R)$. The ring R is noetherian with 4 minimal primes by Goldie's Theorem.

Note that, in R, a and b are common lower bounds of r and s, but a and b are not **rr**-related. The only candidate for a lower bound of r and s which is larger than both a and b is $(0, y, y, y)$. The aim is to show that $(0, y, y, y)$ is not in R. Suppose on the contrary that it is. Then, there is a polynomial expression F, with coefficients in $K[x, y, z]$, in 7 variables with zero constant term which is evaluated at the vector $\mathbf{v} = (r, s, a, b, b_1, b_2, b_3)$ to yield $F(\mathbf{v}) + C = (0, y, y, y)$, where C is a constant from $K[x, y, z]$.

Since z does not appear in an essential way in the main equation and is only used to obtain idempotents in $Q_{\mathrm{cl}}(R)$ it is possible to reduce the problem by setting $z = 0$. Once this has been done there is a polynomial f in four variables u_1, u_2, u_3, u_4 with coefficients in $K[x, y]$ and zero constant term. Set $\mathbf{w} = (r, s, a, b)$ and then $f(\mathbf{w}) + c = (0, y, y, y)$ for some constant $c \in K[x, y]$. The aim is to show that such an equation leads to a contradiction. Subscripts will indicate the four components.

The polynomial f can be written $f = h + f_a + f_b + f_{ab}$ where h has monomials only in u_1 and/or u_2, f_a has monomials with factor u_3 but not u_4, f_b has monomials with factor u_4 but not u_3, and f_{ab} has monomials with factors both u_3 and u_4. Recall that all these polynomials have 0 constant term.

Note first that only h influences the first component. Set $h(\mathbf{w})_1 = l$. Then, $c = (-l, -l, -l, -l)$. Moreover, l is divisible by x. Now consider the other components.

Notice the following: $f_a(\mathbf{w})_1 = f_a(\mathbf{w})_4 = 0$, $f_b(\mathbf{w})_1 = f_b(\mathbf{w})_2 = 0$, $f_{ab}(\mathbf{w})_1 = f_{ab}(\mathbf{w})_2 = f_{ab}(\mathbf{w})_4 = 0$, and

$$\begin{array}{llclcl} (i) & h(\mathbf{w})_2 & = & h(\mathbf{w})_3 & = & h(\mathbf{w})_4 \\ (ii) & f_a(\mathbf{w})_2 & = & f_a(\mathbf{w})_3 & & \\ (iii) & f_b(\mathbf{w})_3 & = & f_b(\mathbf{w})_4 & & \end{array}$$

The elements $h(\mathbf{w})_2, f_a(\mathbf{w})_2, f_b(\mathbf{w})_4$ are divisible by y and $f_{ab}(\mathbf{w})_3$ is divisible by y^2. The components can now be computed.

$$\begin{array}{rclcl} (f(\mathbf{w}) + c)_2 & = & h(\mathbf{w})_2 + f_a(\mathbf{w})_2 - l & = & y \\ (f(\mathbf{w}) + c)_4 & = & h(\mathbf{w})_4 + f_b(\mathbf{w})_4 - l & = & y \\ (f(\mathbf{w}) + c)_3 & = & h(\mathbf{w})_3 + f_a(\mathbf{w})_3 + f_b(\mathbf{w})_3 + f_{ab}(\mathbf{w})_3 - l & = & y \end{array}$$

It follows that $h(\mathbf{w})_2 + f_a(\mathbf{w})_2 = l + y$ and $h(\mathbf{w})_4 + f_b(\mathbf{w})_4 = l + y$ and, using (i), $f_b(\mathbf{w})_4 = f_a(\mathbf{w})_2$. In the expression for $(f(\mathbf{w}) + c)_3$ the sum of the first two terms equals $l + y$ showing that $f_b(\mathbf{w})_3 + f_{ab}(\mathbf{w})_3 = 0$. Since $f_{ab}(\mathbf{w})_3$ is divisible by y^2, so is $f_b(\mathbf{w})_3$ and, from that, so is $f_a(\mathbf{w})_3$. This last is because $f(\mathbf{w})_2 = f(\mathbf{w})_4$ and the various equalities with (ii) and (iii) yield $f_a(\mathbf{w})_3 = f_b(\mathbf{w})_3$.

In $(*)$ $h(\mathbf{w})_3 + f_a(\mathbf{w})_3 + f_b(\mathbf{w})_3 + f_{ab}(\mathbf{w})_3 = l + y$ all the terms on the left are divisible by y. Hence, l is divisible by y. To proceed, a more detailed look at $h(u_1, u_2)$ is needed. Let A be a finite set of pairs of non-negative integers i, j with $i+j \geq 1$ and, for $i, j \in A$, $a_{ij} \in K[x, y]$. Each a_{ij} can be written $a'_{ij} + a''_{ij}$, where a'_{ij} contains all the terms of a_{ij} with factor y and a''_{ij} all the remaining terms. Write

$$(**) \quad l = h(x^2 + x, x) = \sum_A a_{ij}(x^2 + x)^i x^j = \sum_A a'_{ij}(x^2 + x)^i x^j + \sum_A a''_{ij}(x^2 + x)^i x^j \,.$$

Since $y|l$, the last term in $(**)$ is zero.

From the expression for $f(\mathbf{w})_3$, it follows that $f_a(\mathbf{w})_3 + f_b(\mathbf{w})_3 + f_{ab}(\mathbf{w})_3 = l + y - h(\mathbf{w})_3$ is divisible by y^2. Since $x|l$ and $y|l$, one can write $l = l'xy$. Now,

$$h(\mathbf{w})_3 = \sum_A a'_{ij} y^{i+j} + \sum_{i+j \geq 2} a''_{ij} y^{i+j} + (a''_{10} + a''_{01})y \,.$$

The first two terms on the right of this expression are divisible by y^2 and, hence, y^2 divides $l'xy + y - (a''_{10} + a''_{01})y$ and y divides $l'x + 1 - (a''_{10} + a''_{01})$. It follows that $a''_{10} + a''_{01} = px + 1$ where $p \in K[x]$. Now return to $(**)$ to look at the last terms which are $a''_{10}(x^2 + x) + a''_{01}x = a''_{10}x^2 + (px + 1)x$. There are no other degree 1 terms in x. This contradicts the fact that $\sum_A a''_{ij}(x^2 + x)^i x^j = 0$ and concludes the proof. $\square$

The remainder of this section is devoted to showing how to embed each commutative reduced ring into a wB ring in various ways. Every such ring will be seen to have an essential wB extension by using the complete ring of quotients in the next proposition. It does not require commutativity.

PROPOSITION 2.14. *Let $R \subseteq S$ be reduced rings where S is wB. Let $T \subseteq S$ be the ring generated by R and those idempotents from $\mathbf{B}(S)$ which generate the annihilators, in S, of the elements of R. Then, T is wB.*

PROOF. The following notation will be used. For $r \in R$, let $\operatorname{ann}_S r = e(r)S$, where $e(r) \in \mathbf{B}(S)$. Set $\mathcal{S} = \{e(r) \mid r \in R\}$ and $T = R[\mathcal{S}]$. It needs to be shown that the annihilator in T of an element of T is generated by an idempotent. A typical non-zero element $t \in T$ can be written $t = \sum_{i=1}^k r_i e_i$ where the $0 \neq r_i \in R$ and the e_i are idempotents which, after the standard manipulations on idempotents, can be assumed to be orthogonal and derived from elements of $\mathcal{S}$ (and are thus in T). Put $\varepsilon = 1 - (e_1 + \cdots + e_k)$. The claim is that $\operatorname{ann}_T t$ is generated by $e = \sum_{i=1}^k e(r_i)e_i + \varepsilon$. It is clear that $te = 0$. Let $a \in \operatorname{ann}_T t$. Then a can be written in the form $a = \sum_{j=1}^m s_j f_j$, with each $s_j \in R$ and the f_j a set of orthogonal idempotents derived from those in $\mathcal{S}$. For each $i = 1, \ldots, k$, the orthogonality of the e_i shows that $0 = r_i e_i \sum_{j=1}^m s_j f_j$ so that for each $j = 1, \ldots, m$, $s_j f_j e_i \in e(r_i)T$ and $s_j f_j e_i \in e_i e(r_i)T$. From this, $0 = (\sum_{i=1}^k r_i e_i)(\sum_{j=1}^m s_j f_j) = \sum_{j=1}^m \sum_{i=1}^k r_i s_j e_i f_j =$

$\sum_{i=1}^{m} r_i e_i a$. By the orthogonality of the e_i, each $r_i e_i a = 0$, which shows that $e_i a = e_i a e(r_i)$. From this, $a = a(\sum_{i=1}^{k} e_i e(r_i)) + a\varepsilon \in (\sum_{i=1}^{k} e_i e(r_i) + \varepsilon)T$, as required. □

Before applying this proposition it should be noted that neither the category of commutative wB rings, $\mathcal{WB}$, nor the category of commutative **rr**-good rings, $\mathcal{RG}$, is a reflective subcategory of the category of commutative reduced rings $\mathcal{CR}$. (See [**1**, Chapter V].) Both $\mathcal{WB}$ and $\mathcal{RG}$ are closed under products but neither is closed under equalizers as would be required for a reflective subcategory. Hence, in both cases, the inclusion of it into $\mathcal{CR}$ cannot have a left adjoint. This is seen using the following examples. The first simple example, pointed out by J. Kennison, illustrates this for $\mathcal{WB}$.

EXAMPLE 2.15. *Let $R = \mathbf{Z}\times\mathbf{Z}$ and $S = Z_2$ with $\phi\colon R \to S$ defined by $\phi(a,b) = \bar{a}$ and $\psi\colon R \to S$ by $\psi(a,b) = \bar{b}$. Then while R and S are wB, the equalizer E of ϕ and ψ is not wB.*

PROOF. Note that $(0,2) \in E$ but its annihilator is not generated by an idempotent in E, which has no non-trivial idempotents. □

The same remark applies to the category $\mathcal{RG}$ in $\mathcal{CR}$. Here the example is a bit more complex.

EXAMPLE 2.16. *Let $R = \prod_{\mathbf{N}} \mathbf{Z}$ and $S = \mathbf{Z}_2$. Let $\mathcal{U}$ be the set of all free ultrafilters on $\mathbf{N}$. For each $U \in \mathcal{U}$, let the ideal I_U of $\prod_{\mathbf{N}} \mathbf{Z}_2$ be the set of those elements 0 over a set in U. Define $\phi_U\colon R \to S$ by $(a_n) \mapsto (\overline{a_n}) \mapsto (\overline{a_n}) + I_U$. Then the equalizer of all the ϕ_U is not **rr**-good.*

PROOF. The ring in Example 2.12 with $p = 2$, which is not **rr**-good, will used. (Any prime p could be used.) Suppose $r = (a_n) \in R$ is eventually constant modulo 2. Then, since the ultrafilters are free, r will be in the equalizer. On the other hand, if $r = (a_n) \in R$ is not eventually constant modulo 2 the set, X, of components where r is even and the set, Y, where they are odd are disjoint infinite sets. The sets X and Y are not both in all $U \in \mathcal{U}$. Thus, r will not have the same image under all of the ϕ_U. □

In the case of a commutative reduced ring R (i.e., commutative semiprime rings) there are various standard ways of embedding R into a regular ring. Each will yield a wB ring according to Proposition 2.14. The first, which is mentioned in passing, is the *universal regular ring* studied in [**10**] and [**18**]. The functor $T\colon \mathcal{CR} \to \mathcal{VNR}$, where $\mathcal{VNR}$ is the category of commutative regular rings, embeds each $R \in \mathcal{CR}$ in $T(R) \in \mathcal{VNR}$ so that the T is the left adjoint of the inclusion functor $\mathcal{VNR} \to \mathcal{CR}$. If Proposition 2.14 is used, it can be shown that the construction is functorial, say $U\colon \mathcal{CR} \to \mathcal{WB}$. However, U does not preserve rings which are already wB and, hence, details will be omitted.

Another choice of regular ring would be the complete ring of quotients, $Q(R)$ of $R \in \mathcal{CR}$, always regular in this case. This works well but is not functorial. There is another important regular ring between R and $Q(R)$, namely the *epimorphic hull*, the intersection of all regular rings between R and $Q(R)$ (see [**17**, Satz 11.3]). However, the next lemma will show that the construction of Proposition 2.14 gives the same result for any regular ring S, $R \subseteq S \subseteq Q(R)$.

DEFINITION 2.17. Let $R \in \mathcal{CR}$ and $Q(R)$ its complete ring of quotients. The wB ring constructed from R and $Q(R)$ using Proposition 2.14 is denoted $\mathbf{W}(R)$.

It will be seen that an annihilator which is generated by an idempotent behaves well when passing from a ring to an essential extension ring. If $R \subseteq S$ are reduced rings then the extension is called *(right) essential* if every non-zero right ideal of S has non-zero intersection with R. A right essential extension is also called a *right intrinsic extension* in [**8**]. (Note that this is weaker than saying that R_R is essential in S_R.)

LEMMA 2.18. *Let $S \subseteq T$ be reduced rings where T is a right essential extension of S. Suppose that, for $s \in S$, there are idempotents $e \in S$ and $f \in T$ such that* $\operatorname{ann}_S s = eS$ *and* $\operatorname{ann}_T s = fT$. *Then, $e = f$.*

PROOF. Since $es = 0$, $e = ef$. If $f \neq ef$ then there is $t \in T$ such that $0 \neq (f - ef)t \in S$. However, $s(f - ef)t = sf(1 - e)t = 0$ so that $f(1 - e)t = f(1 - e)te = 0$, a contradiction. Hence, $e = f$. □

COROLLARY 2.19. *Let $R \in \mathcal{CR}$ and $S \in \mathcal{CR}$ an essential extension of R which is wB. Let V be the wB ring generated in S as in Proposition* 2.14. *If $r \in R$ is such that* $\operatorname{ann}_R r = eR$ *for some idempotent $e \in R$ then $eS =$* $\operatorname{ann}_S r$. *Thus, in the construction of V, the idempotent for $r \in R$ is already in R.*

PROOF. This is a direct application of Lemma 2.18. □

Everything is now in place to show that the construction of $\mathbf{W}(R)$ not only gives an efficient way of constructing a wB ring from R but also that if R is already wB then $\mathbf{W}(R) = R$.

COROLLARY 2.20. *If $R \in \mathcal{CR}$ is a wB ring then $\mathbf{W}(R) = R$.*

PROPOSITION 2.21. *Let R, S and V be as in Corollary* 2.19. *Then, $V \cong \mathbf{W}(R)$.*

PROOF. There is a copy of $Q(R)$ in $Q(S)$ ([**17**, Satz 10.1]). The wB ring $\mathbf{W}(R)$ can be calculated inside this copy of $Q(R)$. Now consider $r \in R$ and the idempotents e and f where $eV = \operatorname{ann}_V r$ and $fQ(R) = \operatorname{ann}_{Q(R)} r$. Then, $\operatorname{ann}_{Q(S)} r = eQ(S)$ and Lemma 2.18 shows that $e = f$. Hence, V and $\mathbf{W}(R)$ are generated over R by the same set of idempotents. □

The construction of $\mathbf{W}(R)$ can be seen, using Lemma 2.18, to be an example of a wB *absolute to $Q(R)$ (right) ring hull* of R for any commutative reduced ring, in the language of [**4**, Definition 8.2.1(i)]. In the case of a general reduced ring R, the construction of Proposition 2.14 along with Lemma 2.18 will yield a wB right ring hull of R ([**4**, Definition 8.2.1(iii)]) *if there is any* essential right wB extension T of R.

The process of constructing $\mathbf{W}(R)$ is illustrated by the following.

PROPOSITION 2.22. *Suppose $R \in \mathcal{CR}$ is such that its classical ring of quotients $Q = Q_{cl}(R)$ is regular. Then $\mathbf{W}(R)$ is the ring generated over R by $\mathbf{B}(Q)$.*

PROOF. In this case for $r \in R$, $\operatorname{ann}_Q r = (\operatorname{ann}_R r)Q$ and for each $a/b \in Q$, $\operatorname{ann}_Q a/b = (\operatorname{ann}_R a)Q$. Hence, in Proposition 2.14, all the idempotents of Q are adjoined. □

3. Rings of continuous functions. In this section examples of **rr**-good rings will be found which are quite unlike wB rings. The basic reference is [**9**].

Suppose X is a completely regular space and $\mathrm{C}(X)$ is considered with the **rr**-order. What does it mean for $\mathrm{C}(X)$ to be **rr**-good? Let $f, g \in \mathrm{C}(X)$ and suppose $h = f \wedge_{\mathrm{rr}} g$ exists. Then where h is non-zero it must coincide with f and g. This means that $\operatorname{coz} h \subseteq \mathrm{z}(f-g) \cap \operatorname{coz} f$. (Here $\operatorname{coz} f$ and $\mathrm{z}(f)$ have their usual meanings in rings of functions.) Moreover, on $\operatorname{coz} h$, h coincides with f and with g. These conditions only make $h \leq_{\mathrm{rr}} f$ and $h \leq_{\mathrm{rr}} g$. To have the infimum there must be a *unique* largest cozero set $U \subseteq \mathrm{z}(f-g) \cap \operatorname{coz} f$ and h with (i) $\operatorname{coz} h = U$, (ii) h and f (and g) coincide on U, and (iii) for every net $\Lambda = \{\lambda_\alpha\}$ in U converging to a point y on the boundary of U, $\{f(\lambda_\alpha)\} \to f(y) = 0$ and $\{g(\lambda_\alpha)\} \to g(y) = 0$. Notation: For $Y \subseteq X$ the boundary of Y is denoted $\mathbf{bd}\, Y = \overline{Y} \setminus Y$.

Definition 3.1. A completely regular topological space X is called **rr**-*good* if the ring $\mathrm{C}(X)$ is **rr**-good.

As an example consider a pair of functions from $\mathrm{C}(\mathbf{R})$. (It turns out that $\mathbf{R}$ is an **rr**-good space.) Let $f(x) = \sin x$ and $g(x) = |\sin x|$. The function $h(x)$ which is $f(x)$ where the two functions coincide and zero where they differ is continuous. Hence, $h = f \wedge_{\mathrm{rr}} g$.

The purpose here is to find examples of **rr**-good spaces. It will be seen later that it is easy to construct spaces which are not **rr**-good. Before looking at details it is worthwhile to note a class of **rr**-good spaces; more will follow. Recall that a space X is *basically disconnected* if the closure of a cozero set is open. As already mentioned these are precisely the spaces where $\mathrm{C}(X)$ is wB and hence **rr**-good by Theorem 2.6. These spaces include the P-spaces ([**9**, 4J]) where $\mathrm{C}(X)$ is von Neumann regular. Note that a space can be totally disconnected without being **rr**-good: the space $X = \beta\mathbf{N} \setminus \mathbf{N}$ is an example where $\mathrm{C}(X)$ is awB but not wB ([**14**, Example 3.2]) and, by Theorem 2.6 again, X is not **rr**-good.

A complete characterization of all **rr**-good spaces as a class of well-known spaces is elusive since examples include basically disconnected spaces and, as will be seen, locally connected spaces. However, Theorem 3.2 does characterize **rr**-good spaces although by properties not easily translated into global language. It makes more explicit what it means for f and g in $\mathrm{C}(X)$ to have a non-zero infimum. The description will later be used to get sufficient conditions for X to be **rr**-good.

Theorem 3.2. *Let X be a completely regular topological space. Fix $f, g \in$ $C(X)$, $f \neq g$, and suppose that f and g have a non-zero* **rr**-*lower bound. Let $\{h_\alpha\}_{\alpha \in A}$ be the set of non-zero* **rr**-*lower bounds of f and g, $U_\alpha = \operatorname{coz} h_\alpha$, $\alpha \in A$ and $U = \bigcup_{\alpha \in A} U_\alpha$*

(i) Then, $f \wedge_{rr} g$ exists if and only if U has the property: $(*)$ if $x \in \mathbf{bd} U$ and N is any neighbourhood of x then, for some $\beta \in A$, $N \cap \mathbf{bd} U_\beta \neq \emptyset$.

(ii) Moreover, $f \wedge_{rr} g$ exists if and only if U has the property: $(**)$ if $x \in \mathbf{bd} U$ then for each $\varepsilon > 0$ there is a neighbourhood N_ε of x such that for all $\alpha \in A$, $N_\varepsilon \cap U_\alpha \subseteq \{y \in X \mid |h_\alpha(y)| < \varepsilon\}$.

Proof. (i) Note, with f and g as above, that if $\mathbf{bd}\, U = \emptyset$ then U is clopen and there is $e^2 = e \in \mathrm{C}(X)$ such that $\operatorname{coz} e = U$. In this case $f \wedge_{\mathrm{rr}} g = ef = eg$ and $(*)$ is vacuously satisfied.

Notice that, as in the statement, h_α coincides with f and with g on $\operatorname{coz} h_\alpha$; hence, $\operatorname{coz} h_\alpha \subseteq \mathrm{z}(f-g) \cap \operatorname{coz} f$. Moreover, if $\alpha \neq \beta$ in A, then $(\mathbf{bd}\, U_\alpha) \cap U_\beta = \emptyset$.

This is because for $y \in \mathbf{bd}\, U_\alpha$, $h_\alpha(y) = f(y) = 0$ while for $x \in U_\beta$, $f(x) = h_\beta(x) \neq 0$.

First assume property $(*)$ for f and g which have a non-zero lower bound. The proof will show that, among the lower bounds, there is a maximal one. With the h_α as in the statement, define h by $h(x) = f(x)$ for $x \in U$ and $h(x) = 0$ for $x \in X \setminus U$. The continuity of h on $\mathbf{bd}\, U$ must be shown. The condition $(*)$ says that for every neighbourhood N of $x \in \mathbf{bd}\, U$ there is some $\beta \in A$ with $N \cap \mathbf{bd}\, U_\beta \neq \emptyset$. Thus there is $y \in N$ with $f(y) = 0$. By continuity of f it follows that $f(x) = 0$; the fact that h is zero on $X \setminus U$ shows that h is continuous at x and that h coincides with f (and with g) on U. By construction $h \leq_{\mathrm{rr}} f$ and $h \leq_{\mathrm{rr}} g$ and, also, for each $\alpha \in A$, $h_\alpha \leq_{\mathrm{rr}} h$.

In the other direction, if $h = f \wedge_{\mathrm{rr}} g$ exists it is one of the h_α showing that $\operatorname{coz} h = U$. By its continuity, the property $(*)$ follows automatically.

The proof of part (ii) is along the same lines as that of (i). □

The next result gives a sufficient condition on a space X to be an **rr**-good space. For convenience, the condition will be given a name.

DEFINITION 3.3. Let X be a completely regular space. Let $\{U_\alpha\}_{\alpha \in A}$ be any family of non-empty cozero sets in X with the following property: for $\alpha \neq \beta$ in A, $(\mathbf{bd}\, U_\alpha) \cap U_\beta = \emptyset$. The space X is said to satisfy the *B-property* (for *boundary property*) if the following holds for each such family of cozero sets. Let $z \in \mathbf{bd}\,(\bigcup_{\alpha \in A} U_\alpha)$ then for every neighbourhood N of z there is $\beta \in A$ such that $N \cap \mathbf{bd}\, U_\beta \neq \emptyset$.

COROLLARY 3.4. *Let X be a completely regular topological space which satisfies the B-property. Then X is an **rr**-good space.*

PROOF. The B-property is a stronger form (strictly stronger by Example 3.8, below) of the property $(*)$ in Theorem 3.2. This guarantees the existance of $f \wedge_{\mathrm{rr}} g$ whenever f and g have a non-zero lower bound. Otherwise, $f \wedge_{\mathrm{rr}} g = 0$. □

The next task is to find examples of spaces satisfying the conditions of Corollary 3.4. Recall the definition of a locally connected space (e.g., [**7**, page 374]): X is *locally connected* if for each $x \in X$ and a neighbourhood N of x there is a connected subset $C \subseteq N$ such that $x \in \mathbf{int}(C)$.

THEOREM 3.5. *Let X be a completely regular space. (1) If X is locally connected then X is an **rr**-good space. (2) A connected space X need not be **rr**-good even when X is a compact metric space.*

PROOF. (1) It will be shown that X satisfies the B-property. Suppose $\{U_\alpha\}_{\alpha \in A}$ is a family of non-empty open sets in X as in the first part of Definition 3.3. Let $z \in \mathbf{bd}\,(\bigcup_{\alpha \in A} U_\alpha)$. It may be supposed that $z \notin \mathbf{bd}\, U_\beta$ for any $\beta \in A$. Choose a neighbourhood N of z. By [**7**, 6.3.3], there is a connected subset C of N with $z \in \mathbf{int} C$. Then for some $\beta \in A$, $\mathbf{int} C \cap U_\beta \neq \emptyset$. If $C \cap \mathbf{bd}\, U_\beta \neq \emptyset$ then there is nothing to prove. Otherwise, $C \cap \overline{U_\beta} = C \cap U_\beta$. This means that $C \cap U_\beta$ is clopen in C. Moreover, $C \cap U_\beta \neq \emptyset$ and $C \cap U_\beta \neq C$ since $z \notin U_\beta$. This contradicts the fact that C is connected. Hence, $N \cap \mathbf{bd}\, U_\beta \neq \emptyset$.

(2) For the construction, [**19**, Example 27.8(b)] is used but truncated to make it compact. Define $X \subseteq \mathbf{R}^2$ as follows. It consists of the vertical line segments $\{(0, y) \mid -1 \leq y \leq 1\}$ and $\{(1, y) \mid -1 \leq y \leq 1\}$ along with horizontal line segments

$\{(x, 1/n) \mid 0 \leq x \leq 1\}$, for $n \in \mathbf{Z} \setminus \{0\}$, and $\{(x, 0) \mid 0 \leq x \leq 1\}$. Elements $f, g \in \mathrm{C}(X)$ will be defined so that they do not have an **rr**-inf.

$$f(x,y) = \begin{cases} 2x & \text{if } 0 \leq x \leq 1/2, y = 0 \text{ or } y = 1/n, n \in \mathbf{Z} \setminus \{0\} \\ -2x+2 & \text{if } 1/2 \leq x \leq 1, y = 0 \text{ or } y = 1/n, n \in \mathbf{Z} \setminus \{0\} \\ 0 & \text{otherwise} \end{cases}$$

$$g(x,y) = \begin{cases} f(x,y) & \text{if } x = 0, x = 1, \text{ or } y = 0 \text{ or } y = 1/n,\ n \in \mathbf{Z} \setminus \{0\}, n \text{ even} \\ \frac{2|n|}{|n|+1} x & \text{if } 0 \leq x \leq 1/2, y = 1/n, n \in \mathbf{Z} \text{ odd} \\ \frac{-2|n|}{|n|+1} x + \frac{2|n|}{|n|+1} & \text{if } 1/2 \leq x \leq 1, y = 1/n, n \in \mathbf{Z} \text{ odd} \end{cases}$$

A typical **rr**-lower bound of f and g coincides with f and g on finitely many of the horizontal lines, say with $y \in \{1/n_1, \ldots, 1/n_k\}$, the $n_i \neq 0$ and even, and zero elsewhere. However, continuity on the line where $y = 0$ precludes a greatest **rr**-lower bound. □

The next result shows how to construct examples of spaces which are not **rr**-good.

PROPOSITION 3.6. *Let X be a completely regular topological space with the following property: there is a sequence of disjoint clopen subsets $\{C_n\}_{n \in \mathbf{N}}$ such that $U = \bigcup_{n \in \mathbf{N}} C_n$ is not closed in X and, for some $x \in \mathbf{bd}\, U$, every neighbourhood of x meets all but finitely many of the C_n. Then, X is not an **rr**-good space.*

PROOF. The proof of Theorem 3.5 is a model. Define two elements of $\mathrm{C}(X)$ as follows: f is constantly 1 on X, and g is constantly 1 except on C_n, n odd, where it is constantly $1 + 1/n$. It will be shown that g is continuous. Put $U = \bigcup_{n \in \mathbf{N}} C_n$ and $V = \mathbf{int}(X \setminus U)$. Clearly g is continuous on $U \cup V$. Let $z \in \mathbf{bd}\, U$. Given $\varepsilon > 0$ there is $m \in \mathbf{N}$ such that for all $n > m$, $1/n < \varepsilon$. Then for $x \in X \setminus \bigcup_{1 \leq p \leq m} C_p$, $|g(x) - g(z)| < \varepsilon$. This shows continuity at all points of X.

It will be shown that $f \wedge_{\mathrm{rr}} g$ does not exist. There are lower bounds. In fact, any function h which is 0 everywhere except on finitely many C_n, n even, where it is 1, is continuous and a lower bound. However, since, for each $n \in \mathbf{N}$, $\mathbf{bd}\, C_n = \emptyset$, condition $(*)$ of Theorem 3.2 fails at x (as does condition $(**)$). □

Proposition 3.6 has many corollaries. Some of them are now listed.

EXAMPLES 3.7. *The following are examples of spaces X which are not **rr**-good. (1) Let X be the one-point compactification of a disjoint union of an infinite collection of compact spaces; e.g., the one-point compactification of an infinite discrete space. (2) Let X be any dense subset of $\mathbf{R}$ such that $\mathbf{R} \setminus X$ has an accumulation point in X. If X is as stated then X^n is also not **rr**-good. (3) Let $\{F_n\}_{n \in \mathbf{N}}$ be a set of closed subsets of $\mathbf{R}^n$ which are pairwise separated by open sets, $U = \bigcup_{n \in \mathbf{N}} F_n$ is not closed, and for $x \in \mathbf{bd}\, U$ there is a sequence $\{x_n\}_{n \in \mathbf{N}}$ converging to x with $x_n \in F_n$; then let $X = \overline{\bigcup_{n \in \mathbf{N}} F_n}$ as a subspace of $\mathbf{R}^n$.*

PROOF. These can be seen to be special cases of Proposition 3.6. □

It is next seen that the B-condition is not necessary for a space to be **rr**-good.

EXAMPLE 3.8. *Let Y be an uncountable discrete space and $X = Y \cup \{s\}$ where the neighbourhoods of s are co-countable in X. Then, X is an* **rr**-*good space which does not satisfy the B-condition.*

PROOF. According to [**9**, 4N], X is a P-space ($\mathrm{C}(X)$ is regular) and, hence, **rr**-good. Now partition Y into countable subsets $\{Y_\alpha\}_{\alpha \in A}$ and, for each $\alpha \in A$, let e_α be the idempotent whose support is Y_α. Each Y_α is thus a cozero set in X with empty boundary. However, $\bigcup_A Y_\alpha = Y$ whose boundary is $\{s\}$. Hence, the B-condition fails. □

Recall that for a completely regular space X, $C^*(X)$ is the ring of bounded real-valued continuous functions on X.

PROPOSITION 3.9. *Let X be a completely regular topological space. Then X is* **rr**-*good if and only if $C^*(X)$ is an* **rr**-*good ring.*

PROOF. If X is **rr**-good and $f, g \in \mathrm{C}^*(X)$ and $h = f \wedge_{\mathrm{rr}} g$, calculated in $\mathrm{C}(X)$, then $h \in \mathrm{C}^*(X)$.

In the other direction, let $f, g \in \mathrm{C}(X)$. Put $f_b = f/(1+f^2)$ and $g_b = g/(1+g^2)$. Then $f_b, g_b \in \mathrm{C}^*(X)$, as are $f_{bb} = f_b/(1+g^2)$ and $g_{bb} = g_b/(1+f^2)$. Let $h = f_{bb} \wedge_{\mathrm{rr}} g_{bb}$. Then $h^2 = hf_{bb} = hg_{bb}$. Hence, $h^2(1+f^2)^2(1+g^2)^2 = h(1+f^2)(1+g^2)f = h(1+f^2)(1+g^2)g$ showing that $h(1+f^2)(1+g^2)$ is a lower bound for f and g. Let $k \in \mathrm{C}(X)$ be a lower bound for f and g. Then, $k^2 = kf = kg$ and, hence, $k/(1+f^2)(1+g^2)$ is a lower bound for f_{bb} and g_{bb}. It follows that $k/(1+f^2)(1+g^2) \in \mathrm{C}^*(X)$ and $k/(1+f^2)(1+g^2) \leq_{\mathrm{rr}} h$. Then, by Proposition 2.1 (i), $k \leq_{\mathrm{rr}} h(1+f^2)(1+g^2)$ showing that $h(1+f^2)(1+g^2) = f \wedge_{\mathrm{rr}} g$ in $\mathrm{C}(X)$. □

Since for any space X, $C^*(X) \cong C(\beta X)$, where βX is the Čech-Stone compactification of X, Proposition 3.9 shows that if X is **rr**-good so is βX. It is not hard to show that if X is **rr**-good so is any space V, $X \subseteq V \subseteq \beta X$. Moreover, since for any space Y between X and its realcompactification υX, $C(Y) = \mathrm{C}(X)$, if X is **rr**-good so is V. (See [**9**, Chapter 8].)

It is next shown that cozero sets in **rr**-good spaces are also **rr**-good, for example, any open set in an **rr**-good metric space.

PROPOSITION 3.10. *Let X be a completely regular topological space which is* **rr**-*good. Let $Y = \operatorname{coz} k$ where $k \in C(X)$. The space Y is* **rr**-*good.*

PROOF. First note that k may be assumed to be bounded. If $f \in \mathrm{C}^*(V)$ then fk can be extended to $\tilde{f} \in \mathrm{C}(X)$ by defining $\tilde{f}(x) = 0$ if $x \in X \setminus Y$ (see [**5**, Proposition 1.1]). Notice also that an **rr**-lower bound of two bounded functions is necessarily bounded.

It will first be shown that $\mathrm{C}^*(Y)$ is an **rr**-lower semi-lattice; the result for $\mathrm{C}(Y)$ follows from Proposition 3.9. Consider $f, g \in \mathrm{C}^*(V)$. Extend fk and gk as above to get $\tilde{f}$ and $\tilde{g}$ in $\mathrm{C}(X)$. Let $H = \tilde{f} \wedge_{\mathrm{rr}} \tilde{g}$. Then, in particular, $H^2 = \tilde{f}H = \tilde{g}H$. Set $h = H_{|_V}$ showing that $h^2 = fkh = gkh$. Consider $1/k \in \mathrm{C}(Y)$. Then, $(h/k)^2 = f(h/k) = g(h/k)$. Thus h/k is a lower bound for f and g. Since $f, g \in \mathrm{C}^*(V)$, it follows that $h/k \in \mathrm{C}^*(V)$.

Suppose next that a is also a lower bound for f and g. Then, $ak \leq_{\mathrm{rr}} fk$ and $ak \leq_{\mathrm{rr}} gk$. Consider $\tilde{a}$. It follows that $(\tilde{a})^2 = \tilde{a}\tilde{f} = \tilde{a}\tilde{g}$ since the equality holds on Y and all the functions are 0 on $X \setminus Y$. Therefore, $\tilde{a} \leq H$ and $\tilde{a}_{|_Y}$ is a lower bound of fk and gk, showing that $ak \leq_{\rm rr} h$. Then, upon multiplication by $1/k$, $a \leq h/k$. Thus, $h/k = f \wedge_{\rm rr} g$. □

It is not true that the restriction homomorphism $\mathrm{C}(X) \to \mathrm{C}(Y)$ in Proposition 3.10 preserves $\wedge_{\rm rr}$.

EXAMPLE 3.11. *Let $X = \mathbf{R}$ and $Y = (0,1)$, a cozero set, and $\rho\colon C(X) \to C(Y)$. Then, ρ does not preserve **rr**-infima.*

PROOF. Consider two function f and g in $\mathrm{C}(X)$ which are never 0 except at $x = 1/2$ and only coincide on $[0, 1/2]$. Then, $f \wedge_{\rm rr} g = 0$ in $\mathrm{C}(X)$ but $\rho(f) \wedge_{\rm rr} \rho(g) = h \neq 0$ where h coincides with f and g on $(0, 1/2)$ and is 0 on $[1/2, 1)$. □

In the context of $\mathrm{C}(X)$ it is possible to construct other examples of ring surjections where countable **rr**-orthogonal sets lift to **rr**-orthogonal sets. (See also Theorem 2.10.) A subspace V of X is C-embedded if every element of $\mathrm{C}(V)$ extends to an element of $\mathrm{C}(X)$, i.e., the restriction map $\rho\colon \mathrm{C}(X) \to \mathrm{C}(V)$ is surjective.

PROPOSITION 3.12. *Let X be an **rr**-good space, V a C-embedded subspace and $\rho\colon C(X) \to C(V)$ the restriction homomorphism. If $\{f_n\}_{n\in\mathbf{N}}$ is an **rr**-orthogonal set in $C(V)$ then there is an **rr**-orthogonal set $\{F_n\}_{n\in\mathbf{N}}$ in $C(X)$ such that for all $n \in \mathbf{N}$, $\rho(F_n) = f_n$.*

PROOF. The proof will be by induction. For F_1, pick any pre-image of f_1. Let $n \geq 1$ and suppose that an **rr**-orthogonal set $F_1, \dots, F_n$ has been found with, for $1 \leq i \leq n$, $\rho(F_i) = f_i$. Choose $G \in \mathrm{C}(X)$ to be any pre-image of f_{n+1}.

For $1 \leq i \leq n$, let $U_i = \mathrm{coz}(G \wedge_{\rm rr} F_i)$. If all the U_i are empty then put $G = F_{n+1}$. In any case, notice that $U_i \cap V = \emptyset$ since $G|_V \wedge_{\rm rr} F_i|_V = 0$ for $1 \leq i \leq n$.

It will be shown that, in any case, the U_i are pairwise disjoint. Now suppose that for some $1 \leq i < j \leq n$ that $U_i \cap U_j \neq \emptyset$. On $U_i \cap U_j$, all of G, F_i and F_j coincide. Suppose that $\mathbf{bd}\,(U_i \cap U_j) \neq \emptyset$ and that all three functions are 0 on it. Then let H be such that H coincides with G, F_i and F_j on $U_i \cap U_j$ but is 0 elsewhere. This function is continuous. Moreover, $H \leq_{\rm rr} F_i$ and $H \leq_{\rm rr} F_j$. This is impossible since $H \neq 0$ and $F_i \wedge_{\rm rr} F_j = 0$. If follows that for some $x_0 \in \mathbf{bd}\,(U_i \cap U_j)$ that $G(x_0) = F_i(x_0) = F_j(x_0) \neq 0$ (they coincide by continuity). In this case $x_0 \in U_i \cap U_j$. To see this consider a net $\{\lambda_\alpha\}_{\alpha\in A}$ in $U_i \cap U_j$ which converges to x_0. Think of this as a net in U_i; since $F_i(x_0) \neq 0$, the limit of the net, x_0, is in U_i and not in $\mathbf{bd}\,U_i$. Similarly, x_0 is in U_j. Thus, $x_0 \in U_i \cap U_j$. This is impossible by the definition of the boundary. The only remaining possibility is that $\mathbf{bd}\,(U_i \cap U_j) = \emptyset$. This means that $U_i \cap U_j$ is clopen and thus the cozero set of an idempotent, say $e \in \mathrm{C}(X)$. But then eG will be a lower bound for F_i and for F_j. This cannot happen unless $e = 0$ and then $U_i \cap U_j = \emptyset$.

Hence, for each $1 \leq i < j \leq n$, $U_i \cap U_j = \emptyset$. Notice also that each $U_i \subseteq X \setminus V$. Now G can be modified as follows: Define G' by

$$G'(x) = \begin{cases} 0, & \text{if } x \in \bigcup_{i=1}^n \overline{U_i} \\ G(x), & \text{otherwise} \end{cases}$$

The function G is zero on each $\mathbf{bd}\,U_i$ which shows that $G' \in \mathrm{C}(X)$. Since G and G' coincide on V, they have the same image in $\mathrm{C}(V)$. It remains to be shown that F_i and G' are **rr**-orthogonal for $i = 1, \dots n$. Suppose, for some i, that H is an

rr-lower bound of F_i and of G'. Put $W = X \setminus \bigcup_{j=1}^n \overline{U_j}$. Since G' is 0 on $\bigcup_{j=1}^n \overline{U_i}$, $\operatorname{coz} H \subseteq W$. Recall that $\operatorname{coz}(F_i \wedge_{\mathrm{rr}} G) = U_i$. Define a new function K as follows:

$$K(x) = \begin{cases} (F_i \wedge_{\mathrm{rr}} G)(x), & x \in U_i \\ H(x), & x \in \operatorname{coz} H \\ 0, & \text{otherwise} \end{cases}$$

The function K is continuous and is an **rr**-lower bound of F_i and of G. However, $F_i \wedge_{\mathrm{rr}} G \leq_{\mathrm{rr}} K$ showing that $K = F_i \wedge_{\mathrm{rr}} G$. But then, $\operatorname{coz} H = \emptyset$ and $H = 0$.

□

As examples where Proposition 3.12 can be used, recall that any compact subspace V of X is C-embedded ([**9**, 6J]) and, hence, if X is **rr**-good the proposition applies. It is also easy to see that if, in the above, the subspace V is only C*-embedded then countable **rr**-orthogonal sets of bounded elements of $C(V)$ lift to **rr**-orthogonal sets in $C(X)$.

References

[1] J. Adámek, H. Herrlich and G.E. Strecker, *Abstract and Concrete Categories.* Springer Verlag, Online Edition, 2004.

[2] V.A. Andrunakievič and J.M. Rjabuhin, *Rings without nilpotent elements and completely prime ideals*, Soviet Math. Dokl. **9** (1968), 565–568. MR02030760

[3] G. M. Bergman, *Hereditary commutative rings and centres of hereditary rings*, Proc. London Math. Soc. (3) **23** (1971), 214–236. MR0309918

[4] G. F. Birkenmeier, J. K. Park, and S. T. Rizvi, *Extensions of rings and modules*, Birkhäuser/Springer, New York, 2013. MR3099829

[5] R. L. Blair and A. W. Hager, *Extensions of zero-sets and of real-valued functions*, Math. Z. **136** (1974), 41–52. MR0385793

[6] W. D. Burgess and R. Raphael, *Abian's order relation and orthogonal completions for reduced rings*, Pacific J. Math. **54** (1974), 55–64. MR0360708

[7] R. Engelking, *General topology*, 2nd ed., Sigma Series in Pure Mathematics, vol. 6, Heldermann Verlag, Berlin, 1989. Translated from the Polish by the author. MR1039321

[8] C. Faith and Y. Utumi, *Intrinsic extensions of rings*, Pacific J. Math. **14** (1964), 505–512. MR0161873

[9] L. Gillman and M. Jerison, *Rings of continuous functions*, Springer-Verlag, New York-Heidelberg, 1976. Reprint of the 1960 edition; Graduate Texts in Mathematics, No. 43. MR0407579

[10] M. Hochster, *Prime ideal structure in commutative rings*, Trans. Amer. Math. Soc. **142** (1969), 43–60. MR0251026

[11] P. T. Johnstone, *Stone spaces*, Cambridge Studies in Advanced Mathematics, vol. 3, Cambridge University Press, Cambridge, 1982. MR698074

[12] J. L. Kelley, *General topology*, Springer-Verlag, New York-Berlin, 1975. Reprint of the 1955 edition [Van Nostrand, Toronto, Ont.]; Graduate Texts in Mathematics, No. 27. MR0370454

[13] T. Y. Lam, *A first course in noncommutative rings*, 2nd ed., Graduate Texts in Mathematics, vol. 131, Springer-Verlag, New York, 2001. MR1838439

[14] S. B. Niefield and K. I. Rosenthal, *Sheaves of integral domains on Stone spaces*, J. Pure Appl. Algebra **47** (1987), no. 2, 173–179. MR906969

[15] R. Raphael, *On the countable orthogonal lifting property*, in preparation.

[16] P. N. Stewart, *Semi-simple radical classes*, Pacific J. Math. **32** (1970), 249–254. MR0255593

[17] H. H. Storrer, *Epimorphismen von kommutativen Ringen* (German), Comment. Math. Helv. **43** (1968), 378–401. MR0242810

[18] R. Wiegand, *Modules over universal regular rings*, Pacific J. Math. **39** (1971), 807–819. MR0379472

[19] S. Willard, *General topology*, Addison-Wesley Publishing Co., Reading, Mass.-London-Don Mills, Ont., 1970. MR0264581

Department of Mathematics and Statistics, University of Ottawa, Ottawa, Canada, K1N 6N5

Email address: `wburgess@uottawa.ca`

Department of Mathematics and Statistics, Concordia University, Montréal, Canada, H4B 1R6

Email address: `raphael@alcor.concordia.ca`

Contemporary Mathematics
Volume **715**, 2018
https://doi.org/10.1090/conm/715/14407

On a category of extensions whose endomorphism rings have at most four maximal ideals

Federico Campanini and Alberto Facchini

Abstract. We describe the endomorphism ring of a short exact sequence $0 \longrightarrow A_R \longrightarrow B_R \longrightarrow C_R \longrightarrow 0$ with A_R and C_R uniserial modules and the behavior of these short exact sequences as far as their direct sums are concerned.

1. Introduction

Two right modules A_R and B_R are said to have: (1) the same *monogeny class*, denoted by $[A_R]_m = [B_R]_m$, if there exist a monomorphism $A_R \to B_R$ and a monomorphism $B_R \to A_R$; (2) the same *epigeny class*, denoted by $[A_R]_e = [B_R]_e$, if there exist an epimorphism $A_R \to B_R$ and an epimorphism $B_R \to A_R$. A module A_R is *uniserial* if the lattice of its submodules is linearly ordered under inclusion. That is, if, for any submodules V and W of A_R, either $V \subseteq W$ or $W \subseteq V$. Finite direct sums of uniserial modules are classified via their monogeny class and their epigeny class [**F1**]. These two invariants are sufficient because the endomorphism ring of a uniserial module has at most two maximal ideals. Other classes of modules whose endomorphism rings have at most two maximal ideals have similar behaviors [**AAF**, **FEK**, **FG**].

In this article, we focus our attention to short exact sequences

$$0 \longrightarrow A_R \longrightarrow B_R \longrightarrow C_R \longrightarrow 0$$

of right R-modules. They form the class of objects of a category $\mathcal{E}$, which is additive, but is not abelian. If we consider the objects of $\mathcal{E}$ for which A_R and C_R are uniserial R-modules, we find a behavior very similar to that of uniserial modules mentioned in the previous paragraph. In this case, we have that finite direct sums of short exact sequences with A_R and C_R uniserial modules are classified via four invariants, which are the natural generalizations of monogeny class and epigeny class. Four invariants are necessary now because the endomorphism ring in $\mathcal{E}$ of a short exact sequence with A_R and C_R uniserial modules has at most four maximal ideals.

The main results in this paper are Theorem 3.2, in which we describe the endomorphism ring of a short exact sequence $0 \longrightarrow A_R \longrightarrow B_R \longrightarrow C_R \longrightarrow 0$

2010 *Mathematics Subject Classification.* Primary 16D70, 16D80.

Partially supported by Dipartimento di Matematica "Tullio Levi-Civita" of Università di Padova (Project BIRD163492/16 "Categorical homological methods in the study of algebraic structures" and Research program DOR1690814 "Anelli e categorie di moduli").

with A_R and C_R uniserial modules, and Theorem 6.3, in which we describe the behavior of the short exact sequences with A_R and C_R uniserial as far as their direct-sum decompositions in $\mathcal{E}$ are concerned.

In this article, modules are right unitary modules.

2. Modules with a semilocal endomorphism ring

Let R be a ring, associative and with an identity 1. We consider the following category $\mathcal{E}$ of extensions. The objects of $\mathcal{E}$ are the short exact sequences

$$0 \longrightarrow A_R \xrightarrow{\alpha} B_R \xrightarrow{\beta} C_R \longrightarrow 0,$$

where A_R, B_R, C_R are right R-modules. A morphism in $\mathcal{E}$ between two such exact sequences

$$0 \longrightarrow A_R \xrightarrow{\alpha} B_R \xrightarrow{\beta} C_R \longrightarrow 0 \tag{2.1}$$

and

$$0 \longrightarrow A'_R \xrightarrow{\alpha'} B'_R \xrightarrow{\beta'} C'_R \longrightarrow 0 \tag{2.2}$$

is a right R-module morphism $f\colon B_R \to B'_R$ that induces a commutative diagram

$$\begin{array}{ccccccccc} 0 & \longrightarrow & A_R & \xrightarrow{\alpha} & B_R & \xrightarrow{\beta} & C_R & \longrightarrow & 0 \\ & & \downarrow{\scriptstyle f|^{A}_{A'}} & & \downarrow{\scriptstyle f} & & \downarrow{\scriptstyle \overline{f}} & & \\ 0 & \longrightarrow & A'_R & \xrightarrow[\alpha']{} & B'_R & \xrightarrow[\beta']{} & C'_R & \longrightarrow & 0. \end{array}$$

Equivalently, $f\colon B_R \to B'_R$ is a right R-module morphism such that $f(\alpha(A_R)) \subseteq \alpha'(A'_R)$. We will denote by E_B the endomorphism ring of the object

$$0 \longrightarrow A_R \xrightarrow{\alpha} B_R \xrightarrow{\beta} C_R \longrightarrow 0 \tag{2.3}$$

in the category $\mathcal{E}$. Notice that in this notation E_B for the endomorphism ring of the exact sequence (2.3), there is a slight abuse of notation, because we assume that the modules A_R and C_R and the morphisms α and β are fixed once for all, which will be true in the rest of this article.

Clearly, the zero object of the category $\mathcal{E}$ is the exact sequence

$$0 \longrightarrow 0 \longrightarrow 0 \longrightarrow 0 \longrightarrow 0.$$

It will be denoted by 0.

If (2.1) and (2.2) are two objects of $\mathcal{E}$ and $f\colon B_R \to B'_R$ is a morphism in $\mathcal{E}$, that is, a right R-module morphism with $f(\alpha(A_R)) \subseteq \alpha'(A'_R)$, then f is an isomorphism in $\mathcal{E}$ if and only if $f\colon B_R \to B'_R$ is a right R-module isomorphism and $f(\alpha(A_R)) = \alpha'(A'_R)$. Also, the category $\mathcal{E}$ is not an abelian category. In fact, it is easy to construct a morphism f in $\mathcal{E}$ for which $f\colon B_R \to B'_R$ is a right R-module isomorphism and $f(\alpha(A_R)) \subset \alpha'(A'_R)$. Such a morphism f in $\mathcal{E}$ is clearly both right cancellable and left cancellable, that is, is both monic and epi, but is not an isomorphism in $\mathcal{E}$. Thus $\mathcal{E}$ is not abelian.

A ring R is *semilocal* if the factor ring $R/J(R)$, where $J(R)$ denotes the Jacobson radical of R, is a semisimple artinian ring. A ring R is semilocal if and only if its dual dimension codim(R) is finite [**F2**, Proposition 2.43]. If R and S are rings, a ring morphism $\varphi\colon R \to S$ is *local* if, for every $r \in R$, $\varphi(r)$ invertible in S implies r invertible in R [**CD**].

PROPOSITION 2.1. [**CD**, Corollary 2] If $R \to S$ is a local morphism between two rings R and S, then $\operatorname{codim}(R) \leq \operatorname{codim}(S)$.

Recall that a ring S *has type* n [**FP3**], if the factor ring $S/J(S)$ is a direct product of n division rings, and that S is a *ring of finite type* if it has type n for some integer $n \geq 1$. If a ring S has finite type, then the type n of S coincides with the dual Goldie dimension $\operatorname{codim}(S)$ of S [**F2**, Proposition 2.43]. A module M_R has *type* n (*finite type*) if its endomorphism ring $\operatorname{End}(M_R)$ is a ring of type n (of finite type, respectively). The zero module is the unique module of type 0.

PROPOSITION 2.2. [**FP3**, Proposition 2.1] Let S be a ring and $n \geq 1$ be an integer. The following conditions are equivalent:

(a) The ring S has type n.
(b) n is the smallest of the positive integers m for which there is a local morphism of S into a direct product of m division rings.
(c) The ring S has exactly n distinct maximal right ideals, and they are all two-sided ideals in S.
(d) The ring S has exactly n distinct maximal left ideals, and they are all two-sided ideals in S.

PROPOSITION 2.3. Let $0 \longrightarrow A_R \xrightarrow{\alpha} B_R \xrightarrow{\beta} C_R \longrightarrow 0$ be a short exact sequence of right R-modules. Then:

(a) $\operatorname{codim}(E_B) \leq \operatorname{codim}(\operatorname{End}(A_R)) + \operatorname{codim}(\operatorname{End}(C_R))$.
(b) If A_R and C_R are modules whose endomorphism ring is semilocal, then the endomorphism ring E_B of the short exact sequence in the category $\mathcal{E}$ is also a semilocal ring.
(c) If A_R and C_R are modules of type n and m respectively, then the endomorphism ring E_B of the short exact sequence is of type $\leq n+m$.

PROOF. Without loss of generality, we can suppose that A_R is a submodule of B_R, $\alpha\colon A_R \to B_R$ is the embedding, $C_R = B_R/A_R$ and β is the canonical projection. There is a ring morphism $\varphi\colon E_B \to \operatorname{End}(A_R) \times \operatorname{End}(B_R/A_R)$ defined by $f \mapsto (f|_A^A, \overline{f})$, which is a local morphism by the Snake Lemma. Now (a) follows from Proposition 2.1, and (b) follows from the fact that a ring is semilocal if and only if its dual Goldie dimension is finite [**F2**, Proposition 2.43].

As far as (c) is concerned, suppose that A_R and C_R are modules of type n and m respectively. By Proposition 2.2(b), there are local morphisms $\operatorname{End}(A_R) \to D_1 \times \cdots \times D_n$ and $\operatorname{End}(B_R/A_R) \to D_1' \times \cdots \times D_m'$ for suitable division rings $D_1, \ldots, D_n, D_1', \ldots, D_m'$. Composing with φ, we get a local morphism $E_B \to D_1 \times \cdots \times D_n \times D_1' \times \cdots \times D_m'$. Thus E_B has type $\leq n+m$ by Proposition 2.2(b). □

REMARK 2.4. In the particular case where either A_R is zero or C_R is zero, their endomorphism ring is the trivial ring with one element. If we define such a ring to be of dual Goldie dimension 0 and we consider it to be a semilocal ring of finite type 0, then Proposition 2.3 also holds in this case.

3. Uniserial modules

The endomorphism ring $E := \mathrm{End}(U_R)$ of a non-zero uniserial module U_R [**F1**, Theorem 1.2] has at most two maximal right ideals: the two-sided completely prime ideals $I_U := \{ f \in E \mid f \text{ is not injective} \}$ and $K_U := \{ f \in E \mid f \text{ is not surjective} \}$, or only one of them. Here, a *completely prime ideal* P of a ring S is a proper ideal P of S such that, for every $x, y \in S$, $xy \in P$ implies that either $x \in P$ or $y \in P$.

REMARK 3.1. If S is a ring, $P_1, \dots, P_n$ are completely prime two-sided ideals of S and L is a right ideal of S contained in $\bigcup_{i=1}^n P_i$, then $L \subseteq P_i$ for some $i = 1, 2, \dots, n$.

THEOREM 3.2. *Let $A < B_R$ be right R-modules with $A_R \neq 0$. Suppose A_R and $C_R := B_R/A$ uniserial modules. Let E_B be the endomorphism ring of the short exact sequence*

$$0 \longrightarrow A_R \longrightarrow B_R \longrightarrow B_R/A \longrightarrow 0,$$

viewed as an object of the category $\mathcal{E}$. Set

$$I_{B,m,l} := \{ f \in E_B \mid \ker(f) \cap A \neq 0 \},$$
$$I_{B,e,l} := \{ f \in E_B \mid f(A) \subset A \},$$
$$I_{B,m,u} := \{ f \in E_B \mid A \subset f^{-1}(A) \}$$

and

$$I_{B,e,u} := \{ f \in E_B \mid f(B) + A \subset B \}.$$

Then $I_{B,m,l}, I_{B,e,l}, I_{B,m,u}$ and $I_{B,e,u}$ are four two-sided completely prime ideals of E_B, and every proper right ideal of E_B and every proper left ideal of E_B is contained in one of these four ideals of E_B. Moreover, $E_B/J(E_B)$ is isomorphic to the direct product of n division rings $D_1, \dots, D_n$ for some n with $1 \leq n \leq 4$, where $\{D_1, \dots, D_n\} \subseteq \{E_B/I_{B,m,l}, E_B/I_{B,e,l}, E_B/I_{B,m,u}, E_B/I_{B,e,u}\}$.

PROOF. As in the proof of Proposition 2.3, there is a canonical local morphism $\varphi\colon E_B \to \mathrm{End}(A_R) \times \mathrm{End}(B_R/A_R)$. Since A_R is a uniserial module, the canonical projections define a local morphism $\mathrm{End}(A_R) \to \mathrm{End}(A_R)/I_A \times \mathrm{End}(A_R)/K_A$, where $\mathrm{End}(A_R)/I_A$ and $\mathrm{End}(A_R)/K_A$ are integral domains. More precisely, three cases can occur:

(1) $I_A \subseteq K_A$. In this case, $\mathrm{End}(A_R)$ is a local ring with maximal ideal K_A, $\mathrm{End}(A_R)/K_A$ is a division ring, and the canonical projection

$$\mathrm{End}(A_R) \to \mathrm{End}(A_R)/K_A$$

is a local morphism.

(2) $K_A \subseteq I_A$. In this case, $\mathrm{End}(A_R)$ is a local ring with maximal ideal I_A, $\mathrm{End}(A_R)/I_A$ is a division ring, and the canonical projection

$$\mathrm{End}(A_R) \to \mathrm{End}(A_R)/I_A$$

is a local morphism.

(3) $I_A \nsubseteq K_A$ and $K_A \nsubseteq I_A$. In this case, $\mathrm{End}(A_R)$ has two maximal right ideal I_A and K_A, which are two-sided ideals, $\mathrm{End}(A_R)/I_A$ and $\mathrm{End}(A_R)/K_A$ are two division rings, and the canonical mapping

$$\mathrm{End}(A_R) \to \mathrm{End}(A_R)/I_A \times \mathrm{End}(A_R)/K_A$$

is a local morphism.

Similarly for the uniserial module $C_R := B_R/A$. Thus if $D_1, \dots, D_n$ are those n rings between the rings $\mathrm{End}(A_R)/I_A$, $\mathrm{End}(A_R)/K_A$, $\mathrm{End}(C_R)/I_C$ and $\mathrm{End}(C_R)/K_C$ that are division rings, we get a local morphism $E_B \to D_1 \times \cdots \times D_n$ of E_B into $n \le 4$ division rings. The four ideals $I_{B,m,l}, I_{B,e,l}, I_{B,m,u}$ and $I_{B,e,u}$ of E_B are the kernels of the canonical morphisms $E_B \to \mathrm{End}(A_R)/I_A$, $E_B \to \mathrm{End}(A_R)/K_A$, $E_B \to \mathrm{End}(C_R)/I_C$, $E_B \to \mathrm{End}(C_R)/K_C$, respectively. It is therefore immediate that they are completely prime two-sided ideals of E_B.

The non-invertible elements of $\mathrm{End}(A_R)$ are exactly the elements of $I_A \cup K_A$. Similarly, the non-invertible elements of $\mathrm{End}(C_R)$ are exactly the elements of $I_C \cup K_C$. Since $\varphi\colon E_B \to \mathrm{End}(A_R) \times \mathrm{End}(B_R/A_R)$ is a local morphism, the non-invertible elements of E_B are exactly the elements of $I_{B,m,l} \cup I_{B,e,l} \cup I_{B,m,u} \cup I_{B,e,u}$. But these four ideals of E_B are completely prime, so that every proper right (or left) ideal of E_B, which necessarily consists of non-invertible elements, must be contained in one of them (Remark 3.1). In particular, every maximal two-sided ideal of E_B must be one of $I_{B,m,l}, I_{B,e,l}, I_{B,m,u}$ or $I_{B,e,u}$.

The last assertion follows by applying the Chinese Remainder Theorem. □

REMARK 3.3. Note that in Theorem 3.2, without the assumption $0 \ne A_R < B_R$, the sets $I_{B,m,l}$, $I_{B,e,l}$, $I_{B,m,u}$ and $I_{B,e,u}$ could be empty. For example, for $A_R = 0$, we get that $I_{B,m,l}$ and $I_{B,e,l}$ are empty sets, while $I_{B,m,u}$ and $I_{B,e,u}$ coincide with the set of R-module morphisms of B_R that are not injective or not surjective respectively. Similarly for $C_R = 0$. Anyway, if A or C is equal to zero, B turns out to be a uniserial module and the corresponding description of the endomorphism ring E_B of B is well known [**F1**, Theorem 1.2]. Of course, these are the only two cases in which one of the four sets is empty, because otherwise the zero morphism belongs to $I_{B,m,l}, I_{B,e,l}, I_{B,m,u}$ and $I_{B,e,u}$. Indeed, in the hypothesis of Theorem 3.2, these four sets are always proper ideals of E_B.

Let (2.1) and (2.2) be two objects of $\mathcal{E}$. We will write:
$[B]_{m,l} = [B']_{m,l}$ if there exist two morphisms $f\colon B_R \to B'_R$ and $g\colon B'_R \to B_R$ in the category $\mathcal{E}$ such that $f|^{A}_{A'}$ and $g|^{A'}_{A}$ are injective right R-module morphisms.
$[B]_{e,l} = [B']_{e,l}$ if there exist two morphisms $f\colon B_R \to B'_R$ and $g\colon B'_R \to B_R$ in the category $\mathcal{E}$ such that $f(A) = A'$ and $g(A') = A$.
$[B]_{m,u} = [B']_{m,u}$ if there exist two morphisms $f\colon B_R \to B'_R$ and $g\colon B'_R \to B_R$ in the category $\mathcal{E}$ such that $\overline{f}\colon C \to C'$ and $\overline{g}\colon C' \to C$ are injective right R-module morphisms.
$[B]_{e,u} = [B']_{e,u}$ if there exist two morphisms $f\colon B_R \to B'_R$ and $g\colon B'_R \to B_R$ in the category $\mathcal{E}$ such that $\overline{f}\colon C \to C'$ and $\overline{g}\colon C' \to C$ are surjective right R-module morphisms.
(Here the four letters m, e, l, u stand for "monogeny", "epigeny", "lower" and "upper", respectively.)

If $[B]_{a,b} = [B']_{a,b}$ for some $a = m, e$ and $b = l, u$, we will say that the two objects (2.1) and (2.2) have the same *(a, b)-class*.

In this notation, $[B]_{m,l} = [0]_{m,l}$ if and only if $[B]_{e,l} = [0]_{e,l}$, if and only if A_R is the zero module. Similarly, $[B]_{m,u} = [0]_{m,u}$ if and only if $[B]_{e,u} = [0]_{e,u}$, if and only if C_R is the zero module.

4. Isomorphism

In this section, we will consider the full subcategory $\mathcal{U}$ of the category $\mathcal{E}$ whose elements are all short exact sequences $0 \longrightarrow A_R \longrightarrow B_R \longrightarrow C_R \longrightarrow 0$ with A_R, C_R non-zero uniserial right R-modules. In particular, every element of $\mathcal{U}$ has a semilocal endomorphism ring (Proposition 2.3(b)).

REMARK 4.1. Let (2.1) and (2.2) be two objects of $\mathcal{U}$. By [**F1**, Lemma 1.1], $[B]_{a,b} = [B']_{a,b}$ if and only if there exist two morphisms $f\colon B \to B'$ and $g\colon B' \to B$ in the category $\mathcal{U}$ such that $gf \notin I_{B,a,b}$ (or, equivalently, such that $fg \notin I_{B',a,b}$), where $a = m, e$ and $b = l, u$.

LEMMA 4.2. Let (2.1) and (2.2) be two objects in the category $\mathcal{U}$. Fix $a = m, e$ and $b = l, u$. Suppose that $[B]_{a,b} = [B']_{a,b}$. Then the following properties hold:

(a) For every $c = m, e$ and $d = l, u$, one has that $I_{B,c,d} \subseteq I_{B,a,b}$ if and only if $I_{B',c,d} \subseteq I_{B',a,b}$.

(b) For every $c = m, e$ and $d = l, u$, $I_{B,c,d} \subseteq I_{B,a,b}$ implies $[B]_{c,d} = [B']_{c,d}$.

PROOF. (a) It suffices to show that if $c = m, e$, $d = l, u$ and $I_{B,c,d} \subseteq I_{B,a,b}$, then $I_{B',c,d} \subseteq I_{B',a,b}$. Fix $c = m, e$ and $d = l, u$ and suppose that $I_{B',c,d} \nsubseteq I_{B',a,b}$. Let $\varphi\colon B'_R \to B'_R$ be a morphism in $I_{B',c,d}$ not in $I_{B',a,b}$. Since $[B]_{a,b} = [B']_{a,b}$, there exist two morphisms $f\colon B \to B'$ and $g\colon B' \to B$ in the category $\mathcal{U}$ such that $gf \notin I_{B,a,b}$. In particular, by [**F1**, Lemma 1.1], $g\varphi f \in I_{B,c,d}$ and $g\varphi f \notin I_{B,a,b}$, which implies that $I_{B,c,d} \nsubseteq I_{B,a,b}$.

(b) Suppose that $I_{B,c,d} \subseteq I_{B,a,b}$ for some $c = m, e$ and $d = l, u$. Since $[B]_{a,b} = [B']_{a,b}$, there exist two morphisms $f\colon B \to B'$ and $g\colon B' \to B$ in the category $\mathcal{U}$ such that $gf \notin I_{B,a,b}$. In particular $gf \notin I_{B,c,d}$, and therefore $[B]_{c,d} = [B']_{c,d}$ by Remark 4.1. □

COROLLARY 4.3. Let (2.1) and (2.2) be two objects of $\mathcal{U}$ with $[B]_{a,b} = [B']_{a,b}$ for every $a = m, e$ and every $b = l, u$. Consider the sets $\mathcal{S}_B := \{I_{B,m,l}, I_{B,e,l}, I_{B,m,u}, I_{B,e,u}\}$ and $\mathcal{S}_{B'} := \{I_{B',m,l}, I_{B',e,l}, I_{B',m,u}, I_{B',e,u}\}$ partially ordered by set inclusion. Then the canonical mapping $\Phi\colon \mathcal{S}_B \to \mathcal{S}_{B'}$, defined by $I_{B,a,b} \mapsto I_{B',a,b}$, is a partially ordered set isomorphism.

PROOF. It immediately follows from Lemma 4.2 (a). □

PROPOSITION 4.4. Let (2.1) and (2.2) be two objects of $\mathcal{U}$. Then (2.1) and (2.2) are isomorphic in $\mathcal{U}$ if and only if $[B]_{a,b} = [B']_{a,b}$ for every $a = m, e$ and $b = l, u$.

PROOF. Assume that $[B]_{a,b} = [B']_{a,b}$ for every $a = m, e$ and $b = l, u$. By Theorem 3.2, the maximal right ideals of E_B are the ideals $I_{B,a,b}$, where (a, b) ranges in a non-empty subset S of $\{(m, l), (e, l), (m, u), (e, u)\}$. Consequently, the maximal right ideals of $E_{B'}$ are the ideals $I_{B',a,b}$ for the same pairs (a, b) in S (Corollary 4.3).

By hypothesis, there are morphisms $f_{(a,b)}\colon B \to B'$ and $f'_{(a,b)}\colon B' \to B$ such that $f'_{(a,b)} f_{(a,b)} \notin I_{B,a,b}$ for every $(a, b) \in S$. Moreover, since $E_B/J(E_B)$ and $\prod_{(a,b)\in S} E_B/I_{B,a,b}$ are canonically isomorphic, for every $(a, b) \in S$ there exists $\varepsilon_{(a,b)} \in E_B$ such that $\varepsilon_{(a,b)} \equiv \delta_{(a,b),(c,d)} \pmod{I_{B,c,d}}$ for every $(c, d) \in S$. Similarly, for every $(a, b) \in S$ there exists $\varepsilon'_{(a,b)} \in E_{B'}$ such that $\varepsilon'_{(a,b)} \equiv \delta_{(a,b),(c,d)}$ (mod

$I_{B',c,d})$ for every $(c,d) \in S$. The two morphisms $\varphi\colon = \sum_{(a,b)\in S} \varepsilon'_{(a,b)} f_{(a,b)} \varepsilon_{(a,b)}$ and $\psi\colon = \sum_{(a,b)\in S} \varepsilon_{(a,b)} f'_{(a,b)} \varepsilon'_{(a,b)}$ are such that $\psi\varphi \in E_B$ is invertible modulo $J(E_B)$ and $\varphi\psi \in E_{B'}$ is invertible modulo $J(E_{B'})$. Hence $\psi\varphi$ and $\varphi\psi$ are invertible in E_B and $E_{B'}$ respectively, and therefore $\varphi\colon B_R \to B'_R$ is an isomorphism.

The inverse implication is clear. □

Let $\mathcal{C}$ be any preadditive category. An *ideal* of $\mathcal{C}$ assigns to every pair A, B of objects of $\mathcal{C}$ a subgroup $\mathcal{I}(A,B)$ of the abelian group $\mathrm{Hom}_{\mathcal{C}}(A,B)$ with the property that, for all morphisms $\varphi\colon C \to A$, $\psi\colon A \to B$ and $\omega\colon B \to D$ with $\psi \in \mathcal{I}(A,B)$, one has that $\omega\psi\varphi \in \mathcal{I}(C,D)$ [**M**, p. 18].

A *completely prime ideal* $\mathcal{P}$ of $\mathcal{C}$ consists of a subgroup $\mathcal{P}(A,B)$ of $\mathrm{Hom}_{\mathcal{C}}(A,B)$ for every pair of objects of $\mathcal{C}$, such that: (1) for every objects A, B, C of $\mathcal{C}$, for every $f : A \to B$ and for every $g : B \to C$, one has that $gf \in \mathcal{P}(A,C)$ if and only if either $f \in \mathcal{P}(A,B)$ or $g \in \mathcal{P}(B,C)$, and (2) $\mathcal{P}(A,A)$ is a proper subgroup of $\mathrm{Hom}_{\mathcal{C}}(A,A)$ for every object A of $\mathcal{C}$ [**FP4**, p. 565].

Let A be an object of $\mathcal{C}$ and I be a two-sided ideal of the ring $\mathrm{End}_{\mathcal{C}}(A)$. Let $\mathcal{I}$ be the ideal of the category $\mathcal{C}$ defined as follows. A morphism $f\colon X \to Y$ in $\mathcal{C}$ belongs to $\mathcal{I}(X,Y)$ if $\beta f\alpha \in I$ for every pair of morphisms $\alpha\colon A \to X$ and $\beta\colon Y \to A$ in $\mathcal{C}$. The ideal $\mathcal{I}$ is called the *ideal of* $\mathcal{C}$ *associated to* I [**FP2, FP3**]. It is the greatest of the ideals $\mathcal{I}'$ of $\mathcal{C}$ with $\mathcal{I}'(A,A) \subseteq I$. It is easily seen that $\mathcal{I}(A,A) = I$. If A is an object of $\mathcal{C}$, the ideals associated to two distinct ideals of $\mathrm{End}_{\mathcal{C}}(A)$ are obviously two distinct ideals of the category $\mathcal{C}$.

The endomorphism ring E_B of any object (2.1) of $\mathcal{U}$ has n maximal right ideals $I_{B,a,b}$, where $n \leq 4$ (Theorem 3.2). Let $\mathcal{I}_{B,a,b}$ be the corresponding n ideals of $\mathcal{E}$ associated to the maximal ideals of E_B. Set $V(B)\colon = \{\mathcal{I}_{B,a,b} \mid I_{B,a,b}$ is a maximal ideal of $E_B\}$, so that $V(B)$ has $n \leq 4$ elements.

Now set $V(\mathcal{U})\colon = \bigcup_{B\in \mathrm{Ob}(\mathcal{U})} V(B)$. Thus $V(\mathcal{U})$ is a class of ideals of the category $\mathcal{E}$.

LEMMA 4.5. *For every $\mathcal{P} \in V(\mathcal{U})$ and every non-zero object (2.2) of $\mathcal{E}$ such that $E_{B'}$ is semilocal, either $\mathcal{P}(B',B') = E_{B'}$ or $\mathcal{P}(B',B')$ is a maximal ideal of $E_{B'}$. In this second case, the ideal of $\mathcal{E}$ associated to $\mathcal{P}(B',B')$ is equal to $\mathcal{P}$.*

PROOF. The lemma is an immediate consequence of [**FP1**, Lemma 2.1(ii)]. □

For any ideal $\mathcal{P}$ of $\mathcal{E}$, $F\colon \mathcal{E} \to \mathcal{E}/\mathcal{P}$ will denote the canonical functor. Recall that the factor category $\mathcal{E}/\mathcal{P}$ has the same objects as $\mathcal{E}$ and, for $B, B' \in \mathrm{Ob}(\mathcal{E}) = \mathrm{Ob}(\mathcal{E}/\mathcal{P})$, the group of morphisms $B \to B'$ in $\mathcal{E}/\mathcal{P}$ is defined to be the factor group $\mathrm{Hom}_{\mathcal{E}}(B,B')/\mathcal{P}(B,B')$.

LEMMA 4.6. *Let (2.1) be an object of $\mathcal{U}$, $I_{B,a,b}$ be a maximal ideal of E_B, $\mathcal{I}_{B,a,b}$ its associated ideal in $\mathcal{E}$ and $F\colon \mathcal{E} \to \mathcal{E}/\mathcal{I}_{B,a,b}$ be the canonical functor. Then, for any object (2.2) of $\mathcal{E}$ such that $E_{B'}$ is of finite type, either $F(B') = 0$ or $F(B') \cong F(B)$. Moreover, if $[B]_{a,b} = [B']_{a,b}$, then $F(B') \cong F(B)$.*

PROOF. From Lemma 4.5, we know that either $F(B') = 0$ or the endomorphism ring of $F(B')$ is a division ring. Let us consider the latter case. As $1_{B'} \notin \mathcal{I}_{B,a,b}(B',B')$, there are $\alpha\colon B \to B'$ and $\beta\colon B' \to B$ such that $\beta\alpha \notin I_{B,a,b}$. Thus $\beta(\alpha\beta)\alpha \notin I_{B,a,b}$. It follows that $\alpha\beta \notin \mathcal{I}_{B,a,b}(B',B')$, Therefore $F(\beta)F(\alpha)$ and $F(\alpha)F(\beta)$ are automorphisms, so $F(B) \cong F(B')$ follows. For the last assertion, note that, also in the case $[B]_{a,b} = [B']_{a,b}$, there are morphisms $\alpha\colon B \to B'$

and $\beta\colon B' \to B$ such that $\beta\alpha \notin I_{B,a,b}$. So, as before, we can conclude that $F(B) \cong F(B')$. □

In the following corollary, we collect some basic properties that can be deduced easily from the previous two lemmas.

COROLLARY 4.7. Let (2.1) be an object of $\mathcal{U}$, $I_{B,a,b}$ be a maximal ideal of E_B, $\mathcal{I}_{B,a,b}$ its associated ideal in $\mathcal{E}$ and $F\colon \mathcal{E} \to \mathcal{E}/\mathcal{I}_{B,a,b}$ be the canonical functor.

(1) For any object (2.2) of $\mathcal{U}$, the following conditions are equivalent:

(a) $F(B') = 0$ in $\mathcal{E}/\mathcal{I}_{B,a,b}$.
(b) $\mathcal{I}_{B,a,b}(B', B') = E_{B'}$.
(c) $I_{B',a,b} \subset \mathcal{I}_{B,a,b}(B', B')$.
(d) $[B]_{a,b} \neq [B']_{a,b}$.

(2) For any object (2.2) of $\mathcal{U}$, the following conditions are equivalent:

(a) $F(B) \cong F(B')$ in $\mathcal{E}/\mathcal{I}_{B,a,b}$.
(b) $\mathcal{I}_{B,a,b}(B', B')$ is a proper ideal of $E_{B'}$.
(c) $I_{B',a,b} = \mathcal{I}_{B,a,b}(B', B')$.
(d) $[B]_{a,b} = [B']_{a,b}$.

PROOF. Proof of (1). (a) $\Leftrightarrow$ (b) $F(B') = 0$ if and only if the endomorphism ring $E_{B'}/\mathcal{I}_{B,a,b}(B', B')$ of $F(B')$ in the factor category $\mathcal{E}/\mathcal{I}_{B,a,b}$ is the zero ring.

(b) $\Rightarrow$ (c) It suffices to note that $I_{B',a,b}$ is always a proper ideal of $E_{B'}$.

(c) $\Rightarrow$ (d) Let φ be an element in $\mathcal{I}_{B,a,b}(B', B')$ not in $I_{B',a,b}$. For any two morphisms $f\colon B_R \to B'_R$ and $g\colon B'_R \to B_R$ in the category $\mathcal{E}$, one has $g\varphi f \in I_{B,a,b}$, by definition of associated ideal. Since $\varphi \notin I_{B',a,b}$, it follows from [**F1**, Lemma 1.1] that $gf \in I_{B,a,b}$. This means that $[B]_{a,b} \neq [B']_{a,b}$.

(d) $\Rightarrow$ (b) By Remark 4.1, if $[B]_{a,b} \neq [B']_{a,b}$, then, for any morphism $f\colon B \to B'$ and $g\colon B' \to B$ in the category $\mathcal{U}$, we have that $g1_{B'}f = gf \in I_{B,a,b}$. Therefore $1_{B'} \in \mathcal{I}_{B,a,b}(B', B')$, so that $\mathcal{I}_{B,a,b}(B', B') = E_{B'}$.

Proof of (2). First notice that, by Lemma 4.6, either $F(B) = 0$ or $F(B) \cong F(B')$ in $\mathcal{E}/\mathcal{I}_{B,a,b}$. Moreover, $I_{B',a,b}$ is always contained in $\mathcal{I}_{B,a,b}(B', B')$. As a matter of fact, suppose $\varphi \in I_{B',a,b}$. By [**F1**, Lemma 1.1], for any $f\colon B \to B'$ and any $g\colon B' \to B$, we have that $g\varphi f \in I_{B,a,b}$, so $\varphi \in I_{B,a,b}(B', B')$. Now all the implications follow from part (1). □

COROLLARY 4.8. Let (2.1) and (2.2) be two objects in the category $\mathcal{U}$. Fix $a = m, e$ and $b = l, u$. Suppose that $[B]_{a,b} = [B']_{a,b}$. Then the following properties hold:

(a) $I_{B,a,b}$ is a maximal right ideal of E_B if and only if $I_{B',a,b}$ is a maximal right ideal of $E_{B'}$.
(b) Suppose that $I_{B,a,b}$ is a maximal right ideal of E_B. Then, for every $c = m, e$ and $d = l, u$, $I_{B,c,d} = I_{B,a,b}$ if and only if $I_{B',c,d} = I_{B',a,b}$.

PROOF. (a) It suffices to show that $I_{B,a,b}$ maximal implies $I_{B',a,b}$ maximal. From Corollary 4.7 (2), $I_{B',a,b} = \mathcal{I}_{B,a,b}(B', B')$ is a proper ideal of $E_{B'}$, which is a maximal right ideal of $E_{B'}$ by Lemma 4.5.

(b) By (a), the hypotheses on (2.1) and (2.2) are symmetrical. Therefore it suffices to show that $I_{B,c,d} = I_{B,a,b}$ implies $I_{B',c,d} = I_{B',a,b}$. The inclusion $I_{B',c,d} \subseteq I_{B',a,b}$ follows from Lemma 4.2(a). Moreover, by Lemma 4.2 (b), we can interchange the role of (a, b) and (c, d) and deduce the opposite inclusion applying Lemma 4.2 (a) again. □

COROLLARY 4.9. Let (2.1) and (2.2) be two objects of $\mathcal{U}$. Then $V(B)\cap V(B')\neq\emptyset$ if and only if there exists a pair $(p,q)\in\{m,e\}\times\{l,u\}$ such that $I_{B,p,q}$ is a maximal right ideal of E_B, $I_{B',p,q}$ is a maximal ideal of $E_{B'}$ and $\mathcal{I}_{B,p,q}=\mathcal{I}_{B',p,q}$. Moreover, for such a pair, $[B]_{p,q}=[B']_{p,q}$.

PROOF. Let $\mathcal{P}\in V(B)\cap V(B')$. Then there exist two pairs (p,q) and (r,s) in $\{m,e\}\times\{l,u\}$ such that $I_{B,p,q}$ is a maximal right ideal of E_B, $I_{B',r,s}$ is a maximal ideal of $E_{B'}$ and $\mathcal{P}=\mathcal{I}_{B,p,q}=\mathcal{I}_{B',r,s}$ is the ideal of $\mathcal{E}$ associated both to $I_{B,p,q}$ and $I_{B',r,s}$. We have $I_{B',p,q}\subseteq\mathcal{I}_{B,p,q}(B',B')=\mathcal{I}_{B',r,s}(B',B')=I_{B',r,s}$. If the inclusion is proper, then $I_{B',r,s}=E_{B'}$ by Corollary 4.7 (1), a contradiction. It follows that $I_{B',p,q}=I_{B',r,s}$, so we can replace the pair (r,s) with (p,q). Clearly, the converse holds as well. Finally, the last assertion follows from Corollary 4.7 (2). □

According to [**FP1**], we say that an ideal $\mathcal{M}$ of a preadditive category $\mathcal{C}$ is *maximal* if the improper ideal of $\mathcal{C}$ is the unique ideal of the category $\mathcal{C}$ properly containing $\mathcal{M}$. Moreover, we say that a preadditive category $\mathcal{C}$ is *simple* [**FP1**] if it has exactly two ideals, necessarily the trivial ones. Hence, a simple category has non-zero objects.

In the category $\mathcal{U}$, $V(\mathcal{U})$ turns out to be a class of maximal ideals of $\mathcal{U}$, by [**FP1**, Lemma 2.4].

PROPOSITION 4.10. Let (2.1) be an object of $\mathcal{U}$, $I_{B,a,b}$ be a maximal ideal of E_B and $\mathcal{I}_{B,a,b}$ its associated ideal in $\mathcal{U}$. Then the factor category $\mathcal{U}/\mathcal{I}_{B,a,b}$ is simple and the endomorphism ring of all its non-zero objects are division rings.

PROOF. The Proposition is an immediate consequence of [**FP1**, Theorem 3.2]. □

COROLLARY 4.11. Let (2.1) be an object of $\mathcal{U}$, $I_{B,a,b}$ be a maximal ideal of E_B and $\mathcal{I}_{B,a,b}$ its associated ideal in $\mathcal{E}$. If $F(B)^t\cong F(B)^s$ in the factor category $\mathcal{E}/\mathcal{I}_{B,a,b}$ for integers $t,s\geq 0$, then $t=s$.

PROOF. Let D be the endomorphism ring of the object $F(B)$ of $\mathcal{E}/\mathcal{I}_{B,a,b}$. Apply the functor $\operatorname{Hom}(F(B),-)\colon\mathcal{E}/\mathcal{I}_{B,a,b}\to\text{Mod-}D$. □

Notice that we are using the following notations:

$I_{B,a,b}$ is an ideal of E_B, the endomorphism ring of an object of $\mathcal{E}$;

$\mathcal{I}_{B,a,b}$ is the ideal of $\mathcal{E}$, or possibly a full subcategory of $\mathcal{E}$, associated to $I_{B,a,b}$;

$\mathcal{K}_{a,b}$ is one of four completely prime ideals of $\mathcal{U}$, defined, for every pair of objects (2.1), (2.2) of $\mathcal{U}$, by

$\mathcal{K}_{m,l}((2.1),(2.2)):=\{\,f\colon B\to B'\mid \ker(f)\cap A\neq 0\,\}$,

$\mathcal{K}_{e,l}((2.1),(2.2)):=\{\,f\colon B\to B'\mid f(A)\subset A'\,\}$,

$\mathcal{K}_{m,u}((2.1),(2.2)):=\{\,f\colon B\to B'\mid A\subset f^{-1}(A')\,\}$ and

$\mathcal{K}_{e,u}((2.1),(2.2)):=\{\,f\colon B\to B'\mid f(B)+A'\subset B'\,\}$.

LEMMA 4.12. Let

$$0\longrightarrow A_i\longrightarrow B_i\longrightarrow C_i\longrightarrow 0,\qquad i=1,2,\dots,n,$$

be objects of $\mathcal{U}$. Then every maximal two-sided ideal of $E_{\oplus_{j=1}^n B_j}$ is of the form $\mathcal{I}_{B_k,a,b}(\oplus_{j=1}^n B_j,\oplus_{j=1}^n B_j)$ for some $k=1,\dots,n$, $a=m,e$ and $b=l,u$ such that $I_{B_k,a,b}$ is a maximal right ideal of E_{B_k}. Conversely, if (k,a,b) is a triple such that $k=1,\dots,n$, $a=m,e$, $b=l,u$ and $I_{B_k,a,b}$ is a maximal right ideal of E_{B_k}, then $\mathcal{I}_{B_k,a,b}(\oplus_{j=1}^n B_j,\oplus_{j=1}^n B_j)$ is a maximal two-sided ideal of $E_{\oplus_{j=1}^n B_j}$.

PROOF. Let I be a maximal two-sided ideal of $E_{\oplus_{j=1}^n B_j}$ and let $\mathcal{I}$ be the ideal of $\mathcal{E}$ associated to I. Using [**FP1**, Lemma 2.1(ii)], we get that for any $k = 1, \dots, n$, either $\mathcal{I}(B_k, B_k) = E_{B_k}$ or $\mathcal{I}(B_k, B_k)$ is a maximal two-sided ideal of E_{B_k}. If $\mathcal{I}(B_k, B_k) = E_{B_k}$ for all $k = 1, \dots, t$, by the definition of associated ideal it follows that $\varepsilon_k \pi_k \in I$ for every $k = 1, \dots, n$, where $\varepsilon_k \colon B_k \to \oplus_{j=1}^n B_j$ and $\pi_k \colon \oplus_{j=1}^n B_j \to B_k$ are the embedding and the canonical projection respectively. In particular $1_{\oplus_{j=1}^n B_j} = \sum_{k=1}^n \varepsilon_k \pi_k \in I$, which is a contradiction. It follows that there exists a triple (k, a, b) such that $\mathcal{I}(B_k, B_k) = I_{B_k,a,b}$ is a maximal right ideal of E_{B_k}. By [**FP1**, Lemma 2.1(i)], $\mathcal{I}_{B_k,a,b} = \mathcal{I}$, so that, in particular, $I = \mathcal{I}(\oplus_{j=1}^n B_j, \oplus_{j=1}^n B_j) = \mathcal{I}_{B_k,a,b}(\oplus_{j=1}^n B_j, \oplus_{j=1}^n B_j)$.

Conversely, let (k, a, b) be a triple such that $k = 1, \dots, n$, $a = m, e$, $b = l, u$ and $I_{B_k,a,b}$ is a maximal right ideal of E_{B_k}. By [**FP1**, Lemma 2.1(ii)], either $\mathcal{I}_{B_k,a,b}(\oplus_{j=1}^n B_j, \oplus_{j=1}^n B_j)$ is a maximal ideal of $E_{\oplus_{j=1}^n B_j}$ or

$$\mathcal{I}_{B_k,a,b}(\oplus_{j=1}^n B_j, \oplus_{j=1}^n B_j) = E_{\oplus_{j=1}^n B_j}.$$

In this second case, the identity of $\oplus_{j=1}^n B_j$ is in $\mathcal{I}_{B_k,a,b}(\oplus_{j=1}^n B_j, \oplus_{j=1}^n B_j)$, so that the identity of B_k is in $\mathcal{I}_{B_k,a,b}(B_k, B_k) = I_{B_k,a,b}$, a contradiction. □

REMARK 4.13. If T is any semilocal ring and M is a maximal two-sided ideal of T, then there exists an element $\delta_M \in T$ such that $\delta_M \equiv 1_T \pmod{M}$ and $\delta_M \equiv 0 \pmod{N}$ for every other maximal two-sided ideal N of T different from M. This follows from the fact that $T/J(T)$ is a direct product of finitely many simple rings.

PROPOSITION 4.14. Let

$$0 \longrightarrow A_i \longrightarrow B_i \longrightarrow C_i \longrightarrow 0, \qquad i = 1, 2, \dots, n, \tag{4.1}$$

and

$$0 \longrightarrow A'_j \longrightarrow B'_j \longrightarrow C'_j \longrightarrow 0, \qquad j = 1, 2, \dots, m, \tag{4.2}$$

be $n + m$ objects in the category $\mathcal{E}$ with all the modules A_i, C_i, A'_j, C'_j non-zero uniserial modules. For every $a = m, e$ and every $b = l, u$, set

$$X_{a,b} := \{\, i \mid i = 1, 2, \dots, n,\ I_{B_i,a,b} \text{ is a maximal ideal of } E_{B_i} \,\}$$

and

$$Y_{a,b} := \{\, j \mid j = 1, 2, \dots, m,\ I_{B'_j,a,b} \text{ is a maximal ideal of } E_{B'_i} \,\}.$$

Then $\bigoplus_{i=1}^n B_i \cong \bigoplus_{j=1}^m B'_j$ in the category $\mathcal{E}$ if and only if there exist four bijections $\varphi_{a,b} \colon X_{a,b} \to Y_{a,b}$, $a = m, e$ and $b = l, u$, with $[B_i]_{a,b} = [B'_{\varphi_{a,b}(i)}]_{a,b}$ for every $i \in X_{a,b}$.

PROOF. ($\Rightarrow$) Suppose there exists an isomorphism $\bigoplus_{i=1}^n B_i \cong \bigoplus_{j=1}^m B'_j$ in $\mathcal{E}$. Fix $a = m, e$ and $b = l, u$. Apply Proposition 4.10 to any fixed object B'' of $\mathcal{U}$ for which $I_{B'',a,b}$ is a maximal ideal of $E_{B''_i}$. By Proposition 4.10, the corresponding factor category $\mathcal{U}/\mathcal{I}_{B'',a,b}$ is simple. Also, either $[B_i]_{a,b} = [B'']_{a,b}$ or $[B_i]_{a,b} \neq [B'']_{a,b}$. By Corollary 4.7, we have that either $F(B_i) \cong F(B'')$ or $F(B_i) = 0$ in the category $\mathcal{U}/\mathcal{I}_{B'',a,b}$. Thus, in the category $\mathcal{E}/\mathcal{I}_{B'',a,b}$, we have that $F(\bigoplus_{i=1}^n B_i) \cong F(B'')^{t_{B'',a,b}}$, where $t_{B'',a,b} = |\{\, i \mid i = 1, 2, \dots, n,\ [B_i]_{a,b} = [B'']_{a,b} \,\}|$. Similarly, $F(\bigoplus_{j=1}^m B'_j) \cong F(B'')^{s_{B'',a,b}}$, where $s_{B'',a,b} = |\{\, j \mid i = 1, 2, \dots, m,\ [B_j]_{a,b} = [B'']_{a,b} \,\}|$. Then, by Corollary 4.11, $t_{B'',a,b} = s_{B'',a,b}$ for every object B'' of $\mathcal{U}$ for which $I_{B'',a,b}$ is a maximal ideal of $E_{B''_i}$. Thus there is a bijection $\varphi_{B'',a,b}$ between $\{\, i \mid i = 1, 2, \dots, n,\ [B_i]_{a,b} = [B'']_{a,b} \,\}$ and $\{\, j \mid i = 1, 2, \dots, m,\ [B_j]_{a,b} = [B'']_{a,b} \,\}$

for every B'' in $\mathcal{U}$ and every a,b with $I_{B'',a,b}$ a maximal ideal of $E_{B''}$. Now, for $[B_i]_{a,b} = [B'']_{a,b}$, $I_{B'',a,b}$ is a maximal ideal of $E_{B''}$ if and only if $I_{B_i,a,b}$ is a maximal ideal of E_{B_i} (Corollary 4.8 (a)). Hence $\varphi_{B'',a,b}$ is a bijection between $\{\, i \mid i = 1, 2, \dots, n,\ [B_i]_{a,b} = [B'']_{a,b}$ and $I_{B_i,a,b}$ is a maximal ideal of $E_{B_i}\,\}$ and $\{\, j \mid i = 1, 2, \dots, m,\ [B'_j]_{a,b} = [B'']_{a,b}$ and $I_{B'_j,a,b}$ is a maximal ideal of $E_{B'_j}\,\}$ for every B'' in $\mathcal{U}$. Glueing together all these bijections $\varphi_{B'',a,b}$, we get the required bijection $\varphi_{a,b}\colon X_{a,b} \to Y_{a,b}$.

($\Leftarrow$) Conversely, suppose that four bijections $\varphi_{a,b}$ with the property in the statement of the Proposition exist. Fix any object B'' in $\mathcal{U}$ with $I_{B'',a,b}$ a maximal ideal of $E_{B''}$. Then the number of indices i with $i = 1, 2, \dots, n$, $I_{B_i,a,b}$ a maximal ideal of E_{B_i} and $[B_i]_{a,b} = [B'']_{a,b}$ is equal to the number of indices j with $j = 1, 2, \dots, m$, $I_{B'_j,a,b}$ a maximal ideal of $E_{B'_j}$ and $[B'_j]_{a,b} = [B'']_{a,b}$. By Corollary 4.8, the number of indices i with $i = 1, 2, \dots, n$ and $[B_i]_{a,b} = [B'']_{a,b}$ is equal to the number of indices j with $j = 1, 2, \dots, m$ and $[B'_j]_{a,b} = [B'']_{a,b}$. Call $n_{B'',a,b}$ this number. Let $\mathcal{I}_{B'',a,b}$ be the ideal in the category $\mathcal{E}$ associated to $I_{B'',a,b}$ and $F\colon \mathcal{E} \to \mathcal{E}/\mathcal{I}_{B'',a,b}$ be the canonical functor. Then, for every short exact sequence B in $\mathcal{U}$, $F(B) = 0$ if $[B]_{a,b} \neq [B'']_{a,b}$, and $F(B) \cong F(B'')$ if $[B]_{a,b} = [B'']_{a,b}$ (Corollary 4.7). Thus $F(\bigoplus_{i=1}^n B_i) \cong F(B'')^{n_{B'',a,b}}$ and $F(\bigoplus_{j=1}^m B'_j) \cong F(B'')^{n_{B'',a,b}}$, so that $F(\bigoplus_{i=1}^n B_i) \cong F(\bigoplus_{j=1}^m B'_j)$ in $\mathcal{E}/\mathcal{I}_{B'',a,b}$. In particular, this occurs for all the short exact sequences $B'' \in \{B_1, \dots, B_n, B'_1, \dots, B'_m\}$ and the pairs (a,b) with $I_{B'',a,b}$ is a maximal ideal of $E_{B''}$.

From the isomorphism $F(\bigoplus_{i=1}^n B_i) \cong F(B'')^{n_{B'',a,b}}$, we get that the ring $\mathrm{End}_{\mathcal{E}/\mathcal{I}_{B'',a,b}}(F(\bigoplus_{i=1}^n B_i))$ is isomorphic to the ring $M_{n_{B'',a,b}}(E_{B''}/I_{B'',a,b})$ of all $n_{B'',a,b} \times n_{B'',a,b}$ matrices with entries in the division ring $E_{B''}/I_{B'',a,b}$. This ring of matrices is a simple artinian ring. Thus the canonical functor F induces a surjective ring morphism $\mathrm{End}_{\mathcal{E}}(\bigoplus_{i=1}^n B_i) \to \mathrm{End}_{\mathcal{E}/\mathcal{I}_{B'',a,b}}(F(\bigoplus_{i=1}^n B_i))$ onto a simple ring, and its kernel is therefore a maximal two-sided ideal of $\mathrm{End}_{\mathcal{E}}(\bigoplus_{i=1}^n B_i)$. This kernel is $\mathcal{I}_{B'',a,b}(\bigoplus_{i=1}^n B_i, \bigoplus_{i=1}^n B_i)$, which is therefore a maximal two-sided ideal of the semilocal ring $\mathrm{End}_{\mathcal{E}}(\bigoplus_{i=1}^n B_i)$ (cf. Lemma 4.12). By Remark 4.13, there exists an endomorphism $\delta_{[B'']_{a,b}}$ of $\bigoplus_{i=1}^n B_i$ such that $\delta_{[B'']_{a,b}} \equiv 1$ modulo this maximal two-sided ideal and $\delta_{[B'']_{a,b}}$ belongs to all the other maximal two-sided ideals of $\mathrm{End}_{\mathcal{E}}(\bigoplus_{i=1}^n B_i)$. Similarly, we find endomorphisms $\delta'_{[B'']_{a,b}}$ of $\bigoplus_{j=1}^m B'_j$.

Since $F(\bigoplus_{i=1}^n B_i) \cong F(\bigoplus_{j=1}^m B'_j)$, there exists a morphism $f_{B'',a,b}\colon \bigoplus_{i=1}^n B_i \to \bigoplus_{j=1}^m B'_j$ in $\mathcal{E}$, which becomes an isomorphism in $\mathcal{E}/\mathcal{I}_{B'',a,b}$. Consider the morphism $f := \sum \delta'_{[B'']_{a,b}} f_{B'',a,b} \delta_{[B'']_{a,b}}\colon \bigoplus_{i=1}^n B_i \to \bigoplus_{j=1}^m B'_j$, where the sum ranges over the maximal two-sided ideals in the endomorphism ring (Lemma 4.12), so that the number of summands in the definition of f is equal to the number of maximal two-sided ideals of $E_{\bigoplus_{i=1}^n B_i} \cong E_{\bigoplus_{j=1}^m B'_j}$. Thus f becomes an isomorphism in the factor category $\mathcal{E}/\mathcal{I}_{B'',a,b}$ for any triple (B'',a,b) such that $B'' \in \{B_1, \dots, B_n, B'_1, \dots, B'_m\}$, $a \in \{m, e\}$, $b \in \{l, u\}$ and $I_{B'',a,b}$ is a maximal ideal of $E_{B''}$. By [**AF**, Proposition 5.2], f is an isomorphism in the category $\mathcal{E}/\mathcal{I}$, where $\mathcal{I}$ is the intersection of these finitely many ideals $\mathcal{I}_{B'',a,b}$. The restriction of $\mathcal{I}$ to the full additive subcategory $\mathcal{S} := \mathrm{add}(B_1 \oplus \cdots \oplus B'_m)$ of $\mathcal{E}$ is the Jacobson radical $\mathcal{J}(\mathcal{S})$ of $\mathcal{S}$. Hence $\bigoplus_{i=1}^n B_i$ and $\bigoplus_{j=1}^m B'_j$ are isomorphic in the category $\mathcal{E}/\mathcal{I}$. Thus there are morphisms $\alpha\colon \bigoplus_{i=1}^n B_i \to \bigoplus_{j=1}^m B'_j$ and $\beta\colon \bigoplus_{j=1}^m B'_j \to \bigoplus_{i=1}^n B_i$ in $\mathcal{E}$ such that $\alpha\beta \equiv 1$ and $\beta\alpha \equiv 1$ modulo $\mathcal{I}$. Equivalently, the endomorphisms $\alpha\beta$ and $\beta\alpha$ in $\mathcal{S}$ are congruent to 1 modulo $J(\mathrm{End}_{\mathcal{S}}(\bigoplus_{j=1}^m B'_j))$ and $J(\mathrm{End}_{\mathcal{S}}(\bigoplus_{i=1}^n B_i))$, respectively. Thus

$\alpha\beta$ and $\beta\alpha$ are invertible in the rings $\mathrm{End}_{\mathcal{S}}(\bigoplus_{j=1}^m B'_j)$ and $\mathrm{End}_{\mathcal{S}}(\bigoplus_{i=1}^n B_i)$, respectively. In particular, α is both right invertible and left invertible in the category $\mathcal{E}$. It follows that α is an isomorphism in $\mathcal{E}$. □

REMARK 4.15. In the statement of the previous proposition, we have used the condition "$I_{B,a,b}$ is a maximal ideal of E_B". This condition can be expressed very easily in terms relative to B. For instance, suppose we want to express the condition $I_{B,m,l}$ is a maximal ideal of E_B. By Theorem 3.2, the ideals $I_{B,m,l}, I_{B,e,l}, I_{B,m,u}$ and $I_{B,e,u}$ are completely prime ideals, and the maximal ideals of E_B are some of them. Thus, by Remark 3.1, $I_{B,m,l}$ is a maximal ideal of E_B if and only if $I_{B,m,l} \nsubseteq I_{B,m,u} \cup I_{B,e,l} \cup I_{B,e,u}$, that is, if and only if there exists an endomorphism f of (2.1) with $\ker(f) \cap A \neq 0$, $f(A) = A$, $A = f^{-1}(A)$ and $f(B) + A = B$.

In the final result of this section, we will make use of some techniques, notations and ideas taken from [**DF**]. If X and Y are finite disjoint sets, we will denote by $D(X, Y; E)$ the bipartite digraph (= directed graph) having X and Y as disjoint sets of non-adjacent vertices and E as set of edges. Equivalently, $V = X \cup Y$ is the vertex set of $D(X, Y; E)$, $E \subseteq X \times Y \cup Y \times X$ is the set of its edges, and $X \cap Y = \emptyset$. For every subset $T \subseteq V$, let $N^+(T) = \{ w \in V \mid (v, w) \in E \text{ for some } v \in T \}$ be the *out-neighborhood* of T. Define an equivalence relation $\sim_s$ on V by $v \sim_s w$ if there are both an oriented path from v to w and an oriented path from w to v $(v, w \in V)$.

PROPOSITION 4.16. ([**DF**, Lemma 2.1], Krull-Schmidt Theorem for bipartite digraphs). *Let X and Y be disjoint sets of cardinality n and m, respectively. Set $V := X \cup Y$. Let $D = D(X, Y; E)$ be a bipartite digraph having X and Y as disjoint sets of non-adjacent vertices. If $|T| \leq |N^+(T)|$ for every subset T of V, then $n = m$ and, after a suitable relabeling of the indices of the elements $x_1, \dots, x_n$ of X and $y_1, \dots, y_n$ of Y, $x_i \sim_s y_i$ for every $i = 1, \dots, n$.*

PROPOSITION 4.17. *Let*

$$(4.3)\qquad 0 \longrightarrow A_i \longrightarrow B_i \longrightarrow C_i \longrightarrow 0, \qquad i = 1, 2, \dots, n,$$

and

$$(4.4)\qquad 0 \longrightarrow A'_j \longrightarrow B'_j \longrightarrow C'_j \longrightarrow 0, \qquad j = 1, 2, \dots, m,$$

be $n + m$ objects in the category $\mathcal{E}$ with all the modules A_i, C_i, A'_j, C'_j non-zero uniserial modules. Then $\bigoplus_{i=1}^n B_i \cong \bigoplus_{j=1}^m B'_j$ in the category $\mathcal{E}$ if and only if $n = m$ and there exist four permutations $\varphi_{a,b}$ of $\{1, 2, \dots, n\}$, $(a, b) \in \{m, e\} \times \{l, u\}$, with $[B_i]_{a,b} = [B'_{\varphi_{a,b}(i)}]_{a,b}$ for every $i = 1, 2, \dots, n$, every $a = m, e$ and every $b = l, u$.

PROOF. Suppose $\bigoplus_{i=1}^n B_i \cong \bigoplus_{j=1}^m B'_j$ in the category $\mathcal{E}$. Then $\bigoplus_{i=1}^n A_i \cong \bigoplus_{j=1}^m A'_j$ in the category Mod-R, so that, taking the Goldie dimension, we get that $n = m$. Let $\alpha \colon \bigoplus_{i=1}^n B_i \to \bigoplus_{j=1}^m B'_j$ be an isomorphism in $\mathcal{E}$ with inverse $\beta \colon \bigoplus_{j=1}^m B'_j \to \bigoplus_{i=1}^n B_i$. Denote by $\varepsilon_h \colon B_h \to \bigoplus_{i=1}^n B_i$, $\pi_h \colon \bigoplus_{i=1}^n B_i \to B_h$, $\varepsilon'_k \colon B'_k \to \bigoplus_{j=1}^n B'_j$ and $\pi'_k \colon \bigoplus_{j=1}^n B'_j \to B'_k$ the embeddings and the canonical projections and consider the composite morphisms $\chi_{h,k} := \pi'_k \alpha \varepsilon_h \colon B_h \to B'_k$ and $\chi'_{k,h} := \pi_h \beta \varepsilon'_k \colon B'_k \to B_h$. We will prove the existence of the permutation $\varphi_{e,u}$. The proof of the existence of the other three permutations $\varphi_{a,b}$ is similar. Define a bipartite digraph $D = D(X, Y; E)$ having $X = \{B_1, \dots, B_n\}$ and $Y = \{B'_1, \dots, B'_n\}$ as disjoint sets of non-adjacent vertices, and the set E of edges defined as follows: one edge from B_h to B'_k for each h and k such that the induced mapping $\overline{\chi_{h,k}} \colon C_h \to$

C'_k is surjective, and one edge from B'_k to B_h for each h and k such that the induced mapping $\overline{\chi'_{k,h}}\colon C'_k \to C_h$ is surjective. We want to show that $|T| \leq |N^+(T)|$ for every subset $T \subseteq X \cup Y$ of vertices. Since the digraph is bipartite, we can suppose that $T \subseteq X$. If $r = |T|$ and $s = |N^+(T)|$, relabeling the indices we may assume that $T = \{B_1, \dots, B_r\}$ and $N^+(T) = \{B'_1, \dots, B'_s\}$. This means that the induced morphisms $\overline{\chi_{h,k}}$ are not surjective for every $h = 1, \dots, r$ and every $k = s+1, \dots, n$. Since the modules C'_k are all uniserial, we have that $L_k := \sum_{h=1}^r \overline{\chi_{h,k}}(C_h) \subset C'_k$ for every $k = s+1, \dots, n$, and therefore the quotient modules C'_k/L_k are non-zero for every $k = s+1, \dots, n$. Let $\pi\colon \bigoplus_{j=1}^n C'_j \to \bigoplus_{k=s+1}^n C'_k/L_k$ be the canonical projection. For every $h = 1, \dots, r$ and for every $k = s+1, \dots, n$, the composite morphism

$$C_h \xrightarrow{\overline{\varepsilon_h}} \bigoplus_{i=1}^n C_i \xrightarrow{\overline{\alpha}} \bigoplus_{j=1}^n C'_j \xrightarrow{\overline{\pi'_k}} C'_k \to C'_k/L_k$$

is zero because $\overline{\pi'_k}\overline{\alpha}\overline{\varepsilon_h}(C_h) = \overline{\chi_{h,k}}(C_h) \subseteq L_k$. It follows that, for every $h = 1, \dots, r$, $\overline{\varepsilon_h}(C_h)$ is contained in the kernel of the composite mapping

$$\bigoplus_{i=1}^n C_i \xrightarrow{\overline{\alpha}} \bigoplus_{j=1}^n C'_j \xrightarrow{\overline{\pi'_k}} C'_k \to C'_k/L_k$$

for every $k = s+1, \dots, n$. Since $\sum_{h=1}^r \overline{\varepsilon_h}(C_h) = \bigoplus_{h=1}^r C_h$, it follows that there exists a morphism $\bigoplus_{i=1}^n C_i / \bigoplus_{h=1}^r C_h \cong \bigoplus_{\ell=r+1}^n C_\ell \to C'_k/L_k$ making the following diagram

$$\begin{array}{ccc} \bigoplus_{i=1}^n C_i & \longrightarrow & \bigoplus_{\ell=r+1}^n C_\ell \\ {\scriptstyle\overline{\alpha}}\downarrow & & \downarrow \\ \bigoplus_{j=1}^n C'_j & \longrightarrow & C'_k/L_k \end{array}$$

commute for every $k = s+1, \dots, n$. Therefore there exist a morphism

$$\gamma\colon \bigoplus_{\ell=r+1}^n C_\ell \to \bigoplus_{k=s+1}^n C'_k/L_k$$

and a commutative diagram

$$\begin{array}{ccc} \bigoplus_{i=1}^n C_i & \longrightarrow & \bigoplus_{\ell=r+1}^n C_\ell \\ {\scriptstyle\overline{\alpha}}\downarrow & & \downarrow{\scriptstyle\gamma} \\ \bigoplus_{j=1}^n C'_j & \xrightarrow{\pi} & \bigoplus_{k=s+1}^n C'_k/L_k. \end{array}$$

Since the horizontal arrows are the canonical projections and $\overline{\alpha}$ is an isomorphism, the morphism γ must be surjective. Taking the dual Goldie dimension of the domain and the codomain of γ, we get that $n-r \geq n-s$, that is, $|T| = r \leq s = |N^+(T)|$, as we wanted to prove. Now apply Proposition 4.16 to the digraph D. This concludes the proof of one of the implications in the statement of our proposition.

Conversely, assume that $n = m$ and there exist four permutations $\varphi_{a,b}$ of $\{1, 2, \dots, n\}$, $(a,b) \in \{m, e\} \times \{l, u\}$, with $[B_i]_{a,b} = [B'_{\varphi_{a,b}(i)}]_{a,b}$ for every $i = 1, 2, \dots, n$, every $a = m, e$ and every $b = l, u$. By Corollary 4.8 (a), $\varphi_{a,b}(X_{a,b}) = Y_{a,b}$. Thus $\varphi_{a,b}$ restricts to a bijection $\varphi'_{a,b}$ of $X_{a,b}$ onto $Y_{a,b}$. We can now apply

Proposition 4.14, getting that $\bigoplus_{i=1}^n B_i$ and $\bigoplus_{j=1}^m B'_j$ are isomorphic objects in the category $\mathcal{E}$. □

5. Examples

EXAMPLE 5.1. Let R be a ring having two non-isomorphic simple right R-modules S and S'. Consider the following two objects of $\mathcal{U}$:

$$0 \longrightarrow S \xrightarrow{\varepsilon_1} S \oplus S' \xrightarrow{\pi_1} S' \longrightarrow 0$$

and

$$0 \longrightarrow S' \xrightarrow{\varepsilon_2} S \oplus S' \xrightarrow{\pi_2} S \longrightarrow 0,$$

where ε_i and π_j are the embeddings and the canonical projections, respectively. These two objects have the same endomorphism ring in $\mathcal{E}$:

$$E_B = \begin{pmatrix} \mathrm{End}_R(S) & 0 \\ 0 & \mathrm{End}_R(S') \end{pmatrix} = E_{B'}.$$

Moreover, the maximal right ideals of $E_B = E_{B'}$ are:

$$I_{B,m,u} = I_{B,e,u} = \begin{pmatrix} \mathrm{End}_R(S) & 0 \\ 0 & 0 \end{pmatrix} = I_{B',m,l} = I_{B',e,l}$$

and

$$I_{B,m,l} = I_{B,e,l} = \begin{pmatrix} 0 & 0 \\ 0 & \mathrm{End}_R(S') \end{pmatrix} = I_{B',m,u} = I_{B',e,u}.$$

So, for any fixed $(a, b) \in \{m, e\} \times \{l, u\}$, $I_{B,a,b}$ is a maximal right ideal of E_B and $I_{B',a,b}$ is a maximal ideal of $E_{B'}$, but $[B]_{a,b} \neq [B']_{a,b}$, because otherwise S and S' would be isomorphic, contrary to the hypothesis. In particular, the two objects are not isomorphic in $\mathcal{E}$ (Proposition 4.4). Notice that, in the notation of Proposition 4.14, the unique bijections $\varphi_{a,b}\colon X_{a,b} \to Y_{a,b}$ do not satisfy the "compatibility" conditions $[B_i]_{a,b} = [B'_{\varphi_{a,b}(i)}]_{a,b}$.

EXAMPLE 5.2. Let R be a ring and let U, V be two non-zero uniserial right R-modules. Consider the following object of $\mathcal{U}$:

$$0 \longrightarrow U \xrightarrow{\varepsilon} U \oplus V \xrightarrow{\pi} V \longrightarrow 0,$$

where $\varepsilon\colon u \mapsto (u, 0)$ is the embedding and $\pi\colon (u, v) \mapsto v$ is the canonical projection. Its endomorphism ring in $\mathcal{E}$ is

$$E_B = \begin{pmatrix} \mathrm{End}_R(U) & 0 \\ \mathrm{Hom}_R(V, U) & \mathrm{End}_R(V) \end{pmatrix}.$$

For any endomorphism $f = \left(\begin{smallmatrix} f_{1,1} & 0 \\ f_{2,1} & f_{2,2} \end{smallmatrix}\right) \in E_B$, we have $f\,|_U^U = f_{1,1}$ and $\overline{f} = f_{2,2}$. So, using the notation at the beginning of Section 3, $f \in I_{B,m,l}$ (resp. $f \in I_{B,e,l}$) if and only if $f_{1,1} \in I_U$ (resp. $f_{1,1} \in K_U$). Similarly, $f \in I_{B,m,u}$ (resp. $f \in I_{B,e,u}$) if and only if $f_{2,2} \in I_V$ (resp. $f_{2,2} \in K_V$). In particular, the type of E_B is exactly the sum of the types of $\mathrm{End}_R(U)$ and $\mathrm{End}_R(V)$. Therefore, choosing suitable uniserials R-modules, we can produce objects of $\mathcal{U}$ whose endomorphism ring has exactly 2, 3 or 4 maximal right ideals.

EXAMPLE 5.3. In [**F1**, Example 2.1], the second author constructed examples of uniserial right R-modules $U_{i,j}$ to show that a module that is a direct sum of n uniserial modules can have $n!$ pair-wise non-isomorphic direct-sum decompositions into indecomposables. All those modules $U_{i,j}$ have exactly two maximal ideals and we will now show that they are extensions of two uniserial modules with local endomorphism ring. The construction of the ring R and the uniserial right R-modules $U_{i,j}$ was the following.

Let $\mathbf{M}_n(\mathbb{Q})$ be the ring of all $n\times n$-matrices over the field $\mathbb{Q}$ of rational numbers. Let $\mathbb{Z}$ be the ring of integers and let $\mathbb{Z}_p, \mathbb{Z}_q$ be the localizations of $\mathbb{Z}$ at two distinct maximal ideals (p) and (q) of $\mathbb{Z}$ (here $p, q \in \mathbb{Z}$ are distinct prime numbers). Let Λ_p denote the subring of $\mathbf{M}_n(\mathbb{Q})$ whose elements are the $n \times n$-matrices with entries in $\mathbb{Z}_p$ on and above the diagonal and entries in $p\mathbb{Z}_p$ under the diagonal, that is,

$$\Lambda_p := \begin{pmatrix} \mathbb{Z}_p & \mathbb{Z}_p & \dots & \mathbb{Z}_p \\ p\mathbb{Z}_p & \mathbb{Z}_p & \dots & \mathbb{Z}_p \\ \vdots & & \ddots & \\ p\mathbb{Z}_p & p\mathbb{Z}_p & \dots & \mathbb{Z}_p \end{pmatrix} \subseteq \mathbf{M}_n(\mathbb{Q}).$$

Similarly, set

$$\Lambda_q := \begin{pmatrix} \mathbb{Z}_q & \mathbb{Z}_q & \dots & \mathbb{Z}_q \\ q\mathbb{Z}_q & \mathbb{Z}_q & \dots & \mathbb{Z}_q \\ \vdots & & \ddots & \\ q\mathbb{Z}_q & q\mathbb{Z}_q & \dots & \mathbb{Z}_q \end{pmatrix} \subseteq \mathbf{M}_n(\mathbb{Q}).$$

If

$$R := \begin{pmatrix} \Lambda_p & 0 \\ \mathbf{M}_n(\mathbb{Q}) & \Lambda_q \end{pmatrix},$$

then R is a subring of the ring $\mathbf{M}_{2n}(\mathbb{Q})$ of $2n \times 2n$-matrices with rational entries.

For every $i = 1, 2, \dots, n$, set

$$V_i := (\underbrace{\mathbb{Q}, \dots, \mathbb{Q}}_{n}, \underbrace{q\mathbb{Z}_q, \dots, q\mathbb{Z}_q}_{i-1}, \underbrace{\mathbb{Z}_q, \dots, \mathbb{Z}_q}_{n-i+1})$$

and

$$X_i := (\underbrace{p\mathbb{Z}_p, \dots, p\mathbb{Z}_p}_{i-1}, \underbrace{\mathbb{Z}_p, \dots, \mathbb{Z}_p}_{n-i+1}, \underbrace{0, \dots, 0}_{n}).$$

Then V_i and the X_j are right R-modules with right R-module structure given by matrix multiplication. Moreover, the unique infinite composition series of V_1 is

$$\begin{aligned} & V_1 \supset V_2 \supset \dots \supset V_n \supset qV_1 \supset \dots \supset qV_n \\ & \supset q^2V_1 \supset \dots \supset q^2V_n \supset \dots \supset p^{-2}X_1 \supset \dots \supset p^{-2}X_n \\ & \supset p^{-1}X_1 \supset \dots \supset p^{-1}X_n \supset X_1 \supset X_2 \supset \dots \supset X_n \\ & \supset pX_1 \supset \dots pX_n \supset p^2X_1 \supset \dots p^2X_n \supset \dots \supset 0 \end{aligned}$$

The right R-modules X_i/X_{i+1} and V_i/V_{i+1} are simple. Define the n^2 uniserial R-modules $U_{i,j} := V_j/X_i$. Now notice that

$$X := (\underbrace{\mathbb{Q}, \dots, \mathbb{Q}}_{n}, \underbrace{0, \dots, 0}_{n}) = \bigcap_{m \geq 0} q^m V_j = \sum_{m \geq 0} p^{-m} X_i$$

for every $i, j = 1, 2, \dots, n$. Thus we have a short exact sequence

$$0 \longrightarrow X/X_i \longrightarrow U_{i,j} = V_j/X_i \longrightarrow V_j/X \longrightarrow 0. \tag{5.1}$$

The right R-module X/X_i is an artinian uniserial module whose lattice of submodules is order isomorphic to the ordinal number $\omega + 1$, where ω denotes the first infinite ordinal. Since X/X_i is an artinian uniserial R-module, its endomorphism ring $\mathrm{End}(X/X_i)$ is a local ring whose maximal ideal consist of all the endomorphisms of X/X_i that are zero on the socle of X/X_i.

The right R-module V_j/X is a noetherian uniserial module, whose lattice of submodules is order antiisomorphic to the ordinal number $\omega+1$. The cyclic uniserial module V_j/X is annihilated by the two-sided ideal $I := \begin{pmatrix} \Lambda_p & 0 \\ \mathbf{M}_n(\mathbb{Q}) & 0 \end{pmatrix}$ of R, so that it is a module over $R/I \cong \Lambda_q$. In particular, $\mathrm{End}_R(V_j/X) = \mathrm{End}_{\Lambda_q}(V_j/X)$. As a module over Λ_q, V_j/X turns out to be a cyclic projective uniserial module, hence $V_j/X \cong e\Lambda_q$ for some idempotent $e \in \Lambda_q$. Thus $\mathrm{End}_R(V_j/X) = \mathrm{End}_{\Lambda_q}(V_j/X) \cong \mathrm{End}_{\Lambda_q}(e\Lambda_q) \cong e\Lambda_q e \cong \mathbb{Z}_q$ is a local ring.

Finally, every endomorphism of the R-module $U_{i,j} := V_j/X_i$ maps X/X_i into X/X_i, i.e., X/X_i is a fully invariant submodule of V_j/X_i, because X/X_i is the Loewy submodule of V_j/X_i. Thus the endomorphism ring of the extension (5.1) in the category $\mathcal{E}$ is canonically isomorphic to the endomorphism ring of the right R-module $U_{i,j}$, which is a ring of type 2.

EXAMPLE 5.4. Let $R = \mathbb{Z}$ be the ring of integers and let $p \in \mathbb{Z}$ be a prime. Consider the following object of $\mathcal{U}$:

$$0 \longrightarrow \mathbb{Z}/p\mathbb{Z} \xrightarrow{\ \iota\ } \mathbb{Z}(p^\infty) \xrightarrow{\ \beta\ } \mathbb{Z}(p^\infty) \longrightarrow 0,$$

where $\mathbb{Z}(p^\infty)$ denotes the Prüfer group and β is any surjective endomorphism of $\mathbb{Z}(p^\infty)$ having its kernel equal to $\mathrm{soc}(\mathbb{Z}(p^\infty)) \cong \mathbb{Z}/p\mathbb{Z}$. It is easy to check that $I_{B,e,u} \subsetneq I_{B,m,u} \subsetneq I_{B,m,l} = I_{B,e,l}$, and therefore E_B has exactly one maximal right ideal.

6. Main Theorem

Let $\mathcal{U}_0$ denote the full subcategory of $\mathcal{E}$ whose objects are of the form

$$0 \longrightarrow 0 \longrightarrow C' \longrightarrow C \longrightarrow 0, \tag{6.1}$$

for some non-zero uniserial right R-module C. Similarly, define $\mathcal{U}^0$ to be the full subcategory of $\mathcal{E}$ whose objects are of the form

$$0 \longrightarrow A \longrightarrow A' \longrightarrow 0 \longrightarrow 0, \tag{6.2}$$

for some non-zero uniserial right R-module A.

REMARK 6.1. A short exact sequence

$$0 \longrightarrow A \xrightarrow{\ \alpha\ } B \xrightarrow{\ \beta\ } C \longrightarrow 0 \tag{6.3}$$

splits if and only if it is the direct sum in the category $\mathcal{E}$ of two short exact sequences

$$0 \longrightarrow A \xrightarrow{\ \gamma\ } A' \longrightarrow 0 \longrightarrow 0 \tag{6.4}$$

and

$$0 \longrightarrow 0 \longrightarrow C' \longrightarrow C \longrightarrow 0. \tag{6.5}$$

Now suppose we have a split short exact sequence (6.3) with both A and C non-zero modules. The relations between the (a,b)-class of (6.3) and the (a,b)-classes of (6.4) and (6.5) are as follows:

(1) The (a,b)-classes of (6.4) are $[A']_{m,l} \neq [0]_{m,l}$, $[A']_{e,l} \neq [0]_{e,l}$, $[0]_{m,u}$, $[0]_{e,u}$.

(2) The (a,b)-classes of (6.5) are $[0]_{m,l}$, $[0]_{e,l}$, $[C']_{m,u} \neq [0]_{m,u}$, $[C']_{e,u} \neq [0]_{e,u}$.

(3) The (a,b)-classes of (6.3) are $[B]_{m,l} = [A']_{m,l}$, $[B]_{e,l} = [A']_{e,l}$, $[B]_{m,u} = [C']_{m,u}$, $[B]_{e,u} = [C']_{e,u}$.

LEMMA 6.2. The following conditions are equivalent for an object (6.3) of $\mathcal{U}$:

(a) The short exact sequence (6.3) splits.

(b) There exists an object (6.4) of $\mathcal{U}^0$ such that $[B]_{a,l} = [A']_{a,l}$ for both $a = m, e$.

(c) There exists an object (6.5) of $\mathcal{U}_0$ such that $[B]_{a,u} = [C']_{a,u}$ for both $a = m, e$.

PROOF. (a)⇒(b) and (a)⇒(c) easily follow from Remark 6.1.

(b)⇒(a) By hypothesis, there exist $f, g\colon B \to A'$ such that $f|_A^A$ is injective and $g|_A^A$ is surjective. From [**F1**, Lemma 1.4 (a)], one of $f|_A^A$, $g|_A^A$ or $(f+g)|_A^A$ is an isomorphism. So we can find a morphism $h\colon B \to A'$ in $\mathcal{E}$ for which $h|_A^A$ is an isomorphism. Now it is immediate to see that $(h|_A^A)^{-1}\gamma^{-1}h\alpha = id_A$, where $\gamma\colon A \to A'$ is the isomorphism in the short exact sequence (6.4). This means that (6.3) splits. Similarly for (c)⇒ (a). □

THEOREM 6.3. *Let*

$$0 \longrightarrow A_i \longrightarrow B_i \longrightarrow C_i \longrightarrow 0, \qquad i = 1, 2, \dots, n, \tag{6.6}$$

and

$$0 \longrightarrow A'_j \longrightarrow B'_j \longrightarrow C'_j \longrightarrow 0, \qquad j = 1, 2, \dots, m, \tag{6.7}$$

be $n+m$ non-zero objects in the category $\mathcal{E}$ with all the modules A_i, C_i, A'_j, C'_j uniserial modules. Set

$$\begin{aligned} X_l &:= \{\, i \mid i = 1, 2, \dots, n,\ A_i \neq 0 \,\}, \\ X_u &:= \{\, i \mid i = 1, 2, \dots, n,\ C_i \neq 0 \,\}, \\ X'_l &:= \{\, j \mid j = 1, 2, \dots, m,\ A'_j \neq 0 \,\}, \\ X'_u &:= \{\, j \mid j = 1, 2, \dots, m,\ C'_j \neq 0 \,\}. \end{aligned}$$

Then $\bigoplus_{i=1}^n B_i \cong \bigoplus_{j=1}^m B'_j$ in the category $\mathcal{E}$ if and only if there exist four bijections $\varphi_{a,b}\colon X_b \to X'_b$, $a = m, e$ and $b = l, u$, such that $[B_i]_{a,b} = [B'_{\varphi_{a,b}(i)}]_{a,b}$ for every $a = m, e$, $b = l, u$ and $i \in X_b$.

PROOF. (⇒) Suppose $\bigoplus_{i=1}^n B_i \cong \bigoplus_{j=1}^m B'_j$ in the category $\mathcal{E}$. Then $\bigoplus_{i=1}^n A_i \cong \bigoplus_{j=1}^m A'_j$ and $\bigoplus_{i=1}^n C_i \cong \bigoplus_{j=1}^m C'_j$ in Mod-R. Taking the Goldie dimensions of these two pairs of isomorphic modules, we get that $|X_l| = |X'_l|$ and $|X_u| = |X'_u|$. Set $r := |X_l| = |X'_l|$ and $s := |X_u| = |X'_u|$. Suppose for instance $r \leq s$, so that $r + t = s$ for some integer $t \geq 0$. The set $\{1, 2, \dots, n\}$ is partitioned into three pair-wise disjoint subsets $X_l \cap X_u$, $X_l \setminus X_u$ and $X_u \setminus X_l$, so that $\{1, 2, \dots, n\} = (X_l \cap X_u) \dot\cup (X_l \setminus X_u) \dot\cup (X_u \setminus X_l)$. Then $|X_l \setminus X_u| + t = |X_u \setminus X_l|$. Now set $u := |X_l \setminus X_u|$, so that $|X_l \cap X_u| = n - 2u - t$. Let $i_1, \dots, i_u$ be the elements of $X_l \setminus X_u$ and fix u elements $k_1, \dots, k_u$ in $X_u \setminus X_l$. In the direct sum $\bigoplus_{i=1}^n B_i$, replace the $2u$ sequences

$$0 \longrightarrow A_{i_\ell} \longrightarrow B_{i_\ell} \longrightarrow 0 \longrightarrow 0, \qquad \ell = 1, \dots, u, \tag{6.8}$$

and

$$0 \longrightarrow 0 \longrightarrow B_{k_\ell} \longrightarrow C_{k_\ell} \longrightarrow 0, \qquad \ell = 1, \dots, u, \tag{6.9}$$

with the u split sequences

$$0 \longrightarrow A_{i_\ell} \longrightarrow B_{i_\ell} \oplus B_{k_\ell} \longrightarrow C_{k_\ell} \longrightarrow 0, \qquad \ell = 1, \dots, u. \tag{6.10}$$

In each of these replacements, the (a,b)-classes $[B_{i_\ell}]_{m,l} \neq [0]_{m,l}$, $[B_{i_\ell}]_{e,l} \neq [0]_{e,l}$, $[0]_{m,u}$, $[0]_{e,u}$ of (6.8) and $[0]_{m,l}$, $[0]_{e,l}$, $[B_{k_\ell}]_{m,u} \neq [0]_{m,u}$, $[B_{k_\ell}]_{e,u} \neq [0]_{e,u}$ of (6.9) are replaced with the (a,b)-classes $[B_{i_\ell} \oplus B_{k_\ell}]_{m,l} = [B_{i_\ell}]_{m,l}$, $[B_{i_\ell} \oplus B_{k_\ell}]_{e,l} = [B_{i_\ell}]_{e,l}$, $[B_{i_\ell} \oplus B_{k_\ell}]_{m,u} = [B_{k_\ell}]_{m,u}$, $[B_{i_\ell} \oplus B_{k_\ell}]_{e,u} = [B_{k_\ell}]_{e,u}$ of (6.10). That is, only some zero (a,b)-classes are deleted in the replacement.

Now fix a simple right R-module S and replace each of the remaining t sequences indexed in $(X_u \setminus X_l) \setminus \{k_1, \dots, k_u\}$ with the direct sum of the sequence and the sequence $0 \longrightarrow S \longrightarrow S \longrightarrow 0 \longrightarrow 0$. After these substitutions, the object $(\bigoplus_{i=1}^n B_i) \oplus S^t$ of $\mathcal{E}$ is the direct sum of the $n - 2u - t$ sequences (6.6) indexed in $X_l \cap X_u$ and $u + t$ split sequences. All these $n - u$ summands are objects of $\mathcal{U}$.

Similarly, the object $(\bigoplus_{j=1}^m B'_j) \oplus S^t$ of $\mathcal{E}$ can be written as the direct sum of $m - 2u' - t$ sequences indexed in $X'_l \cap X'_u$, plus u' split sequences of the form

$$0 \longrightarrow A'_{j_v} \longrightarrow B'_{j_v} \oplus B'_{j'_v} \longrightarrow C'_{j'_v} \longrightarrow 0, \qquad v = 1, \dots, u', \tag{6.11}$$

plus t split sequences of the form $0 \longrightarrow S \longrightarrow B'_j \oplus S \longrightarrow C'_j \longrightarrow 0$.

Moreover, $\bigoplus_{i=1}^n B_i \cong \bigoplus_{j=1}^m B'_j$ implies that $(\bigoplus_{i=1}^n B_i) \oplus S^t \cong (\bigoplus_{j=1}^m B'_j) \oplus S^t$ in $\mathcal{E}$, so that we can apply Proposition 4.17 to this isomorphism. Thus, fix $a = m, e$ and any short exact sequence

$$0 \longrightarrow U \longrightarrow V \longrightarrow W \longrightarrow 0 \tag{6.12}$$

in $\mathcal{U}$. Then the number of indices $i \in X_l \cap X_u$ with $[B_i]_{a,l} = [V]_{a,l}$ plus the number of indices $\ell = 1, \dots, u$ with $[B_{i_\ell} \oplus B_{k_\ell}]_{a,l} = [V]_{a,l}$ plus the number of indices $i \in (X_u \setminus X_l) \setminus \{k_1, \dots, k_u\}$ such that $[B_i \oplus S]_{a,l} = [V]_{a,l}$ is equal to the number of indices $j \in X'_l \cap X'_u$ with $[B'_j]_{a,l} = [V]_{a,l}$ plus the number of indices $v = 1, \dots, u'$ with $[B'_{j_v} \oplus B'_{k'_v}]_{a,l} = [V]_{a,l}$ plus the number of indices $j \in (X'_u \setminus X'_l) \setminus \{k'_1, \dots, k'_{u'}\}$ such that $[B'_j \oplus S]_{a,l} = [V]_{a,l}$. From Remark 6.1, it follows that the number of indices $i \in X_l \cap X_u$ with $[B_i]_{a,l} = [V]_{a,l}$ plus the number of indices $\ell = 1, \dots, u$ with $[B_{i_\ell}]_{a,l} = [V]_{a,l}$ plus the number of indices $i \in (X_u \setminus X_l) \setminus \{k_1, \dots, k_u\}$ such that $[S]_{a,l} = [V]_{a,l}$ (that is, t if $[S]_{a,l} = [V]_{a,l}$ and 0 if $[S]_{a,l} \neq [V]_{a,l}$) is equal to the number of indices $j \in X'_l \cap X'_u$ with $[B'_j]_{a,l} = [V]_{a,l}$ plus the number of indices $v = 1, \dots, u'$ with $[B'_{j_v}]_{a,l} = [V]_{a,l}$ plus the number of indices $j \in (X'_u \setminus X'_l) \setminus \{k'_1, \dots, k'_{u'}\}$ with $[S]_{a,l} = [V]_{a,l}$ (that is, t if $[S]_{a,l} = [V]_{a,l}$ and 0 if $[S]_{a,l} \neq [V]_{a,l}$).

This is true for every sequence (6.12) in $\mathcal{U}$, so that, by Remark 6.1, it is also true for every sequence (6.12) in $\mathcal{U}^0$. Notice that, for every sequence (6.12) in $\mathcal{U}$ or $\mathcal{U}^0$, one has that $[B'_j]_{a,l} = [V]_{a,l}$ implies $A'_j \neq 0$. Therefore, the number of indices $i = 1, 2, \dots, n$ with $[B_i]_{a,l} = [V]_{a,l}$ is equal to the number of indices $j = 1, 2, \dots, m$ with $[B'_j]_{a,l} = [V]_{a,l}$.

Similarly for the case $r \geq s$ and the (a,b)-classes with $b = u$.

($\Leftarrow$) Suppose there exist four bijections $\varphi_{a,b} \colon X_b \to X'_b$, $a = m, e$ and $b = l, u$, such that $[B_i]_{a,b} = [B'_{\varphi_{a,b}(i)}]_{a,b}$ for every $a = m, e$, $b = l, u$ and $i \in X_b$. Set $r := |X_l| = |X'_l|$ and $s := |X_u| = |X'_u|$. Suppose $r \leq s$, so that $r + t = s$ for some integer $t \geq 0$. The set $\{1, 2, \dots, n\}$ is partitioned into three pair-wise disjoint subsets

$X_l \cap X_u$, $X_l \setminus X_u$ and $X_u \setminus X_l$, so that $\{1, 2, \dots, n\} = (X_l \cap X_u) \dot\cup (X_l \setminus X_u) \dot\cup (X_u \setminus X_l)$. Then $|X_l \setminus X_u| + t = |X_u \setminus X_l|$. Now set $u := |X_l \setminus X_u|$, so that $|X_l \cap X_u| = n - 2u - t$. Let $i_1, \dots, i_u$ be the elements of $X_l \setminus X_u$ and fix u elements $k_1, \dots, k_u$ in $X_u \setminus X_l$. Fix a simple right R-module S and consider the sequence

$$0 \longrightarrow S \longrightarrow S \longrightarrow 0 \longrightarrow 0. \tag{6.13}$$

Let $i_1, \dots, i_u$ be the elements of $X_l \setminus X_u$ and fix u elements $k_1, \dots, k_u$ in $X_u \setminus X_l$. Like in the proof of the previous implication, we have that, of the n sequences (6.6), $n - 2u - t$ are in $\mathcal{U}$, u are in $\mathcal{U}_0$, and $u + t$ are in $\mathcal{U}^0$. Replace the n sequences (6.6) with $n - u$ sequences, which are:

(1) the $n - 2u - t$ sequences (6.6) that are in $\mathcal{U}$;

(2) u sequences that are the direct sum of the sequences (6.6) in $\mathcal{U}_0$ plus u of the sequences (6.6) in $\mathcal{U}^0$;

(3) t sequences that are the direct sum of the remaining t sequences (6.6) in $\mathcal{U}^0$ plus the sequence (6.13).

These new $n - u$ sequences are indexed in X_u and are in $\mathcal{U}$.

The (a, b)-classes of the new $n - u$ sequences with their multiplicity are the same as the non-zero (a, b)-classes of the old n sequences (6.6) for $b = u$, and are the same as the non-zero (a, b)-classes of the old n sequences (6.6) plus the (a, b)-class $[S]_{a,b}$ with multiplicity t for for $b = l$.

Similarly for the m sequences (6.7). We can replace them with $m - u'$ sequences, which are:

(1) the $m - 2u' - t$ sequences (6.7) that are in $\mathcal{U}$;

(2) u' sequences that are the direct sum of the sequences (6.7) in $\mathcal{U}_0$ plus u' of the sequences (6.7) in $\mathcal{U}^0$;

(3) t sequences that are the direct sum of the remaining t sequences (6.7) in $\mathcal{U}^0$ plus the sequence (6.13).

These $m - u'$ sequences are indexed in X'_u and are all in $\mathcal{U}$. The (a, b)-classes of these new $m - u'$ sequences with their multiplicity are the same as the non-zero (a, b)-classes of the old m sequences (6.7) for $b = u$, and are the same as the non-zero (a, b)-classes of the old m sequences (6.7) plus the (a, b)-class $[S]_{a,b}$ with multiplicity t for for $b = l$.

The existence of the four bijections $\varphi_{a,b} \colon X_b \to X'_b$, $a = m, e$ and $b = l, u$ with $[B_i]_{a,b} = [B'_{\varphi_{a,b}(i)}]_{a,b}$ for every $a = m, e$, $b = l, u$ and $i \in X_b$ implies that $|X_u| = |X'_u|$, that is, $n - u = m - u'$, and that there are four bijections between the new $n - u$ sequences and the new $m - u'$ sequences that preserve the (a, b)-classes. Since all the new sequences are in $\mathcal{U}$, we can apply Proposition 4.17 to the new $n - u + m - u'$ sequences, getting that $\bigoplus_{i=1}^n B_i \oplus S^t \cong \bigoplus_{j=1}^m B'_j \oplus S^t$ in the category $\mathcal{E}$. Now the sequence S cancels from direct sums in $\mathcal{E}$ because its endomorphism ring is a division ring [**F2**, Theorem 4.5]. (Notice that Evans' result is stated in [**F2**, Theorem 4.5] only for modules, but its proof shows that it holds in any additive category. Moreover, the endomorphism ring of the sequence (6.13) in the category $\mathcal{E}$ is the endomorphism ring of the simple module S, which is a division ring.) Thus $\bigoplus_{i=1}^n B_i \cong \bigoplus_{j=1}^m B'_j$ in $\mathcal{E}$. Similarly for the case $r \geq s$, making use of the exact sequence $0 \longrightarrow 0 \longrightarrow S \longrightarrow S \longrightarrow 0$. $\square$

The authors are very grateful to the referee of the paper, who discovered a mistake hidden in the proof of one of the implications of Proposition 4.17.

References

[AF] A. Alahmadi and A. Facchini, *Some remarks on categories of modules modulo morphisms with essential kernel or superfluous image*, J. Korean Math. Soc. **50** (2013), no. 3, 557–578. MR3056922

[AAF] B. Amini, A. Amini, and A. Facchini, *Equivalence of diagonal matrices over local rings*, J. Algebra **320** (2008), no. 3, 1288–1310. MR2427644

[CD] R. Camps and W. Dicks, *On semilocal rings*, Israel J. Math. **81** (1993), no. 1-2, 203–211. MR1231187

[DF] L. Diracca and A. Facchini, *Uniqueness of monogeny classes for uniform objects in abelian categories*, J. Pure Appl. Algebra **172** (2002), no. 2-3, 183–191. MR1906873

[F1] A. Facchini, *Krull-Schmidt fails for serial modules*, Trans. Amer. Math. Soc. **348** (1996), no. 11, 4561–4575. MR1376546

[F2] A. Facchini, *Module theory. Endomorphism rings and direct sum decompositions in some classes of modules*, Progress in Mathematics, vol. 167, Birkhäuser Verlag, Basel, 1998. MR1634015

[FEK] A. Facchini, Ş. Ecevit, and M. T. Koşan, *Kernels of morphisms between indecomposable injective modules*, Glasg. Math. J. **52** (2010), no. A, 69–82. MR2669097

[FG] A. Facchini and N. Girardi, *Couniformly presented modules and dualities*, Advances in ring theory, Trends Math., Birkhäuser/Springer Basel AG, Basel, 2010, pp. 149–164. MR2664670

[FP1] A. Facchini and M. Perone, *Maximal ideals in preadditive categories and semilocal categories*, J. Algebra Appl. **10** (2011), no. 1, 1–27. MR2784751

[FP2] A. Facchini and P. Příhoda, *Factor categories and infinite direct sums*, Int. Electron. J. Algebra **5** (2009), 135–168. MR2471385

[FP3] A. Facchini and P. Příhoda, *Endomorphism rings with finitely many maximal right ideals*, Comm. Algebra **39** (2011), no. 9, 3317–3338. MR2845575

[FP4] A. Facchini and P. Příhoda, *The Krull-Schmidt theorem in the case two*, Algebr. Represent. Theory **14** (2011), no. 3, 545–570. MR2785922

[M] B. Mitchell, *Rings with several objects*, Advances in Math. **8** (1972), 1–161. MR0294454

Dipartimento di Matematica, Università di Padova, 35121 Padova, Italy
Email address: federico.campanini@math.unipd.it
Email address: facchini@math.unipd.it

Contemporary Mathematics
Volume **715**, 2018
https://doi.org/10.1090/conm/715/14408

Tilting cotorsion pairs and pure semisimple rings

Nguyen Viet Dung and José Luis García

ABSTRACT. A result of L. Angeleri Hügel in [A key module over pure-semisimple hereditary rings, J. Algebra **307** (2007), 361-376] asserts that if R is a left pure semisimple hereditary ring, then every cotorsion pair in R-Mod is generated by a finitely generated tilting module. In this paper, we obtain a new characterization of left pure semisimple hereditary rings by proving that the converse of Angeleri Hügel's result is true, namely that if R is a ring such that every cotorsion pair in R-Mod is generated by a finitely generated tilting module, then R is left pure semisimple hereditary.

1. Introduction

A ring R is called *left pure semisimple* if every left R-module is a direct sum of finitely generated left R-modules, or equivalently, if R has left pure global dimension zero [**19**, **31**, **32**]. It is well known that a ring R is left and right pure semisimple if and only if R is of *finite representation type*, i.e. R is left (and right) artinian with only finitely many non-isomorphic finitely generated indecomposable left (and right) modules (see [**8**, **20**, **28**]). It is still an open problem, known as the *pure semisimplicity conjecture*, whether left pure semisimple rings are always of finite representation type. It is also known that the pure semisimplicity conjecture is true if and only if it is true for all left hereditary rings (see [**24**]).

Tilting modules have played an important role in the study of hereditary pure semisimple rings. Early results in this direction include Simson's work [**33**] (see also [**34**]) using tilting modules, in the form of the reflection functors, to study pure semisimple triangular matrix rings induced by bimodules over division rings. Several recent results in the literature have shown that, if R is a left pure semisimple hereditary (indecomposable) ring, there is an abundance of tilting modules in the category R-mod of finitely generated left R-modules (see, e.g., [**2**, **4**, **6**, **16**, **17**]). Tilting modules are also crucial in the study of potential counter-examples to the pure semisimplicity conjecture (see, e.g., [**21**, **22**, **35**, **37**]).

Cotorsion pairs in module categories originated in the work by Salce [**29**], and have been studied extensively by many authors (see, e.g., the monograph by Göbel and Trlifaj [**23**] and the references therein). Among cotorsion pairs in R-Mod for a ring R, those that are generated by tilting modules (also called *tilting cotorsion*

2010 *Mathematics Subject Classification.* 16G10, 16G60, 16D70, 16D90.
Key words and phrases. Pure semisimple ring, cotorsion pair, tilting module.

pairs), are of particular interest. Cotorsion pairs were used by Angeleri Hügel [**2**] to study left pure semisimple hereditary rings, and it was shown in [**2**, Proposition 4.2] that, over such a ring R, every cotorsion pair in R-Mod is generated by a finitely generated tilting module.

One of the points of this paper is to show that there is a close relationship between the existence of a "sufficiently large" number of tilting modules in R-Mod and the pure semisimplicity of a ring R. As the main result, we will show that the converse of Angeleri Hügel's result is true, namely that if R is a ring such that every cotorsion pair in R-Mod is generated by a finitely generated tilting module, then R is left pure semisimple hereditary. We thus obtain a new criterion, in terms of tilting cotorsion pairs, for a ring to be left pure semisimple hereditary.

Our paper is organized as follows. Section 2 gives basic definitions and some preliminary results that will be needed later on. In Section 3, we focus on rings R such that every cotorsion pair in R-Mod is generated by a finitely generated tilting module, and present the proof of the main result (Theorem 3.7). We conclude the paper with some related remarks, in particular regarding triangular matrix rings induced by Cohn's construction [**11**, Sec. 5.6] of the quadratic division ring extension $G \subseteq F$ (solution of Artin's problem) and their relation to the pure semisimplicity problem (see Remark 3.9).

2. Definitions and Preliminaries

Throughout this paper, R is an associative ring with identity. We denote by R-mod the category of all finitely generated left R-modules, and by R-Mod the category of all left R-modules.

If $\mathcal{B}$ is a family of finitely generated left R-modules, ind$(\mathcal{B})$ will stand for the set of all non-isomorphic indecomposable summands of modules belonging to $\mathcal{B}$. In particular, we set R-ind = ind(R-mod). For any module $M \in R$-Mod, by Add(M) we denote the class consisting of all left R-modules that are isomorphic to direct summands of direct sums of copies of M.

Let $\mathcal{M}$ be a full subcategory of R-Mod. We denote $\mathcal{M}^{\perp} = \{X \in R\text{-Mod} \mid \operatorname{Ext}^1_R(M, X) = 0 \text{ for each } M \in \mathcal{M}\}$, and ${}^{\perp}\mathcal{M} = \{X \in R\text{-Mod} \mid \operatorname{Ext}^1_R(X, M) = 0 \text{ for each } M \in \mathcal{M}\}$. Let $\mathcal{A}, \mathcal{B} \subseteq R$-Mod be classes of modules. Then $(\mathcal{A}, \mathcal{B})$ is said to be a *cotorsion pair* in R-Mod provided $\mathcal{A} = {}^{\perp}\mathcal{B}$ and $\mathcal{B} = \mathcal{A}^{\perp}$. If $\mathcal{C}$ is a class of modules, then $({}^{\perp}(\mathcal{C}^{\perp}), \mathcal{C}^{\perp})$ is a cotorsion pair, and we say it is the cotorsion pair *generated by* $\mathcal{C}$. We say that the cotorsion pair generated by $\mathcal{C} = \{W\}$ is generated by W. If $(\mathcal{A}, \mathcal{B})$ is a cotorsion pair, the class $\mathcal{K} = \mathcal{A} \cap \mathcal{B}$ is called the *kernel* of $(\mathcal{A}, \mathcal{B})$.

A module T in R-Mod is called a *tilting module* if the following conditions are satisfied: (1) T has projective dimension at most one; (2) $\operatorname{Ext}^1_R(T, T^{(I)}) = 0$ for all sets I; (3) there is an exact sequence $0 \to R \to T' \to T'' \to 0$ such that T' and T'' are in Add(T). It is well known that T is a tilting module if and only if the class Gen(T) of all T-generated modules coincides with the class $T^{\perp} = \{X \in R\text{-Mod} \mid \operatorname{Ext}^1_R(T, X) = 0\}$ (see [**14**, Proposition 1.3]).

We will need the following result, due to Angeleri Hügel and Coelho [**3**], on the kernel of a tilting cotorsion pair over an arbitrary ring.

LEMMA 2.1. *Let R be a ring and $(\mathcal{A}, \mathcal{B})$ a cotorsion pair in R-Mod generated by a tilting module T. Then $\mathcal{A} \cap \mathcal{B} = Add(T)$.*

PROOF. See [**3**, Lemma 2.4] (cf. [**39**, Lemma 2.7]). □

If $\mathcal{C}$ is a class of modules in R-Mod, a left R-module X in $\mathcal{C}$ is said to be an *Ext-injective module* in $\mathcal{C}$ if $\mathrm{Ext}^1_R(C, X) = 0$ for all $C \in \mathcal{C}$. Dually, a left R-module Y in $\mathcal{C}$ is said to be an *Ext-projective module* in $\mathcal{C}$ if $\mathrm{Ext}^1_R(Y, C) = 0$ for all $C \in \mathcal{C}$.

Recall that a pair $(\mathcal{T}, \mathcal{F})$ of subclasses of R-Mod is called a *torsion theory* if the following conditions are satisfied:

a) $\mathrm{Hom}_R(X, Y) = 0$ for any $X \in \mathcal{T}$ and $Y \in \mathcal{F}$.

b) $\mathcal{T} = \{X \in R\text{-Mod} \mid \mathrm{Hom}_R(X, Y) = 0 \text{ for all } Y \in \mathcal{F}\}$

c) $\mathcal{F} = \{Y \in R\text{-Mod} \mid \mathrm{Hom}_R(X, Y) = 0 \text{ for all } X \in \mathcal{T}\}$

$\mathcal{T}$ is called the *torsion class* and $\mathcal{F}$ is the *torsionfree class* of this torsion theory. If $(\mathcal{T}, \mathcal{F})$ is a torsion theory, then for each module M in R-Mod, there is a largest submodule $t(M) \in \mathcal{T}$ and then $M/t(M) \in \mathcal{F}$, because t is an idempotent radical [**38**]. A torsion theory $(\mathcal{T}, \mathcal{F})$ is called *splitting* if for any module M in R-Mod, $t(M)$ is a direct summand of M. It is clear that a pair $(\mathcal{T}, \mathcal{F})$ of classes of modules is a splitting torsion theory if and only if $\mathrm{Hom}_R(X, Y) = 0$ for any $X \in \mathcal{T}$ and $Y \in \mathcal{F}$ and for any module M in R-Mod, there is a direct sum decomposition $M = X \oplus Y$ with $X \in \mathcal{T}$ and $Y \in \mathcal{F}$.

The following lemma is essentially due to Hoshino [**25**].

LEMMA 2.2. *Let R be a ring and $(\mathcal{T}, \mathcal{F})$ be a torsion theory in R-Mod. Let $0 \to A \to B \to C \to 0$ be an almost split sequence in R-Mod. Then C is Ext-projective in $\mathcal{T}$ if and only if A is Ext-injective in $\mathcal{F}$.*

PROOF. See [**6**, Lemma 4.2]. □

The following concept was recently introduced in [**15, 16**].

DEFINITION 2.3. *Let R be any ring and $\mathcal{C}$ a family of non-isomorphic finitely generated indecomposable left R-modules. We say that $\mathcal{C}$ has an Ext-injective partition if $\mathcal{C} = \bigcup_{\alpha<\rho} \mathcal{U}_\alpha$, for an ordinal ρ, with the following properties:*

(i) $\mathcal{U}_\alpha \cap \mathcal{U}_\beta = \emptyset$ whenever $\alpha \neq \beta$.

(ii) Each $\mathcal{U}_\alpha$ is a finite and non-empty set.

(iii) For each ordinal $\alpha < \rho$, $\mathcal{U}_\alpha$ is the set of all Ext-injective modules of $\bigcup_{\beta\geq\alpha} \mathcal{U}_\beta$.

We will make use of the following result, showing that, for a left artinian hereditary ring R, the existence of the Ext-injective partition in R-mod is equivalent to the left pure semisimplicity of the ring R.

PROPOSITION 2.4. *Let R be a left artinian hereditary ring. Then the following conditions are equivalent:*

(a) The family of all finitely generated indecomposable left R-modules has an Ext-injective partition R-ind $= \bigcup_{\alpha<\rho} \mathcal{U}_\alpha$.

(b) R is left pure semisimple.

PROOF. See [**16**, Proposition 2.2] □

We refer the reader to [**1, 38**] for general properties of rings and modules, and to [**7, 10**] for basic representation-theoretic notions and results. The reader is referred to [**26, 27, 36**] for the history and basic background on the pure semisimplicity conjecture, and to [**12, 23, 39**] for basic results and concepts on tilting modules and cotorsion pairs.

3. Main Results

In this section we will prove the main result stated in the abstract, establishing a converse to Angeleri Hügel's result [**2**, Proposition 4.2] (see also [**5**, Proposition 6.2]). We will need some preparatory results for the proof of the main theorem.

Recall that a module M is said to have the *exchange property* provided that whenever there is a direct sum decomposition $A = \oplus_{i\in I} A_i = M \oplus N$, there are submodules $B_i \subseteq A_i$ for all $i \in I$ such that $A = M \oplus (\oplus_{i\in I} B_i)$. For a ring R, let $\mathcal{I}$ denote the class of all injective left R-modules, then (R-Mod, $\mathcal{I}$) is a cotorsion pair in R-Mod.

PROPOSITION 3.1. *A ring R is left artinian hereditary with a left Morita duality if and only if the cotorsion pair (R-Mod, $\mathcal{I}$) is generated by a finitely generated tilting module.*

PROOF. Assume first that the cotorsion pair (R-Mod, $\mathcal{I}$) is generated by a finitely generated tilting module T. It then follows by [**39**, Theorem 2.8] that each module in R-Mod has projective dimension at most one, so R is left hereditary. Also, we know by Lemma 2.1 that the kernel of this cotorsion pair (which is $\mathcal{I}$) coincides with Add(T). Hence each injective left R-module E is isomorphic to a direct summand of a direct sum of copies of T. Since it is well known that each injective module has the exchange property (see [**41**, Lemma 2]), it is clear that E is isomorphic to a direct sum of direct summands of T, so in particular E is a direct sum of finitely generated modules. By a result due to Faith and Walker [**18**, Theorem 3.1], if R is a ring such that each injective left R-module is a direct sum of finitely generated modules, then R is left artinian. It is clear by the above argument that if E is an indecomposable injective left R-module, then E is finitely generated, hence R has a left Morita duality. Conversely, suppose that R is left artinian hereditary with a left Morita duality. Let M be the minimal injective cogenerator of R-Mod, i.e. M is the finite direct sum of all nonisomorphic indecomposable injective left R-modules. Then it is clear that M is a finitely generated tilting module which generates the cotorsion pair (R-Mod, $\mathcal{I}$). □

We thank the referee for communicating to us the remark below, showing that a ring R, over which every cotorsion pair in R-Mod is generated by a (not necessarily finitely generated) tilting module, is left artinian hereditary (see also [**5**, Theorem 6.1]).

REMARK 3.2. Let R be a ring such that every cotorsion pair in R-Mod is generated by a tilting module. Note that if $(\mathcal{A}, \mathcal{B})$ is any cotorsion pair in R-Mod, then $(\mathcal{A}, \mathcal{B})$ is a tilting cotorsion pair, hence $\mathcal{B}$ is a *definable class*, i.e. it is closed under direct products, direct limits and pure submodules (see, e.g., [**23**, Theorems 13.26 and 13.46]). In particular, the definability of $\mathcal{I}$ in the tilting cotorsion pair (R-Mod, $\mathcal{I}$) implies that R is left noetherian. Also, because (R-Mod, $\mathcal{I}$) is a tilting cotorsion pair, it follows by [**39**, Theorem 2.8] that R is left hereditary. A left R-module M is called *cotorsion* if $\mathrm{Ext}^1_R(F, M) = 0$ for every flat left R-module. It is well-known that any pure-injective R-module is cotorsion. Let R-$\mathcal{FLAT}$ and R-$\mathcal{COT}$ be the classes of all flat and cotorsion left R-modules, respectively. Then (R-$\mathcal{FLAT}$, R-$\mathcal{COT}$) is a tilting cotorsion pair, hence R-$\mathcal{COT}$ is definable. Since each left R-module M is a pure submodule of a pure-injective module N which belongs to R-$\mathcal{COT}$, M itself belongs to R-$\mathcal{COT}$. Therefore every left R-module is cotorsion, implying that each flat left R-module is projective, hence it is well-known that R is a left perfect ring. As R is left noetherian, it follows that R is left artinian.

The next lemma will give some useful information on the kernel of a cotorsion pair generated by a finitely generated tilting module, over a left artinian hereditary ring.

LEMMA 3.3. *Let R be a left artinian hereditary ring and $(\mathcal{C}, \mathcal{D})$ be a cotorsion pair in R-Mod generated by a finitely generated tilting module T. Then each module in $\mathcal{C} \cap \mathcal{D}$ is a direct sum of indecomposable modules each isomorphic to a direct summand of T. Moreover, a finitely generated indecomposable left R-module M is Ext-injective in $\mathcal{C}$ if and only if it is Ext-injective in $\mathcal{C} \cap R$-mod.*

PROOF. Since $(\mathcal{C}, \mathcal{D})$ is a tilting cotorsion pair, it follows by Lemma 2.1 that $\mathcal{C} \cap \mathcal{D} = \text{Add}(T)$. Hence, for any module $M \in \mathcal{C} \cap \mathcal{D}$, M is a direct summand of a direct sum of copies of T, which in turn is a finite direct sum of finitely generated indecomposable left R-modules with local endomorphism rings. It follows, by the Crawley-Jønsson-Warfield theorem and Azumaya's theorem (see [**1**, Theorem 26.5, Corollary 26.6]), that M is a direct sum of indecomposable modules each isomorphic to a direct summand of T.

For the second statement, it is clear that if an indecomposable module M in R-mod is Ext-injective in $\mathcal{C}$, then it is also Ext-injective in $\mathcal{C} \cap R$-mod. Conversely, let M be a finitely generated indecomposable left R-module in $\mathcal{C}$ and assume that M is Ext-injective in $\mathcal{C} \cap R$-mod. In particular, M will be Ext-injective relative to the finitely generated summands of the tilting module T, i.e., $\text{Ext}^1_R(T, M) = 0$. Now, since T is tilting, we have that $\text{Gen}(T)$ coincides with the class $T^{\perp}$, and hence M is generated by T. Since M is finitely generated, there is an exact sequence $0 \to K \to T^n \to M \to 0$ for some positive integer n. Thus, for each module $C \in \mathcal{C}$, we get the induced exact sequence $\text{Ext}^1_R(C, T^n) \to \text{Ext}^1_R(C, M) \to \text{Ext}^2_R(C, K) = 0$ because R is left hereditary. Since $\text{Ext}^1_R(C, T^n) = 0$, it follows that $\text{Ext}^1_R(C, M) = 0$, showing that M is Ext-injective in $\mathcal{C}$. □

The next two results, dealing with successor and limit ordinals respectively, will be essential in the proof of our main theorem.

PROPOSITION 3.4. *Let R be a left artinian hereditary ring and let $(\mathcal{C}, \mathcal{D})$ be a cotorsion pair of R-Mod which is generated by a finitely generated tilting module T. Then there is a torsion theory $(\mathcal{D}, \mathcal{C}_1)$ and if $\mathcal{C}_1 \subseteq \mathcal{C}$, then the following hold:*

(i) $(\mathcal{D}, \mathcal{C}_1)$ is a splitting torsion theory;

(ii) $\mathcal{C}_1$ consists of all modules of $\mathcal{C}$ which have no nonzero direct summands belonging to Add(T);

(iii) Let $\mathcal{D}_1$ denote the class of all modules that are the direct sum of a module of $\mathcal{D}$ and an Ext-injective module of $\mathcal{C}_1$, and let $\hat{\mathcal{C}}_1$ be the class of all modules that are direct sums of a module of $\mathcal{C}_1$ and a projective module. Then $(\hat{\mathcal{C}}_1, \mathcal{D}_1)$ is a cotorsion pair and, if it is generated by a (finitely generated) tilting module, then the associated torsion theory $(\mathcal{D}_1, \mathcal{C}_2)$ satisfies the condition $\mathcal{C}_2 \subseteq \mathcal{C}_1$.

PROOF. Suppose that $(\mathcal{C}, \mathcal{D})$ is a cotorsion pair generated by the finitely generated tilting module T. Then $\mathcal{D}$ is closed under direct sums (see, e.g., [**39**, Theorem 2.8]). Moreover, $\mathcal{D}$ is closed under epimorphic images and extensions due to the fact that R is left hereditary and $(\mathcal{C}, \mathcal{D})$ is a cotorsion pair. Then $\mathcal{D}$ is the torsion class for a torsion theory of R-Mod. By hypothesis, the torsion theory is $(\mathcal{D}, \mathcal{C}_1)$ with

$\mathcal{C}_1 \subseteq \mathcal{C}$. We show next that this is a splitting torsion theory. Take any module M and consider the short exact sequence $0 \to t(M) \to M \to \overline{M} \to 0$, where $t(M) \in \mathcal{D}$ and $\overline{M} \in \mathcal{C}_1$. Since $\mathcal{C}_1 \subseteq \mathcal{C}$, we have $\mathrm{Ext}^1_R(\overline{M}, t(M)) = 0$ and hence the given short exact sequence is split and $M \cong t(M) \oplus \overline{M}$, which shows that the torsion theory is splitting. This proves (i).

We set now to prove (ii). It is clear that all modules in $\mathcal{C}_1$ are modules of $\mathcal{C}$ which have no nonzero direct summands in $\mathcal{D}$, hence they have no nonzero direct summands in $\mathrm{Add}(T) = \mathcal{C} \cap \mathcal{D}$. Conversely, let $M \in \mathcal{C}$ be a module having no nonzero direct summands in $\mathrm{Add}(T)$. By the property (i), $M = M_1 \oplus M_2$ with $M_1 \in \mathcal{C}_1$ and $M_2 \in \mathcal{D}$. Since $\mathcal{C}$ is closed under direct summands, $M_2 \in \mathcal{C} \cap \mathcal{D} = \mathrm{Add}(T)$, hence $M_2 = 0$, from which it follows that $M \in \mathcal{C}_1$. This is (ii).

Let us set $\mathcal{D}' = (\mathcal{C}_1)^{\perp}$. Since $\mathcal{C}_1 \subseteq \mathcal{C}$ and $\mathcal{C}^{\perp} = \mathcal{D}$, we deduce that $\mathcal{D} \subseteq \mathcal{D}'$. Now, let $M \in \mathcal{D}'$. Since the torsion theory $(\mathcal{D}, \mathcal{C}_1)$ is splitting, we have $M = M_1 \oplus M_2$ where $M_1 \in \mathcal{C}_1$ and $M_2 \in \mathcal{D}$. Since $M \in (\mathcal{C}_1)^{\perp}$, we see that M_1 is a module of $\mathcal{C}_1$ which is Ext-injective in that class. This shows that $\mathcal{D}' = \mathcal{D}_1$, where $\mathcal{D}_1$ is the class of all modules that are the direct sum of a module of $\mathcal{D}$ and an Ext-injective module of $\mathcal{C}_1$. To show that $(\hat{\mathcal{C}}_1, \mathcal{D}_1)$ is a cotorsion pair, it remains only to see that ${}^{\perp}(\mathcal{D}_1) = \hat{\mathcal{C}}_1$.

By Lemma 2.1, $\mathcal{C} \cap \mathcal{D} = \mathrm{Add}(T)$ since the cotorsion pair $(\mathcal{C}, \mathcal{D})$ is generated by the tilting module T. Also, every module $M \in \mathcal{C}$ is the direct sum of a module in $\mathcal{C}_1$ and a module in $\mathcal{C} \cap \mathcal{D}$. Now, since $\mathcal{D} \subseteq \mathcal{D}_1$ and ${}^{\perp}\mathcal{D} = \mathcal{C}$, we have that ${}^{\perp}(\mathcal{D}_1) \subseteq \mathcal{C}$. On the other hand, $\mathcal{C}_1 \subseteq {}^{\perp}(\mathcal{D}_1)$ by the definition of $\mathcal{D}' = \mathcal{D}_1$. If ${}^{\perp}(\mathcal{D}_1) \neq \hat{\mathcal{C}}_1$, then there exists some module $X \in {}^{\perp}(\mathcal{D}_1)$ with $X \notin \hat{\mathcal{C}}_1$. Since $X = N \oplus L$ with $N \in \mathcal{D}, L \in \mathcal{C}_1$, it follows that $N \neq 0$ belongs to ${}^{\perp}(\mathcal{D}_1) \setminus \hat{\mathcal{C}}_1$ and $N \in \mathcal{C} \cap \mathcal{D}$. Therefore N belongs to the kernel $\mathcal{C} \cap \mathcal{D}$ of the cotorsion pair $(\mathcal{C}, \mathcal{D})$, hence by Lemma 3.3, N is the direct sum of finitely generated indecomposable modules which are isomorphic to direct summands of T. Thus we can assume that $N \in \mathcal{D}$ is nonzero indecomposable and not projective (since $N \notin \hat{\mathcal{C}}_1$). Then by Auslander's theorem (see [**9**, Theorem I.3.9]) there exists an almost split sequence in R-Mod

$$0 \to A \to X \to N \to 0$$

and N is an Ext-projective module of the class $\mathcal{D}$ (because $\mathcal{D} \subseteq \mathcal{D}_1$). This entails, by Lemma 2.2, that A is Ext-injective in the class $\mathcal{C}_1$, since $(\mathcal{D}, \mathcal{C}_1)$ is a torsion theory of R-Mod. Therefore $A \in \mathcal{D}_1$ and $N \in {}^{\perp}(\mathcal{D}_1)$, hence $\mathrm{Ext}^1_R(N, A) = 0$, which contradicts the existence of the almost split sequence above. This contradiction shows that ${}^{\perp}(\mathcal{D}_1) = \hat{\mathcal{C}}_1$.

To prove (iii) it remains to see that, if the cotorsion pair $(\hat{\mathcal{C}}_1, \mathcal{D}_1)$ is generated by a (finitely generated) tilting module, then $(\mathcal{D}_1, \mathcal{C}_2)$ is a torsion theory with $\mathcal{C}_2 \subseteq \mathcal{C}_1$. But this is trivial because $\mathcal{D} \subseteq \mathcal{D}_1$ and $(\mathcal{D}, \mathcal{C}_1)$ is a torsion theory, hence $\mathcal{C}_2 \subseteq \mathcal{C}_1$. This completes the proof of the proposition. □

PROPOSITION 3.5. *Let R be a left artinian hereditary ring such that every cotorsion pair in R-Mod is generated by a finitely generated tilting module. Let λ be a nonzero limit ordinal and suppose that there exist classes $\mathcal{C}_\beta, \mathcal{D}_\beta$ of left R-modules for each $\beta < \lambda$, so that the following conditions hold for any $\beta < \lambda$:*

(1) $(\mathcal{D}_\beta, \mathcal{C}_\beta)$ is a splitting torsion theory of R-Mod.

(2) If $\gamma \leq \beta$, then $\mathcal{C}_\beta \subseteq \mathcal{C}_\gamma$ and $\mathcal{D}_\gamma \subseteq \mathcal{D}_\beta$.

(3) If $\hat{\mathcal{C}}_\beta$ denotes the class of left R-modules which are the direct sum of a module of $\mathcal{C}_\beta$ and a projective module, then $(\hat{\mathcal{C}}_\beta, \mathcal{D}_{\beta+1})$ is a cotorsion pair.

(4) For any $M \in R$-Mod, let $t_\beta(M)$ denote the torsion submodule of M in the torsion theory $(\mathcal{D}_\beta, \mathcal{C}_\beta)$. There exist submodules U_β^M of M such that: (a) $U_\beta^M \in \mathcal{C}_\beta \cap \mathcal{D}_{\beta+1}$; (b) $t_{\beta+1}(M) = t_\beta(M) \oplus U_\beta^M$; (c) $t_\beta(M) = \oplus_{\gamma<\beta} U_\gamma^M$.

Then, if $\mathcal{C}_\lambda = \cap_{\beta<\lambda}\mathcal{C}_\beta$ and $\mathcal{D}_\lambda$ is the smallest torsion class containing $\cup_{\beta<\lambda}\mathcal{D}_\beta$, there exists a torsion class $\mathcal{D}_{\lambda+1}$ of left R-modules such that properties (1) to (4) hold also for $\beta = \lambda$.

PROOF. We define $\mathcal{C}$ as the intersection of all the classes $\mathcal{C}_\beta$, and $\mathcal{D}$ as the union of all the classes $\mathcal{D}_\beta$, with $\beta < \lambda$. Let $\overline{\mathcal{D}}$ be the smallest torsion class that contains $\mathcal{D}$. We denote by $\hat{\mathcal{C}}$ the class of all modules that are the direct sum of a module in $\mathcal{C}$ and a projective module. Then we will show that the following hold: (i) $(\overline{\mathcal{D}}, \mathcal{C})$ is a torsion theory of R-Mod. (ii) There is a class $\mathcal{D}'$ of left R-modules such that $(\hat{\mathcal{C}}, \mathcal{D}')$ is a cotorsion pair of R-Mod.

Let $Y \in R$-Mod be such that $\operatorname{Hom}_R(D, Y) = 0$ for every $D \in \mathcal{D}$. Thus $\operatorname{Hom}_R(D, Y) = 0$ for every $D \in \mathcal{D}_\beta$ and $\beta < \lambda$, so that $Y \in \mathcal{C}_\beta$ by assumption (1), and thus $Y \in \mathcal{C}$. Conversely, if $Y \in \mathcal{C}$, then $Y \in \mathcal{C}_\beta$ for each $\beta < \lambda$ so that $\operatorname{Hom}_R(D, Y) = 0$ for any $D \in \mathcal{D}_\beta$, from which it follows that $\operatorname{Hom}_R(D, Y) = 0$ for any $D \in \mathcal{D}$. By [**38**, Ch. VI, 2, Definition], $\mathcal{C}$ is a torsionfree class, whose corresponding torsion class is the smallest torsion class containing $\mathcal{D}$, i.e., $\overline{\mathcal{D}}$. This gives (i).

Starting now with the class ${}^\perp\mathcal{D}$ and using assumption (3) instead of (1), one may show in an analogous manner that $\operatorname{Ext}_R^1(X, D) = 0$ for every $D \in \mathcal{D}$ if and only if $X \in \cap_{\beta<\lambda}\hat{\mathcal{C}}_\beta$. Therefore $\mathcal{C} \subseteq {}^\perp\mathcal{D}$, whence $\hat{\mathcal{C}} \subseteq {}^\perp\mathcal{D}$. We want to show that ${}^\perp\mathcal{D} = \hat{\mathcal{C}}$, thus it remains to see that any module of ${}^\perp\mathcal{D}$ is the direct sum of a module of $\mathcal{C}$ and a projective module.

To this end, suppose that $X \in {}^\perp\mathcal{D}\backslash\mathcal{C}$. Since $X \notin \mathcal{C}$, there is by (2) a smallest $\beta < \lambda$ such that $X \notin \mathcal{C}_\beta$. But $X \in \hat{\mathcal{C}}_\beta$ from which it follows that $X = X_1 \oplus P_1$ with $X_1 \in \mathcal{C}_\beta$ and P_1 nonzero projective; moreover, no nonzero direct summand of P_1 belongs to $\mathcal{C}_\beta$. If $X_1 \notin \mathcal{C}$, then there exists a smallest $\gamma < \lambda$ such that $X_1 \notin \mathcal{C}_\gamma$ (hence $\beta < \gamma$ by (2)), but $X_1 = X_2 \oplus P_2$ with $X_2 \in \mathcal{C}_\gamma$ and P_2 nonzero projective and with each nonzero direct summand of P_2 not in $\mathcal{C}_\gamma$. Thus $X = X_2 \oplus (P_1 \oplus P_2)$ and since $P_2 \in \mathcal{C}_\beta$, each nonzero direct summand of P_2 is not isomorphic to any direct summand of P_1. Since every projective module is a direct sum of finitely presented modules and the number of isomorphism classes of indecomposable projective modules is finite, the above process is necessarily finite, and there is some decomposition $X = X_h \oplus P$ with P projective and $X_h \in \mathcal{C}$. Therefore, every module of ${}^\perp\mathcal{D}$ is the direct sum of a module in $\mathcal{C}$ and a projective module. This means that ${}^\perp\mathcal{D} = \hat{\mathcal{C}}$ and $(\hat{\mathcal{C}}, \mathcal{D}')$ is a cotorsion pair, with $\mathcal{D} \subseteq \mathcal{D}'$, showing (ii).

Now, we take $\mathcal{C}_\lambda = \mathcal{C}$, $\mathcal{D}_\lambda = \overline{\mathcal{D}}$ and $\mathcal{D}_{\lambda+1} = \mathcal{D}'$ so that conditions (2) and (3) are satisfied for all $\beta \leq \lambda$. We next show (1), i.e., that the torsion theory $(\overline{\mathcal{D}}, \mathcal{C})$ is splitting.

By [**39**, Theorem 2.8], $\mathcal{D}'$ is the torsion class for a torsion theory $(\mathcal{D}', \mathcal{C}')$ of R-Mod because $(\hat{\mathcal{C}}, \mathcal{D}')$ is a cotorsion pair which is generated by a tilting module. Since $\mathcal{D} \subseteq \mathcal{D}'$, we have then $\overline{\mathcal{D}} \subseteq \mathcal{D}'$.

Let M be any module. For any $\beta < \lambda$, $t_\beta(M)$ is the torsion submodule of M in the torsion theory $(\mathcal{D}_\beta, \mathcal{C}_\beta)$. By (1) there are split short exact sequences

$0 \to t_\beta(M) \to M \to C_\beta^M \to 0$, where $C_\beta^M \in \mathcal{C}_\beta$. Moreover, the inclusions $t_\beta(M) \to t_\alpha(M)$ for $\beta < \alpha$ (due to (2)) determine projections $C_\beta^M \to C_\alpha^M$ for $\beta < \alpha$; this results in a direct system of split short exact sequences. By the exactness of the direct limit functor, we get a short exact sequence

$$0 \to \varinjlim_{\beta<\lambda} t_\beta(M) \to M \to \varinjlim_{\beta<\lambda} C_\beta^M \to 0 \tag{3.1}$$

which is pure because it is a direct limit of split sequences.

Let $\theta < \lambda$, $X \in \mathcal{D}_\theta$ and let $f : X \to \varinjlim C_\beta^M$ be a homomorphism. By (c) of assumption (4), $X = t_\theta(X)$ is a direct sum of modules belonging to the sets $\mathcal{U}_\rho = \mathcal{C}_\rho \cap \mathcal{D}_{\rho+1} \subseteq \mathrm{Add}(W_\rho)$ provided W_ρ is the finitely generated tilting module generating the cotorsion pair $(\hat{\mathcal{C}}_\rho, \mathcal{D}_{\rho+1})$. By Lemma 3.3 each of these modules U_ρ^M is pure-projective and consequently X is pure-projective. Since the sequence (3.1) is pure, f can be lifted to $g : X \to M$. Since $M = t_\theta(M) \oplus C_\theta^M$ and X is torsion in the theory $(\mathcal{D}_\theta, \mathcal{C}_\theta)$, it follows that the image of g is contained in $t_\theta(M)$. Therefore, $g : X \to M$ factors through $\varinjlim t_\beta(M) = \sum_{\beta<\lambda} t_\beta(M)$ and consequently $f = 0$. This means that $\varinjlim C_\beta^M \in \mathcal{C}$ because it belongs to $\mathcal{C}_\theta$ for any $\theta < \lambda$. But $\varinjlim t_\beta(M) \in \overline{\mathcal{D}}$ because any torsion class is closed under direct limits. Since $(\hat{\mathcal{C}}, \mathcal{D}')$ is a cotorsion pair, $\overline{\mathcal{D}} \subseteq \mathcal{D}'$ and $\varinjlim C_\beta^M \in \mathcal{C} \subseteq \hat{\mathcal{C}}$, we infer that $\mathrm{Ext}_R^1(\varinjlim C_\beta^M, \varinjlim t_\beta(M)) = 0$ and hence the sequence (3.1) is split. This gives that M is the direct sum of a module of $\overline{\mathcal{D}}$ and a module of $\mathcal{C}$, hence the torsion theory $(\overline{\mathcal{D}}, \mathcal{C})$ is splitting, proving condition (1). This also shows that $t_\lambda(M) = \varinjlim t_\beta(M) = \sum_{\beta<\lambda} t_\beta(M)$. But then $t_\lambda(M) = \sum_{\beta<\lambda}(\oplus_{\gamma<\beta} U_\gamma^M) = \oplus_{\beta<\lambda} U_\beta^M$, by assumption (4).

We set now to prove (4) for M and λ. By (1), $M = t_\lambda(M) \oplus C$ with $C \in \mathcal{C}_\lambda$. Since $\mathcal{D}_\lambda \subseteq \mathcal{D}' = \mathcal{D}_{\lambda+1}$, $t_\lambda(M) \subseteq t_{\lambda+1}(M)$ and hence we may find U_λ^M so that $t_{\lambda+1}(M) = t_\lambda(M) \oplus U_\lambda^M$. We choose the collection of all U_β^M for $\beta \leq \lambda$ by adding U_λ^M to the sequence existing already by the inductive assumption. With this choice, the above two equalities prove (4)(b) directly and (4)(a) because U_λ^M is a direct summand of $t_{\lambda+1}(M)$ and is isomorphic to a submodule of C. Finally, (4)(c) is the last equality of the preceding paragraph. □

The next result follows from a construction by transfinite recursion.

PROPOSITION 3.6. *Let R be a left artinian hereditary ring such that every cotorsion pair in R-Mod is generated by a finitely generated tilting module. For each ordinal α, there is a 3-tuple $(\mathcal{D}_\alpha, \mathcal{C}_\alpha, \mathcal{D}_{\alpha+1})$ of classes of left R-modules, so that the following properties are satisfied.*

(1) $(\mathcal{D}_\alpha, \mathcal{C}_\alpha)$ is a splitting torsion theory of R-Mod for any α.

(2) For any pair of ordinals $\beta \leq \alpha$, $\mathcal{D}_\beta \subseteq \mathcal{D}_\alpha$ and $\mathcal{C}_\alpha \subseteq \mathcal{C}_\beta$.

(3) For any α, $(\hat{\mathcal{C}}_\alpha, \mathcal{D}_{\alpha+1})$ is a cotorsion pair, where $\hat{\mathcal{C}}_\alpha$ is the class of left R-modules which are the direct sum of a module of $\mathcal{C}_\alpha$ and a projective module.

(4) For every left R-module M and ordinal α, let $t_\alpha(M)$ denote the torsion submodule of M in the torsion theory $(\mathcal{D}_\alpha, \mathcal{C}_\alpha)$. Then there exists a submodule U_α^M of M for each α, so that: (a) $U_\alpha^M \in \mathcal{C}_\alpha \cap \mathcal{D}_{\alpha+1}$; (b) $t_{\alpha+1}(M) = t_\alpha(M) \oplus U_\alpha^M$; and (c) $t_\alpha(M) = \oplus_{\beta<\alpha} U_\beta^M$.

PROOF. The proof will consist in defining, by transfinite induction, the classes $\mathcal{D}_\alpha, \mathcal{C}_\alpha, \mathcal{D}_{\alpha+1}$ for each ordinal α; in such a way that, if we admit inductively that the classes exist and satisfy conditions (1) to (4) for all ordinals $\beta < \alpha$, then $\mathcal{D}_\alpha, \mathcal{C}_\alpha, \mathcal{D}_{\alpha+1}$ are defined and conditions (1) to (4) are satisfied for all ordinals $\leq \alpha$ (but we must understand that the meaning of "condition (2) is satisfied for all ordinals $\leq \alpha$" is: if $\gamma < \beta \leq \alpha + 1$, then $\mathcal{D}_\gamma \subseteq \mathcal{D}_\beta$; and if $\gamma < \beta \leq \alpha$, then $\mathcal{C}_\beta \subseteq \mathcal{C}_\gamma$).

(i) Let $\alpha = 0$. Then we set $\mathcal{C}_0 = R$-Mod, $\mathcal{D}_0 = \{0\}$ and $\mathcal{D}_1$ is the class of injective modules. We must show that properties (1) to (4) hold for all ordinals ≤ 0. This is obvious for (1) and (3). The condition (2) reduces to $\mathcal{D}_0 \subseteq \mathcal{D}_1$ and is equally obvious. For (4), note that $t_0(M) = 0$ and $t_1(M)$ is the largest injective submodule of M, so $t_1(M) = U_0^M$.

(ii) Let $\alpha = \beta + 1$ a successor ordinal and assume that the classes $\mathcal{D}_\rho, \mathcal{C}_\rho, \mathcal{D}_{\rho+1}$ are defined up to $\rho = \beta$ and satisfy the corresponding properties (1) to (4). Then $(\hat{\mathcal{C}}_\beta, \mathcal{D}_\alpha)$ is a cotorsion pair which is generated by a finitely generated tilting module by hypothesis. As before, $\mathcal{D}_\alpha$ is a torsion class and there is a torsion theory $(\mathcal{D}_\alpha, \mathcal{C}_\alpha)$ so that we define $\mathcal{C}_\alpha$ from this property. Since $\mathcal{D}_\beta \subseteq \mathcal{D}_\alpha$ by assumption (2), $\mathcal{C}_\alpha \subseteq \mathcal{C}_\beta \subseteq \hat{\mathcal{C}}_\beta$ and we may apply Proposition 3.4 to obtain $\mathcal{D}_{\alpha+1}$. This construction is functional and it defines the inductive step for successor ordinals. Then properties (1) and (3) follow by Proposition 3.4 and the induction assumptions. For (2) we have to check that $\mathcal{C}_\alpha \subseteq \mathcal{C}_\beta$ which is known, and consequently $\mathcal{D}_\alpha \subseteq \mathcal{D}_{\alpha+1}$. The condition (4) now reduces to the equalities $t_{\alpha+1}(M) = t_\alpha(M) \oplus U_\alpha^M$ and $t_\alpha(M) = \oplus_{\gamma \leq \beta} U_\gamma^M$, since (a) is a consequence of these as seen in the proof of this item for Proposition 3.5. As before, the first equation follows because $t_\alpha(M)$ is a direct summand of M. Also, $t_\alpha(M) = t_\beta(M) \oplus U_\beta^M = (\oplus_{\gamma<\beta} U_\gamma^M) \oplus U_\beta^M$, so (c) follows.

(iii) Let α be a nonzero limit ordinal and assume that the classes $\mathcal{C}_\beta, \mathcal{D}_\beta$ are defined for $\beta < \alpha$ with properties (1) to (4). Then Proposition 3.5 shows that we can define the classes $\mathcal{D}_\lambda, \mathcal{C}_\lambda, \mathcal{D}_{\lambda+1}$ so that (1) to (4) hold for all $\beta \leq \lambda$. □

We can now state and prove the main result of our paper, giving necessary and sufficient conditions for a ring to be left pure semisimple hereditary. The necessity part of the result is due to Angeleri Hügel [**2**].

THEOREM 3.7. *Let R be a ring. Then R is left pure semisimple hereditary if and only if every cotorsion pair in R-Mod is generated by a finitely generated tilting module.*

PROOF. The necessity is proven in [**2**, Proposition 4.2] (the ring R in [**2**, Proposition 4.2] is assumed to be left pure semisimple hereditary indecomposable, but the indecomposable condition of the ring R is not needed in the proof in [**2**] of this result).

So we have to prove only the sufficiency. Assume that every cotorsion pair in R-Mod is generated by a finitely generated tilting module. First, it follows from Proposition 3.1 (cf. Remark 3.2) that R is left artinian hereditary. Our goal is to show that the family R-ind of all finitely generated indecomposable left R-modules has an Ext-injective partition, and the left pure semisimplicity of R would follow from Proposition 2.4. To this end, we consider the classes $\mathcal{C}_\alpha, \mathcal{D}_\alpha$ for ordinals α of Proposition 3.6 satisfying properties (1) to (4). We will also use the notation introduced in that proposition.

Take $\mathcal{U}_\alpha = \mathcal{C}_\alpha \cap \mathcal{D}_{\alpha+1}$ for each ordinal α, and set $\mathcal{W}_\alpha = \text{ind}(\mathcal{U}_\alpha)$. We note first that $\mathcal{W}_\alpha \cap \mathcal{W}_\beta = \emptyset$ if $\alpha \neq \beta$. For suppose $\beta < \alpha$ and $M \in \mathcal{W}_\alpha \cap \mathcal{W}_\beta$. Then $M \in \mathcal{C}_\alpha$ so that $M \in \mathcal{C}_{\beta+1}$ by condition (3). Also, $M \in \mathcal{U}_\beta$ and therefore $M \in \mathcal{D}_{\beta+1}$. But $\mathcal{C}_{\beta+1} \cap \mathcal{D}_{\beta+1} = \{0\}$, which gives a contradiction.

We know that $(\hat{\mathcal{C}}_\alpha, \mathcal{D}_{\alpha+1})$ is a cotorsion pair generated by a finitely generated tilting module T, and then $\mathcal{U}_\alpha \subseteq \mathrm{Add}(T)$ by Lemma 2.1 and every module of $\mathcal{U}_\alpha$ is a direct sum of finitely generated indecomposable modules which are isomorphic to direct summands of T. Therefore each module of $\mathcal{W}_\alpha$ is isomorphic to a direct summand of T and $\mathcal{W}_\alpha$ is a finite set.

We consider next when is $\mathcal{W}_\alpha = \emptyset$. Since each module of $\mathcal{U}_\alpha$ is a direct sum of modules belonging to $\mathcal{W}_\alpha$, $\mathcal{W}_\alpha = \emptyset$ if and only if $\mathcal{U}_\alpha$ has only the zero module. But $\mathcal{U}_\alpha = \mathcal{C}_\alpha \cap \mathcal{D}_{\alpha+1} \subseteq \mathrm{Add}(T) = \hat{\mathcal{C}}_\alpha \cap \mathcal{D}_{\alpha+1}$, hence $\mathcal{W}_\alpha = \emptyset$ if and only if no indecomposable direct summand of T belongs to $\mathcal{C}_\alpha$, i.e., T is projective. By Valenta's result ([**40**]; cf. [**12**, Corollary 3.7.5]), T is projective if and only if T is a projective generator. On the other hand, an indecomposable direct summand of T belongs either to $\mathcal{C}_\alpha$ or to $\mathcal{D}_\alpha$. Therefore $\mathcal{W}_\alpha = \emptyset$ if and only if $R \in \mathcal{D}_\alpha$ and, since $\mathcal{D}_\alpha$ is a torsion class, $\mathcal{W}_\alpha = \emptyset \Leftrightarrow \mathcal{D}_\alpha = R\text{-Mod}$. Since R-ind is a set and the sets $\mathcal{W}_\alpha$ are pairwise disjoint, $\mathcal{W}_\alpha$ is empty for some α. There is then some first ordinal ρ satisfying $\mathcal{D}_\rho = R\text{-Mod}$, and $\mathcal{D}_\alpha = R\text{-Mod}$ if $\rho \leq \alpha$. Thus if $M \in R\text{-ind}$, $M = t_\rho(M)$ and, by (4) of Proposition 3.6, $M = U_\alpha^M \in \mathcal{U}_\alpha$ for some $\alpha < \rho$. This proves that $\bigcup_{\beta<\rho} \mathcal{W}_\beta = R\text{-ind}$.

Since $\mathcal{W}_\alpha \neq \emptyset$ if $\alpha < \rho$, it only remains to show that $\mathcal{W}_\alpha$ is the set of all indecomposable finitely generated Ext-injective modules of $\bigcup_{\beta \geq \alpha} \mathcal{W}_\beta$ in order to use Definition 2.3 and prove that R-ind has an Ext-injective partition. By Proposition 2.4, it will follow that R is left pure semisimple.

First, if $M \in \mathcal{W}_\alpha$, then $M \in \mathcal{D}_{\alpha+1}$ hence M is Ext-injective in $\mathcal{C}_\alpha$ and, forcely, Ext-injective in $\bigcup_{\beta \geq \alpha} \mathcal{W}_\beta$. Conversely, let $X \in \mathcal{W}_\delta$ for some $\delta > \alpha$. Then $X \in \mathcal{C}_\delta \subseteq \mathcal{C}_\alpha$, but $X \notin \mathcal{D}_{\alpha+1}$ (because otherwise $X \in \mathcal{W}_\alpha$) and consequently X is not Ext-injective in $\mathcal{C}_\alpha$. Therefore there exists $Y \in \mathcal{C}_\alpha$ such that $\mathrm{Ext}^1_R(Y, X) \neq 0$. Let $Y = t_\delta(Y) \oplus C$ with $C \in \mathcal{C}_\delta$. Since X is Ext-injective in $\mathcal{C}_\delta$, $\mathrm{Ext}^1_R(C, X) = 0$ and hence $\mathrm{Ext}^1_R(t_\delta(Y), X) \neq 0$. By (c) of condition (4) of Proposition 3.6, there is $\beta < \delta$ such that $\mathrm{Ext}^1_R(U^Y_\beta, X) \neq 0$. If $\beta < \alpha$ then $U^Y_\beta \in \mathcal{D}_\alpha$ and we would have a nonzero direct summand of Y which belongs to $\mathcal{D}_\alpha \cap \mathcal{C}_\alpha$ and this is absurd. Hence $\beta \geq \alpha$ and $U^Y_\beta \in \bigcup_{\gamma \geq \alpha} \mathcal{U}_\gamma$. Any indecomposable direct summand of U^Y_β gives then an element of $\bigcup_{\gamma \geq \alpha} \mathcal{W}_\gamma$ and hence X is not Ext-injective in $\bigcup_{\gamma \geq \alpha} \mathcal{W}_\gamma$, which shows our claim. This completes the proof. □

We conclude the paper with some remarks.

REMARK 3.8. Left hereditary rings R over which each cotorsion pair in R-Mod is generated by a (possibly infinitely generated) tilting module have recently been studied by Angeleri Hügel, Šaroch and Trlifaj [**5**, Section 6]. It was remarked in [**5**, p. 13] that it was not known if the condition that every cotorsion pair in R-Mod is generated by a finitely generated tilting module is left-right symmetric, and it was shown in [**5**, Proposition 6.3] that the left-right symmetry of this condition over left hereditary rings actually would imply the validity of the pure semisimplicity conjecture. Our result indeed shows that the left-right symmetry of this condition is in fact *equivalent* to the pure semisimplicity conjecture.

REMARK 3.9. By results of Herzog [**24**], the pure semisimplicity conjecture is true if and only if it is true for all hereditary rings. It was proven by Simson [**33**] (cf. [**34**]) that the pure semisimplicity conjecture is true for all hereditary rings if and only if it is true for all triangular matrix rings $R = \begin{pmatrix} F & 0 \\ B & G \end{pmatrix}$ where F, G are division rings and B is a G-F-bimodule.

As pointed out by Simson [**34, 35**], if there exists an extension $G \subsetneq F$ of division rings such that the right G-dimension of F is infinite, the left G-dimension of F is 2, and the sequence of left dimensions of the successive left dual bimodules $F^* = \mathrm{Hom}_G(F, G)$, $F^{**} = \mathrm{Hom}_F(F^*, F)$, etc., is constantly equal to 2, then the triangular matrix ring $R = \begin{pmatrix} F & 0 \\ F & G \end{pmatrix}$ is a counter-example to the pure semisimplicity conjecture. The problem of the existence of such an extension is a (more difficult) version of Artin's problem, which was solved by Cohn [**11**] (see also Schofield [**30**]). Let $G \subsetneq F$ be the quadratic division ring extension, constructed by Cohn [**11**, Sec. 5.6], such that $\dim({}_G F) = 2$ and $\dim(F_G)$ is infinite, and let $R = \begin{pmatrix} F & 0 \\ F & G \end{pmatrix}$.

Then, as shown by Colpi and Trlifaj [**14**, Example 2.12] (see also [**13**, Sec. 3]), there is an injective left R-module P over the left artinian hereditary ring R such that $P^\perp$ is not generated by a finitely generated tilting module. Thus $({}^\perp(P^\perp), P^\perp)$ is a cotorsion pair, which is not generated by a finitely generated tilting module, hence R is not a left pure semisimple ring by Theorem 3.7 (indeed this follows already from [**2**, Proposition 4.2]).

Acknowledgements. A part of the results in this paper was obtained during the visit of the first-named author at University of Murcia in April 2016. He would like to express his warmest thanks to his host, Professor José Luis García, the members of the Algebra group of the Department of Mathematics, and University of Murcia for their kind invitation, hospitality, and support. The second-named author thanks the Spanish Ministry of Education and Science for support (MTM2013-46837-P). Both authors would like to thank the referee for his/her careful reading of the paper, and for several helpful comments and suggestions (particularly for suggesting Remark 3.2).

References

[1] F. W. Anderson and K. R. Fuller, *Rings and categories of modules*, 2nd ed., Graduate Texts in Mathematics, vol. 13, Springer-Verlag, New York, 1992. MR1245487

[2] L. Angeleri Hügel, *A key module over pure-semisimple hereditary rings*, J. Algebra **307** (2007), no. 1, 361–376. MR2278060

[3] L. Angeleri Hügel and F. U. Coelho, *Infinitely generated tilting modules of finite projective dimension*, Forum Math. **13** (2001), no. 2, 239–250. MR1813669

[4] L. Angeleri Hügel and D. Herbera, *Auslander-Reiten components over pure-semisimple hereditary rings*, J. Algebra **331** (2011), no. 1, 285–303. MR2774659

[5] L. Angeleri Hügel, J. Šaroch and J. Trlifaj, *Approximations and Mittag-Leffler conditions—the applications*, preprint 2016 (arXiv:1612.01140) (to appear in Israel J. Math.).

[6] L. Angeleri Hügel and H. Valenta, *A duality result for almost split sequences*, Colloq. Math. **80** (1999), no. 2, 267–292. MR1703897

[7] I. Assem, D. Simson, and A. Skowroński, *Elements of the representation theory of associative algebras. 1: Techniques of representation theory*, London Mathematical Society Student Texts, vol. 65, Cambridge University Press, Cambridge, 2006. MR2197389

[8] M. Auslander, *Representation theory of Artin algebras. I, II*, Comm. Algebra **1** (1974), no. 3, 177–268; ibid. 1 (1974), 269–310. MR0349747

[9] M. Auslander, *Functors and morphisms determined by objects*, Representation theory of algebras (Proc. Conf., Temple Univ., Philadelphia, Pa., 1976), Dekker, New York, 1978, pp. 1–244. Lecture Notes in Pure Appl. Math., Vol. 37. MR0480688

[10] M. Auslander, I. Reiten, and S. O. Smalø, *Representation theory of Artin algebras*, Cambridge Studies in Advanced Mathematics, vol. 36, Cambridge University Press, Cambridge, 1995. MR1314422

[11] P. M. Cohn, *Skew field constructions*, London Mathematical Society Lecture Note Series, vol. 27, Cambridge University Press, Cambridge-New York-Melbourne, 1977. MR0463237

[12] R. R. Colby and K. R. Fuller, *Equivalence and duality for module categories with tilting and cotilting for rings*, Cambridge Tracts in Mathematics, vol. 161, Cambridge University Press, Cambridge, 2004. MR2048277

[13] R. Colpi and J. Trlifaj, *Classes of generalized $*$-modules*, Comm. Algebra **22** (1994), no. 10, 3985–3995. MR1280103

[14] R. Colpi and J. Trlifaj, *Tilting modules and tilting torsion theories*, J. Algebra **178** (1995), no. 2, 614–634. MR1359905

[15] N. V. Dung and J. L. García, *Definable subcategories over pure semisimple rings*, J. Algebra Appl. **11** (2012), no. 5, 1250099, 24 pp. MR2983191

[16] N. V. Dung and J. L. García, *Indecomposable modules over pure semisimple hereditary rings*, J. Algebra **371** (2012), no. 1, 577–595. MR2975414

[17] N. V. Dung and J. L. García, *Splitting torsion pairs over pure semisimple rings*, J. Pure Appl. Algebra **219** (2015), no. 7, 2637–2657. MR3313500

[18] C. Faith and E. A. Walker, *Direct-sum representations of injective modules*, J. Algebra **5** (1967), no. 2, 203–221. MR0207760

[19] K. R. Fuller, *On rings whose left modules are direct sums of finitely generated modules*, Proc. Amer. Math. Soc. **54** (1976), no. 1, 39–44. MR0393133

[20] K. R. Fuller and I. Reiten, *Note on rings of finite representation type and decompositions of modules*, Proc. Amer. Math. Soc. **50** (1975), no. 1, 92–94. MR0376768

[21] J. L. García, *On hereditary Artinian rings and the pure semisimplicity conjecture: rigid tilting modules and a weak conjecture*, Colloq. Math. **135** (2014), no. 2, 227–262. MR3229419

[22] J. L. García, *On hereditary rings and the pure semisimplicity conjecture II: sporadic potential counterexamples*, Colloq. Math. **139** (2015), no. 1, 55–93. MR3332734

[23] R. Göbel and J. Trlifaj, *Approximations and endomorphism algebras of modules*, De Gruyter Expositions in Mathematics, vol. 41, Walter de Gruyter GmbH & Co. KG, Berlin, 2006. MR2251271

[24] I. Herzog, *A test for finite representation type*, J. Pure Appl. Algebra **95** (1994), no. 2, 151–182. MR1293051

[25] M. Hoshino, *On splitting torsion theories induced by tilting modules*, Comm. Algebra **11** (1983), no. 4, 427–439. MR689417

[26] B. Huisgen-Zimmermann, *Purity, algebraic compactness, direct sum decompositions, and representation type*, Infinite length modules (Bielefeld, 1998), Trends Math., Birkhäuser, Basel, 2000, pp. 331–367. MR1789225

[27] B. Zimmermann-Huisgen and W. Zimmermann, *On the sparsity of representations of rings of pure global dimension zero*, Trans. Amer. Math. Soc. **320** (1990), no. 2, 695–711. MR965304

[28] C. M. Ringel and H. Tachikawa, QF $-$ 3 *rings*, J. Reine Angew. Math. **272** (1974), 49–72. MR0379578

[29] L. Salce, *Cotorsion theories for abelian groups*, Symposia Mathematica, Vol. XXIII (Conf. Abelian Groups and their Relationship to the Theory of Modules, INDAM, Rome, 1977), Academic Press, London-New York, 1979, pp. 11–32. MR565595

[30] A. H. Schofield, *Representation of rings over skew fields*, London Mathematical Society Lecture Note Series, vol. 92, Cambridge University Press, Cambridge, 1985. MR800853

[31] D. Simson, Pure semisimple categories and rings of finite representation type, J. Algebra **48** (1977), no. 2, 290-296. MR0460386. Corrigendum J. Algebra **67** (1980), no. 1, 254-256. MR595031

[32] D. Simson, *On pure semisimple Grothendieck categories. II*, Fund. Math. **110** (1980), no. 2, 107–116. MR600584

[33] D. Simson, *Partial Coxeter functors and right pure semisimple hereditary rings*, J. Algebra **71** (1981), no. 1, 195–218. MR627434

[34] D. Simson, *On right pure semisimple hereditary rings and an Artin problem*, J. Pure Appl. Algebra **104** (1995), no. 3, 313–332. MR1361578

[35] D. Simson, *An Artin problem for division ring extensions and the pure semisimplicity conjecture. I*, Arch. Math. (Basel) **66** (1996), no. 2, 114–122. MR1367152

[36] D. Simson, *A class of potential counterexamples to the pure semisimplicity conjecture*, Advances in Algebra and Model Theory (Essen, 1994; Dresden, 1995), Algebra Logic Appl., vol. 9, Gordon and Breach, Amsterdam, 1997, pp. 345–373. MR1683552

[37] D. Simson, *An Artin problem for division ring extensions and the pure semisimplicity conjecture. II*, J. Algebra **227** (2000), no. 2, 670–705. MR1759841

[38] B. Stenström, *Rings of quotients. An introduction to methods of ring theory*, Die Grundlehren der Mathematischen Wissenschaften, Band 217, Springer-Verlag, New York-Heidelberg, 1975. MR0389953.

[39] J. Trlifaj, *Infinite dimensional tilting modules and cotorsion pairs*, Handbook of Tilting Theory, London Math. Soc. Lecture Note Ser., vol. 332, Cambridge Univ. Press, Cambridge, 2007, pp. 279–321. MR2384614

[40] H. Valenta, *Existence criteria and construction methods for certain classes of tilting modules*, Comm. Algebra **22** (1994), no. 15, 6047–6071. MR1302993

[41] R. B. Warfield Jr., *Decompositions of injective modules*, Pacific J. Math. **31** (1969), no. 1, 263–276. MR0249467

Department of Mathematics, Ohio University-Zanesville, Zanesville, OH 43701, USA

Email address: nguyend2@ohio.edu

Department of Mathematics, University of Murcia, 30071 Murcia, Spain

Email address: jlgarcia@um.es

Contemporary Mathematics
Volume **715**, 2018
https://doi.org/10.1090/conm/715/14409

Twist invariants of graded algebras

K. R. Goodearl and M. T. Yakimov

ABSTRACT. We define two invariants for (semiprime right Goldie) algebras, one for algebras graded by arbitrary abelian groups, which is unchanged under twists by 2-cocycles on the grading group, and one for $\mathbb{Z}$-graded or $\mathbb{Z}_{\geq 0}$-filtered algebras. The first invariant distinguishes quantum algebras which are "truly multiparameter" apart from ones that are "essentially uniparameter", meaning cocycle twists of uniparameter algebras. We prove that both invariants are stable under adjunction of polynomial variables. Methods for computing these invariants for large families of algebras are given, including quantum nilpotent algebras and algebras admitting one quantum cluster, and applications to non-isomorphism theorems are obtained.

1. Introduction

We introduce and investigate two interrelated invariants for graded or filtered algebras. The first concerns algebras A graded by abelian groups Γ, and takes account of the related algebras obtained by twisting A with respect to 2-cocycles on Γ. The resulting invariant, which we denote $\mathrm{tw}_\Gamma(A)$, is a cocycle twist invariant, and helps to separate algebras which are not twist equivalent. We also apply it to distinguish between quantum algebras which are basically uniparameter – meaning the scalars involved in commutation relations reduce to powers of a single scalar – and those which are truly multiparameter, in that their commutation scalars under any cocycle twist do not form a cyclic group. Our second invariant concerns $\mathbb{Z}$-graded or $\mathbb{Z}_{\geq 0}$-filtered algebras A and applies the invariant tw_Γ to a suitably maximal grading group Γ, obtained as the character group of a maximal torus in an affine algebraic group of graded or filtered automorphisms. The resulting invariant, denoted $\mathrm{tw}(A)$, is used to help separate algebras which are not isomorphic as $\mathbb{Z}$-graded or $\mathbb{Z}_{\geq 0}$-filtered algebras, respectively. These invariants, and a more basic invariant of Alev and Dumas [**1**] on which they are founded, are in many circumstances stable under polynomial extensions. Consequently, they allow to prove stability results saying that certain algebras cannot be made isomorphic even after adjoining arbitrarily many polynomial indeterminates.

2010 *Mathematics Subject Classification.* Primary 16K40; Secondary 16S85, 16T20, 17B37, 13F60.

Key words and phrases. Locally finite graded and filtered algebras, 2-cocycle twists of algebras, Alev-Dumas invariants, stable invariants, maximal tori of automorphism groups.

The research of K.R.G. was partially supported by NSF grant DMS-1601184, and that of M.T.Y. by NSF grant DMS-1601862.

Our original motivation for these invariants was to attack the question "uniparameter or multiparameter?" for classes of quantum algebras, such as quantum nilpotent algebras. We do so by adapting an invariant that was originally introduced by Alev and Dumas in [**1**] to tell quantum affine spaces and their division rings of fractions apart. We put their invariant into the following context: For any semiprime right Goldie algebra A over a field $\mathbb{K}$, we define $\mathrm{AD}(A)$ to be the intersection of $\mathbb{K}^*$ with the commutator subgroup of the group of units of the right Goldie quotient ring Fract A. If, additionally, A is graded by an abelian group Γ, we define the "Γ-twist invariant" $\mathrm{tw}_\Gamma(A)$ to be the intersection of the AD-invariants of all cocycle twists of A. For instance, consider a quantum affine space $A = \mathcal{O}_{\boldsymbol{q}}(\mathbb{K}^n)$, for some multiplicatively skew-symmetric matrix $\boldsymbol{q} = (q_{ij}) \subseteq M_n(\mathbb{K}^*)$. In this case, $\mathrm{AD}(A)$ is the subgroup of $\mathbb{K}^*$ generated by all q_{ij} [**1**, Proposition 3.9]. However, when A is given its natural $\mathbb{Z}^n$-grading, $\mathrm{tw}_{\mathbb{Z}^n}(A)$ is the trivial group, due to the fact that A can be written as a cocycle twist of a commutative polynomial ring. This is an extreme case, since quantum affine spaces and their graded subalgebras are essentially the only quantum algebras that can be obtained from commutative algebras via cocycle twists. In the case of multiparameter quantum matrix algebras $A = \mathcal{O}_{\lambda,\boldsymbol{p}}(M_n(\mathbb{K}))$, the invariant $\mathrm{AD}(A)$ again records all the relevant commutation scalars (λ and the entries of $\boldsymbol{p}$) whereas $\mathrm{tw}_\Gamma(A)$ is the cyclic group generated by λ (assuming A is graded by $\Gamma = \mathbb{Z}^{2n}$ in the natural way); see Example 5.8. On the other hand, the multiparameter quantized Weyl algebras $A = A_n^{Q,P}(\mathbb{K})$ are typically truly multiparameter: with respect to the natural $\mathbb{Z}^n$-grading, $\mathrm{tw}_{\mathbb{Z}^n}(A)$ is the group generated by the entries of the vector Q (Example 4.7).

From the perspective of $\mathbb{Z}$-graded algebras, the twist invariant may not provide a lot of new information, due to the paucity of "new commutators" when twisting by a cocycle on $\mathbb{Z}$ – in particular, multiplicative commutators of homogeneous elements do not change under cocycle twists in this setting. However, there are many algebras for which $\mathbb{Z}$-gradings can be extended to larger grading groups. We develop a procedure for obtaining gradings by abelian groups which are maximal in a suitable sense. Specifically, if an algebra A is affine (finitely generated as an algebra) and has a locally finite dimensional $\mathbb{Z}$-grading, the group of graded algebra automorphisms of A has a natural structure of affine algebraic group, and A receives a grading from the character group $X(\mathcal{H})$ of any maximal torus $\mathcal{H}$ of $\mathrm{Aut}_{\mathrm{gr}} A$. We define the "general twist invariant" of A to be $\mathrm{tw}(A) = \mathrm{tw}_{X(\mathcal{H})}(A)$ which, as we observe, is independent of the choice of $\mathcal{H}$ (Lemma 6.5). A similar procedure allows to define $\mathrm{tw}(A)$ for algebras A equipped with a locally finite dimensional $\mathbb{Z}_{\geq 0}$-filtration, and this coincides with the previous definition if A is $\mathbb{Z}$-graded and the filtration is the natural one associated with the $\mathbb{Z}$-grading (Lemma 6.9).

We prove that tw is unchanged in the passage from a $\mathbb{Z}$-graded algebra A to a polynomial ring $A[x]$, where the grading on A is extended to $A[x]$ so that x is homogeneous with degree 1, and similarly for $\mathbb{Z}_{\geq 0}$-filtered algebras (Theorem 8.2). This allows to show that certain algebras cannot become isomorphic after making polynomial extensions. For instance, consider quantized Weyl algebras $A = A_n^{Q,P}(\mathbb{K})$ and $A' = A_{n'}^{Q',P'}(\mathbb{K})$. If $\langle q_1, \dots, q_n \rangle \neq \langle q_1', \dots, q_n' \rangle$, then no cocycle twisted polynomial algebra $\big(A[X_1, \dots, X_s]\big)^{\mathbf{c}}$ is isomorphic, as a filtered $\mathbb{K}$-algebra, to any cocycle twisted polynomial algebra $\big(A'[X_1, \dots, X_t]\big)^{\mathbf{c}'}$ (Example 8.4). Here A is filtered so that the standard generators x_i, y_j for $A_n^{Q,P}(\mathbb{K})$ and the polynomial variables X_k all have degree 1.

Notation Fix a base field $\mathbb{K}$ throughout. Starting at the end of Subsection 6.1, we will assume that $\mathbb{K}$ is infinite, but otherwise $\mathbb{K}$ is arbitrary. We assume that all algebras are unital $\mathbb{K}$-algebras, all automorphisms are $\mathbb{K}$-algebra automorphisms, and that all skew derivations are $\mathbb{K}$-linear.

When we use the group $\mathbb{Z}^n$, we denote its standard basis by $e_1, \dots, e_n$.

2. 2-cocycles and twists

In this section we gather some background material on 2-cocycle twists of algebras.

DEFINITION 2.1. Let Γ be an (additive) abelian group, and view $\mathbb{K}^*$ as a Γ-module with the trivial action. Then a *$\mathbb{K}^*$-valued 2-cocycle on* Γ is a map $\mathbf{c} : \Gamma \times \Gamma \to \mathbb{K}^*$ such that

$$\mathbf{c}(s, t+u)\mathbf{c}(t, u) = \mathbf{c}(s+t, u)\mathbf{c}(s, t), \quad \forall\, s, t, u \in \Gamma. \tag{2.1}$$

It is *normalized* if $\mathbf{c}(0,0) = 1$. Among these maps are the *bicharacters* on Γ, that is, the maps $\Gamma \times \Gamma \to \mathbb{K}^*$ which are homomorphisms in each variable. A bicharacter $\mathbf{c}$ is *alternating* provided $\mathbf{c}(s,s) = 1$ for all $s \in \Gamma$, which implies skew-symmetry: $\mathbf{c}(t,s) = \mathbf{c}(s,t)^{-1}$ for all $s, t \in \Gamma$.

Note that (2.1) with $t = u = 0$ yields $\mathbf{c}(s,0)\mathbf{c}(0,0) = \mathbf{c}(s,0)^2$. Thus, $\mathbf{c}(s,0) = \mathbf{c}(0,0)$ for all $s \in \Gamma$, and similarly $\mathbf{c}(0,s) = \mathbf{c}(0,0)$. Consequently, if $\mathbf{c}$ is normalized, then $\mathbf{c}(s,0) = \mathbf{c}(0,s) = 1$ for all $s \in \Gamma$.

Denote by $Z^2(\Gamma, \mathbb{K}^*)$ the set of all $\mathbb{K}^*$-valued 2-cocycles on Γ, and note that $Z^2(\Gamma, \mathbb{K}^*)$ is an abelian group under pointwise multiplication.

Recall that a matrix (α_{ij}) of scalars is *multiplicatively skew-symmetric* provided $\alpha_{ii} = 1$ and $\alpha_{ij}\alpha_{ji} = 1$ for all indices i, j.

PROPOSITION 2.2. *Suppose Γ is free abelian with a basis $\{e_i \mid i \in I\}$.*

(a) *The rule $\rho(\mathbf{c}) = \big(\mathbf{c}(e_i, e_j)\mathbf{c}(e_j, e_i)^{-1}\big)$ defines a surjection*

$$\rho : Z^2(\Gamma, \mathbb{K}^*) \longrightarrow \{\textit{multiplicatively skew-symmetric } I \times I \textit{ matrices over } \mathbb{K}^*\}.$$

(b) *For $\mathbf{c}, \mathbf{c}' \in Z^2(\Gamma, \mathbb{K}^*)$, one has $\rho(\mathbf{c}) = \rho(\mathbf{c}')$ if and only if there is a function $f : \Gamma \to \mathbb{K}^*$ such that*

$$\mathbf{c}'(s,t) = \frac{f(s)f(t)}{f(s+t)}\,\mathbf{c}(s,t), \quad \forall\, s, t \in \Gamma. \tag{2.2}$$

(c) *If $\mathbf{c} \in Z^2(\Gamma, \mathbb{K}^*)$ and $\mathbf{c}^\# : \Gamma \times \Gamma \to \mathbb{K}^*$ is defined by*

$$\mathbf{c}^\#(s,t) = \mathbf{c}(s,t)\mathbf{c}(t,s)^{-1}, \quad \forall\, s, t \in \Gamma,$$

then $\mathbf{c}^\#$ is an alternating bicharacter on Γ.

PROOF. [**2**, Proposition 1]. □

COROLLARY 2.3. *Suppose Γ is free abelian with a basis $\{e_i \mid i \in I\}$, and put a total order on I. Let $\mathbf{c} \in Z^2(\Gamma, \mathbb{K}^*)$, and assume there exists $p_{ij} = \sqrt{\mathbf{c}(e_i, e_j)\mathbf{c}(e_j, e_i)^{-1}} \in \mathbb{K}^*$ for all $i < j$. Let $\mathbf{b} : \Gamma \times \Gamma \to \mathbb{K}^*$ be the alternating bicharacter such that $\mathbf{b}(e_i, e_j) = p_{ij}$ for all $i < j$. Then there is a function $f : \Gamma \to \mathbb{K}^*$ such that*

$$\mathbf{c}(s,t) = \frac{f(s)f(t)}{f(s+t)}\,\mathbf{b}(s,t), \quad \forall\, s, t \in \Gamma.$$

PROOF. By construction, $\mathbf{b}(e_i, e_j)\mathbf{b}(e_j, e_i)^{-1} = p_{ij}^2 = \mathbf{c}(e_i, e_j)\mathbf{c}(e_j, e_i)^{-1}$ for all $i < j$. Since $\mathbf{b}$ is alternating, it follows that $\rho(\mathbf{b}) = \rho(\mathbf{c})$ for ρ as in Proposition 2.2(a). Apply Proposition 2.2(b) to $\mathbf{c}$ and $\mathbf{b}$. □

DEFINITION 2.4. Suppose A is a Γ-graded $\mathbb{K}$-algebra and $\mathbf{c} \in Z^2(\Gamma, \mathbb{K}^*)$. Define a bilinear multiplication operation $*_{\mathbf{c}}$ on A by the rule

$$x *_{\mathbf{c}} y := \mathbf{c}(s,t)xy, \quad \forall x \in A_s, \ \ y \in A_t, \ \ s,t \in \Gamma. \tag{2.3}$$

Then $(A, *_{\mathbf{c}})$ is a Γ-graded $\mathbb{K}$-algebra, called the *twist of A by $\mathbf{c}$* [**2**]. We shall denote this algebra by $A^{\mathbf{c}}$.

OBSERVATIONS 2.5. Let A be a Γ-graded $\mathbb{K}$-algebra.

(1) If $\mathbf{c} \in Z^2(\Gamma, \mathbb{K}^*)$, then the identity element in $A^{\mathbf{c}}$ is $\mathbf{c}(0,0)^{-1} \cdot 1$, so if $\mathbf{c}$ is normalized, this identity equals the identity of A. Moreover, in this case the subalgebras $A_0 \subseteq A$ and $A_0^{\mathbf{c}} \subseteq A^{\mathbf{c}}$ are identical.

(2) Suppose $\mathbf{c}, \mathbf{c}' \in Z^2(\Gamma, \mathbb{K}^*)$ and there is a function $f : \Gamma \to \mathbb{K}^*$ such that (2.2) holds. Then $A^{\mathbf{c}}$ and $A^{\mathbf{c}'}$ are isomorphic, via the Γ-graded linear isomorphism given by

$$x \mapsto f(s)^{-1}x, \quad \forall x \in A_s, \quad s \in \Gamma$$

[**2**]. Thus, starting with $\mathbf{c}$, choosing f such that $f(0) = \mathbf{c}(0,0)^{-1}$, and defining $\mathbf{c}'$ by (2.2), we obtain $\mathbf{c}'(0,0) = 1$. This yields a twist $A^{\mathbf{c}'}$, isomorphic to $A^{\mathbf{c}}$, in which the identity is the identity of A.

(3) For any $\mathbf{c}, \mathbf{d} \in Z^2(\Gamma, \mathbb{K}^*)$,

$$A^{\mathbf{cd}} = (A^{\mathbf{c}})^{\mathbf{d}}. \tag{2.4}$$

Hence, the relation of *cocycle twist equivalence*, where Γ-graded $\mathbb{K}$-algebras A and B are cocycle twist equivalent if and only if $B \cong A^{\mathbf{c}}$ (as Γ-graded algebras) for some $\mathbf{c} \in Z^2(\Gamma, \mathbb{K}^*)$, is an equivalence relation.

3. Twist invariants for Γ-graded algebras

In this section we review the Alev-Dumas invariant of noncommutative algebras and define a twist invariant of graded algebras. We also provide various examples to illustrate both notions.

DEFINITION 3.1. Suppose A is a semiprime right Goldie $\mathbb{K}$-algebra. The *Alev-Dumas invariant* of A [**1**, Définition 3.8] is the group

$$\mathrm{AD}(A) := \mathrm{GL}_1(\mathrm{Fract}\, A)' \cap \mathbb{K}^*,$$

where $\mathrm{GL}_1(\mathrm{Fract}\, A)'$ is the derived subgroup (= commutator subgroup) of the group of units of the classical right quotient ring $\mathrm{Fract}\, A$.

OBSERVATIONS 3.2. (1) If A and B are semiprime right Goldie $\mathbb{K}$-algebras with $\mathrm{Fract}\, A \cong \mathrm{Fract}\, B$, then $\mathrm{AD}(A) = \mathrm{AD}(B)$.

(2) If $A_1, \dots, A_k$ are semiprime right Goldie $\mathbb{K}$-algebras, then

$$\mathrm{AD}(A_1 \times \cdots \times A_k) = \mathrm{AD}(A_1) \cap \cdots \cap \mathrm{AD}(A_k). \tag{3.1}$$

(3) For all positive integers n,

$$\mathrm{AD}(M_n(\mathbb{K})) = \{\alpha \in \mathbb{K} \mid \alpha^n = 1\}. \tag{3.2}$$

For $n \geq 3$, this follows from the well known fact that $\mathrm{GL}_n(\mathbb{K})' = \mathrm{SL}_n(\mathbb{K})$ (e.g., [**15**, Theorem XIII.9.2]). For $n = 2$, it follows from that fact that $\mathrm{GL}_2(\mathbb{K})' \subseteq \mathrm{SL}_2(\mathbb{K})$ together with the observation $\left[\begin{smallmatrix} -1 & 0 \\ 0 & 1 \end{smallmatrix}\right]\left[\begin{smallmatrix} 0 & 1 \\ 1 & 0 \end{smallmatrix}\right]\left[\begin{smallmatrix} -1 & 0 \\ 0 & 1 \end{smallmatrix}\right]\left[\begin{smallmatrix} 0 & 1 \\ 1 & 0 \end{smallmatrix}\right] = \left[\begin{smallmatrix} -1 & 0 \\ 0 & -1 \end{smallmatrix}\right]$.

Consequently, the n-th roots of unity in K are contained in $\mathrm{AD}(M_n(A))$ for any semiprime right Goldie $\mathbb{K}$-algebra A.

(4) The AD-invariant obviously depends on the choice of base field. For instance, if we view $M_3(\mathbb{C})$ as a $\mathbb{C}$-algebra, then $\mathrm{AD}(M_3(\mathbb{C}))$ has order 3, while as an $\mathbb{R}$-algebra, $\mathrm{AD}(M_3(\mathbb{C}))$ is the trivial group.

For any multiplicatively skew-symmetric matrix $\boldsymbol{q} = (q_{ij}) \in M_n(\mathbb{K}^*)$, the corresponding *quantum affine space* is the $\mathbb{K}$-algebra

$$\mathcal{O}_{\boldsymbol{q}}(\mathbb{K}^n) := \mathbb{K}\langle y_1, \ldots, y_n \mid y_i y_j = q_{ij} y_j y_i \ \forall\, i, j \in [1, n]\rangle.$$

The standard (one-parameter) quantum affine space, denoted $\mathcal{O}_q(\mathbb{K}^n)$ for $q \in \mathbb{K}^*$, is the case of $\mathcal{O}_{\boldsymbol{q}}(\mathbb{K}^n)$ where $q_{ij} = q$ for $1 \le i < j \le n$. The localization of $\mathcal{O}_{\boldsymbol{q}}(\mathbb{K}^n)$ at the multiplicative subset generated by $y_1, \ldots, y_n$ is the *quantum torus*

$$\mathcal{T}_{\boldsymbol{q}}^n(\mathbb{K}) = \mathcal{O}_{\boldsymbol{q}}((\mathbb{K}^*)^n) := \mathbb{K}\langle y_1^{\pm 1}, \ldots, y_n^{\pm 1} \mid y_i y_j = q_{ij} y_j y_i \ \forall\, i, j \in [1, n]\rangle.$$

By way of abbreviation, we write $\langle \boldsymbol{q} \rangle$ for the group $\langle q_{ij} \mid i, j \in [1, n]\rangle \subseteq \mathbb{K}^*$.

PROPOSITION 3.3. [**1**, Proposition 3.9] *Let $n \in \mathbb{Z}_{>0}$ and $\boldsymbol{q} \in M_n(\mathbb{K}^*)$ a multiplicatively skew-symmetric matrix. Then $\mathrm{AD}(\mathcal{O}_{\boldsymbol{q}}(\mathbb{K}^n)) = \langle \boldsymbol{q} \rangle$.*

Of course, $\mathrm{AD}(\mathcal{T}_{\boldsymbol{q}}^n(\mathbb{K})) = \langle \boldsymbol{q} \rangle$ as well.

PROPOSITION 3.4. *Let A be a prime affine PI $\mathbb{K}$-algebra of PI degree n. Then*

$$\mathrm{AD}(A) \subseteq \{\alpha \in \mathbb{K} \mid \alpha^n = 1\}.$$

PROOF. Denote by Q the field of fractions of $Z := Z(A)$. By Posner's theorem, $B := A \otimes_Z Q$ is a central simple algebra over Q, and by definition, $n = \sqrt{\dim_Q B}$. This algebra splits over the algebraic closure $\overline{Q}$,

$$A \otimes_Z \overline{Q} \cong M_n(\overline{Q}),$$

see [**18**, Corollary 2.3.25]. Therefore A is isomorphic to a $\mathbb{K}$-subalgebra of $M_n(\overline{Q})$, and by Observation 3.2(1),

$$\mathrm{AD}(A) \subseteq \mathrm{AD}(M_n(\overline{Q})) \cap \mathbb{K}^* = \{\alpha \in \mathbb{K} \mid \alpha^n = 1\}. \qquad \square$$

Further examples appear in the following section.

DEFINITION 3.5. Suppose A is a Γ-graded semiprime right Goldie $\mathbb{K}$-algebra. The *twisted commutation invariant* of A as a Γ-graded $\mathbb{K}$-algebra is the group

$$\mathrm{tw}_\Gamma(A) := \bigcap_{\mathbf{c} \in Z^2(\Gamma, \mathbb{K}^*)} \mathrm{AD}(A^{\mathbf{c}}) \subseteq \mathbb{K}^*.$$

EXAMPLE 3.6. For any multiplicatively skew-symmetric matrix $\boldsymbol{q} = (q_{ij}) \in M_n(\mathbb{K}^*)$, the quantum affine space $A = \mathcal{O}_{\boldsymbol{q}}(\mathbb{K}^n)$ has a natural grading by the group $\Gamma = \mathbb{Z}^n$ with $\deg y_i = e_i$ for all i. This graded algebra can be expressed as a cocycle twist

$$A = \mathcal{O}(\mathbb{K}^n)^{\mathbf{c}},$$

where $\mathcal{O}(\mathbb{K}^n) = \mathbb{K}[x_1, \ldots, x_n]$ is a polynomial algebra and $\mathbf{c}$ is the 2-cocycle such that

$$\mathbf{c}(s, t) = \prod_{1 \le i < j \le n} q_{ij}^{s_i t_j}, \quad \forall\, s = (s_1, \ldots, s_n),\ t = (t_1, \ldots, t_n) \in \Gamma.$$

Since $\mathrm{AD}(\mathcal{O}(\mathbb{K}^n))$ is trivial (because $\mathcal{O}(\mathbb{K}^n)$ is commutative), we conclude that

$$\mathrm{tw}_\Gamma(\mathcal{O}_{\boldsymbol{q}}(\mathbb{K}^n)) = \{1\}.$$

EXAMPLE 3.7. The $n \times n$ *multiparameter quantum matrix algebra* $\mathcal{O}_{\lambda,\boldsymbol{p}}(M_n(\mathbb{K}))$ is given by generators X_{ij} for $i, j \in [1, n]$ and relations

$$X_{lm}X_{ij} = \begin{cases} p_{li}p_{jm}X_{ij}X_{lm} + (\lambda - 1)p_{li}X_{im}X_{lj}, & (l > i,\ m > j), \\ \lambda p_{li}p_{jm}X_{ij}X_{lm}, & (l > i,\ m \leq j), \\ p_{jm}X_{ij}X_{lm}, & (l = i,\ m > j), \end{cases} \tag{3.3}$$

where $\lambda \in \mathbb{K}^*$ and $\boldsymbol{p} = (p_{ij}) \in M_n(\mathbb{K}^*)$ is multiplicatively skew-symmetric. The *standard* $n \times n$ *quantum matrix algebra* $\mathcal{O}_q(M_n(\mathbb{K}))$, for $q \in \mathbb{K}^*$, is the case of $\mathcal{O}_{\lambda,\boldsymbol{p}}(M_n(\mathbb{K}))$ for which $\lambda = q^{-2}$ and $p_{ij} = q$ for all $i > j$.

It is obvious from (3.3) that $\mathrm{AD}(\mathcal{O}_{\lambda,\boldsymbol{p}}(M_n(\mathbb{K})))$ contains λ and all p_{ij}. We show below (5.16) that if λ is not a root of unity, then $\mathrm{AD}(\mathcal{O}_{\lambda,\boldsymbol{p}}(M_n(\mathbb{K}))) = \langle \lambda, \boldsymbol{p} \rangle$. In particular, $\mathrm{AD}(\mathcal{O}_q(M_n(\mathbb{K}))) = \langle q \rangle$ if q is not a root of unity.

There is a standard grading of $\mathcal{O}_{\lambda,\boldsymbol{p}}(M_n(\mathbb{K}))$ by $\Gamma = \mathbb{Z}^n \times \mathbb{Z}^n$, under which each X_{ij} is homogeneous of degree (e_i, e_j). If $\mathbf{c}$ is the 2-cocycle on Γ given by

$$\mathbf{c}\big((s,t),(s',t')\big) = \prod_{1 \leq i < j \leq n} p_{ij}^{s_j s'_i - t_j t'_i},$$

then we find that

$$\mathcal{O}_{\lambda,\boldsymbol{p}}(M_n(\mathbb{K}))^{\mathbf{c}} \cong \mathcal{O}_{\lambda,\mathbf{1}}(M_n(\mathbb{K})),$$

where $\mathbf{1}$ denotes the matrix in $M_n(\mathbb{K}^*)$ with all entries equal to 1. Thus, by the result mentioned above, $\mathrm{AD}(\mathcal{O}_{\lambda,\boldsymbol{p}}(M_n(\mathbb{K}))^{\mathbf{c}}) = \langle \lambda \rangle$, assuming that λ is a non-root of unity. In fact, $\mathrm{tw}_\Gamma(\mathcal{O}_{\lambda,\boldsymbol{p}}(M_n(\mathbb{K}))) = \langle \lambda \rangle$, as we show in (5.16).

4. The twist invariant for algebras with one quantum cluster

In this section we prove a general theorem that can be used to effectively compute the twist invariants of algebras that have one quantum cluster, that is, algebras squeezed between a quantum affine space algebra and the corresponding quantum torus. The theorem is applied to compute the twist invariants of the multiparameter quantized Weyl algebras in full generality.

4.1. The AD invariant for quantum cluster algebras. A *quantum cluster algebra* $\mathcal{A}$ is generated by a collection of cluster variables, whose different clusters are obtained from each other by mutation. Each quantum seed of $\mathcal{A}$ contains elements $y_1, \dots, y_n \in \mathcal{A}$ such that $y_i y_j = q_{ij} y_j y_i$ for some $q_{ij} \in \mathbb{K}^*$ and, in addition, these elements give rise to embeddings

$$\mathcal{O}_{\boldsymbol{q}}(\mathbb{K}^n) \subseteq \mathcal{A} \subset \mathcal{T}_{\boldsymbol{q}}^n(\mathbb{K}), \tag{4.1}$$

where $\boldsymbol{q} = (q_{ij})$ is a multiplicatively skew-symmetric matrix. We refer the reader to [**4**] for details on the uniparameter case when the scalars q_{ij} are powers of a single scalar $q \in \mathbb{K}^*$ and to [**11**, Ch. 2] for the general case. To each quantum cluster algebra we associate the abelian group

$$q(\mathcal{A}) := \langle \boldsymbol{q} \rangle.$$

This is independent of the choice of seed due to the mutation formula [**11**, Eq. (2.19)] for the matrices $\boldsymbol{q}$ of the quantum seeds of $\mathcal{A}$. The *upper quantum cluster algebra* $\mathcal{U}$ associated to $\mathcal{A}$ (with or without inverted frozen variables, see [**4**, **11**]) also satisfies

$$\mathcal{O}_{\boldsymbol{q}}(\mathbb{K}^n) \subseteq \mathcal{U} \subset \mathcal{T}_{\boldsymbol{q}}^n(\mathbb{K}). \tag{4.2}$$

Set also $q(\mathcal{U}) := q(\mathcal{A}) = \langle \boldsymbol{q} \rangle$. Analogously to the case for $\mathcal{A}$, this is independent of the choice of seed.

A $\mathbb{K}$-algebra $\mathcal{A}$ that satisfies (possibly only one) embedding of the form (4.1) is called *an algebra with a quantum cluster*. The class of algebras with one quantum cluster is strictly larger than that of quantum cluster algebras.

PROPOSITION 4.1. (a) *If A is a $\mathbb{K}$-algebra with a quantum cluster*

$$\mathcal{O}_{\boldsymbol{q}}(\mathbb{K}^n) \subseteq A \subseteq \mathcal{T}^n_{\boldsymbol{q}}(\mathbb{K})$$

for some multiplicatively skew-symmetric matrix $\boldsymbol{q} \in M_n(\mathbb{K}^*)$, then $\mathrm{AD}(A) = \langle \mathbf{q} \rangle$.

(b) *For every quantum cluster algebra $\mathcal{A}$ and the associated upper quantum cluster algebra $\mathcal{U}$,*

$$\mathrm{AD}(\mathcal{A}) = \mathrm{AD}(\mathcal{U}) = q(\mathcal{A}).$$

The statement follows from Proposition 3.3, Observation 3.2(1) and the embeddings (4.1)-(4.2).

Propositions 3.4 and 4.1 have the following interesting corollary.

COROLLARY 4.2. *Assume that A is a $\mathbb{K}$-algebra with a quantum cluster*

$$\mathcal{O}_{\boldsymbol{q}}(\mathbb{K}^n) \subseteq A \subseteq \mathcal{T}^n_{\boldsymbol{q}}(\mathbb{K})$$

for some multiplicatively skew-symmetric matrix $\boldsymbol{q} \in M_n(\mathbb{K}^*)$. Then A is a PI algebra if and only if $\mathrm{AD}(A)$ *is finite, in which case*

$$\text{PI-deg}\, A \geq \operatorname{card} \mathrm{AD}(A)\,.$$

PROOF. It is well known that $\mathcal{O}_{\boldsymbol{q}}(\mathbb{K}^n)$ is a PI algebra if and only if all q_{ij} are roots of unity (e.g., [**14**, Theorem 7]), i.e., if and only if $\langle \boldsymbol{q} \rangle$ is finite. Since $\mathcal{O}_{\boldsymbol{q}}(\mathbb{K}^n) \subseteq A \subseteq \operatorname{Fract} \mathcal{O}_{\boldsymbol{q}}(\mathbb{K}^n)$, one of these algebras is PI if and only if the other is. The first result thus follows from Proposition 4.1(a).

If A is PI with PI-deg $A = d$, then $\operatorname{card} \mathrm{AD}(A) \leq d$ by Proposition 3.4. $\square$

The inequality in Corollary 4.2 is by no means sharp, as can be seen from the formula for PI-deg $\mathcal{O}_{\boldsymbol{q}}(\mathbb{K}^n)$ in [**7**, Proposition 7.1(c)]. For instance, if $\boldsymbol{q}$ is given by $q_{ij} = q^{a_{ij}}$ where $q \in \mathbb{K}^*$ is a primitive ℓ-th root of unity and $(a_{ij}) \in M_n(\mathbb{Z})$ is an invertible antisymmetric matrix, then PI-deg $\mathcal{O}_{\boldsymbol{q}}(\mathbb{K}^n) = \ell^{n/2}$.

4.2. The twist invariant for algebras with one quantum cluster.

DEFINITION 4.3. Let $\chi : \mathbb{Z}^n \times \mathbb{Z}^n \to \mathbb{K}^*$ be an alternating bicharacter. Set

$$\boldsymbol{q}(\chi) := (\chi(e_i, e_j)) \in M_n(\mathbb{K}^*),$$

which is a multiplicatively skew-symmetric matrix, and

$$\mathcal{O}_\chi(\mathbb{K}^n) := \mathcal{O}_{\boldsymbol{q}(\chi)}(\mathbb{K}^n), \qquad \mathcal{T}^n_\chi(\mathbb{K}) := \mathcal{T}^n_{\boldsymbol{q}(\chi)}(\mathbb{K}).$$

The next theorem provides an effective way for computing twist invariants of algebras with one quantum cluster.

THEOREM 4.4. *Let $\chi : \mathbb{Z}^n \times \mathbb{Z}^n \to \mathbb{K}^*$ be an alternating bicharacter and*

$$\phi : \mathbb{Z}^n \to \Gamma$$

a surjective group homomorphism, where Γ is a free abelian group. Assign to $\mathcal{O}_\chi(\mathbb{K}^n)$ and $\mathcal{T}_\chi(\mathbb{K})$ the Γ-gradings induced from their $\mathbb{Z}^n$-gradings along ϕ. For every inclusion of Γ-graded $\mathbb{K}$-algebras

$$\mathcal{O}_\chi(\mathbb{K}^n) \subseteq A \subseteq \mathcal{T}^n_\chi(\mathbb{K}),$$

we have

$$\mathrm{tw}_\Gamma(A) = \langle \chi(\ker\phi,\, \mathbb{Z}^n)\rangle \subseteq \mathbb{K}^*. \tag{4.3}$$

PROOF. For $\mathbf{c} \in Z^2(\Gamma, \mathbb{K}^*)$, let $\mathbf{c}^\#$ be the alternating bicharacter on Γ obtained from $\mathbf{c}$ as in Proposition 2.2(c), and define the alternating bicharacter $\chi_{\mathbf{c}}$ on $\mathbb{Z}^n$ by

$$\chi_{\mathbf{c}}(s,t) = \chi(s,t)\,\mathbf{c}^\#(\phi(s), \phi(t)). \tag{4.4}$$

It is easy to see that there are compatible $\mathbb{K}$-algebra isomorphisms

$$\mathcal{O}_\chi(\mathbb{K}^n)^{\mathbf{c}} \to \mathcal{O}_{\chi_{\mathbf{c}}}(\mathbb{K}^n) \qquad \text{and} \qquad \mathcal{T}^n_\chi(\mathbb{K})^{\mathbf{c}} \to \mathcal{T}^n_{\chi_{\mathbf{c}}}(\mathbb{K}).$$

The second isomorphism sends $A^{\mathbf{c}}$ to a subalgebra of $\mathcal{T}^n_{\chi_{\mathbf{c}}}(\mathbb{K})$ containing $\mathcal{O}_{\chi_{\mathbf{c}}}(\mathbb{K}^n)$, and so Proposition 4.1(a) implies that $\mathrm{AD}(A^{\mathbf{c}}) = \langle \boldsymbol{q}(\chi_{\mathbf{c}})\rangle$. Therefore

$$\mathrm{tw}_\Gamma(A) = \bigcap_{\mathbf{c}\in Z^2(\Gamma,\mathbb{K}^*)} \langle \boldsymbol{q}(\chi_{\mathbf{c}})\rangle.$$

Given $s \in \ker\phi$ and $t \in \mathbb{Z}^n$, we see that $\chi_{\mathbf{c}}(s,t) = \chi(s,t)$ for any $\mathbf{c} \in Z^2(\Gamma, \mathbb{K}^*)$. Thus,

$$\mathrm{tw}_\Gamma(A) \supseteq \langle \chi(\ker\phi,\, \mathbb{Z}^n)\rangle.$$

We are left with proving the opposite inclusion.

Since Γ is free abelian, there is a basis $b_1, \dots, b_n$ for $\mathbb{Z}^n$ such that some initial sublist $b_1, \dots, b_m$ is a basis for $\ker\phi$ and $\phi(b_{m+1}), \dots, \phi(b_n)$ is a basis for Γ. Define $\mathbf{c} \in Z^2(\Gamma, \mathbb{K}^*)$ by

$$\mathbf{c}\Big(\sum_{i=m+1}^n u_i\phi(b_i),\ \sum_{j=m+1}^n v_j\phi(b_j)\Big) = \prod_{m+1\le i<j\le n} \chi(b_i,b_j)^{-u_iv_j}.$$

Then $\mathbf{c}^\#(\phi(b_i), \phi(b_j)) = \chi(b_i,b_j)^{-1}$ for $i,j \in [m+1,n]$, and so

$$\chi_{\mathbf{c}}(b_i,b_j) = \begin{cases} \chi(b_i,b_j) & (i \le m \text{ or } j \le m) \\ 1 & (i,j > m) \end{cases}$$

for $i,j \in [1,n]$. It follows that

$$\mathrm{tw}_\Gamma(A) \subseteq \mathrm{AD}(A^{\mathbf{c}}) = \langle \boldsymbol{q}(\chi_{\mathbf{c}})\rangle = \langle \chi(b_i,b_j) \mid i \le m \text{ or } j \le m\rangle = \langle \chi(\ker\phi,\, \mathbb{Z}^n)\rangle,$$

which completes the proof of the proposition. □

If the homomorphism ϕ in Theorem 4.4 is an isomorphism, equation (4.3) says that $\mathrm{tw}_\Gamma(A)$ is trivial. This yields the following.

COROLLARY 4.5. *Let $\boldsymbol{q} \in M_n(\mathbb{K}^*)$ be a multiplicatively skew-symmetric matrix, and give $\mathcal{O}_{\boldsymbol{q}}(\mathbb{K}^n)$ and $\mathcal{T}^n_{\boldsymbol{q}}(\mathbb{K})$ the standard $\mathbb{Z}^n$-gradings for which the canonical generators y_i are homogeneous of degree e_i. For every inclusion $\mathcal{O}_{\boldsymbol{q}}(\mathbb{K}^n) \subseteq A \subseteq \mathcal{T}^n_{\boldsymbol{q}}(\mathbb{K})$ of $\mathbb{Z}^n$-graded $\mathbb{K}$-algebras, we have*

$$\mathrm{tw}_{\mathbb{Z}^n}(A) = \langle 1\rangle.$$

Many algebras A with a quantum cluster $\mathcal{O}_{\boldsymbol{q}}(\mathbb{K}^n) \subseteq A \subseteq \mathcal{T}^n_{\boldsymbol{q}}(\mathbb{K})$ are not graded by $\mathbb{Z}^n$ but by a proper homomorphic image Γ of $\mathbb{Z}^n$. We shall see that $\mathrm{tw}_\Gamma(A)$ is typically nontrivial in such cases. In particular, in the case when $\Gamma = \mathbb{Z}$, the invariant $\mathrm{tw}_\Gamma(A)$ reduces to $\mathrm{AD}(A)$, as follows.

PROPOSITION 4.6. *Let $\boldsymbol{q} \in M_n(\mathbb{K}^*)$ be a multiplicatively skew-symmetric matrix, and suppose $\mathcal{O}_{\boldsymbol{q}}(\mathbb{K}^n)$ and $\mathcal{T}_{\boldsymbol{q}}^n(\mathbb{K})$ have $\mathbb{Z}$-gradings such that the canonical generators y_i are homogeneous. For every inclusion $\mathcal{O}_{\boldsymbol{q}}(\mathbb{K}^n) \subseteq A \subseteq \mathcal{T}_{\boldsymbol{q}}^n(\mathbb{K})$ of $\mathbb{Z}$-graded $\mathbb{K}$-algebras, we have*

$$\mathrm{tw}_{\mathbb{Z}}(A) = \langle \boldsymbol{q} \rangle.$$

PROOF. View the given $\mathbb{Z}$-gradings on $\mathcal{O}_{\boldsymbol{q}}(\mathbb{K}^n)$ and $\mathcal{T}_{\boldsymbol{q}}^n(\mathbb{K})$ as induced from the standard $\mathbb{Z}^n$-gradings along a suitable homomorphism $\phi : \mathbb{Z}^n \to \mathbb{Z}$. As in the proof of Theorem 4.4,

$$\mathrm{tw}_{\mathbb{Z}}(A) = \bigcap_{\mathbf{c} \in Z^2(\mathbb{Z}, \mathbb{K}^*)} \langle \boldsymbol{q}(\chi_{\mathbf{c}}) \rangle,$$

where χ is the alternating bicharacter on $\mathbb{Z}^n$ such that $\chi(e_i, e_j) = q_{ij}$ for all i, j and the alternating bicharacters $\chi_{\mathbf{c}}$ are defined as in (4.4).

Observe that any alternating bicharacter $\mathbf{d}$ on $\mathbb{Z}$ is trivial, since $\mathbf{d}(1,1) = 1$ and $\mathbf{d}(s,t) = \mathbf{d}(1,1)^{st}$ for all $s, t \in \mathbb{Z}$. Hence, $\chi_{\mathbf{c}} = \chi$ for all $\mathbf{c} \in Z^2(\mathbb{Z}, \mathbb{K}^*)$, and thus $\mathrm{tw}_{\mathbb{Z}}(A) = \langle \boldsymbol{q}(\chi) \rangle = \langle \boldsymbol{q} \rangle$. □

EXAMPLE 4.7. The *multiparameter quantized Weyl algebra* $A := A_n^{Q,P}(\mathbb{K})$ is given by generators $x_1, y_1, \dots, x_n, y_n$ satisfying the following relations:

$$\begin{aligned}
&y_i y_j = p_{ij} y_j y_i\,, && (\text{all } i, j),\\
&x_i x_j = q_i p_{ij} x_j x_i\,, && (i < j),\\
&x_i y_j = p_{ji} y_j x_i\,, && (i < j),\\
&x_i y_j = q_j p_{ji} y_j x_i\,, && (i > j),\\
&x_j y_j = 1 + q_j y_j x_j + \sum_{l<j} (q_l - 1) y_l x_l\,, && (\text{all } j),
\end{aligned} \tag{4.5}$$

where $Q = (q_1, \dots, q_n) \in (\mathbb{K}^*)^n$ and $P = (p_{ij}) \in M_n(\mathbb{K}^*)$ is a multiplicatively skew-symmetric matrix. There are commuting normal elements $z_k := x_k y_k - y_k x_k \in A$ for $k \in [1, n]$, satisfying

$$z_k y_j = \begin{cases} q_j y_j z_k & (j \le k), \\ y_j z_k & (j > k), \end{cases} \qquad z_k x_j = \begin{cases} q_j^{-1} x_j z_k & (j \le k), \\ x_j z_k & (j > k), \end{cases} \qquad \forall\, j, k \in [1, n]$$

[**12**, §§2.8, 3.1]. Moreover, A has a quantum cluster of the form

$$\mathcal{O}_{\boldsymbol{q}}(\mathbb{K}^{2n}) \subseteq A \subseteq \mathcal{T}_{\boldsymbol{q}}^{2n}(\mathbb{K}) \tag{4.6}$$

where $\mathcal{O}_{\boldsymbol{q}}(\mathbb{K}^{2n})$ is generated by $z_1, \dots, z_n, y_1, \dots, y_n$ [**12**, §3.1].

We may grade A by $\Gamma := \mathbb{Z}^n$, with each x_k homogeneous of degree e_k and each y_k homogeneous of degree $-e_k$. If we assign to $\mathcal{O}_{\boldsymbol{q}}(\mathbb{K}^{2n})$ and $\mathcal{T}_{\boldsymbol{q}}^{2n}(\mathbb{K})$ the $\mathbb{Z}^n$-gradings induced from their standard $\mathbb{Z}^{2n}$-gradings along the homomorphism

$$\phi : \mathbb{Z}^{2n} \longrightarrow \Gamma, \quad \phi(e_j) = \begin{cases} 0 & (j \le n), \\ -e_{j-n} & (j > n), \end{cases} \quad \forall\, j \in [1, 2n],$$

then (4.6) is an inclusion of Γ-graded $\mathbb{K}$-algebras. We may write $\mathcal{O}_{\boldsymbol{q}}(\mathbb{K}^{2n}) = \mathcal{O}_{\chi}(\mathbb{K}^{2n})$ and $\mathcal{T}_{\boldsymbol{q}}^{2n}(\mathbb{K}) = \mathcal{T}_{\chi}^{2n}(\mathbb{K})$ where χ is the alternating bicharacter on $\mathbb{Z}^{2n}$ such that

$$\chi(e_k, e_j) = 1, \qquad \chi(e_{n+k}, e_{n+j}) = p_{kj}\,, \qquad \chi(e_k, e_{n+j}) = \begin{cases} q_j & (j \le k), \\ 1 & (j > k), \end{cases}$$

for all $j, k \in [1, n]$. Since $\ker \phi = \bigoplus_{k=1}^{n} \mathbb{Z}e_k$, we conclude from (4.3) of Theorem 4.4 that

$$\operatorname{tw}_\Gamma(A_n^{Q,P}(\mathbb{K})) = \langle q_1, \dots, q_n \rangle. \tag{4.7}$$

5. The twist invariant for quantum nilpotent algebras and quantum Schubert cell algebras

In this section we recall results on the structure of CGL extensions (also known as quantum nilpotent algebras), which form a large class of algebras that contains as special cases various families of algebras from the theory of quantum groups. A theorem is proved for the effective calculation of their twist invariants and is illustrated for the algebras of quantum matrices and more generally the quantum Schubert cell algebras for all symmetrizable Kac–Moody algebras.

We apply the twist invariant to prove that many quantized Weyl algebras are not uniparameter CGL extensions. This means that the scalars in the leading terms of the commutation relations of these CGL extensions do not form a cyclic group.

5.1. Quantum nilpotent algebras and their AD invariants. The class of *quantum nilpotent algebras* consists of certain iterated skew polynomial extensions of the form

$$A := \mathbb{K}[x_1][x_2; \sigma_2, \delta_2] \cdots [x_N; \sigma_N, \delta_N], \tag{5.1}$$

where the σ_k are $\mathbb{K}$-algebra automorphisms and the δ_k are $\mathbb{K}$-linear σ_k-derivations. The precise conditions were formulated by Launois-Lenagan-Rigal in [**16**, Definition 3.1] and baptized "CGL extensions".

DEFINITION 5.1. An iterated skew polynomial extension (5.1) is called a *CGL extension* if it is equipped with a rational action of a $\mathbb{K}$-torus $\mathcal{H}$ by $\mathbb{K}$-algebra automorphisms satisfying the following conditions:

(i) The elements $x_1, \dots, x_N$ are $\mathcal{H}$-eigenvectors.

(ii) For every $k \in [2, N]$, δ_k is a locally nilpotent σ_k-derivation of the algebra

$$A_{k-1} = \mathbb{K}[x_1][x_2; \sigma_2, \delta_2] \cdots [x_{k-1}; \sigma_{k-1}, \delta_{k-1}] \tag{5.2}$$

(iii) For every $k \in [1, N]$, there exists $h_k \in \mathcal{H}$ such that $\sigma_k = (h_k \cdot)|_{A_{k-1}}$ and the h_k-eigenvalue of x_k, to be denoted by λ_k, is not a root of unity.

Conditions (i) and (iii) imply that

$$\sigma_k(x_j) = \lambda_{kj} x_j \text{ for some } \lambda_{kj} \in \mathbb{K}^*, \ \ \forall 1 \le j < k \le N.$$

We then set $\lambda_{kk} := 1$ and $\lambda_{jk} := \lambda_{kj}^{-1}$ for $j < k$, obtaining a multiplicatively skew-symmetric matrix $\boldsymbol{\lambda} := (\lambda_{kj}) \in M_N(\mathbb{K}^*)$.

Denote the character group of the torus $\mathcal{H}$ by $X(\mathcal{H})$ and treat it as an additive abelian group. The action of $\mathcal{H}$ on A gives rise to an $X(\mathcal{H})$-grading of A (e.g., [**6**, Lemma II.2.11]), and the $\mathcal{H}$-eigenvectors in A are precisely the nonzero homogeneous elements with respect to this grading.

For a function $\eta : [1, N] \to \mathbb{Z}$, define the predecessor function for its level sets, $p = p_\eta : [1, N] \to [1, N] \sqcup \{-\infty\}$, by

$$p(k) := \max\{j < k \mid \eta(j) = \eta(k)\},$$

where we set $\max \varnothing = -\infty$. For $k \in [1, N]$, denote

$$O_-(k) := \max\{l \in \mathbb{Z}_{\ge 0} \mid p^l(k) \ne -\infty\}.$$

An element y of an algebra A is called *prime* if it is a nonzero nonunit,

$$Ay = yA, \quad \text{and} \quad A/Ay \text{ is a domain.}$$

We will need the following theorem for the homogeneous prime elements of the algebras A_k from (5.2), which are CGL extensions for the restriction of the $\mathcal{H}$-action to them. Set $A_0 := \mathbb{K}$.

THEOREM 5.2. [**10**, Theorems 4.3, 4.6] *Let A be a CGL extension as in Definition* 5.1.

(a) *Each of the algebras A_k contains a unique (up to scalar multiples) $X(\mathcal{H})$-homogeneous prime element that does not belong to A_{k-1}. This element will be denoted by y_k.*

(b) *There exists a function $\eta : [1, N] \to \mathbb{Z}$ such that (up to scalar multiples) the elements $y_1, \dots, y_N$ are given by*

$$y_k = \begin{cases} y_{p(k)}x_k - c_k, & \text{for some } c_k \in A_{k-1}, & \text{if } p(k) \neq -\infty \\ x_k, & & \text{if } p(k) = -\infty. \end{cases}$$

(c) *The elements $y_1, \dots, y_N$ define embeddings*

$$\mathcal{O}_{\boldsymbol{q}}(\mathbb{K}^N) \subseteq A \subset \mathcal{T}_{\boldsymbol{q}}^N(\mathbb{K}),$$

where the entries q_{jk} of the matrix $\boldsymbol{q}$ are given by

$$q_{jk} := \prod_{l=0}^{O_-(j)} \prod_{m=0}^{O_-(k)} \lambda_{p^l(j), p^m(k)} \, .$$

It follows from part (b) of the theorem that

(5.3) $$\delta_k \neq 0 \quad \text{if and only if} \quad p(k) = -\infty,$$

see [**11**, Theorem 3.6] for details.

PROPOSITION 5.3. *If A is a CGL extension as in Definition* 5.1, *then* $\mathrm{AD}(A) = \langle \boldsymbol{\lambda} \rangle$.

PROOF. It is easy to see that $\langle \boldsymbol{q} \rangle = \langle \boldsymbol{\lambda} \rangle$. Hence, it follows from Theorem 5.2(c), Observation 3.2(1) and Proposition 3.3 that $\mathrm{AD}(A) = \langle \boldsymbol{\lambda} \rangle$. □

5.2. The twist invariant for quantum nilpotent algebras. Consider a CGL extension A as in Definition 5.1. For $h \in \mathcal{H}$ and $\alpha \in X(\mathcal{H})$, we will denote by $h^\alpha \in \mathbb{K}^*$ the corresponding eigenvalue. Consider the homomorphism

(5.4) $$\pi : \mathbb{Z}^N \to X(\mathcal{H}) \quad \text{given by} \quad \pi(e_i) := \deg x_i \in X(\mathcal{H}), \ \ \forall\, i \in [1, N].$$

Denote the subsets of $[1, N]$ indexing the variables x_k with vanishing and non-vanishing derivations by

$$V(A) := \{k \in [1, N] \mid \delta_k = 0\}, \quad N(A) := [1, N] \backslash V(A).$$

LEMMA 5.4. *Keep the above notation.*

(a) *The set $\{\pi(e_i) \mid i \in V(A)\}$ generates* $\operatorname{im} \pi$.

(b) *For each $k \in N(A)$, there exists $b_k \in \ker \pi$ such that $b_k - e_k \in \sum_{j=1}^{k-1} \mathbb{Z} e_j$. If such elements b_k have been chosen, and if*

(5.5) *The elements $\pi(e_j)$ for $j \in V(A)$ are $\mathbb{Z}$-linearly independent,*

then $(b_k)_{k \in N(A)}$ is a basis for $\ker \pi$.

PROOF. (a) For $k \in N(A)$, choose $j_k \in [1, k-1]$ and $(m_{k,1}, \dots, m_{k,k-1}) \in \mathbb{Z}_{\geq 0}^{k-1}$ such that $\delta_k(x_j) \neq 0$ and the monomial $x_1^{m_{k,1}} \cdots x_{k-1}^{m_{k,k-1}}$ appears in $\delta_k(x_{j_k})$ (when expressed in the standard PBW basis of A) with a nonzero coefficient. Comparing the $X(\mathcal{H})$-degrees of the two sides of the relation $x_k x_{j_k} - (h_k \cdot x_{j_k})x_k = \delta_k(x_{j_k})$, we obtain

$$\pi(e_k) + \pi(e_{j_k}) = \sum_{i=1}^{k-1} m_{k,i}\pi(e_i).$$

Now part (a) follows by induction on $k \in N(A)$.

(b) Keeping the notation above, we can choose

$$b_k := e_k + e_{j_k} - \sum_{i=1}^{k-1} m_{k,i} e_i \in \ker \pi\,, \quad \forall\, k \in N(A).$$

Now assume just that for all $k \in N(A)$, we have chosen $b_k \in \ker \pi$ with $b_k - e_k \in \sum_{j=1}^{k-1} \mathbb{Z}e_j$. Regardless of (5.5), the elements b_k are $\mathbb{Z}$-linearly independent, and

$$\mathbb{Z}^N = \sum_{j \in V(A)} \mathbb{Z}e_j + \sum_{k \in N(A)} \mathbb{Z}b_k\,.$$

The remainder of part (b) follows. □

The condition (5.5) does not always hold. For instance, if $A = \mathcal{O}_q(\mathbb{K}^N)$ is the standard quantum affine space with q not a root of unity, then A is a CGL extension with respect to the action of $\mathcal{H} = \mathbb{K}^*$ under which $\alpha.x_i = \alpha x_i$ for all $\alpha \in \mathcal{H}$ and $i \in [1, N]$. Here $V(A) = [1, N]$ and $X(\mathcal{H}) \cong \mathbb{Z}$, so the $\pi(e_i)$ are not $\mathbb{Z}$-linearly independent if $N > 1$.

For any CGL extension A, there is a canonical choice of a torus, acting rationally on A in the manner required by Definition 5.1, for which (5.5) does hold. This torus, which is maximal in a suitable sense, was denoted $\mathcal{H}_{\max}(A)$ in [**10**, Eq. (5.3)], and may be described as follows. For each $k \in N(A)$, choose $j_k \in [1, k-1]$ and $(m_{k,1}, \dots, m_{k,k-1}) \in \mathbb{Z}_{\geq 0}^{k-1}$ as in the proof of Lemma 5.4. Then $\mathcal{H}_{\max}(A)$ is the following subgroup of $(\mathbb{K}^*)^N$:

$$\mathcal{H}_{\max}(A) = \{(\psi_1, \dots, \psi_N) \in (\mathbb{K}^*)^N \mid \psi_k \psi_{j_k} = \prod_{i=1}^{k-1} \psi_i^{m_{k,i}},\ \forall\, k \in N(A)\}. \tag{5.6}$$

It acts on A so that

$$(\psi_1, \dots, \psi_N).x_i = \psi_i x_i\,, \quad \forall\, (\psi_1, \dots, \psi_N) \in \mathcal{H}_{\max}(A),\ i \in [1, N].$$

The action of the given torus $\mathcal{H}$ (as in Definition 5.1) on A factors through $\mathcal{H}_{\max}(A)$ via a morphism of algebraic groups, and so A is also a CGL extension with respect to the $\mathcal{H}_{\max}(A)$-action [**10**, §5.2]. The character group $X(\mathcal{H}_{\max}(A))$ may be identified with $\mathbb{Z}^{V(A)}$, and the degree homomorphism $\pi : \mathbb{Z}^N \to X(\mathcal{H}_{\max}(A))$ of (5.4) is then determined recursively by

$$\pi(e_j) = e_j\,, \qquad j \in V(A),$$

$$\pi(e_k) = -\pi(e_{j_k}) + \sum_{i=1}^{k-1} m_{k,i}\pi(e_i), \qquad k \in N(A).$$

In particular, the $\pi(e_j)$ for $j \in V(A)$ are the standard basis vectors for $\mathbb{Z}^{V(A)}$.

For an algebra A as in (5.1), denote the *interval subalgebra*

$$A_{[j,k]} := \mathbb{K}\langle x_i \mid j \le i \le k \rangle, \quad \forall j,k \in [1,N].$$

In particular, $A_{[j,k]} = \mathbb{K}$ if $j \not\le k$.

DEFINITION 5.5. We call a CGL extension A as in Definition 5.1 *symmetric* if the following two conditions hold:

(i) For all $1 \le j < k \le N$,

$$\delta_k(x_j) \in A_{[j+1,k-1]}.$$

(ii) For all $j \in [1,N]$, there exists $h_j^* \in \mathcal{H}$ such that

$$h_j^* \cdot x_k = \lambda_{kj}^{-1} x_k = \lambda_{jk} x_k, \quad \forall k \in [j+1,N]$$

and $h_j^* \cdot x_j = \lambda_j^* x_j$ for some $\lambda_j^* \in \mathbb{K}^*$ which is not a root of unity.

A symmetric CGL extension A has a presentation as a CGL extension with the variables x_k in descending order:

$$A = \mathbb{K}[x_N][x_{N-1}; \sigma_{N-1}^*, \delta_{N-1}^*] \cdots [x_1; \sigma_1^*, \delta_1^*],$$

see [**10**, Corollary 6.4]. It follows from condition (i) in Definition 5.1 and condition (i) in Definition 5.5 that, for $j < k$, the interval subalgebras $A_{[j,k]}$ also have presentations as CGL extensions

$$A_{[j,k]} = \mathbb{K}[x_j][x_{j+1}; \sigma_{j+1}, \delta_{j+1}] \cdots [x_k; \sigma_k, \delta_k]$$

for the appropriate restrictions of the maps $\sigma_{j+1}, \delta_{j+1}, \dots, \sigma_k, \delta_k$.

THEOREM 5.6. *Let A be a CGL extension as in Definition 5.1, and assume that (5.5) holds (for instance, in case $\mathcal{H} = \mathcal{H}_{\max}(A)$). Let χ' be the alternating bicharacter on $\mathbb{Z}^N$ such that $\chi'(e_i, e_j) = \lambda_{ij}$ for $i,j \in [1,N]$.*

(a) *Choose $b_k \in \mathbb{Z}^N$ as in Lemma 5.4(b). With the above notation,*

$$\mathrm{tw}_{X(\mathcal{H})}(A) = \langle \lambda_k, \ \chi'(b_k, e_i) \mid k \in N(A), \ i \in V(A), \ i < k \rangle.$$

(b) *If A is a symmetric CGL extension then*

$$\mathrm{tw}_{X(\mathcal{H})}(A) = \langle \lambda_k \mid k \in N(A) \rangle.$$

PROOF OF THEOREM 5.6(a). There is an automorphism β of $\mathbb{Z}^N$ such that

$$\beta(e_k) = \sum_{l=0}^{O_-(k)} e_{p^l(k)}, \quad \forall k \in [1,N],$$

and $\chi := \chi' \circ (\beta \times \beta)$ is an alternating bicharacter on $\mathbb{Z}^n$ such that

$$\chi(e_k, e_j) = q_{kj}, \quad \forall k,j \in [1,N].$$

In the notation of Definition 4.3, Theorem 5.2(c) says that

$$\mathcal{O}_\chi(\mathbb{K}^N) \subseteq A \subset \mathcal{T}_\chi^N(\mathbb{K}). \tag{5.7}$$

Giving A its $X(\mathcal{H})$-grading (from the $\mathcal{H}$-action) and $\mathcal{O}_\chi(\mathbb{K}^N)$ and $\mathcal{T}_\chi(\mathbb{K})$ the $X(\mathcal{H})$-gradings induced from their canonical $\mathbb{Z}^N$-gradings along $\phi := \pi\beta$, we see that (5.7) is an inclusion of $X(\mathcal{H})$-graded $\mathbb{K}$-algebras. Thus, Theorem 4.4 implies that

$$\mathrm{tw}_{X(\mathcal{H})}(A) = \langle \chi(\ker \phi, \ \mathbb{Z}^N) \rangle.$$

Since β is an isomorphism, this yields

$$\mathrm{tw}_{X(\mathcal{H})}(A) = \langle \chi'(\ker \pi, \mathbb{Z}^N) \rangle = \langle \chi'(b_k, e_j) \mid k \in N(A), \ j \in [1, N] \rangle, \tag{5.8}$$

taking account of Lemma 5.4(b).

Conditions (i) and (iii) in Definition 5.1 imply the following important property of χ':

$$\chi'(e_j, a) = h_j^{\pi(a)}, \ \ \forall j \in [2, N], \ a \in \bigoplus_{i<j} \mathbb{Z} e_i \,. \tag{5.9}$$

It follows from this property that

$$\chi'(b_k, e_k) = \chi'(e_k, e_k - b_k) = h_k^{\pi(e_k - b_k)} = h_k^{\pi(e_k)} = \lambda_k \,, \ \ \forall\, k \in N(A), \tag{5.10}$$

and that

$$\chi'(b_k, e_j) = h_j^{-\pi(b_k)} = h_j^0 = 1, \ \ \forall k \in N(A), \ j \in [k+1, N]. \tag{5.11}$$

For $k \in N(A)$, write

$$b_k = e_k + a_k \quad \text{with} \quad a_k \in \bigoplus_{i=1}^{k-1} \mathbb{Z} e_i.$$

For $k, j \in N(A)$ with $k > j$, we have,

$$\begin{aligned} \chi'(b_k, e_j) &= \chi'(b_k, b_j - a_j) = \chi'(b_k, b_j)\chi'(b_k, a_j)^{-1} = \chi'(e_k, b_j)\chi'(a_k, b_j)\chi'(b_k, a_j)^{-1} \\ &= \chi'(b_j, a_k)^{-1}\chi'(b_k, a_j)^{-1} \in \langle \lambda_j, \chi'(b_k, e_s), \chi'(b_j, e_t), \ s, t < j \rangle, \end{aligned}$$

where in the last equality we used (5.10)–(5.11). An inductive argument yields that for $k, j \in N(A)$ with $k > j$

$$\chi'(b_k, e_j) \in \langle \lambda_l, \ \chi'(b_l, e_i) \mid l \in N(A), \ i \in V(A), \ i < l \leq k \rangle.$$

Part (a) of the theorem follows from this property and Eqs. (5.8), (5.10)–(5.11). $\square$

5.3. Proof of Theorem 5.6(b). Before we proceed with the proof of part (b) of Theorem 5.6 we establish a lemma. Continue with the previous notation.

LEMMA 5.7. *Let A be a symmetric CGL extension. For each $k \in N(A)$, there exists $a_k \in \bigoplus_{i=p(k)+1}^{k-1} \mathbb{Z} e_i$ such that*

$$b_k := e_{p(k)} - a_k + e_k \in \ker \pi \tag{5.12}$$

$$\chi'(b_k, e_j) = \begin{cases} 1, & \forall j \in [1, N] \setminus \{p(k), k\}, \\ \lambda_k, & j = k, \\ \lambda_k^{-1}, & j = p(k). \end{cases} \tag{5.13}$$

PROOF. Fix $k \in N(A)$. Then $p(k) \neq -\infty$ and the interval subalgebra $A' := A_{[p(k),k]}$ is a CGL extension (with the restricted action of $\mathcal{H}$) of the form

$$A' = \mathbb{K}[x'_1][x'_2; \sigma'_2, \delta'_2] \cdots [x'_n; \sigma'_n, \delta'_n]$$

where $n := k - p(k) + 1$ and $x'_i := x_{p(k)+i-1}$ for $i \in [1, n]$, while σ'_i and δ'_i are the restrictions of $\sigma_{p(k)+i-1}$ and $\delta_{p(k)+i-1}$ to $A_{[p(k),p(k)+i-2]}$ for $i \in [2, n]$. The η-function for A' given by Theorem 5.2(b), call it η', can be chosen as follows:

$$\eta'(i) = \eta(p(k) + i - 1), \ \ \forall i \in [1, n],$$

see [**11**, Corollary 5.6(b)]. If p' is the corresponding predecessor function, then $p'(n) = 1$. Hence, up to a scalar multiple the final $X(\mathcal{H})$-homogeneous prime element of A' in Theorem 5.2(a)(b) is

$$y'_n = x'_1 x'_n - c'_n \tag{5.14}$$

where $c'_n \in A_{[p(k),k-1]}$ and all terms in (5.14) are homogeneous of the same $X(\mathcal{H})$-degree. In fact, c'_n is a scalar multiple of $\delta'_n(x'_1) = \delta_k(x_{p(k)})$ [**10**, Proposition 4.7(b)], which means that

$$c'_n \in A_{[p(k)+1,k-1]}\,,$$

due to the symmetry assumption on A. Moreover, by [**10**, Corollary 4.8],

$$y'_n x_j = (\lambda_{jk}\lambda_{j,p(k)})^{-1} x_j y'_n\,, \quad \forall j \in [p(k), k]. \tag{5.15}$$

The monomials

$$x_{p(k)}^{n_{p(k)}} \dots x_k^{n_k}, \quad (n_{p(k)}, \dots, n_k) \in \mathbb{Z}^n_{\geq 0}$$

form a $\mathbb{K}$-basis of A', and the vector of exponents for the leading term of y'_n with respect to the right-to-left lexicographic order on $\mathbb{Z}^n_{\geq 0}$ is $(1, 0, \dots, 0, 1)$. Let $(0, m_{p(k)+1}, \dots, m_{k-1}, 0)$ be the vector of exponents for the leading term of c'_n, and set

$$a_k := m_{p(k)+1} e_{p(k)+1} + \dots + m_{k-1} e_{k-1}.$$

It follows from the $X(\mathcal{H})$-homogeneity of (5.14) that (5.12) holds, and from (5.10), (5.11) we obtain (5.13) for $j \in [k, N]$.

As far as the CGL extension $A_{[p(k)+1,k]}$ is concerned, the index for x_k has no predecessor, and so x_k is a homogeneous normal element of that algebra. Similarly, $x_{p(k)}$ is a homogeneous normal element of the algebra $A_{[p(k),k-1]}$. Hence,

$$x_{p(k)} x_k x_j = \lambda_{p(k),j} \lambda_{kj} x_j x_{p(k)} x_k\,, \quad \forall j \in [p(k)+1, k-1].$$

Taking (5.15) into account, we see that

$$c'_n x_j = \lambda_{p(k),j} \lambda_{kj} x_j c'_n\,, \quad \forall j \in [p(k)+1, k-1].$$

This equation, together with the relation [**10**, Eq. (4.12)] for leading terms of products of monomials in CGL extensions, implies

$$\prod_{i=j+1}^{k-1} \lambda_{ij}^{m_i} = \lambda_{p(k),j} \lambda_{kj} \prod_{i=p(k)+1}^{j-1} \lambda_{ji}^{m_i}\,, \quad \forall j \in [p(k)+1, k-1],$$

which yields

$$\chi'(b_k, e_j) = 1, \quad \forall j \in [p(k)+1, k-1].$$

Condition (ii) of Definition 5.5 implies the following analog of (5.9):

$$\chi'(e_j, a) = (h_j^*)^{\pi(a)}\,, \quad \forall j \in [1, p(k)-1], \quad a \in \bigoplus_{i>j} \mathbb{Z} e_i\,.$$

As with (5.10) and (5.11), it follows that

$$\begin{aligned} \chi'(b_k, e_{p(k)}) &= \lambda^*_{p(k)} = \lambda_k^{-1}\,, \\ \chi'(b_k, e_j) &= 1, \qquad \forall j \in [1, p(k)-1], \end{aligned}$$

where in the first equality we used [**11**, Proposition 5.8]. This completes the proof of the lemma. □

PROOF OF THEOREM 5.6(b). For $k \in N(A)$, let $b_k := e_{p(k)} - a_k + e_k$ as in (5.12), and note that these elements satisfy the hypotheses of Lemma 5.4(b). In view of (5.13),

$$\chi'(b_k, e_j) \in \langle \lambda_k \rangle, \ \ \forall k \in N(A), \ \ j \in [1, N].$$

Part (b) of the theorem follows from this and part (a). □

EXAMPLE 5.8. It is well known that $A := \mathcal{O}_{\lambda,\boldsymbol{p}}(M_n(\mathbb{K}))$ can be expressed as a CGL extension provided λ is a non-root of unity (cf. [**8**, §5.6] and [**16**, Corollary 3.8]) with the generators X_{ij} adjoined in lexicographic order and the torus $\mathcal{H} := (\mathbb{K}^*)^{2n}$ acting so that

$$\alpha.X_{ij} = \alpha_i \alpha_{n+j} X_{ij}\,, \ \ \forall \alpha = (\alpha_1, \dots, \alpha_{2n}) \in \mathcal{H}, \ \ i, j \in [1, n].$$

Equally, A can be expressed as a CGL extension with the generators adjoined in reverse lexicographic order and the same $\mathcal{H}$-action, which shows that this algebra is a symmetric CGL extension. We may identify $X(\mathcal{H})$ with $\mathbb{Z}^{2n}$, and we note that the $X(\mathcal{H})$-grading of A equals the $\mathbb{Z}^{2n}$-grading given in Example 3.7. Observe that

$$V(A) = [1, n] \cup \{in + 1 \mid i \in [1, n-1]\},$$

and that the degree homomorphism $\pi : \mathbb{Z}^{n^2} \to X(\mathcal{H})$ satisfies $\pi(j) = e_1 + e_{n+j}$ for $j \in [1, n]$ while $\pi(in + 1) = e_{i+1} + e_{n+1}$ for $i \in [1, n-1]$. Hence, (5.5) holds.

The calculations in [**8**, §5.6] show that $\lambda_k = \lambda$ for all $k \in N(A)$. Therefore, by Proposition 5.3 and Theorem 5.6,

$$\text{(5.16)} \quad \mathrm{AD}(\mathcal{O}_{\lambda,\boldsymbol{p}}(M_n(\mathbb{K}))) = \langle \lambda, \boldsymbol{p} \rangle \quad \text{and} \quad \mathrm{tw}_{X(\mathcal{H})}(\mathcal{O}_{\lambda,\boldsymbol{p}}(M_n(\mathbb{K}))) = \langle \lambda \rangle,$$

assuming λ is a non-root of unity. Specializing to the standard quantum matrix algebra yields

$$\text{(5.17)} \quad \mathrm{AD}(\mathcal{O}_q(M_n(\mathbb{K}))) = \langle q \rangle \quad \text{and} \quad \mathrm{tw}_{X(\mathcal{H})}(\mathcal{O}_q(M_n(\mathbb{K}))) = \langle q^2 \rangle$$

provided q is not a root of unity.

5.4. Uniparameter CGL extensions.

DEFINITION 5.9. We will say that a CGL extension A as in Definition 5.1 is *uniparameter* if there is some $q \in \mathbb{K}^*$ such that all $\lambda_{ij} \in \langle q \rangle$, that is, $\langle \boldsymbol{\lambda} \rangle$ is a cyclic group.

For example, the standard quantum matrix algebras $\mathcal{O}_q(M_n(\mathbb{K}))$ are uniparameter, and all the multiparameter quantum matrix algebras $\mathcal{O}_{\lambda,\boldsymbol{p}}(M_n(\mathbb{K}))$ are cocycle twists of uniparameter ones, as shown in Example 3.7. On the other hand, many quantum algebras cannot be so reduced to uniparameter algebras, as we now illustrate.

EXAMPLE 5.10. Let $A := A_n^{Q,P}(\mathbb{K})$ as in Example 4.7, and assume that all the q_k are non-roots of unity. As is well known, A is then a CGL extension with variables in the order $x_1, y_1, \dots, x_n, y_n$, equipped with the action of $\mathcal{H} := (\mathbb{K}^*)^n$ by algebra automorphisms such that

$$\text{(5.18)} \quad (\alpha_1, \dots, \alpha_n).x_i = \alpha_i x_i \ \text{and} \ (\alpha_1, \dots, \alpha_n).y_i = \alpha_i^{-1} y_i\,, \ \ \forall i \in [1, n]$$

(cf. [**8**, §5.10], [**16**, Corollary 3.8]). We can identify the $X(\mathcal{H})$-grading of A with the $\mathbb{Z}^n$-grading of Example 4.7, and so by (4.7),

$$\mathrm{tw}_{X(\mathcal{H})}(A_n^{Q,P}(\mathbb{K})) = \langle q_1, \dots, q_n \rangle.$$

(This can also be obtained from Theorem 5.6 after rewriting A as a symmetric CGL extension, which can be done by adjoining the generators in the order $y_n, \ldots, y_1, x_1, \ldots, x_n$.) It follows that whenever the group $\langle q_1, \ldots, q_n \rangle$ fails to be cyclic, $A_n^{Q,P}(\mathbb{K})$ is not cocycle twist equivalent to a uniparameter CGL extension.

5.5. Quantum Schubert cell algebras. Let $\mathfrak{g}$ be a symmetrizable Kac–Moody algebra with set of simple roots $\alpha_1, \ldots, \alpha_r$, Weyl group W, and root lattice Q. Denote by $U_q(\mathfrak{g})$ the corresponding quantized universal enveloping algebra over $\mathbb{K}$, for a deformation parameter $q \in \mathbb{K}^*$ which is not a root of unity. Let $\{K_i^{\pm 1}, E_i, F_i \mid i \in [1, r]\}$ be the generators of $U_q(\mathfrak{g})$. Denote by

$$\langle .,. \rangle : Q \times Q \longrightarrow \mathbb{Z}$$

the invariant symmetric form, normalized by $\|\alpha\|^2 := \langle \alpha, \alpha \rangle = 2$ for a short root α.

To each element $w \in W$, one associates a quantum Schubert cell subalgebra $U_-^q[w] \subset U_q(\mathfrak{g})$, [**17**, §39.3]. Given a reduced expression $w = s_{i_1} \ldots s_{i_l}$, define the roots

$$\beta_k := s_{i_1} \ldots s_{i_{k-1}}(\alpha_{i_k}), \quad \forall\, k \in [1, l]$$

of $\mathfrak{g}$ and the root vectors

$$F_{\beta_k} := T_{i_1} \ldots T_{i_{k-1}}(E_{i_k}), \quad \forall\, k \in [1, l]$$

of $U_q(\mathfrak{g})$. Here T denotes Lusztig's action of the braid group of $\mathfrak{g}$ on $U_q(\mathfrak{g})$, [**17**, Ch. 5]. The algebra $U_-^q[w]$ is defined as the subalgebra of $U_q(\mathfrak{g})$ generated by $F_{\beta_1}, \ldots, F_{\beta_l}$. This subalgebra is independent of the choice of a reduced expression of w. The algebras of quantum matrices $\mathcal{O}_q(M_n(\mathbb{K}))$ are recovered as special cases for $\mathfrak{g} = \mathfrak{sl}_{2n}$ and a particular choice of $w \in S_{2n}$, [**19**, Sect. 4].

The algebra $U_-^q[w]$ is Q-graded by

$$\deg F_{\beta_k} := \beta_k\,, \quad \forall\, k \in [1, l].$$

The grading gives rise to a rational action of a torus $\mathcal{H} := (\mathbb{K}^*)^r$ by algebra automorphisms obtained by identifying $X(\mathcal{H}) \cong Q$. The algebra $U_-^q[w]$ is a symmetric CGL extension when its generators are adjoined in the order $F_{\beta_1}, \ldots, F_{\beta_l}$, [**11**, Lemma 9.1]. The corresponding scalars $\lambda_1, \ldots, \lambda_l$ are given by

$$\lambda_k = q^{-\|\beta_k\|^2} = q^{-\|\alpha_{i_k}\|^2}, \quad \forall\, k \in [1, l], \tag{5.19}$$

see [**11**, Eq. (9.15)]. These facts were stated in [**11**] in the case where $\mathfrak{g}$ is finite dimensional, but the proofs in the symmetrizable Kac–Moody case are identical.

Denote the support of $w \in W$ by

$$\operatorname{supp}(w) := \{j \in [1, r] \mid j = i_k \text{ for some } k \in [1, l]\},$$

and set

$$d(w) := \gcd(\|\alpha_j\|^2,\ j \in \operatorname{supp}(w)).$$

Theorem 5.6(b) and (5.19) imply at once the following result.

PROPOSITION 5.11. *For all symmetrizable Kac–Moody algebras $\mathfrak{g}$ with root lattice Q, Weyl group elements w, and non-roots of unity $q \in \mathbb{K}^*$,*

$$\operatorname{tw}_Q(U_-^q[w]) = \langle q^{d(w)} \rangle.$$

6. General twist invariants

For a $\mathbb{Z}$-graded algebra A, the invariant $\mathrm{tw}_{\mathbb{Z}}(A)$ may not distinguish A from twists of A by 2-cocycles on finer grading groups, as illustrated by Proposition 4.6. In order to recover cocycle twist invariant information within a $\mathbb{Z}$-graded invariant, we work with gradings of A by character groups of maximal tori of suitable automorphism groups of A with algebraic group structures. For this process to function, the homogeneous components of the $\mathbb{Z}$-grading on A need to be finite dimensional. We can enlarge the setting to include filtered algebras, and we begin in that setting.

6.1. Filtered automorphism groups.

DEFINITION 6.1. All filtered $\mathbb{K}$-algebras A that we consider will be nonnegatively filtered, and the filtration will be denoted $F_\bullet A = (F_n A)_{n\geq 0}$. The *filtered automorphism group* of such a filtered algebra is

$$\mathrm{Aut}_{\mathrm{fl}}\, A := \{\phi \in \mathrm{Aut}_{\mathbb{K}}\, A \mid \phi(F_n A) = F_n A, \ \ \forall n \geq 0\},$$

where $\mathrm{Aut}_{\mathbb{K}}\, A$ denotes the group of all $\mathbb{K}$-algebra automorphisms of A. Recall that the filtration $F_\bullet A$ is called *locally finite* if each of the subspaces $F_n A$ is finite dimensional over $\mathbb{K}$.

LEMMA 6.2. *Let A be an affine $\mathbb{K}$-algebra and $V \subseteq W$ finite dimensional $\mathbb{K}$-subspaces of A such that V generates A as an algebra. Set*

$$\mathcal{A}_V := \{\phi \in \mathrm{Aut}_{\mathbb{K}}\, A \mid \phi(V) = V\} \qquad \textit{and} \qquad \mathcal{A}_W := \{\phi \in \mathrm{Aut}_{\mathbb{K}}\, A \mid \phi(W) = W\},$$

and let

$$\rho_V : \mathcal{A}_V \longrightarrow \mathrm{GL}(V) \qquad \textit{and} \qquad \rho_W : \mathcal{A}_W \longrightarrow \mathrm{GL}(W)$$

be the homomorphisms given by restriction to V and W, respectively.

(a) *$\rho_V(\mathcal{A}_V)$ and $\rho_V(\mathcal{A}_V \cap \mathcal{A}_W)$ are (Zariski-)closed subgroups of $\mathrm{GL}(V)$, while $\rho_W(\mathcal{A}_V)$ and $\rho_W(\mathcal{A}_V \cap \mathcal{A}_W)$ are closed subgroups of $\mathrm{GL}(W)$.*

(b) *The map $\rho_{WV} : \rho_W(\mathcal{A}_V \cap \mathcal{A}_W) \to \rho_V(\mathcal{A}_V \cap \mathcal{A}_W)$ given by restriction to V is an isomorphism of algebraic groups.*

PROOF. (a) Let $\widehat{A} := T(V)$ be the tensor algebra of V, graded by $\mathbb{Z}_{\geq 0}$ in the standard way, so that $\widehat{A}_m = V^{\otimes m}$, and let $\pi : \widehat{A} \to A$ be the unique $\mathbb{K}$-algebra homomorphism which is the identity on V. For $g \in \mathrm{GL}(V)$, let $\widehat{g}$ be the unique $\mathbb{K}$-algebra automorphism of $\widehat{A}$ which restricts to g on V. Observe that $g \in \rho_V(\mathcal{A}_V)$ if and only if $\widehat{g}$ induces a $\mathbb{K}$-algebra automorphism of A, if and only if $\widehat{g}(\ker \pi) = \ker \pi$. The last condition is equivalent to $\widehat{g}(K_m) = K_m$ for all $m \geq 0$, where $K_m := \widehat{A}_{\leq m} \cap \ker \pi$.

Now $H_m := \{h \in \mathrm{GL}(\widehat{A}_{\leq m}) \mid h(K_m) = K_m\}$ is a closed subgroup of $\mathrm{GL}(\widehat{A}_{\leq m})$, and the set

$$G_m := \{g \in \mathrm{GL}(V) \mid \widehat{g}(K_m) = K_m\}$$

is the inverse image of H_m under the morphism

$$\tau_m : \mathrm{GL}(V) \longrightarrow \mathrm{GL}(\widehat{A}_{\leq m}), \quad g \longmapsto \bigoplus_{n=0}^{m} (g^{\otimes n})|_{V^{\otimes n}} = \widehat{g}|_{\widehat{A}_{\leq m}},$$

so G_m is closed in $\mathrm{GL}(V)$. Therefore $\rho_V(\mathcal{A}_V) = \bigcap_{m\geq 0} G_m$ is closed in $\mathrm{GL}(V)$.

Similarly, $\rho_W(\mathcal{A}_W)$ is closed in $\mathrm{GL}(W)$. Then also

$$\rho_W(\mathcal{A}_V \cap \mathcal{A}_W) = \{f \in \rho_W(\mathcal{A}_W) \mid f(V) = V\}$$

is closed in $\mathrm{GL}(W)$. Choose $m_0 \in \mathbb{Z}_{\geq 0}$ large enough so that $W \subseteq V^{m_0}$, and observe that

$$\rho_V(\mathcal{A}_V \cap \mathcal{A}_W) = \bigcap_{m \geq m_0} \tau_m^{-1}\big(\{l \in \mathrm{GL}(\widehat{A}_{\leq m}) \mid l(K_m) = K_m \text{ and } l(\pi^{-1}(W)) = \pi^{-1}(W)\}\big).$$

It follows that $\rho_V(\mathcal{A}_V \cap \mathcal{A}_W)$ is closed in $\mathrm{GL}(V)$.

(b) Injectivity of ρ_{WV} is clear, since automorphisms of A are determined by their actions on V. Set

$$E_V := \rho_V(\mathcal{A}_V \cap \mathcal{A}_W) \qquad \text{and} \qquad E_W := \rho_W(\mathcal{A}_V \cap \mathcal{A}_W),$$

and choose $m \in \mathbb{Z}_{\geq 0}$ such that $W \subseteq V^m$. Let $\mu : V^{\otimes m} \to V^m$ be the multiplication map. Set

$$C := \{f \in \mathrm{GL}(V^{\otimes m}) \mid f(\ker \mu) = \ker \mu\} \quad \text{and} \quad D := \{f \in \mathrm{GL}(V^m) \mid f(W) = W\},$$

which are closed subgroups of $\mathrm{GL}(V^{\otimes m})$ and $\mathrm{GL}(V^m)$, respectively. Restriction from V^m to W provides an algebraic group morphism $r : D \to \mathrm{GL}(W)$, and induction from $V^{\otimes m}$ to V^m along μ provides an algebraic group morphism $s : C \to \mathrm{GL}(V^m)$. Finally, we have a morphism $t : \mathrm{GL}(V) \to \mathrm{GL}(V^{\otimes m})$ given by $t(g) = g^{\otimes m}$.

If $\phi \in \mathcal{A}_V \cap \mathcal{A}_W$ and $g = \rho_V(\phi) = \phi|_V$, then since

$$\mu t(g)(v_1 \otimes \cdots \otimes v_m) = g(v_1)g(v_2)\cdots g(v_m) = \phi(v_1 v_2 \cdots v_m)$$

for all $v_1, \ldots, v_m \in V$, we see that $\mu t(g) = \phi\mu$. In particular, it follows that $t(g) \in C$. Hence, s composed with $t|_{E_V}$ provides a morphism $st : E_V \to \mathrm{GL}(V^m)$. For g and ϕ as above, $st(g)\mu = \mu t(g) = \phi\mu$, whence $st(g) = \phi|_{V^m}$. Since $\phi \in \mathcal{A}_W$, it follows that $st(g) \in D$ and $rst(g) = \phi|_W = \rho_W(\phi)$, which means that $rst(g) \in E_W$ and $\rho_{WV} rst(g) = \phi|_V = g$. Therefore ρ_{WV} is bijective and $\rho_{WV}^{-1} = rst|_{E_V}$, which is a morphism. This establishes part (b). □

PROPOSITION 6.3. *Let A be an affine locally finite filtered $\mathbb{K}$-algebra, and for $d \in \mathbb{Z}_{\geq 0}$ let $\rho_d : \mathrm{Aut_{fl}}\, A \to \mathrm{GL}(F_d A)$ be the homomorphism given by restriction to $F_d A$.*

(a) *If $F_d A$ generates A (as an algebra), then ρ_d is a group isomorphism from $\mathrm{Aut_{fl}}\, A$ onto a (Zariski-)closed subgroup of $\mathrm{GL}(F_d A)$.*

(b) *If $F_d A$ generates A and $e \geq d$, then the map $\rho_{ed} : \operatorname{im} \rho_e \to \operatorname{im} \rho_d$ given by restriction to $F_d A$ is an isomorphism of algebraic groups.*

(c) *Choose $d \in \mathbb{Z}_{\geq 0}$ such that $F_d A$ generates A, and make $\mathrm{Aut_{fl}}\, A$ into an affine algebraic group via the isomorphism ρ_d. Then the action of $\mathrm{Aut_{fl}}\, A$ on A is a rational action.*

PROOF. Fix $d \leq e$ in $\mathbb{Z}_{\geq 0}$ such that $F_d A$ generates A. Set $V := F_d A$ and $W := F_e A$. For $n \in \mathbb{Z}_{\geq 0}$, define

$$\mathcal{A}_n := \{\phi \in \mathrm{Aut}_{\mathbb{K}}\, A \mid \phi(F_n A) = F_n A\},$$

and let $\rho'_n : \mathcal{A}_n \to \mathrm{GL}(F_n A)$ be the homomorphism given by restriction to $F_n A$.

(a) By Lemma 6.2(a), $\rho'_d(\mathcal{A}_d)$ is a closed subgroup of $\mathrm{GL}(V)$. It is clear that ρ'_d and ρ_d are injective. Now $\operatorname{im}\rho_d = \rho'_d(\mathrm{Aut_{fl}}\, A)$, so it follows from the injectivity of ρ'_d that

$$\operatorname{im}\rho_d = \bigcap_{n=0}^{\infty} \rho'_d(\mathcal{A}_d \cap \mathcal{A}_n).$$

For $n \geq d$, we have $\rho'_d(\mathcal{A}_d \cap \mathcal{A}_n)$ closed in $\mathrm{GL}(V)$ by Lemma 6.2(a), while for $n < d$, closedness holds because

$$\rho'_d(\mathcal{A}_d \cap \mathcal{A}_n) = \{f \in \rho'_d(\mathcal{A}_d) \mid f(F_n A) = F_n A\}.$$

Therefore $\operatorname{im}\rho_d$ is closed in $\mathrm{GL}(V)$.

(b) By Lemma 6.2(b), the map $\rho'_{ed} : \rho'_e(\mathcal{A}_d \cap \mathcal{A}_e) \to \rho'_d(\mathcal{A}_d \cap \mathcal{A}_e)$ given by restriction to V is an isomorphism of algebraic groups. Since ρ_{ed} is the restriction of ρ'_{ed} to $\operatorname{im}\rho_e$, it follows that ρ_{ed} is an algebraic group isomorphism of $\operatorname{im}\rho_e$ onto a closed subgroup of $\rho'_d(\mathcal{A}_d \cap \mathcal{A}_e)$. But clearly $\rho_{ed}(\operatorname{im}\rho_e) = \operatorname{im}\rho_d$.

(c) For each $e \geq d$, the space $F_e A$ is an $\mathrm{Aut_{fl}}\, A$-stable finite dimensional subspace of A, and the map $\rho_e : \mathrm{Aut_{fl}}\, A \to \mathrm{GL}(F_e A)$ is a morphism by part (b). $\square$

We now impose the assumption that the field $\mathbb{K}$ is infinite, so that rational actions of $\mathbb{K}$-tori on $\mathbb{K}$-algebras correspond to $X(\mathcal{H})$-gradings (e.g., [**6**, Lemma II.2.11]).

DEFINITION 6.4. Suppose A is an affine locally finite filtered $\mathbb{K}$-algebra. Choose $d \in \mathbb{Z}_{\geq 0}$ such that $F_d A$ generates A, and give $\mathrm{Aut_{fl}}\, A$ the structure of an affine algebraic group as in Proposition 6.3(a). By parts (b),(c) of the proposition, this structure is independent of the choice of d, and the action of $\mathrm{Aut_{fl}}\, A$ on A is rational.

6.2. General twist invariants of filtered algebras. Let $\mathcal{H}$ be a maximal torus of $\mathrm{Aut_{fl}}\, A$. The restricted action of $\mathcal{H}$ on A is rational, so A receives a corresponding $X(\mathcal{H})$-grading.

Now suppose A is also a semiprime right Goldie ring. The *general twist invariant* of A is

$$\mathrm{tw}(A) = \mathrm{tw}(A, F_\bullet) := \mathrm{tw}_{X(\mathcal{H})}(A),$$

which is well defined by following lemma:

LEMMA 6.5. *Let A be an affine, locally finite, filtered, semiprime right Goldie $\mathbb{K}$-algebra. Then $\mathrm{tw}(A)$ is independent of the choice of a maximal torus $\mathcal{H}$ of $\mathrm{Aut_{fl}}\, A$.*

PROOF. Any two maximal tori $\mathcal{H}_1$ and $\mathcal{H}_2$ of $\mathrm{Aut_{fl}}\, A$ are conjugate, as follows by applying [**5**, Corollary 11.3] to the connected group $(\mathrm{Aut_{fl}}\, A)^\circ$, so $\mathcal{H}_2 = \phi\mathcal{H}_1\phi^{-1}$ for some $\phi \in \mathrm{Aut_{fl}}\, A$. The isomorphism $\theta : \mathcal{H}_1 \to \mathcal{H}_2$ given by conjugation by ϕ induces an isomorphism $\theta^* : \Gamma_2 := X(\mathcal{H}_2) \to \Gamma_1 := X(\mathcal{H}_1)$. We check that ϕ provides an isomorphism from (A with Γ_1-grading) to (A with Γ_2-grading), in the sense that for any $d \in \Gamma_2$, the automorphism ϕ restricts to a vector space isomorphism of $A_{\theta^*(d)}$ onto A_d. Namely, if $a \in A_{\theta^*(d)}$, then

$$\theta(h_1).\phi(a) = \phi h_1 \phi^{-1}.\phi(a) = \phi(h_1.a) = \phi\big(\big[\theta^*(d)(h_1)\big]a\big) = \big[d(\theta(h_1))\big]\phi(a), \quad \forall\, h_1 \in \mathcal{H}_1\,,$$

whence $\phi(a) \in A_d$. By symmetry, ϕ^{-1} maps A_d to $A_{\theta^*(d)}$, establishing the claim. It follows that $\mathrm{tw}_{\Gamma_1}(A) = \mathrm{tw}_{\Gamma_2}(A)$. $\square$

EXAMPLE 6.6. Consider $A := \mathcal{O}_{\boldsymbol{q}}(\mathbb{K}^n)$, where $\boldsymbol{q} \in M_n(\mathbb{K}^*)$ is an arbitrary multiplicatively skew-symmetric matrix, and give A the standard filtration in which $F_k A$ is the $\mathbb{K}$-span of the monomials $y_1^{m_1} \cdots y_n^{m_n}$ with $m_1 + \cdots + m_n \leq k$. Then

$\operatorname{Aut}_{\text{fl}} A$ becomes an algebraic group isomorphic, by restriction, to a closed subgroup of $\operatorname{GL}(F_1A)$. Identify the torus $\mathcal{H} := (\mathbb{K}^*)^n$ in the standard way with a subgroup of $\operatorname{Aut}_{\text{fl}} A$, where

$$(\alpha_1, \dots, \alpha_n)(y_i) = \alpha_i y_i\,, \quad \forall\,(\alpha_1, \dots, \alpha_n) \in \mathcal{H}, \quad i \in [1, n].$$

it is easily checked that $\mathcal{H}$ is a maximal torus of $\operatorname{Aut}_{\text{fl}} A$. When $X(\mathcal{H})$ is identified with $\mathbb{Z}^n$ in the canonical manner, the corresponding grading of A is the one for which $\deg y_i = e_i$ for all i. Thus, we conclude from Example 3.6 that

$$\operatorname{tw}(A) = \operatorname{tw}_{\mathbb{Z}^n}(A) = \{1\}.$$

THEOREM 6.7. *Let A be an affine, locally finite, filtered, semiprime right Goldie $\mathbb{K}$-algebra and $\mathcal{H} \subseteq \operatorname{Aut}_{\text{fl}} A$ a $\mathbb{K}$-torus acting rationally on A by algebra automorphisms. For every 2-cocycle $\mathbf{c} \in Z^2(X(\mathcal{H}), \mathbb{K}^*)$,*

$$\operatorname{tw}(A) = \operatorname{tw}(A^{\mathbf{c}}).$$

Here $\operatorname{tw}(A^{\mathbf{c}})$ is defined because $F_\bullet A$ is also a filtration of $A^{\mathbf{c}}$, and $\mathcal{H}$ is a subtorus of $\operatorname{Aut}_{\text{fl}} A^{\mathbf{c}}$.

PROOF. The subtorus $\mathcal{H} \subseteq \operatorname{Aut}_{\text{fl}} A$ is contained in a maximal torus $\widetilde{\mathcal{H}} \subseteq \operatorname{Aut}_{\text{fl}} A$. Using the projection $\pi : X(\widetilde{\mathcal{H}}) \to X(\mathcal{H})$ define the 2-cocycle

$$\widetilde{\mathbf{c}} := \mathbf{c} \circ (\pi \times \pi) \in Z^2(X(\widetilde{\mathcal{H}}), \mathbb{K}^*).$$

Clearly $A^{\widetilde{\mathbf{c}}} \cong A^{\mathbf{c}}$ and thus,

$$\operatorname{tw}(A^{\mathbf{c}}) = \operatorname{tw}(A^{\widetilde{\mathbf{c}}}) = \operatorname{tw}_{X(\widetilde{\mathcal{H}})}(A^{\widetilde{\mathbf{c}}}) = \operatorname{tw}_{X(\widetilde{\mathcal{H}})}(A) = \operatorname{tw}(A). \qquad \square$$

6.3. General twist invariants of graded algebras.

OBSERVATIONS 6.8. Recall that nonnegatively $\mathbb{Z}$-graded algebras are turned into nonnegatively filtered algebras in a canonical way. Namely, if $A = \bigoplus_{n=0}^\infty A_n$ is a $\mathbb{Z}_{\geq 0}$-graded $\mathbb{K}$-algebra, we equip A with the associated filtration $F_\bullet A$ where $F_n A := \bigoplus_{k=0}^n A_k$ for all $n \geq 0$. Then $\operatorname{Aut}_{\text{fl}} A$ contains as a subgroup the group

$$\operatorname{Aut}_{\text{gr}} A := \{\phi \in \operatorname{Aut}_{\mathbb{K}} A \mid \phi(A_n) = A_n\,, \ \ \forall\, n \geq 0\}$$

of graded $\mathbb{K}$-algebra automorphisms of A.

Now suppose A is affine and its grading is *locally finite*, meaning that each A_n is finite dimensional over $\mathbb{K}$. If we give $\operatorname{Aut}_{\text{fl}} A$ the structure of an affine algebraic group via Proposition 6.3, then $\operatorname{Aut}_{\text{gr}} A$ is a Zariski-closed subgroup. More precisely, if $d \in \mathbb{Z}_{\geq 0}$ is chosen so that $F_d A$ generates A as an algebra, then restriction to $F_d A$ provides an isomorphism ρ_d of $\operatorname{Aut}_{\text{fl}} A$ onto a closed subgroup of $\operatorname{GL}(F_d A)$, and ρ_d restricts to an isomorphism of $\operatorname{Aut}_{\text{gr}} A$ onto the following closed subgroup of $\operatorname{im} \rho_d$:

$$\{f \in \operatorname{im} \rho_d \mid f(A_k) = A_k\,, \ \ \forall\, k \leq d\} \cap \bigcap_{e > d} \rho_{ed}^{-1}\big(\{g \in \operatorname{im} \rho_e \mid g(A_e) = A_e\}\big),$$

in the notation of Proposition 6.3. Therefore $\operatorname{Aut}_{\text{gr}} A$ becomes an affine algebraic group.

Thus, if A is an affine locally finite $\mathbb{Z}_{\geq 0}$-graded $\mathbb{K}$-algebra and also a semiprime right Goldie ring, $\operatorname{tw}(A)$ is defined. In setting up $\operatorname{tw}(A)$, one may use a maximal torus of $\operatorname{Aut}_{\text{gr}} A$ since any such is also a maximal torus of $\operatorname{Aut}_{\text{fl}} A$, as we now show.

LEMMA 6.9. *Let A be an affine locally finite $\mathbb{Z}_{\geq 0}$-graded $\mathbb{K}$-algebra. Give $\operatorname{Aut}_{\text{fl}} A$ and $\operatorname{Aut}_{\text{gr}} A$ affine algebraic group structures as in Proposition 6.3 and Observations 6.8. Then any maximal torus of $\operatorname{Aut}_{\text{gr}} A$ is also a maximal torus of $\operatorname{Aut}_{\text{fl}} A$.*

PROOF. Let $\mathcal{H}$ be a maximal torus of $\operatorname{Aut}_{\mathrm{gr}} A$ and $\mathcal{H}'$ a subtorus of $\operatorname{Aut}_{\mathrm{fl}} A$ which contains $\mathcal{H}$. We must show $\mathcal{H}' = \mathcal{H}$.

Since the filtration on A is the one induced from the grading, the associated graded algebra of $(A, F_\bullet)$ is canonically isomorphic to A as a graded algebra. Consequently, automorphisms of $(A, F_\bullet)$ canonically induce graded automorphisms of A, yielding a canonical homomorphism

$$\gamma : \operatorname{Aut}_{\mathrm{fl}} A \longrightarrow \operatorname{Aut}_{\mathrm{gr}} A.$$

Observe that γ is a morphism of algebraic groups, and that γ restricts to the identity on $\operatorname{Aut}_{\mathrm{gr}} A$.

Now $\gamma(\mathcal{H}')$ is a subtorus of $\operatorname{Aut}_{\mathrm{gr}} A$ containing $\gamma(\mathcal{H}) = \mathcal{H}$, so $\gamma(\mathcal{H}') = \mathcal{H}$. On the other hand, if we identify $\operatorname{Aut}_{\mathrm{fl}} A$ with the closed subgroup $\operatorname{im} \rho_d \subseteq \mathrm{GL}(F_d A)$ for a suitable d as in Proposition 6.3, then

$$\mathcal{H}' \cap \ker \gamma = \Big\{ f \in \operatorname{im} \rho_d \mid (f - \mathrm{id})\Big(\bigoplus_{k=0}^{n} A_k\Big) \subseteq \bigoplus_{k=0}^{n-1} A_k, \ \ \forall n \in [0, d] \Big\}.$$

It follows that $\mathcal{H}' \cap \ker \gamma$ is unipotent. Since algebraic tori have no nontrivial unipotent subgroups, $\mathcal{H}' \cap \ker \gamma = \{\mathrm{id}\}$. Therefore we conclude from $\gamma(\mathcal{H}') = \gamma(\mathcal{H})$ that $\mathcal{H}' = \mathcal{H}$, as desired. □

We illustrate the use of the above results to recapture the twisted commutation invariant of a quantum matrix algebra from the general twist invariant of the algebra.

EXAMPLE 6.10. Let $A := \mathcal{O}_{\lambda,\mathbf{p}}(M_n(\mathbb{K}))$ as in Example 3.7, and equip A with the natural $\mathbb{Z}$-grading in which all the generators X_{ij} have degree 1. The first stage of the corresponding filtration on A is the subspace $V := \mathbb{K} + \bigoplus_{i,j=1}^{n} \mathbb{K} X_{ij}$, which generates A as a $\mathbb{K}$-algebra. By Observations 6.8, $\operatorname{Aut}_{\mathrm{gr}} A$ becomes an algebraic group isomorphic by the restriction morphism ρ_1 to the closed subgroup $\mathcal{A} := \rho_1(\operatorname{Aut}_{\mathrm{gr}} A)$ of $\mathrm{GL}(V)$.

Now view A as a CGL extension as in Example 5.8. If we express the maps in $\mathrm{GL}(V)$ as matrices with respect to the basis $1, X_{11}, X_{12}, \dots, X_{nn}$, then the action of $\mathcal{H}$ on A corresponds to the morphism $\Delta : \mathcal{H} \to \mathcal{A}$ given by the rule

$$\Delta(\alpha_1, \dots, \alpha_n, \beta_1, \dots, \beta_n) = \\ \operatorname{diag}(1, \alpha_1\beta_1, \alpha_1\beta_2, \dots, \alpha_1\beta_n, \alpha_2\beta_1, \dots, \alpha_n\beta_1, \dots, \alpha_n\beta_n).$$

Our chosen basis vectors $1, X_{11}, X_{12}, \dots, X_{nn}$ for V are $\Delta(\mathcal{H})$-eigenvectors with distinct $\Delta(\mathcal{H})$-eigenvalues, and so the $\Delta(\mathcal{H})$-eigenspaces of V are all 1-dimensional. If $\mathcal{H}'$ is a subtorus of $\mathcal{A}$ containing $\Delta(\mathcal{H})$, all the $\Delta(\mathcal{H})$-eigenspaces of V must be invariant under $\mathcal{H}'$, due to the commutativity of $\mathcal{H}'$. Consequently, each of $1, X_{11}, X_{12}, \dots, X_{nn}$ must be an $\mathcal{H}'$-eigenvector. It is readily checked that the only diagonal matrices in $\mathcal{A}$ are those in $\Delta(\mathcal{H})$, so $\mathcal{H}' = \Delta(\mathcal{H})$. Therefore $\Delta(\mathcal{H})$ is a maximal torus of $\mathcal{A}$, identified via ρ_1^{-1} with a maximal torus of $\operatorname{Aut}_{\mathrm{gr}} A$. By Lemma 6.9, $\rho_1^{-1}(\Delta(\mathcal{H}))$ is also a maximal torus of $\operatorname{Aut}_{\mathrm{fl}} A$. Taking account of (5.16), we conclude that

$$\mathrm{tw}(A) = \mathrm{tw}_{X(\rho_1^{-1}(\Delta(\mathcal{H})))}(A) = \mathrm{tw}_{X(\mathcal{H})}(A) = \langle \lambda \rangle. \tag{6.1}$$

7. Stability of the AD-invariant

This section establishes stability properties of the AD-invariant. The first result is that the invariant remains unchanged under polynomial extensions. The second result establishes invariance under general tensoring with commutative algebras C under natural assumptions on C.

7.1. Polynomial stability.

PROPOSITION 7.1. *If A is a semiprime right Goldie $\mathbb{K}$-algebra, then the polynomial ring $A[x]$ is semiprime right Goldie and*

$$\mathrm{AD}(A[x]) = \mathrm{AD}(A).$$

PROOF. That $A[x]$ is semiprime right Goldie is well known (e.g., [**13**, Corollary 11.19]). Of course, all regular elements of A are regular in $A[x]$, from which it is clear that $\mathrm{AD}(A) \subseteq \mathrm{AD}(A[x])$. It remains to establish the reverse inclusion.

By Observation 3.2(1), $\mathrm{AD}(A) = \mathrm{AD}(\mathrm{Fract}\, A)$ and $\mathrm{AD}(A[x]) = \mathrm{AD}((\mathrm{Fract}\, A)[x])$, so there is no loss of generality in assuming that A is semisimple. Then, due to (3.1), we lose no generality in assuming that $A = M_n(D)$ for some $n \in \mathbb{Z}_{>0}$ and some division $\mathbb{K}$-algebra D. We may identify

$$A((x)) = M_n\big(D((x))\big),$$

where $A((x))$ and $D((x))$ denote the algebras of Laurent series in x over A and D, respectively. Since $D((x))$ is a division algebra, it follows that $A((x))$ is semisimple. Moreover, the inclusion map $D[x] \to D((x))$ extends uniquely to an injective homomorphism $\mathrm{Fract}\, D[x] \to D((x))$, from which we obtain a natural embedding

$$\mathrm{Fract}\, A[x] \equiv M_n(\mathrm{Fract}\, D[x]) \longrightarrow A((x)).$$

Consequently, it suffices to show that $\mathrm{AD}\big(A((x))\big) \subseteq \mathrm{AD}(A)$.

Any scalar $\alpha \in \mathrm{AD}\big(A((x))\big)$ has the form $\alpha = u_1v_1u_1^{-1}v_1^{-1} \cdots u_nv_nu_n^{-1}v_n^{-1}$ for some units $u_i, v_i \in A((x))$. If $a_i, b_i \in A$ denote the leading coefficients of u_i and v_i, respectively (i.e., the coefficients on the lowest powers of x appearing in these series), then $u_iv_iu_i^{-1}v_i^{-1}$ is a power series with constant term $a_ib_ia_i^{-1}b_i^{-1}$. The product of the $u_iv_iu_i^{-1}v_i^{-1}$ is thus a power series whose constant term lies in $\mathrm{GL}(A)'$. Therefore $\alpha \in \mathrm{GL}_1(A)' \cap \mathbb{K}^* = \mathrm{AD}(A)$, as desired. □

Proposition 7.1 immediately implies that various polynomial algebras cannot be isomorphic. In particular:

COROLLARY 7.2. *Let $m, n \in \mathbb{Z}_{>0}$, and let $\boldsymbol{p} \in M_m(\mathbb{K}^*)$, $\boldsymbol{q} \in M_n(\mathbb{K}^*)$ be multiplicatively skew-symmetric matrices. If $\langle \boldsymbol{p} \rangle \neq \langle \boldsymbol{q} \rangle$, then the polynomial algebras*

$$\mathcal{O}_{\boldsymbol{p}}(\mathbb{K}^m)[x_1, \ldots, x_s] \not\cong \mathcal{O}_{\boldsymbol{q}}(\mathbb{K}^n)[x_1, \ldots, x_t]$$

for all $s, t \in \mathbb{Z}_{\geq 0}$.

More generally, for all pairs of quantum cluster algebras $\mathcal{A}_1$ and $\mathcal{A}_2$ (and pairs of upper quantum cluster algebras $\mathcal{U}_1$ and $\mathcal{U}_2$) we have

$$\mathcal{A}_1[x_1, \ldots, x_s] \not\cong \mathcal{A}_2[x_1, \ldots, x_t] \quad \text{if} \quad q(\mathcal{A}_1) \neq q(\mathcal{A}_2)$$

and

$$\mathcal{U}_1[x_1, \ldots, x_s] \not\cong \mathcal{U}_2[x_1, \ldots, x_t] \quad \text{if} \quad q(\mathcal{U}_1) \neq q(\mathcal{U}_2)$$

for all $s, t \in \mathbb{Z}_{\geq 0}$.

7.2. General stability under tensoring with commutative algebras.

LEMMA 7.3. *Let R be a ring and $X \subseteq R$ a right denominator set. Suppose $a_0, \dots, a_n \in R$ and $x_0, \dots, x_n \in X$ such that*

$$a_0 x_0^{-1} = (a_1 x_1^{-1})(a_2 x_2^{-1}) \cdots (a_n x_n^{-1}) \tag{7.1}$$

in RX^{-1}.

(a) *There exist $b_1, \dots, b_n \in R$ and $y_1 = 1, y_2, \dots, y_n, z \in X$ such that*

$$\begin{aligned} a_1 b_2 \cdots b_n z &= a_0 b_1 & x_n y_n z &= x_0 b_1 \\ x_{i-1} y_{i-1} b_i &= a_i y_i & \forall i &\in [2, n]. \end{aligned} \tag{7.2}$$

(b) *Suppose $\phi : R \to S$ is a ring homomorphism which maps $x_0, \dots, x_n, y_2, \dots, y_n, z$ to units of S. Then*

$$\phi(a_0)\phi(x_0)^{-1} = \phi(a_1)\phi(x_1)^{-1}\phi(a_2)\phi(x_2)^{-1} \cdots \phi(a_n)\phi(x_n)^{-1}. \tag{7.3}$$

PROOF. (a) By induction on n, there exist $b_2, \dots, b_n \in R$ and $y_2, \dots, y_n \in X$ such that $x_{i-1} b_i = a_i y_i$ for $i \in [2, n]$ and

$$(a_1 x_1^{-1})(a_2 x_2^{-1}) \cdots (a_n x_n^{-1}) = (a_1 b_2 \cdots b_n)(x_n y_n)^{-1}.$$

There also exist $b_1 \in R$ and $z \in X$ such that $x_n y_n z = x_0 b_1$. Incorporating (7.1) and multiplying on the right by $x_0 b_1$ yields $a_0 b_1 = a_1 b_2 \cdots b_n z$.

(b) Apply ϕ to (7.2) and unwind. □

In the following, we let $\operatorname{u.dim} S$ denote the right uniform dimension of a ring S, i.e., the uniform dimension of the module S_S.

PROPOSITION 7.4. *Let R be a semiprime right noetherian ring and $\mathcal{P}$ a set of prime ideals of R such that $\bigcap \mathcal{P} = 0$ and there is a finite upper bound on $\{\operatorname{u.dim} R/P \mid P \in \mathcal{P}\}$. For any regular element $c \in R$, there is some $P \in \mathcal{P}$ such that c is regular modulo $\mathcal{P}$.*

PROOF. There are only finitely many distinct minimal prime ideals in R (e.g., [**9**, Theorem 3.4]), say $P_1, \dots, P_n$, and c is regular modulo each P_i (e.g., [**9**, Lemma 7.4]). If $\mathcal{P}_i := \{P \in \mathcal{P} \mid P \supseteq P_i\}$, then $\bigcap_{i=1}^n \bigcap \mathcal{P}_i = \bigcap \mathcal{P} = 0$. Consequently, $\bigcap \mathcal{P}_j \subseteq P_1$ for some j. Since $P_j \subseteq \bigcap \mathcal{P}_j$, we must have $j = 1$ and $\bigcap \mathcal{P}_1 = P_1$.

Since $\{\operatorname{u.dim} R/P \mid P \in \mathcal{P}_1\}$ is bounded above, [**9**, Proposition 16.10] implies that there is a neighborhood $\mathcal{U}$ of P_1 in the patch topology on $\operatorname{Spec} R$ such that c is regular modulo P for all $P \in \mathcal{P}_1 \cap \mathcal{U}$. The discussion in [**9**, pp. 275-6] shows that $\mathcal{U}$ contains a patch neighborhood of the form

$$\mathcal{U}' = \{Q \in \operatorname{Spec} R \mid Q \supseteq P_1 \text{ but } Q \not\supseteq I\},$$

for some ideal I properly containing P_1. Since $\bigcap \mathcal{P}_1 = P_1$, we conclude that $\mathcal{P}_1 \cap \mathcal{U}'$ is nonempty. Any $P \in \mathcal{P}_1 \cap \mathcal{U}'$ satisfies the requirements of the proposition. □

THEOREM 7.5. *Assume $\mathbb{K} = \overline{\mathbb{K}}$, and let C be a reduced (i.e., semiprime) commutative $\mathbb{K}$-algebra. If A is a semiprime right Goldie $\mathbb{K}$-algebra and $A \otimes_{\mathbb{K}} C$ is semiprime right Goldie, then*

$$\operatorname{AD}(A \otimes_{\mathbb{K}} C) = \operatorname{AD}(A). \tag{7.4}$$

PROOF. We just need to show that $\mathrm{AD}(A \otimes_{\mathbb{K}} C) \subseteq \mathrm{AD}(A)$, since the reverse inclusion is clear.

If X is the set of regular elements of A, then X is a right denominator set in A and $X \otimes 1$ is a right denominator set of regular elements in $A \otimes_{\mathbb{K}} C$. Consequently, $\mathrm{AD}(A) = \mathrm{AD}(AX^{-1})$ and $\mathrm{AD}(A \otimes_{\mathbb{K}} C) = \mathrm{AD}((A \otimes_{\mathbb{K}} C)(X \otimes 1)^{-1})$, so there is no loss of generality in replacing A by AX^{-1}. Thus, we may now assume that A is semisimple. In view of (3.1), we may further assume that A is simple artinian.

Now C is a directed union of its affine subalgebras C', so $A \otimes_{\mathbb{K}} C$ is a directed union of the subalgebras $A \otimes_{\mathbb{K}} C'$. Since $A \otimes_{\mathbb{K}} C$ is a central extension of $A \otimes_{\mathbb{K}} C'$, semiprimeness descends from $A \otimes_{\mathbb{K}} C$ to $A \otimes_{\mathbb{K}} C'$. Moreover, $A \otimes_{\mathbb{K}} C'$ is a homomorphic image of a polynomial ring $A[x_1, \dots, x_n]$, and so $A \otimes_{\mathbb{K}} C'$ is right noetherian, hence right Goldie. Note that any element of $\mathrm{AD}(A \otimes_{\mathbb{K}} C)$ lies in $\mathrm{AD}(A \otimes_{\mathbb{K}} C')$ for a suitable C'. Thus, it suffices to prove that $\mathrm{AD}(A \otimes_{\mathbb{K}} C') \subseteq \mathrm{AD}(A)$ for all C'. Consequently, without loss of generality, we may assume that C is affine and $A \otimes_{\mathbb{K}} C$ is right noetherian.

Let Φ denote the set of characters $\phi : C \to \mathbb{K}$ and Ψ the set of $\mathbb{K}$-algebra homomorphisms of the form

$$\mathrm{id}_A \otimes \phi : A \otimes_{\mathbb{K}} C \longrightarrow A,$$

for $\phi \in \Phi$. Since C is affine and reduced and $\mathbb{K}$ is algebraically closed, $\bigcap_{\phi \in \Phi} \ker \phi = 0$, from which it follows that $\bigcap_{\phi \in \Phi} \ker(\mathrm{id}_A \otimes \phi) = 0$. In other words, the collection

$$\mathcal{P} := \{\ker \psi \mid \psi \in \Psi\}$$

of maximal ideals of $A \otimes_{\mathbb{K}} C$ has zero intersection. Note that the algebras $(A \otimes_{\mathbb{K}} C)/P$, for $P \in \mathcal{P}$, all have the same uniform dimension, namely $\mathrm{u.dim}\, A$.

Let X denote the set of regular elements of $A \otimes_{\mathbb{K}} C$. Any scalar $\alpha \in \mathrm{AD}(A \otimes_{\mathbb{K}} C)$ has the form

$$\alpha = s_1 t_1^{-1} u_1 v_1^{-1} t_1 s_1^{-1} v_1 u_1^{-1} \cdots s_n t_n^{-1} u_n v_n^{-1} t_n s_n^{-1} v_n u_n^{-1}$$

for some $s_i, t_i, u_i, v_i \in X$. Relabel the sequences

$$\alpha, s_1, u_1, t_1, v_1, \dots, s_n, u_n, t_n, v_n \qquad \text{and} \qquad 1, t_1, v_1, s_1, u_1, \dots, t_n, v_n, s_n, u_n,$$

as $a_0, \dots, a_{4n}$ and $x_0, \dots, x_{4n}$, respectively. By Lemma 7.3(a), there exist $b_1, \dots, b_{4n} \in R$ and $y_1 = 1, y_2, \dots, y_{4n}, z \in X$ such that

$$\begin{aligned} a_1 b_2 \cdots b_{4n} z &= a_0 b_1 & \qquad x_{4n} y_{4n} z &= x_0 b_1 \\ x_{i-1} y_{i-1} b_i &= a_i y_i & \forall\, i &\in [2, 4n]. \end{aligned}$$

Now let c be the product (in some order) of the x_i, y_i, and z, a regular element of $A \otimes_{\mathbb{K}} C$. By Proposition 7.4, there exists $P \in \mathcal{P}$ such that c is regular modulo P. Then $P = \ker \psi$ for some $\psi \in \Psi$, and $\psi(c)$ is a unit in the simple artinian ring A. Consequently, the $\psi(x_i)$, $\psi(y_i)$, and $\psi(z)$ are units in A. Applying Lemma 7.3(b) and rewriting the result in terms of the original s_i, t_i, u_i, v_i, we obtain that the $\psi(s_i)$, $\psi(t_i)$, $\psi(u_i)$, $\psi(v_i)$ are units in A and

$$\begin{aligned} \alpha = \psi(s_1)\psi(t_1)^{-1}\psi(u_1)\psi(v_1)^{-1}\psi(t_1)\psi(s_1)^{-1}\psi(v_1)\psi(u_1)^{-1} \cdots \\ \psi(s_n)\psi(t_n)^{-1}\psi(u_n)\psi(v_n)^{-1}\psi(t_n)\psi(s_n)^{-1}\psi(v_n)\psi(u_n)^{-1}. \end{aligned}$$

Thus $\alpha \in \mathrm{AD}(A)$, as required. □

COROLLARY 7.6. *Assume $\mathbb{K} = \overline{\mathbb{K}}$, and let C be a reduced commutative noetherian $\mathbb{K}$-algebra. If A is an iterated Ore extension of $\mathbb{K}$, then*

$$\mathrm{AD}(A \otimes_{\mathbb{K}} C) = \mathrm{AD}(A).$$

PROOF. If X is the set of regular elements of C, then CX^{-1} is a finite direct product of field extensions K_i of $\mathbb{K}$. Now $1 \otimes X$ is a central multiplicative set in $A \otimes_{\mathbb{K}} C$, and $(A \otimes_{\mathbb{K}} C)(1 \otimes X)^{-1}$ is isomorphic to the finite direct product of the $\mathbb{K}$-algebras $A \otimes_{\mathbb{K}} K_i$. Each of the latter tensor products is an iterated Ore extension of K_i, hence a noetherian domain. Thus, $(A \otimes_{\mathbb{K}} C)(1 \otimes X)^{-1}$ is a semiprime noetherian ring, from which we conclude that $A \otimes_{\mathbb{K}} C$ is semiprime Goldie. The result now follows from Theorem 7.5. □

8. Stability of the twist invariants

In this section we prove that both twist invariants remain unchanged under polynomial extensions. We apply the fact for the general twist invariant to establish non-isomorphisms between twisted polynomial extensions of quantized Weyl algebras.

8.1. Maximal tori of polynomial extensions. First we show that the maximal tori of the automorphism groups of filtered locally finite algebras behave well under polynomial extensions. For any such algebra A, we give the polynomial ring $A[x]$ the natural induced filtration with

$$F_k(A[x]) := \bigoplus_{j=0}^{k} (F_j A) x^{k-j}, \quad \forall\, k \in \mathbb{Z}_{\geq 0}\,. \tag{8.1}$$

PROPOSITION 8.1. *Let A be an affine filtered locally finite algebra over an infinite field $\mathbb{K}$, with $F_0 A = \mathbb{K}$. For every maximal torus $\mathcal{H}$ of $\mathrm{Aut_{fl}}(A)$, the group $\mathcal{H} \times \mathbb{K}^*$ is a maximal torus of $\mathrm{Aut_{fl}}(A[x])$, where $\mathcal{H}$ acts trivially on x and $\mathbb{K}^*$ acts by rescaling x.*

PROOF. Set $\mathcal{H}^+ := \mathcal{H} \times \mathbb{K}^*$, and observe that the $\mathcal{H}^+$-eigenspaces of $A[x]$ are the spaces Ex^n where E is an $\mathcal{H}$-eigenspace of A and $n \in \mathbb{Z}_{\geq 0}$. In particular, $A[x]$ is a direct sum of $\mathcal{H}^+$-eigenspaces.

Suppose $\mathcal{H}'$ is a subtorus of $\mathrm{Aut_{fl}}(A[x])$ containing $\mathcal{H}^+$. Since $\mathcal{H}'$ is abelian, it stabilizes the $\mathcal{H}^+$-eigenspaces of $A[x]$, and so it stabilizes A. Thus, restriction to A provides a group homomorphism $\rho_1 : \mathcal{H}' \to \mathrm{Aut_{fl}}(A)$, and we observe that ρ_1 is a morphism of algebraic groups. Now $\mathcal{H}'$ also stabilizes $F_1(A[x]) = (F_1 A) + \mathbb{K}x$ as well as the $\mathcal{H}^+$-eigenspace $A^{\mathcal{H}}x$, where $A^{\mathcal{H}}$ is the fixed subalgebra of A under $\mathcal{H}$, so $\mathcal{H}'$ stabilizes $\mathbb{K}x$. Consequently, restriction to $\mathbb{K}x$ provides a second algebraic group morphism $\rho_2 : \mathcal{H}' \to \mathrm{GL}(\mathbb{K}x) = \mathbb{K}^*$. The combined morphism $\rho_1 \times \rho_2 : \mathcal{H}' \to \mathrm{Aut_{fl}}(A) \times \mathbb{K}^*$ is injective, because the automorphisms in $\mathcal{H}'$ are determined by their actions on $A \cup \mathbb{K}x$. Obviously $(\rho_1 \times \rho_2)(\mathcal{H}^+) = \mathcal{H}^+$. Hence, once we show that $(\rho_1 \times \rho_2)(\mathcal{H}') = \mathcal{H}^+$, we can conclude that $\mathcal{H}' = \mathcal{H}^+$, establishing the maximality of $\mathcal{H}^+$.

Choose $n > 0$ large enough so that $F_n A$ generates A, whence $F_n(A[x])$ generates $A[x]$. The algebraic group structure on $\mathrm{Aut_{fl}}(A[x])$ thus comes from the isomorphism of this group with a closed subgroup of $\mathrm{GL}(F_n(A[x]))$. Since the group $\mathrm{Aut_{fl}}(A[x])$ acts rationally on $A[x]$, so does $\mathcal{H}'$. This means $\mathcal{H}'$ acts diagonalizably on $F_n(A[x])$. Now $F_n(A[x])$ is $\mathcal{H}^+$-stable, and some of its $\mathcal{H}^+$-eigenspaces sum to

F_nA, from which we see that $\mathcal{H}'$ stabilizes F_nA. Consequently, $\mathcal{H}'$ acts diagonalizably on F_nA. It follows that $\rho_1(\mathcal{H}')$ is a diagonalizable algebraic group. On the other hand, it is connected, being a continuous image of a connected group. Thus, $\rho_1(\mathcal{H}')$ is a torus. Since $\rho_1(\mathcal{H}') \supseteq \mathcal{H}$, we obtain $\rho_1(\mathcal{H}') = \mathcal{H}$ from the maximality of $\mathcal{H}$. Therefore $(\rho_1 \times \rho_2)(\mathcal{H}') \subseteq \mathcal{H}^+$ and consequently $(\rho_1 \times \rho_2)(\mathcal{H}') = \mathcal{H}^+$, as desired. □

8.2. Stability of the twist invariants.

THEOREM 8.2. (a) *Let A be a Γ-graded semiprime right Goldie $\mathbb{K}$-algebra for an abelian group Γ. Then the polynomial algebra $A[x]$ is also semiprime right Goldie and*

$$\mathrm{tw}_{\Gamma\times\mathbb{Z}}(A[x]) = \mathrm{tw}_{\Gamma}(A),$$

where the grading of $A[x]$ is obtained by extending that of A by placing x in degree $(0,1)$.

(b) *Let A be an affine filtered locally finite algebra over an infinite field $\mathbb{K}$, with $F_0A = \mathbb{K}$. Assume in addition that A is a semiprime right Goldie algebra. Then the polynomial algebra $A[x]$ has the same properties with respect to the filtration (8.1) and*

$$\mathrm{tw}(A[x]) = \mathrm{tw}(A).$$

PROOF. (a) The fact that $A[x]$ is also semiprime right Goldie is discussed in Proposition 7.1. Each $\mathbf{c} \in Z^2(\Gamma, \mathbb{K}^*)$ gives rise to

$$\mathbf{c}^+ \in Z^2(\Gamma\times\mathbb{Z}, \mathbb{K}^*) \quad \text{defined by} \quad \mathbf{c}^+((a_1,n_1),(a_2,n_2)) := \mathbf{c}(a_1,a_2)$$

for $a_i \in \Gamma$, $n_i \in \mathbb{Z}$. Obviously,

$$(A[x])^{\mathbf{c}^+} \cong A^{\mathbf{c}}[x],$$

and by Proposition 7.1, $\mathrm{AD}(A^{\mathbf{c}}[x]) = \mathrm{AD}(A^{\mathbf{c}})$. Therefore,

$$\mathrm{tw}_\Gamma(A) = \bigcap_{\mathbf{c}\in Z^2(\Gamma,\mathbb{K}^*)} \mathrm{AD}(A^{\mathbf{c}}) = \bigcap_{\mathbf{c}\in Z^2(\Gamma,\mathbb{K}^*)} \mathrm{AD}((A[x])^{\mathbf{c}^+}) \supseteq$$
$$\supseteq \bigcap_{\mathbf{d}\in Z^2(\Gamma\times\mathbb{Z},\mathbb{K}^*)} \mathrm{AD}((A[x])^{\mathbf{d}}) = \mathrm{tw}_{\Gamma\times\mathbb{Z}}(A[x]).$$

In the opposite direction, for $\mathbf{d} \in Z^2(\Gamma\times\mathbb{Z}, \mathbb{K}^*)$, we have the restricted cocycle

$$\overline{\mathbf{d}} := \mathbf{d}|_{\Gamma\times\Gamma} \in Z^2(\Gamma, \mathbb{K}^*).$$

Since $\mathrm{Fract}(A^{\overline{\mathbf{d}}}) \subset \mathrm{Fract}((A[x])^{\mathbf{d}})$,

$$\mathrm{AD}((A[x])^{\mathbf{d}}) \supseteq \mathrm{AD}(A^{\overline{\mathbf{d}}}).$$

Thus,

$$\mathrm{tw}_{\Gamma\times\mathbb{Z}}(A[x]) = \bigcap_{\mathbf{d}\in Z^2(\Gamma\times\mathbb{Z},\mathbb{K}^*)} \mathrm{AD}((A[x])^{\mathbf{d}}) \supseteq \bigcap_{\mathbf{d}\in Z^2(\Gamma\times\mathbb{Z},\mathbb{K}^*)} \mathrm{AD}(A^{\overline{\mathbf{d}}})$$
$$\supseteq \bigcap_{\mathbf{c}\in Z^2(\Gamma,\mathbb{K}^*)} \mathrm{AD}(A^{\mathbf{c}}) = \mathrm{tw}_\Gamma(A).$$

Combining the two inclusions above gives $\mathrm{tw}_{\Gamma\times\mathbb{Z}}(A[x]) = \mathrm{tw}_\Gamma(A)$.

(b) The properties of $A[x]$ are discussed in the proofs of Propositions 7.1 and 8.1. Let $\mathcal{H}$ be a maximal torus of $\mathrm{Aut}_{\mathrm{fl}}(A)$. By Proposition 8.1, $\mathcal{H}^+ := \mathcal{H} \times \mathbb{K}^*$ is a maximal torus of $\mathrm{Aut}_{\mathrm{fl}}(A[x])$, where $\mathcal{H}$ acts trivially on x and $\mathbb{K}^*$ acts by rescaling

x. We identify $X(\mathcal{H}^+) \cong X(\mathcal{H}) \times \mathbb{Z}$ in the canonical way. Applying the first part of the theorem and taking into account Lemma 6.5, we obtain

$$\operatorname{tw}(A[x]) = \operatorname{tw}_{X(\mathcal{H}^+)}(A[x]) = \operatorname{tw}_{X(\mathcal{H})\times\mathbb{Z}}(A[x]) = \operatorname{tw}_{X(\mathcal{H})}(A) = \operatorname{tw}(A),$$

which completes the proof of the theorem. □

EXAMPLE 8.3. Let $A := \mathcal{O}_{\lambda,\mathbf{p}}(M_n(\mathbb{K}))$ as in Examples 3.7 and 6.10, treated as a $\mathbb{Z}$-graded algebra with $\deg X_{ij} = 1$ for all i, j. Extend this grading to any polynomial algebra $A[x_1, \dots, x_s]$ so that $\deg x_j = 1$ for $j \in [1, s]$. In view of Theorem 8.2(b) and equation (6.1), we obtain

$$\operatorname{tw}(A[x_1, \dots, x_s]) = \langle \lambda \rangle, \quad \forall \text{ polynomial algebras } A[x_1, \dots, x_s]. \tag{8.2}$$

EXAMPLE 8.4. Let $A := A_n^{Q,P}(\mathbb{K})$ as in Example 4.7, treated as a $\mathbb{Z}$-graded algebra with $\deg x_i = 1$ and $\deg y_i = -1$ for all $i \in [1, n]$. This grading is not locally finite, but A has a locally finite filtration based on the generating subspace $V := \mathbb{K} \oplus \bigoplus_{i=1}^n (\mathbb{K}x_i \oplus \mathbb{K}y_i)$, where we take $F_k A = V^k$ for $k \in \mathbb{Z}_{\geq 0}$. By Observations 6.8, $\operatorname{Aut_{fl}} A$ becomes an algebraic group isomorphic by the restriction morphism ρ_1 to the closed subgroup $\mathcal{A} := \operatorname{im} \rho_1$ of $\operatorname{GL}(V)$.

Let the torus $\mathcal{H} := (\mathbb{K}^*)^n$ act on A as in (5.18) (but here we allow the q_i to possibly be roots of unity), and note that the corresponding $X(\mathcal{H})$-grading on A coincides with the $\mathbb{Z}^n$-grading of Example 4.7. Obviously $\mathcal{H}$ acts by filtered automorphisms, and we may identify $\mathcal{H}$ with a closed subgroup of $\operatorname{Aut_{fl}} A$. Note that the $\mathcal{H}$-eigenspaces of V are all 1-dimensional. Hence, it follows as in Example 6.10 that $\mathcal{H}$ is a maximal torus of $\operatorname{Aut_{fl}} A$. By (4.7), we have

$$\operatorname{tw}(A) = \operatorname{tw}_{X(\mathcal{H})}(A) = \langle q_1, \dots, q_n \rangle, \tag{8.3}$$

and Theorem 8.2(b) thus implies

$$\operatorname{tw}(A[X_1, \dots, X_s]) = \langle q_1, \dots, q_n \rangle, \quad \forall \text{ polynomial algebras } A[X_1, \dots, X_s],$$

where $A[X_1, \dots, X_s]$ is filtered by iterating Eqn. (8.1):

$$F_k(A[X_1, \dots, X_s]) := \bigoplus_{j_0 + j_1 + \cdots + j_s \leq k} (F_{j_0} A) X_1^{j_1} \cdots X_s^{j_s}.$$

Taking all polynomial algebras over quantized Weyl algebras to be filtered in this manner, we conclude that

> Given $A_n^{Q,P}(\mathbb{K})$ and $A_{n'}^{Q',P'}(\mathbb{K})$ with $\langle q_1, \dots, q_n \rangle \neq \langle q'_1, \dots, q'_n \rangle$, no cocycle twisted polynomial algebra
>
> $$\big(A_n^{Q,P}(\mathbb{K})[X_1, \dots, X_s]\big)^{\mathbf{c}}$$
>
> is isomorphic, as a filtered $\mathbb{K}$-algebra, to any cocycle twisted polynomial algebra
>
> $$\big(A_{n'}^{Q',P'}(\mathbb{K})[X_1, \dots, X_t]\big)^{\mathbf{c}'}$$
>
> for any two 2-cocycles $\mathbf{c} \in Z^2(\mathbb{Z}^m, \mathbb{K}^*)$ and $\mathbf{c}' \in Z^2(\mathbb{Z}^{m'}, \mathbb{K}^*)$. The twists are performed with respect to any gradings of the two polynomial algebras that respect the filtrations of these algebras.

In the last step we made use of the invariance property of the general twist invariant from Theorem 6.7.

We finish by noting that recently Bell and Zhang [**3**] investigated Zariski cancellative algebras, which are $\mathbb{K}$-algebras A such that $A[x] \cong B[x]$ implies $A \cong B$ for any $\mathbb{K}$-algebra B. They proved that this is the case if A is LND-rigid or if the discriminant of A over its center exists and is cancellative. Currently, there are not many families of algebras that are known to have these properties and even quantum matrices of size $n > 3$ are conjectured not to have the properties. Our stable twist invariant can be explicitly computed from Theorems 4.4 and 5.6 and used to detect non-isomorphisms

$$A[x] \ncong B[x]$$

of filtered algebras when discriminant methods are not applicable.

References

[1] J. Alev and F. Dumas, *Sur le corps des fractions de certaines algèbres quantiques* (French, with English summary), J. Algebra **170** (1994), no. 1, 229–265. MR1302839

[2] M. Artin, W. Schelter, and J. Tate, *Quantum deformations of* GL_n, Comm. Pure Appl. Math. **44** (1991), no. 8-9, 879–895. MR1127037

[3] J. Bell and J. J. Zhang, *Zariski cancellation problem for noncommutative algebras*, Selecta Math. (N.S.) **23** (2017), no. 3, 1709–1737. MR3663593

[4] A. Berenstein and A. Zelevinsky, *Quantum cluster algebras*, Adv. Math. **195** (2005), no. 2, 405–455. MR2146350

[5] A. Borel, *Linear algebraic groups*, 2nd ed., Graduate Texts in Mathematics, vol. 126, Springer-Verlag, New York, 1991. MR1102012

[6] K. A. Brown and K. R. Goodearl, *Lectures on algebraic quantum groups*, Advanced Courses in Mathematics. CRM Barcelona, Birkhäuser Verlag, Basel, 2002. MR1898492

[7] C. De Concini and C. Procesi, *Quantum groups*, D-modules, representation theory, and quantum groups (Venice, 1992), Lecture Notes in Math., vol. 1565, Springer, Berlin, 1993, pp. 31–140. MR1288995

[8] K. R. Goodearl and E. S. Letzter, *The Dixmier-Moeglin equivalence in quantum coordinate rings and quantized Weyl algebras*, Trans. Amer. Math. Soc. **352** (2000), no. 3, 1381–1403. MR1615971

[9] K. R. Goodearl and R. B. Warfield Jr., *An introduction to noncommutative Noetherian rings*, 2nd ed., London Mathematical Society Student Texts, vol. 61, Cambridge University Press, Cambridge, 2004. MR2080008

[10] K. R. Goodearl and M. T. Yakimov, *From quantum Ore extensions to quantum tori via noncommutative UFDs*, Adv. Math. **300** (2016), 672–716. MR3534843

[11] K. R. Goodearl and M. T. Yakimov, *Quantum cluster algebra structures on quantum nilpotent algebras*, Mem. Amer. Math. Soc. **247** (2017), no. 1169, vii+119. MR3633289

[12] D. A. Jordan, *A simple localization of the quantized Weyl algebra*, J. Algebra **174** (1995), no. 1, 267–281. MR1332871

[13] T.-Y. Lam, *Lectures on modules and rings*, Graduate Texts in Mathematics, vol. 189, Springer-Verlag, New York, 1999. MR1653294

[14] A. Leroy and J. Matczuk, *On q-skew iterated Ore extensions satisfying a polynomial identity*, J. Algebra Appl. **10** (2011), no. 4, 771–781. MR2834115

[15] S. Lang, *Algebra*, Third Ed., Reading Mass. (1993), Addison-Wesley.

[16] S. Launois, T. H. Lenagan, and L. Rigal, *Quantum unique factorisation domains*, J. London Math. Soc. (2) **74** (2006), no. 2, 321–340. MR2269632

[17] G. Lusztig, *Introduction to quantum groups*, Progress in Mathematics, vol. 110, Birkhäuser Boston, Inc., Boston, MA, 1993. MR1227098

[18] L. H. Rowen, *Ring theory. Vol. I*, Pure and Applied Mathematics, vol. 127, Academic Press, Inc., Boston, MA, 1988. MR940245

[19] M. Yakimov, *A proof of the Goodearl–Lenagan polynormality conjecture*, Int. Math. Res. Not. IMRN **2013**, no. 9, 2097–2132. MR3053415

Department of Mathematics, University of California, Santa Barbara, CA 93106, U.S.A.

Email address: goodearl@math.ucsb.edu

Department of Mathematics, Louisiana State University, Baton Rouge, LA 70803, U.S.A.

Email address: yakimov@math.lsu.edu

Contemporary Mathematics
Volume **715**, 2018
https://doi.org/10.1090/conm/715/14410

Modules invariant under monomorphisms of their envelopes

Pedro A. Guil Asensio, Derya Keskin Tütüncü, and Ashish K. Srivastava

Dedicated to the memory of Bruno Müller

ABSTRACT. In this paper we prove that a module M is invariant under automorphisms of its $\mathcal{X}$-envelope if and only if M is invariant under monomorphisms of its $\mathcal{X}$-envelope where $\mathcal{X}$ is an enveloping class with the property that for each object X in $\mathcal{X}$, $\mathrm{End}(X)/J(\mathrm{End}(X))$ is a von Neumann regular right self-injective ring and idempotents lift modulo $J(\mathrm{End}(X))$. In particular, invariance under monomorphisms is equivalent to the invariance under automorphisms of $\mathcal{X}$-envelope when $\mathcal{X}$ is the class of injective modules, pure-injective modules or flat cotorsion modules. We finish the paper by proving that an $\mathcal{X}$-automorphism invariant (equivalently, $\mathcal{X}$-monomorphism invariant) module M satisfies the property $\mathcal{P}$ if and only if its $\mathcal{X}$-envelope $X(M)$ satisfies the property $\mathcal{P}$, when $\mathcal{P}$ is direct-finiteness, internal cancellation, cancellation or the substitution property.

1. Introduction

The theory of modules invariant under automorphisms of their envelopes or covers was introduced in [**6**]. Let $\mathcal{X}$ be a class of right R-modules closed under isomorphisms. An $\mathcal{X}$-*preenvelope* of a right module M is a morphism $u : M \to X$ with $X \in \mathcal{X}$ such that any other morphism $g : M \to X'$ with $X' \in \mathcal{X}$ factors through u. A preenvelope $u : M \to X$ is called an $\mathcal{X}$-*envelope* if, moreover, it satisfies that any endomorphism $h : X \to X$ such that $h \circ u = u$ must be an automorphism. An $\mathcal{X}$-(pre)envelope $u : M \to X$ is called *monomorphic* if u is a monomorphism. A class $\mathcal{X}$ of right modules over a ring R, closed under isomorphisms, is called an *enveloping class* if any right R-module M has an $\mathcal{X}$-envelope. A module M having a monomorphic $\mathcal{X}$-envelope $u : M \to X(M)$ is said to be $\mathcal{X}$-*automorphism invariant* (resp., $\mathcal{X}$-*endomorphism invariant*) if for any automorphism (resp., endomorphism) $\varphi : X(M) \to X(M)$, there exists an endomorphism $f : M \to M$ such

2010 *Mathematics Subject Classification.* 16D40, 16D80.

Key words and phrases. $\mathcal{X}$-automorphism invariant module, $\mathcal{X}$-automorphism coinvariant module, $\mathcal{X}$-monomorphism invariant module, directly finite module, unit-regular ring.

The first author has being partially supported by the grant MTM2016-77445-P from the Agencia Española de Investigación and from the Fundación Séneca (part of the sources of these institutions come from the FEDER funds of the European Union). The third author is partially supported by a grant from Simons Foundation (grant number 426367).

that $u \circ f = \varphi \circ u$. In this paper we introduce the notion of $\mathcal{X}$-monomorphism invariant modules. We call a module M to be an *$\mathcal{X}$-monomorphism invariant* module if it is invariant under any monomorphism of its $\mathcal{X}$-envelope.

If $\mathcal{X}$ is the class of injective (pure-injective) modules, then $\mathcal{X}$-automorphism invariant modules are simply called *automorphism-invariant* (*pure-automorphism-invariant*) modules and $\mathcal{X}$-endomorphism invariant modules are called *quasi-injective* (*pure-quasi-injective*) modules. In [**10**] and [**12**], a module M is called *pseudo-injective* if any monomorphism from a submodule of M to M extends to an endomorphism of M. Recently in [**2**] and [**7**], it is proved that a module M is automorphism-invariant if and only if M is pseudo-injective. In [**1**] modules invariant under monomorphisms of their injective envelope were studied and such modules were called *essentially pseudo-injective* modules. In this paper we prove that if M is a module with a monomorphic $\mathcal{X}$-envelope $u : M \longrightarrow X(M)$ such that $\mathrm{End}(X(M))/J(\mathrm{End}(X(M)))$ is a von Neumann regular right self-injective ring and idempotents lift modulo $J(\mathrm{End}(X(M)))$, then M is invariant under automorphisms of its $\mathcal{X}$-envelope if and only if M is invariant under monomorphisms of its $\mathcal{X}$-envelope. In particular, this shows that a module is invariant under automorphisms of its (pure-) injective envelope if and only if it is invariant under monomorphisms of its (pure-) injective envelope. We also investigate the arbitrary direct sums of automorphism-invariant modules. Finally, we obtain some examples of $\mathcal{X}$-automorphism invariant modules that are not $\mathcal{X}$-endomorphism invariant, where the class of injective modules is different from the class of pure-injective modules.

Dually, if $\mathcal{X}$ is a covering class then a module M is called *$\mathcal{X}$-automorphism coinvariant* if it is coinvariant under any automorphism of its $\mathcal{X}$-cover. If $\mathcal{X}$ is the class of projective modules, then $\mathcal{X}$-automorphism-coinvariant modules are simply called *automorphism-coinvariant* modules.

Throughout this paper, R is an associative ring with identity and all the modules are unital right R-modules. Also throughout this paper $\mathcal{X}$ will denote an enveloping (or a covering) class of right R-modules. By $\mathrm{End}(M)$ and $J(\mathrm{End}(M))$, we denote the endomorphism ring of a module M and the Jacobson radical of the ring $\mathrm{End}(M)$, respectively.

Let M be an R-module with an $\mathcal{X}$-envelope $u : M \longrightarrow X(M)$. If $f : M \longrightarrow M$ is an endomorphism, then by the definition of envelope, $u \circ f$ extends to an endomorphism $g : X(M) \longrightarrow X(M)$ such that $g \circ u = u \circ f$. Clearly this extension is unique modulo the Jacobson radical of $\mathrm{End}(X(M))$. Therefore we can define a ring homomorphism $\varphi : \mathrm{End}(M) \longrightarrow \mathrm{End}(X(M))/J(\mathrm{End}(X(M)))$ by $\varphi(f) = g + J(\mathrm{End}(X(M)))$. Let us denote $K = \mathrm{Ker}\varphi$. Then φ induces an injective ring homomorphism $\Psi : \mathrm{End}(M)/K \longrightarrow \mathrm{End}(X(M))/J(\mathrm{End}(X(M)))$ which allows us to identify $\mathrm{End}(M)/K$ with the subring $\mathrm{Im}\Psi \subseteq \mathrm{End}(X(M))/J(\mathrm{End}(X(M)))$ (see [**6**]).

In [**9**] it is shown that if M is an $\mathcal{X}$-automorphism invariant module with a monomorphic $\mathcal{X}$-envelope $u : M \longrightarrow X(M)$ such that $\mathrm{End}(X(M))/J(\mathrm{End}(X(M)))$ is a von Neumann regular right self-injective ring and idempotents lift modulo $J(\mathrm{End}(X(M)))$, then $M = N \oplus L$ where N is a semiboolean module, that is, $\mathrm{End}(N)/J(\mathrm{End}(N))$ is a Boolean ring and L is an $\mathcal{X}$-endomorphism invariant module. Moreover, $\mathrm{End}(L)/J(\mathrm{End}(L))$ is a von Neumann regular right self-injective ring with the property that its each element is a sum of two units.

2. Equivalence of invariance under automorphisms and monomorphisms

In this section we will show that the notions of $\mathcal{X}$-automorphism invariance and $\mathcal{X}$-monomorphism invariance coincide for a module M with a monomorphic $\mathcal{X}$-envelope $u : M \to X(M)$ such that $\mathrm{End}(X(M))/J(\mathrm{End}(X(M)))$ is a von Neumann regular right self-injective ring and idempotents lift modulo $J(\mathrm{End}(X(M)))$. Note that this assumption that $\mathrm{End}(X(M))/J(\mathrm{End}(X(M)))$ is a von Neumann regular right self-injective ring and idempotents lift modulo $J(\mathrm{End}(X(M)))$ is not a restrictive assumption as it holds true for all the enveloping classes that we are interested in. If $\mathcal{X}$ is the class of injective modules or pure-injective modules then $\mathrm{End}(X(M))/J(\mathrm{End}(X(M)))$ is a von Neumann regular right self-injective ring and idempotents lift modulo $J(\mathrm{End}(X(M)))$. Also, if M is a flat module and $\mathcal{X}$ is the class of cotorsion modules then $\mathrm{End}(X(M))/J(\mathrm{End}(X(M)))$ is a von Neumann regular right self-injective ring and idempotents lift modulo $J(\mathrm{End}(X(M)))$.

THEOREM 2.1. *Let M be a module with a monomorphic $\mathcal{X}$-envelope $u : M \to X(M)$ such that $End(X(M))/J(End(X(M)))$ is a von Neumann regular right self-injective ring and idempotents lift modulo $J(End(X(M)))$. Then the following are equivalent:*

(1) *M is an $\mathcal{X}$-automorphism invariant module.*
(2) *M is an $\mathcal{X}$-monomorphism invariant module.*

PROOF. (1) $\Longrightarrow$ (2). Let M be $\mathcal{X}$-automorphism-invariant. Call $X = X(M)$. As X is obviously $\mathcal{X}$-automorphism invariant, we have that $X = X_1 \oplus X_2$, where X_1 is a semiboolean module, X_2 is $\mathcal{X}$-endomorphism invariant, $\mathrm{Hom}(X_1, X_2)$, $\mathrm{Hom}(X_2, X_1) \subseteq J(\mathrm{End}(X))$ and

$$\mathrm{End}(X)/J(\mathrm{End}(X)) = \mathrm{End}(X_1)/J(\mathrm{End}(X_1)) \times \mathrm{End}(X_2)/J(\mathrm{End}(X_2))$$

where $\mathrm{End}(X_1)/J(\mathrm{End}(X_1))$ is a Boolean ring and each element in the ring $\mathrm{End}(X_2)/J(\mathrm{End}(X_2))$ is the sum of two units. Call $S = \mathrm{End}(X)$, $S_1 = \mathrm{End}(X_1)$ and $S_2 = \mathrm{End}(X_2)$. Let $f : X \to X$ be a monomorphism. As $\bar{S} = \bar{S_1} \times \bar{S_2}$, there exist $\bar{f_1} \in \bar{S_1}$ and $\bar{f_2} \in \bar{S_2}$ such that $\bar{f} = \bar{f_1} \times \bar{f_2}$. Moreover, each $\bar{f_i} \in \bar{S_i}$ is a monomorphism. As $\bar{S_1}$ is a Boolean ring, $\bar{f_1} \in \bar{S_1}$ is a unit. On the other hand, $\bar{f_2} \in \bar{S_2}$, so we can write $\bar{f_2}$ as the sum of two units, say $\bar{f_2} = \bar{f_2'} + \bar{f_2''}$. And, $\bar{f_2''}$ is again a sum of two units, say $\bar{f_2''} = \bar{f_3'} + \bar{f_3''}$. Now $\bar{f} = \bar{f_1} \times \bar{f_2} = (\bar{1}_{\bar{S_1}} \times \bar{f_2'}) + (-\bar{1}_{\bar{S_1}} \times \bar{f_3'}) + (\bar{f_1} \times \bar{f_3''})$.

Set $\bar{\gamma_1} = \bar{1}_{\bar{S_1}} \times \bar{f_2'}$, $\bar{\gamma_2} = -\bar{1}_{\bar{S_1}} \times \bar{f_3'}$, and $\bar{\gamma_3} = \bar{f_1} \times \bar{f_3''}$. Then $\bar{f} = \bar{\gamma_1} + \bar{\gamma_2} + \bar{\gamma_3}$. Thus $\bar{f}$ is the sum of three units $\bar{\gamma_1}$, $\bar{\gamma_2}$, and $\bar{\gamma_3}$ in $\bar{S}$. Let us lift the three units $\bar{\gamma_i} \in \bar{S}$ to units $\gamma_i \in S$. Then $f = \gamma_1 + \gamma_2 + \gamma_3 + j$ for some $j \in J(S)$. As M is $\mathcal{X}$-automorphism invariant, $\gamma_i(M) \subseteq M$ for $i = 1, 2, 3$. Also, as $j \in J(S)$, $1 - j$ is an automorphism and consequently, M is invariant under $1 - j$ and hence under j. Thus it follows that M is invariant under f.

(2) $\Longrightarrow$ (1). This implication is obvious. □

In particular, we have

COROLLARY 2.2. *For a module M, we have the following*

(1) *M is automorphism-invariant if and only if M is invariant under monomorphisms of its injective envelope.*

(2) *M is pure-automorphism-invariant if and only if M is invariant under monomorphisms of its pure-injective envelope.*
(3) *If M is flat then M is invariant under automorphisms of its cotorsion envelope if and only if it is invariant under all monomorphisms of its cotorsion envelope.*

Note that in the case of injective envelopes, the above result extends previous results from [**1**] and [**10**]. The first point in the corollary above shows that pseudo-injective modules and essentially pseudo-injective modules studied in [**1**] are not two distinct notions and they are both equivalent to the notion of automorphism-invariant modules.

REMARK 2.3. It follows from the arguments of [**11**, Lemma 1.13] that M is pseudo-N-injective if and only if $f(N) \leq M$ for every monomorphism $f : E(N) \longrightarrow E(M)$.

THEOREM 2.4. *Let $\{M_i \mid i \in I\}$ be a family of modules. Then the following are equivalent:*
(1) *$M = \oplus_{i\in I} M_i$ is automorphism-invariant.*
(2) *$\oplus_{j\neq i} M_j$ is M_i-injective and M_i is automorphism-invariant for every $i \in I$.*

PROOF. (1) $\Rightarrow$ (2). By [**10**, Lemma 1 and Theorem 1].

(2) $\Rightarrow$ (1). It is enough for us to prove that M is invariant under monomorphisms of its injective envelope by Corollary 2.2. Let f be a monomorphism of the injective envelope of M. Since $f(M) = f(\oplus_{i\in I} M_i) = \sum_{i\in I} f(M_i)$, we need to show that $f(M_i) \subseteq M$ for every $i \in I$. Let $i \in I$. Assume that $M_1 = M_i$ and $M_2 = \oplus_{j\neq i} M_j$. Note that M_1 and M_2 are relatively injective. Let E, E_1 and E_2 denote the injective envelopes of M, M_1 and M_2, respectively. Then $E = E_1 \oplus E_2$ and

$$f = \begin{pmatrix} f_{11} & f_{12} \\ f_{21} & f_{22} \end{pmatrix},$$

where $f_{ij} : E_j \longrightarrow E_i$. Since M_2 is M_1-injective, $f_{21}(M_1) \subseteq M_2$ by [**11**, Lemma 1.13]. Therefore $f(M_1) = f_{11}(M_1)+f_{21}(M_1) \subseteq f_{11}(M_1)+M_2$. Now we have to show that $f_{11}(M_1) \subseteq M$. Since f is monic, $\mathrm{Ker}(f_{11})\cap\mathrm{Ker}(f_{21}) = 0$. If $\mathrm{Ker}(f_{11}) = 0$, since M_1 is automorphism-invariant, $f_{11}(M_1) \subseteq M_1$ by Corollary 2.2. If $\mathrm{Ker}(f_{21}) = 0$, then since E_1 is E_2-injective, there exists a homomorphism $g : E_2 \longrightarrow E_1$ such that $g \circ f_{21} = f_{11}$. Now $f_{11}(M_1) = (g \circ f_{21})(M_1) \subseteq g(M_2) \subseteq M_1$. Now assume that $\mathrm{Ker}(f_{11})$ and $\mathrm{Ker}(f_{21})$ are nonzero. Then there exist nonzero orthogonal idempotents α and β such that $E_1 = \alpha(E_1) \oplus \beta(E_1)$ and $\mathrm{Ker}(f_{11}) \subseteq \alpha(E_1)$ and $\mathrm{Ker}(f_{21}) \subseteq \beta(E_1)$.

Let us define the homomorphism $\Psi : f_{21}(E_1) \longrightarrow E_1$ with $\Psi(f_{21}(x_1)) = f_{11}(\alpha(x_1))$ for every $x_1 \in E_1$. Clearly, it is well-defined since α and β are orthogonal idempotents. Since E_1 is E_2-injective, there exists a homomorphism $\eta : E_2 \longrightarrow E_1$ such that $f_{11}(\alpha(M_1)) = (\eta\circ f_{21})(M_1) \subseteq \eta(M_2) \subseteq M_1$. Therefore $f_{11}(\alpha(M_1)) \subseteq M_1$. We also can define a homomorphism $\Gamma : f_{21}(E_1) \longrightarrow E_1$ with $\Gamma(f_{21}(x_1)) = \alpha(x_1)$ for every $x_1 \in E_1$. Then from the same proof as above, $\alpha(M_1) \subseteq M_1$. Hence clearly $\beta(M_1) \subseteq M_1$, too. Assume that $f_{11} \circ \beta(x_1) = f_{11} \circ \beta(y_1)$ for $x_1, y_1 \in E_1$. Then $\beta(x_1 - y_1) \in \mathrm{Ker} f_{11} \subseteq \alpha(E_1)$. Since α and β are orthogonal idempotents, $\beta(x_1) = \beta(y_1)$. Hence f_{11} is one to one on $\beta(E_1)$. On the other hand, there exists a direct summand K of $\beta(E_1)$ such that $\beta(M_1)$ is essential in K. This means that K is the injective envelope of $\beta(M_1)$ in $\beta(E_1)$. Now $f_{11|K} : K = E(\beta(M_1)) \longrightarrow$

$E_1 = E(M_1)$ is a monomorphism. Note that since M_1 is pseudo-injective, M_1 is pseudo-$\beta(M_1)$-injective as we indicated in the Introduction. Then by Remark 2.3, $f_{11}(\beta(M_1)) \subseteq M_1$. Finally, $f_{11}(M_1) = f_{11}((\alpha+\beta)(M_1)) \subseteq M_1$. Therefore M is invariant under monomorphisms of its injective envelope. Hence M is automorphism-invariant. □

EXAMPLE 2.5. Let $M_{\mathbb{Z}} = \mathbb{Z}/2\mathbb{Z} \oplus \mathbb{Z}/8\mathbb{Z}$. Clearly, $\mathbb{Z}/2\mathbb{Z}$ and $\mathbb{Z}/8\mathbb{Z}$ are pseudo-injective $\mathbb{Z}$-modules, but $M_{\mathbb{Z}}$ is not pseudo-injective since $\mathbb{Z}/2\mathbb{Z}$ is not $\mathbb{Z}/8\mathbb{Z}$-injective.

COROLLARY 2.6. *Let R be any ring. $\oplus_{i=1}^n M_i$ is automorphism-invariant if and only if each M_i is automorphism-invariant and M_j-injective for all $j \neq i$.*

Remember that a module M is called *purely infinite* if $M \cong M \oplus M$ (see [**11**, Definition 1.32]).

COROLLARY 2.7. *A purely infinite M is quasi-injective if and only if M is automorphism-invariant.*

By similar arguments, we may obtain the following dual results.

THEOREM 2.8. *Let M be a module with an epimorphic $\mathcal{X}$-cover $p : X(M) \to M$ such that $End(X(M))/J(End(X(M)))$ is a von Neumann regular right self-injective ring and idempotents lift modulo $J(End(X(M)))$. Then the following are equivalent:*

(1) *M is an $\mathcal{X}$-automorphism coinvariant module.*
(2) *M is an $\mathcal{X}$-epimorphism coinvariant module.*

COROLLARY 2.9. *For a module M, we have the following*

(1) *A right module M over a right perfect ring is automorphism-coinvariant if and only if M is coinvariant under epimorphisms of its projective cover.*
(2) *If M is cotorsion then M is coinvariant under automorphisms of its flat cover if and only if it is coinvariant under all epimorphisms of its flat cover.*

THEOREM 2.10. *Let R be a right perfect ring. Let $\{M_i \mid i \in I\}$ be a family of right R-modules. Then the following are equivalent:*

(1) *$M = \oplus_{i\in I} M_i$ is automorphism-coinvariant.*
(2) *M_i is $\oplus_{j\neq i} M_j$-projective and M_i is automorphism-coinvariant for every $i \in I$.*

COROLLARY 2.11. *Let R be a right perfect ring and $\{M_i \mid i \in I\}$ be a family of right R-modules. Then $\oplus_{i=1}^n M_i$ is automorphism-coinvariant if and only if each M_i is automorphism-coinvariant and M_j-projective for all $j \neq i$.*

COROLLARY 2.12. *Let R be a right perfect ring. A purely infinite right R-module M is quasi-projective if and only if M is automorphism-coinvariant.*

3. Properties of $\mathcal{X}$-automorphism invariant modules

In this section, we will obtain directly-finiteness, cancellation and substitution properties for $\mathcal{X}$-automorphism invariant modules. Recall that a module M is called *directly-finite* (or *Dedekind-finite*) if M is not isomorphic to a proper summand of itself. A ring R is called *directly-finite* if $xy = 1$ implies $yx = 1$ for any $x, y \in R$.

It is well-known that a module M is directly-finite if and only if its endomorphism ring is directly-finite (see [**11**, Proposition 1.25]). A right R-module M is said to satisfy the *full exchange property* if for every right R-module A and any two direct sum decompositions $A = M^{'} \oplus N = \oplus_{i \in I} A_i$ with $M^{'} \cong M$, there exist submodules B_i of A_i such that $A = M^{'} \oplus (\oplus_{i \in I} B_i)$. If this holds only for $|I| < \infty$, then M is said to satisfy the *finite exchange property*. A module M is said to have the *cancellation property* if whenever $M \oplus A \cong M \oplus B$, then $A \cong B$. A module M is said to have the *internal cancellation property* if whenever $M = A_1 \oplus B_1 \cong A_2 \oplus B_2$ with $A_1 \cong A_2$, then $B_1 \cong B_2$. A module with the cancellation property always satisfies the internal cancellation property but the converse need not be true, in general. Fuchs [**3**] had shown that if M is a module with the finite exchange property, then M has the cancellation property if and only if M has the internal cancellation property. A module A is said to have the *substitution property* if for every module M with decompositions $M = A_1 \oplus H = A_2 \oplus K$ with $A_1 \cong A \cong A_2$, there exists a submodule C of M such that $M = C \oplus H = C \oplus K$. In general, we have the following hierarchy:

$$\text{substitution} \Rightarrow \text{cancellation} \Rightarrow \text{internal cancellation} \Rightarrow \text{directly-finite}.$$

In [**5**, Lemma 6 and Theorem 7], it is proved that if M is automorphism-invariant and directly-finite, then $E(M)$ is directly-finite and $\mathrm{End}(M)/J(\mathrm{End}(M))$ is unit-regular, where $E(M)$ is the injective envelope of M. In the next theorem we generalize this result for $\mathcal{X}$-automorphism invariant modules.

LEMMA 3.1. *Let M be an $\mathcal{X}$-automorphism invariant module with $u : M \longrightarrow X(M)$, a monomorphic $\mathcal{X}$-envelope such that $End(X(M))/J(End(X(M)))$ is a von Neumann regular right self-injective ring and idempotents lift modulo $J(End(X(M)))$. Then M is directly-finite if and only if $X(M)$ is directly-finite.*

PROOF. Assume M is directly-finite. Let $S = \mathrm{End}(X(M))$ and let $s, t \in S$ such that $t \circ s = 1_{X(M)}$. Then $s : X(M) \longrightarrow X(M)$ is a monomorphism. In the view of Theorem 2.1 we know that M is $\mathcal{X}$-monomorphism invariant. So, there exists $f \in \mathrm{End}(M) = T$ such that $u \circ f = s \circ u$. Thus we have $\Psi(f + K) = s + J(S)$. Therefore, f is a monomorphism. As $K = J(T)$, T/K is von Neumann regular and idempotents lift (see [**6**, Theorem 3.8]), there exists an idempotent $e \in T$ such that $Te + K = Tf + K$. This means that there exists $t' \in T$ such that $f = t' \circ e + k$ with $k \in K$ and thus, $f \circ (1 - e) = k \circ (1 - e)$. Therefore, $\Psi(f + K) \circ \Psi((1 - e) + K) = 0 + J(S)$ and, as $\Psi(f + K)$ is monic in $S/J(S)$, this means that $\Psi((1-e)+K) = 0+J(S)$. Since again Ψ is monic, $(1-e) \in K = J(T)$. Then e is unit in T. Since $M = e(M) \oplus (1 - e)(M)$, $(1 - e)(M) = 0$, and so $e = 1_M$. Therefore, $T = Tf + K$. Now there exists an element $f' \in T$ such that $1_M - f' \circ f \in J(T)$ and thus, $f' \circ f$ is an automorphism. As M is directly-finite, this means that f must be an automorphism. Therefore, $s + J(S) = \Psi(f + K)$ is also an automorphism. Then s is an automorphism. As $t \circ s = 1_{X(M)}$, we get that $s \circ t = 1_{X(M)}$, too. This shows that $X(M)$ is directly-finite.

Conversely, suppose $X(M)$ is directly-finite. We wish to show that M is directly-finite. Assume to the contrary that M is not directly-finite. Then $M \cong M \oplus N$ for some nonzero submodule N. Let $u : M \longrightarrow X(M)$ and $v : N \longrightarrow X(N)$ be $\mathcal{X}$-envelopes of M and N, respectively. Then $u \oplus v : M \oplus N \longrightarrow X(M) \oplus X(N)$ is an $\mathcal{X}$-envelope of $M \oplus N$. Clearly, then $X(M) \cong X(M) \oplus X(N)$. This yields a contradiction to our assumption that $X(M)$ is directly-finite. Thus we have that M is directly-finite if and only if $X(M)$ is directly-finite. □

The following results generalize [**5**, Theorem 7] and [**5**, Corollary 8], respectively.

LEMMA 3.2. *Let M be an $\mathcal{X}$-automorphism invariant module with $u : M \longrightarrow X(M)$ a monomorphic $\mathcal{X}$-envelope such that $End(X(M))/J(End(X(M)))$ is a von Neumann regular right self-injective ring and idempotents lift modulo $J(End(X(M)))$. If M is directly finite, then $End(M)/J(End(M))$ is unit-regular.*

PROOF. We know that $\mathrm{End}(M)/J(\mathrm{End}(M)) = R_1 \times R_2$ where R_1 is a Boolean ring and R_2 is a von Neumann regular right self-injective ring. If M is a directly-finite module then ($\mathrm{End}(M)$ and hence) $\mathrm{End}(M)/J(\mathrm{End}(M))$ is a directly-finite ring. Thus, R_2 is a directly-finite von Neumann regular right self-injective ring and hence a unit-regular ring. Clearly R_1 is a unit-regular ring being a Boolean ring. Since a product of two unit-regular rings is unit-regular, $\mathrm{End}(M)/J(\mathrm{End}(M))$ is unit-regular. □

THEOREM 3.3. *Let M be an $\mathcal{X}$-automorphism invariant module with a monomorphic $\mathcal{X}$-envelope $u : M \longrightarrow X(M)$ such that $End(X(M))/J(End(X(N)))$ is a von Neumann regular right self-injective ring and idempotents lift modulo $J(End(X(M)))$. Then the following are equivalent:*

(1) *M is directly-finite.*
(2) *M has the internal cancellation property.*
(3) *M has the cancellation property.*
(4) *M has the substitution property.*
(5) *$X(M)$ is directly-finite.*
(6) *$X(M)$ has the internal cancellation property.*
(7) *$X(M)$ has the cancellation property.*
(8) *$X(M)$ has the substitution property.*

PROOF. (1) $\Rightarrow$ (2). By Lemma 3.2, $\mathrm{End}(M)/J(\mathrm{End}(M))$ is unit-regular and therefore $\mathrm{End}(M)$ is a ring with internal cancellation. Now by [**4**], M has the internal cancellation property.

(2) $\Rightarrow$ (3). By [**6**, Corollary 3.9], M has the finite exchange property. By [**13**] or [**3**], M has the cancellation property.

(3) $\Rightarrow$ (1). By definitions.

(2) $\Leftrightarrow$ (4). By [**13**], since M has the finite exchange property.

The equivalence of (5), (6), (7) and (8) with rest of the statements follows from Lemma 3.1. □

By similar arguments we can give the following dual results to Lemma 3.1, Lemma 3.2 and Theorem 3.3, respectively.

LEMMA 3.4. *Let M be an $\mathcal{X}$-automorphism coinvariant module with epimorphic $\mathcal{X}$-cover $p : X(M) \longrightarrow M$ such that $End(X(M))/J(End(X(M)))$ is a von Neumann regular right self-injective ring and idempotents lift modulo $J(End(X(M)))$. If M is directly-finite, so is its $\mathcal{X}$-cover.*

LEMMA 3.5. *Let M be an $\mathcal{X}$-automorphism coinvariant module with epimorphic $\mathcal{X}$-cover $p : X(M) \longrightarrow M$ such that $End(X(M))/J(End(X(M)))$ is a von Neumann regular right self-injective ring and idempotents lift modulo $J(End(X(M)))$. If M is directly-finite, then $End(M)/J(End(M))$ is unit-regular.*

THEOREM 3.6. *Let M be an $\mathcal{X}$-automorphism coinvariant module with epimorphic $\mathcal{X}$-cover $p : X(M) \longrightarrow M$ such that $End(X(M))/J(End(X(M)))$ is a von Neumann regular right self-injective ring and idempotents lift modulo $J(End(X(M)))$. Then the following are equivalent:*

(1) *M is directly-finite.*
(2) *M has the internal cancellation property.*
(3) *M has the cancellation property.*
(4) *M has the substitution property.*
(5) *$X(M)$ is directly-finite.*
(6) *$X(M)$ has the internal cancellation property.*
(7) *$X(M)$ has the cancellation property.*
(8) *$X(M)$ has the substitution property.*

4. Examples

In [**6**] an example was constructed of an $\mathcal{X}$-automorphism invariant module that is not $\mathcal{X}$-endomorphism invariant but this example was over a von Neumann regular ring where the class $\mathcal{X}$ of injective modules coincides with the class of pure-injective modules and also with the class of flat cotorsion modules. In this section we will construct some examples where the class of injective modules is different from the class of pure-injective modules. The basic idea behind construction of these examples is similar to the one used in [**8**].

(1) Let R be a von Neumann regular ring which is not semisimple artinian. Then there exists a right R-module A such that A is quasi-injective but not injective. Let S be a ring with a right S-module N' which is pure-injective but not injective. Let $N = E(S) \oplus N'$, $T = R \times S$ and $M_T = A \times N$. Note that the pure-injective envelope of M, $PE(M) = E(A) \times N$. Clearly M_T is invariant under endomorphisms of $PE(M)$. Note that M is not invariant under endomorphisms of its injective envelope because if M is quasi-injective then N would also be quasi-injective and as $S \subset N$, this would mean that N is injective. This gives us a contradiction to the fact that N', a direct summand of N, is not injective.
(2) Let R be the ring of all eventually constant sequences of elements from $\mathbb{F}_2$, the field of two elements. Then R_R is automorphism-invariant but not quasi-injective. Let S be a ring with a right S-module N' which is pure-injective but not injective. Furthermore, assume that S has no homomorphic image isomorphic to $\mathbb{F}_2$. Let $N = E(S) \oplus N'$, $T = R \times S$, $A = R_R$, and $M_T = A \times N$. Note that M_T is invariant under automorphisms of $PE(M)$. It may be observed that M_T is not invariant under endomorphisms of $PE(M)$ because if M is pure-quasi-injective then A would be quasi-injective, a contradiction. Also, M is not automorphism-invariant because if M was automorphism-invariant then it would force N to be automorphism-invariant. But as S has no homormorphic image

isomorphic to $\mathbb{F}_2$, then N would be quasi-injective and hence injective as $S \subseteq N$, a contradiction to our assumption that N' is not injective.

References

[1] A. Alahmadi, N. Er, and S. K. Jain, *Modules which are invariant under monomorphisms of their injective hulls*, J. Aust. Math. Soc. **79** (2005), no. 3, 349–360. MR2190686

[2] N. Er, S. Singh, and A. K. Srivastava, *Rings and modules which are stable under automorphisms of their injective hulls*, J. Algebra **379** (2013), 223–229. MR3019253

[3] L. Fuchs, *On quasi-injective modules*, Ann. Scuola Norm. Sup. Pisa (3) **23** (1969), 541–546. MR0258873

[4] R. Guralnick and C. Lanski, *Pseudosimilarity and cancellation of modules*, Linear Algebra Appl. **47** (1982), 111–115. MR672734

[5] P. A. Guil Asensio and A. K. Srivastava, *Automorphism-invariant modules satisfy the exchange property*, J. Algebra **388** (2013), 101–106. MR3061680

[6] P. A. Guil Asensio, D. Keskin Tütüncü, and A. K. Srivastava, *Modules invariant under automorphisms of their covers and envelopes*, Israel J. Math. **206** (2015), no. 1, 457–482. MR3319648

[7] P. A. Guil Asensio, D. Keskin Tütüncü, B. Kaleboğaz, and A. K. Srivastava, *Modules which are coinvariant under automorphisms of their projective covers*, J. Algebra **466** (2016), 147–152. MR3541679

[8] P. A. Guil Asensio, Berke Kaleboğaz and A. K. Srivastava, Schröder-Bernstein problem for modules, preprint.

[9] P. A. G. Asensio, Truong Cong Quynh, and A. K. Srivastava, *Additive unit structure of endomorphism rings and invariance of modules*, Bull. Math. Sci. **7** (2017), no. 2, 229–246. MR3671737

[10] S. K. Jain and S. Singh, *On pseudo injective modules and self pseudo injective rings*, J. Math. Sci. **2** (1967), 23–31. MR0219569

[11] S. H. Mohamed and B. J. Müller, *Continuous and discrete modules*, London Mathematical Society Lecture Note Series, vol. 147, Cambridge University Press, Cambridge, 1990. MR1084376

[12] M. L. Teply, *Pseudo-injective modules which are not quasi-injective*, Proc. Amer. Math. Soc. **49** (1975), 305–310. MR0366977

[13] H.-P. Yu, *Stable range one for exchange rings*, J. Pure Appl. Algebra **98** (1995), no. 1, 105–109. MR1317002

Departamento de Mathematicas, Universidad de Murcia, Murcia, 30100, Spain
Email address: `paguil@um.es`

Department of Mathematics, Hacettepe University, Ankara, 06800, Turkey
Email address: `keskin@hacettepe.edu.tr`

Department of Mathematics and Statistics, St. Louis University, St. Louis, MO-63103, USA
Email address: `asrivas3@slu.edu`

Contemporary Mathematics
Volume **715**, 2018
https://doi.org/10.1090/conm/715/14411

Some results and questions on left-right symmetry

Dinh Van Huynh and Dinh Duc Tai

In Memory of Professor Bruno Mueller with love and respect!

ABSTRACT. We summarize a left-right symmetric result from [**17**] on prime rings together with applications on related rings. We also ask questions on the left-right symmetry of other properties of rings. Theorem 7.1 is new.

1. Introduction

In this paper we study associative rings R with identity $1 \neq 0$, all modules are unitary modules. If M is a right module over a ring R, we denote it by M_R, while ${}_RM$ stands for left R-module.

A module M_R is defined to be nonsingular if for each nonzero element $m \in M$, the annihilator $r_R(m) = \{x \in R | mx = 0\}$ is not an essential right ideal of R. Otherwise, if each $r_R(m)$ is essential in R_R, then M is called a singular module. Any module M contains a submodule which is singular. The sum of all such submodules is singular and this is the largest singular submodule of M. A ring R is right nonsingular if R_R is nonsingular. Similarly, left nonsingular (resp., singular) modules are defined.

A module M is defined to be a module of finite uniform (or Goldie) dimension if M does not contain an infinite direct sum of nonzero submodules. In this case, there are uniform submodules $U_1, \cdots, U_k$, $k \in \mathbb{N}$ such that $U_1 \oplus \cdots \oplus N_k$ is essential in M, and we will write u-dim$(M) = k$.

The questions on left-right symmetry of non-commutative rings are always interesting and challenging. There are more questions than answers. In this note we summarize some of the results that have been obtained by the first author either solely or in collaboration with others.

In order to do that we start with some definitions: A right R-module M is called CS or extending if every submodule of M is essential in a direct summand. A cyclic right (or left) R-module X is called proper cyclic if $X \ncong R_R$. A module M_R is called a V-module if every simple right R-module is M-injective. A ring R is called a right V-ring if R_R is a V-module. Left V-rings are defined similarly.

2010 *Mathematics Subject Classification.* Primary 16N60, 16P40; Secondary 16L60, 16P20.
The first author is the corresponding author: huynhohio@gmail.com or huynh@ohio.edu.

The reader may refer to [**23**], [**25**], [**24**], [**20**], etc. for an account of V-modules and V-rings.

2. Prime Rings

For a prime right Goldie ring R, the injectivity of R is left-right symmetric, and in this case the ring becomes semisimple artinian. Concerning the weaker condition CS, the following theorem was obtained in [**17**].

THEOREM 2.1. *For a prime ring R with $u - dim(R_R) > 1$, the following are equivalent:*

(1) R is right Goldie and right CS.

(2) R is left Goldie and left CS.

Note that the condition of $u - dim(R_R) > 1$ is unremoveable because, as is well-known, there are right Ore domains which are not left Ore. As an application of Theorem 2.1, the following theorem was obtained in [**15**].

THEOREM 2.2. *For a prime ring R and $0 \neq e^2 = e \in R$, the following conditions are equivalent:*

(i) eR is a CS right R-module and $1 < u - dim(eR) < \infty$.

(ii) Re is a CS left R-module and $1 < u - dim(Re) < \infty$.

(iii) eRe is a left and right Goldie left and right CS ring with left and right u-dimension ≥ 2.

Also here our theorem does not hold if $u - dim(R) = 1$.

For prime rings we would like to mention some other results concerning the left-right symmetry. We need some notation for that.

For any positive integer n, a uniform module U is called a $n - CS^+$ modue if U^n is CS and each uniform direct summand of it is isomorphic to U. $2 - CS^+$-rings and modules were studied in [**2**] using Bezout domains and the main result is the following theorem. A domain R is called right Bezout if every finitely generated right ideal of D is principal. Left Bezout domains are defined similarly. A Bezout domain is a domain which is left and right Bezout.

THEOREM 2.3 ([**2**]). *Let R be a simple ring. Then the following conditions are equivalent:*

(i) R contains a uniform right ideal V such that V is $2 - CS^+$.

(ii) R contains a uniform left ideal T such that T is $2 - CS^+$.

(iii) R is isomorphic to a $k \times k$ matrix ring over a Bezout domain D for some positive integer k.

The following result is obtained as a consequence of Theorem 2.3.

COROLLARY 2.4 ([**2**]). *Let R be a simple ring with acc on principal right ideals and acc on principal left ideals. Then the following conditions are equivalent:*

(i) R contains a uniform right ideal V such that V is $2 - CS^+$.

(ii) R contains a uniform left ideal T such that T is $2 - CS^+$.

(iii) R is isomorphic to an $k \times k$ *matrix ring over a principal right and principal left ideal domain D for some positive integer k.*

3. WV-Rings

A right R-module M is called a V-module if every simple module in the category MOD-R is M-injective.

For $M = R$, we obtain the known definition of V-rings: A ring R is a right V-ring if every simple right R-module is injective. For basic properties of V-rings we refer to [**23**].

Let R be an arbitrary ring. A cyclic right R-module X is called a proper cyclic module if $X_R \not\cong R_R$ (see [**10**]). Recently, using the concept of proper cyclic modules, Holston-Jain-Leroy [**14**] defined a generalization of V-rings: A ring R is called a right weakly V-ring (briefly, right WV-ring) if every simple right R-module is injective relative to proper cyclic right R-modules. Left WV-rings are defined similarly. Using the concept of V-modules, this definition is equivalent to saying that a ring R is a right (left) WV-ring if every proper cyclic right (left) R-module is a V-module. If R is a right (left) WV-ring which is not right (resp., left) V, then we say that R is a strictly right (left) WV-ring.

The structure of a strictly right WV-ring R was described in [**8**] by the following theorem.

THEOREM 3.1. *Let R be a strictly right WV-ring with Jacobson radical J. Set $\overline{R} = R/J$, $S = End_R(J_R)$. Then R has the following properties:*

(I) R has exactly 3 distinct two-sided ideals $0 \subset J \subset R$ such that J is a minimal right ideal of R and R/J is a simple right V-domain. For any nonzero elements $x,\ y \in J$,

$$Rx \cong Ry \cong {}_R\overline{R}.$$

Moreover, the mapping $\lambda : \overline{r} \mapsto \lambda_{\overline{r}},\ \ \forall\ \overline{r} \in \overline{R}$, is a ring embedding $\overline{R} \to S$ where $\lambda_{\overline{r}} \in S$ is defined by $\lambda_{\overline{r}} : x \mapsto \overline{r}x$ for $\forall\ x \in J$.

(II) R is left artinian and right uniserial of right composition length 2 if and only if ${}_RJ$ is finitely generated.

(III) R_R is uniserial of composition length 2 if and only if R has ACC on left annihilators.

(IV) The following conditions are equivalent:

(i) The injective hull $E(J_R)$ of the minimal right ideal J is a uniserial module.

(ii) R is a strictly left WV-ring.

(iii) The monomorphism λ in (I) is an isomorphism.

(iv) R is a Frobenius local ring.

By (I) of Theorem 3.1 we see that if R is also a left strictly WV-ring, then $\overline{R}$ is a left and right V-domain. Moreover, as it has composition length 1, it is a division ring. In this case the sructure of R is completely described, namely:

$$R \cong \left\{ \begin{pmatrix} a & b \\ 0 & a \end{pmatrix} :\ a,\ b \in Q \right\}$$

where Q is a division ring. Thus

QUESTION 3.2 (See [**8**, Remarks 8]). *Is a strictly right WV-ring strictly left WV?*

We feel that the answer for this question would be negative. One of the reasons for this is that, in general, a right V-domain may be not left V, as R/J might not be left V. As discribed above, a two-sided strictly WV-ring is a local Frobenius ring with square zero Jacobson radical.

A special case of strictly right WV-rings was considered in [**8**, Corollary 7]. Now, we assume that R is strictly right WV, and that R/J is finite. Hence R/J must be a finite division ring because it is a domain. Hence by the Wedderburn's Theorem, it must be a field. It follows that the ring is finite. Moreover, as $J^2 = 0$, J is commutative. If we could show that R is then commutative, we would improve [**8**, Corollary 7].

4. Direct sums of CS-modules

In [**16**] it was studied the question of when any direct sum of CS-modules is CS. The structure of this rings is as the following:

THEOREM 4.1. *For a ring R the following statements are equivalent:*
(1) Every direct sum of CS right R-modules is CS;
(2) R is right artinian and every uniform right R-module has composition length at most 2.

Thinking about the left-right symmetry we ask the question:

QUESTION 4.2. *Are the following statements equivalent?*
(a) Every direct sum of CS right R-modules is CS;
(b) Every direct sum of CS left R-modules is CS.

5. All modules are CS

The structure of rings over which every right module is CS was obtained by Dung and Smith (see, [**9**, 13.5]).

THEOREM 5.1. *For a ring R the following statements are equivalent:*

(1) Every right R-module is CS;
(2) $R_R = R_1 \oplus ... \oplus R_n$ where every R_i is uniform of composition length at most 2 and if $l(R_j) = 2$, then R_j is injective.
(3) Every left R-module is CS.

In [**19**] it was shown that

THEOREM 5.2. *If every finitely generated right R-module is CS, then R is right noetherian, and every uniform right R-module has composition length at most 2.*

For the left-right symmetry we ask

QUESTION 5.3. *Are the following statements equivalent for a ring R?*
(a) Every finitely generated right R-module is CS.
(b) Every finitely generated left R-module is CS.

6. Continuous, quasi-continuous modules

A CS R-module M is continuous if any submodule isomorphic to a direct summand is a direct summand. A module M is called a quasi-continuous module if it is CS and whenever A and B are direct summands with $A \cap B = 0$, $A \oplus B$ is also a direct summand.

In [**18**] the following theorem is obtained:

THEOREM 6.1. *If every continuous right R-module is injective, then R is semisimple artinian.*

The following result was shown in [**7**].

THEOREM 6.2. *For a right semiartinian ring R the following conditions are equivalent:*
(a) R is right V;
(b) Every finitely generated continuous right R-module is injective.

QUESTION 6.3. *(a) Find the structure of all rings over which every finitely generated continuous right modules is injective.*

(b) Are the following conditions equivalent for a ring R?[1]
(1) Every finitely generated continuous right R-module is injective.
(2) Every finitely generated continuous left R-module is injective.

7. PCI-Domains

We return to Theorem 2.1. Why did we try to obtain this result? To answer this we need some definitions. A ring R is called a right PCI-ring if every proper cyclic right R-module is injective (Faith, [**10**], [**5**]). These rings were studied by many authors: Faith, Cozzens, Damiano, etc. Among other results they showed that a right PCI-ring is either a semisimple artinian ring or a right noetherian right hereditary domain D such that for any right ideal $0 \neq A \subseteq D$, D/A is a semisimple module. We call D a right PCI-domain, and assume that D is not a division ring. An example of J. Cozzens [**4**] shows that such a domain exists. But the question if a right PCI-domain is left PCI, remains open since m?e than 40 years. Theorem 2.1 or 2.2 was obtained as a "by-product" of the efforts to answer this question. We are not able to remove the additional assumption in (a) or equivalently, in (b) of the following result.

THEOREM 7.1. *For a domain D which is not a division ring, the following conditions are equivalent:*

(a) D is right PCI, and every nonsingular uniform 2-generated right D-module is cyclic.

(b) D is left PCI, and every nonsingular uniform 2-generated left D-module is cyclic.

[1]The referee made a remark that he/she believes the answer to (b) of **6.3** is negative as follows.

Let R be the ring in Example 5.14 of [**5**]. That ring R is semiprime, right but not left V. Futhermore $S = Soc(R_R) = Soc({}_RR)$ is a maximal right ideal of R. Hence R/S is a division ring, therefore R is left semiartinian. But R is not left V. Thus by Theorem 6.2, R satisfies (1) but not (2).

(c) D is a left and right PID which is left and right V.

Proof. (a)$\Rightarrow$(b). Let D be a right PCI-domain, then for any right ideal xD, with nonzero $x \in D$, $xD \cong D$. With the second condition of (a) and the fact that D is right noetherian, we can see that D is a right principal ideal domain. Now let A be a closed uniform submodule of $D^2 = D_D \oplus D_D$. Then D^2/A is a uniform nonsingular 2-generated right D-module. Hence it is cyclic, so it must be isomorphic to D_D. Therefore A splits in D^2, i.e., D^2 is CS.

By 12.8 of [**9**], the 2×2-matrix ring $R = M_2(D)$ is also right CS. By Theorem 2.1, R is left Goldie, and so D is a left Ore domain. Hence, by [**21**, Theorem 4.2], D is left V and left noetherian. In particular, as D is right PCI, D must be two-sided noetherian and two-sided hereditary, and so we obtain the restricted minimum condition on left ideals of D (cf. [**26**]). That implies that D is left PCI.

Moreover, every uniform direct summand of D^2 is 2-generated, hence by the assumption of (a), cyclic, and so D is 2-CS+. It follows, from Corollary 2.4, D is a left principal ideal domain. Let N be a nonsingular 2-generated uniform left D-module. Then we can embed N in the classical left quotient ring Q of D. As Q is also the right quotient ring of D, we can easily show that N is isomorphic to a left ideal of D, and so (b) holds. Here we use the fact that D is a simple ring for applying Corollary 2.4, and Theorem 4.2 of [**21**].

Similarly (b)$\Rightarrow$(a) holds. The equivalence (a)$\Leftrightarrow$ (c) is clear. Namely, assume (c), in particular, if D is a left and right PID, D is left and right hereditary, so D/A is an artinian D-module by [**26**] for any nonzero right or left ideal $A \subseteq D$. As D is a V-domain, it is a right and left PCI domain. The direction (a)$\Rightarrow$(c) is already included in the proof of (a)$\Rightarrow$(b). $\square$

We remark that the Cozzens example is a ring of Theorem 7.1. Of course we want to answer the following long standing question but we are not able to do it now, so it still remains open.

QUESTION 7.2. *Is it true that a domain D is right PCI if and only if D is left PCI ?*

Acknowledgment. We would like to express our thanks to the referee for his/her carefully reading the manuscript and for making many useful comments and suggestions that help us to improve our paper considerably.

References

[1] F. W. Anderson and K. R. Fuller, *Rings and categories of modules*, 2nd ed., Graduate Texts in Mathematics, vol. 13, Springer-Verlag, New York, 1992. MR1245487

[2] J. Clark and D. Van Huynh, *A study of uniform one-sided ideals in simple rings*, Glasg. Math. J. **49** (2007), no. 3, 489–495. MR2371513

[3] J. Clark and D. Van Huynh, *Simple rings with injectivity conditions on one-sided ideals*, Bull. Austral. Math. Soc. **76** (2007), no. 2, 315–320. MR2353217

[4] J. H. Cozzens, *Homological properties of the ring of differential polynomials*, Bull. Amer. Math. Soc. **76** (1970), 75–79. MR0258886

[5] J. Cozzens and C. Faith, *Simple Noetherian rings*, Cambridge University Press, Cambridge-New York-Melbourne, 1975. Cambridge Tracts in Mathematics, No. 69. MR0396660

[6] R. F. Damiano, *A right PCI ring is right Noetherian*, Proc. Amer. Math. Soc. **77** (1979), no. 1, 11–14. MR539620

[7] H. Q. Dinh and P. F. Smith, *A result on semi-Artinian rings*, Proc. Edinb. Math. Soc. (2) **46** (2003), no. 1, 63–66. MR1961171

[8] H. Q. Dinh, C. J. Holston, and D. V. Huynh, *Some results on V-rings and weakly V-rings*, J. Pure Appl. Algebra **217** (2013), no. 1, 125–131. MR2965910

[9] N. V. Dung, D. V. Huynh, P. F. Smith, and R. Wisbauer, *Extending modules*, Pitman Research Notes in Mathematics Series, vol. 313, Longman Scientific & Technical, Harlow; copublished in the United States with John Wiley & Sons, Inc., New York, 1994. MR1312366

[10] C. Faith, *When are proper cyclics injective?*, Pacific J. Math. **45** (1973), 97–112. MR0320069

[11] C. Faith, *Algebra: rings, modules and categories. I*, Springer-Verlag, New York-Heidelberg, 1973. Die Grundlehren der mathematischen Wissenschaften, Band 190. MR0366960

[12] C. Faith, *Rings and things and a fine array of twentieth century associative algebra*, Mathematical Surveys and Monographs, vol. 65, American Mathematical Society, Providence, RI, 1999. MR1657671

[13] K.R. Goodearl, *Singular torsion and the splitting properties,* Mem. Amer. Math. Soc. **124** (1972). MR0340335

[14] C. J. Holston, S. K. Jain, and A. Leroy, *Rings over which cyclics are direct sums of projective and CS or Noetherian*, Glasg. Math. J. **52** (2010), no. A, 103–110. MR2669100

[15] D. V. Huynh, *The symmetry of the CS condition on one-sided ideals in a prime ring*, J. Pure Appl. Algebra **212** (2008), no. 1, 9–13. MR2355030

[16] D. Van Huynh, S. K. Jain, and S. R. López-Permouth, *Rings characterized by direct sums of CS modules*, Comm. Algebra **28** (2000), no. 9, 4219–4222. MR1772501

[17] Dinh Van Huynh, S. K. Jain, and S. R. López-Permouth, *On the symmetry of the Goldie and CS conditions for prime rings*, Proc. Amer. Math. Soc. **128** (2000), no. 11, 3153–3157. MR1670375

[18] D. V. Huynh and S. T. Rizvi, *An approach to Boyle's conjecture*, Proc. Edinburgh Math. Soc. (2) **40** (1997), no. 2, 267–273. MR1454022

[19] D. V. Huynh, S. T. Rizvi, and M. F. Yousif, *Rings whose finitely generated modules are extending*, J. Pure Appl. Algebra **111** (1996), no. 1-3, 325–328. MR1394360

[20] S. K. Jain, T. Y. Lam, and A. Leroy, *Ore extensions and V-domains*, Rings, modules and representations, Contemp. Math., vol. 480, Amer. Math. Soc., Providence, RI, 2009, pp. 249–262. MR2508155

[21] K. A. Kosler, *On hereditary rings and Noetherian V-rings*, Pacific J. Math. **103** (1982), no. 2, 467–473. MR705245

[22] T.Y. Lam, *Lectures on Modules and Rings,* Springer Verlag, Berlin-Heidelberg-New York, GTM 190, 1998.

[23] G. O. Michler and O. E. Villamayor, *On rings whose simple modules are injective*, J. Algebra **25** (1973), 185–201. MR0316505

[24] A. K. Srivastava, *On Σ-V rings*, Comm. Algebra **39** (2011), no. 7, 2430–2436. MR2821721

[25] A. Tuganbaev, *Max rings and V-rings*, Handbook of Algebra, Vol. 3, Elsevier/North-Holland, Amsterdam, 2003, pp. 565–584. MR2035106

[26] D. B. Webber, *Ideals and modules of simple Noetherian hereditary rings*, J. Algebra **16** (1970), 239–242. MR0265395

Department of Mathematics, Ohio University, Athens, Ohio
Email address: huynhohio@gmail.com
Email address: huynh@ohio.edu

Department of Mathematics, Vinh University, Nghe An, Vietnam
Email address: taidd@vinhuni.edu.vn

Contemporary Mathematics
Volume **715**, 2018
https://doi.org/10.1090/conm/715/14412

Rings in which every unit is a sum of a nilpotent and an idempotent

Arezou Karimi-Mansoub, Tamer Koşan, and Yiqiang Zhou

Dedicated to the memory of Bruno J. Müller

ABSTRACT. A ring R is a UU ring if every unit is unipotent, or equivalently if every unit is a sum of a nilpotent and an idempotent that commute. These rings have been investigated in Călugăreanu [**3**] and in Danchev and Lam [**7**]. In this paper, two generalizations of UU rings are discussed. We study rings for which every unit is a sum of a nilpotent and an idempotent, and rings for which every unit is a sum of a nilpotent and two idempotents that commute with one another.

1. Introduction

The motivation of this paper is a recent work of Danchev and Lam [**7**] on UU rings. Throughout, R is an associative ring with identity. We denote by $J(R), U(R)$, $\mathrm{Nil}(R)$ and $\mathrm{idem}(R)$ the Jacobson radical, the unit group, the set of nilpotent elements and the set of idempotents of R, respectively. In [**8**], Diesl introduced (strongly) nil-clean elements and rings as follows. An element a in a ring R is called (strongly) nil-clean if a is the sum of an idempotent and a nilpotent (that commute with each other), and the ring is called (strongly) nil-clean if each of its elements is (strongly) nil-clean. One of the results in [**8**] states that a ring R is strongly nil-clean if and only if R is strongly π-regular with $U(R) = 1 + \mathrm{Nil}(R)$. This motivated Călugăreanu [**3**] to introduce and study UU rings (rings whose units are unipotent). Equivalently, a ring is a UU ring if and only if every unit is strongly nil-clean. These rings have been extensively investigated in Danchev and Lam [**7**], where, among others, it is proved that a ring is strongly nil-clean if and only if it is an exchange (clean) UU ring, and it is asked whether a clean ring R is nil-clean if and only if every unit of R is nil-clean. Here we are motivated to study rings whose units are nil-clean. These rings will be called UNC rings.

In section 2, we first prove several basic properties of UNC rings. Especially it is proved that every semilocal UNC ring is nil-clean. This can be seen as a partial answer to the question of Danchev and Lam. We next show that the matrix ring

---

2010 *Mathematics Subject Classification.* Primary 16E50, 16L30, 16N20, 16S34, 16S50, 16U60.

Key words and phrases. unit, idempotent, nilpotent, nil-clean ring, strongly nil-clean ring, clean ring, semipotent ring, UU ring, UNC ring, group ring.

over a commutative ring R is a UNC ring if and only if $R/J(R)$ is Boolean with $J(R)$ nil. As a consequence, the matrix ring over a UNC ring need not be a UNC ring. We also discuss when a group ring is a UNC (UU) ring. In the last part of this section, it is shown that UU rings are exactly those rings whose units are uniquely nil-clean. As the main result in this section, it is proved that a ring R is strongly nil-clean if and only if R is a semipotent UU ring.

As another natural generalization of UU rings, in section 3 we determine the rings for which every unit is a sum of a nilpotent and two idempotents that commute with one another. We also deal with a special case where every unit of the ring is a sum of two commuting idempotents. These conditions can be compared with the so-called strongly 2-nil-clean rings introduced by Chen and Sheibani [**4**], and the rings for which every element is a sum of two commuting idempotents, studied by Hirano and Tominaga in [**11**].

We write $\mathbb{M}_n(R)$, $\mathbb{T}_n(R)$ and $R[t]$ for the $n \times n$ matrix ring, the $n \times n$ upper triangular matrix ring, and the polynomial ring over R, respectively. For an endomorphism σ of a ring R, let $R[t;\sigma]$ denote the ring of left skew power series over R. Thus, elements of $R[t;\sigma]$ are polynomials in t with coefficients in R written on the left, subject to the relation $tr = \sigma(r)t$ for all $r \in R$. The group ring of a group G over a ring R is denoted by RG.

2. Units being nil-clean

2.1. Basic properties. We present various properties of the rings whose units are nil-clean, and prove that every semilocal ring whose units are nil-clean is a nil-clean ring.

DEFINITION 2.1. A ring R is called a UNC ring if every unit of R is nil-clean.

Every nil-clean ring is a UNC ring. A ring R is called a UU ring if $U(R) = 1 + \mathrm{Nil}(R)$ (see [**3**]).

PROPOSITION 2.2. [**8**] *A unit u of R is strongly nil-clean if and only if $u \in 1 + \mathrm{Nil}(R)$. In particular, R is a UU ring if and only if every unit of R is strongly nil-clean.*

Thus, UU rings can be viewed as the "strong version" of UNC rings.

LEMMA 2.3. *The class of UNC rings is closed under finite direct sums.*

The next lemma was proved for a nil-clean ring in [**8**] and for a UU ring in [**7**].

LEMMA 2.4. *If R is a UNC ring, then $J(R)$ is nil and $2 \in J(R)$.*

PROOF. Let $j \in J(R)$. Then $1 + j = e + b$ where $e^2 = e$ and $b \in \mathrm{Nil}(R)$, and so $e = (1 - b) + j \in U(R)$. It follows that $e = 1$, and hence $j = b \in \mathrm{Nil}(R)$. Write $-1 = e + b$ where $e^2 = e$ and $b \in \mathrm{Nil}(R)$. Then $e = -1 - b \in U(R)$. So $e = 1$ and hence $2 = -b \in \mathrm{Nil}(R)$. □

The next result is basic for studying the structure of a UNC ring.

THEOREM 2.5. *Let R be a ring, and I a nil ideal of R.*

(1) *R is a UNC ring if and only if $J(R)$ is nil and $R/J(R)$ is a UNC ring.*
(2) *R is a UNC ring if and only if R/I is a UNC ring.*

PROOF. (1) The necessity is clear in view of Lemma 2.4. For the sufficiency, let $u \in U(R)$. Then $\bar{u} \in U(R/J(R))$, and write $\bar{u} = \bar{e} + \bar{b}$ where $\bar{e} \in \text{idem}(R/J(R))$ and $\bar{b} \in \text{Nil}(R/J(R))$. As $J(R)$ is nil, idempotents of $R/J(R)$ can be lifted to idempotents of R. So we can assume that $e^2 = e \in R$. Moreover, $b \in R$ is nilpotent. Thus, for some $j \in J(R)$, $u = e + b + j = e + (b + j)$ is nil-clean because $b + j \in \text{Nil}(R)$.

(2) The proof is similar to (1). $\square$

The following corollary can be quickly verified using Theorem 2.5.

COROLLARY 2.6. *Let R, S be rings, M be an (R, S)-bimodule, and N a bimodule over R.*

1. *The trivial extension $R \propto N$ is a UNC ring if and only if R is a UNC ring.*
2. *For $n \geq 2$, $R[t]/(t^n)$ is UNC ring if and only if R is a UNC ring.*
3. *The formal triangular matrix ring $\begin{pmatrix} R & M \\ 0 & S \end{pmatrix}$ is a UNC ring if and only if R, S are UNC rings.*
4. *For $n \geq 1$, $\mathbb{T}_n(R)$ is a UNC ring if and only if R is a UNC ring.*

The easiest way to see a UNC ring that is neither UU nor nil-clean is to form the ring direct sum $R = R_1 \oplus R_2$, where R_1 is a UU ring that is not nil-clean and R_2 is a nil-clean ring that is not UU. For instance, the ring $\mathbb{Z}_2[t] \oplus \mathbb{M}_2(\mathbb{Z}_2)$ is a UNC ring that is neither UU nor nil-clean. The next example gives an indecomposable UNC ring that is neither UU nor nil-clean.

EXAMPLE 2.7. *Let $R = \mathbb{M}_n(\mathbb{Z}_2)$ with $n \geq 2$, $S = \mathbb{Z}_2[t]$ and $M = S^n$. For $(a_{ij}) \in R$, $b \in S$ and $\vec{x} = \begin{bmatrix} x_1 \\ x_2 \\ \vdots \\ x_n \end{bmatrix} \in S^n$, define*

$$(a_{ij})\vec{x} = \begin{bmatrix} a_{11}x_1 + a_{12}x_2 + \cdots + a_{1n}x_n \\ a_{21}x_1 + a_{22}x_2 + \cdots + a_{2n}x_n \\ \vdots \\ a_{n1}x_1 + a_{n2}x_2 + \cdots + a_{nn}x_n \end{bmatrix} \text{ and } \vec{x}b = \begin{bmatrix} x_1 b \\ x_2 b \\ \vdots \\ x_n b \end{bmatrix}.$$

Then M is an (R, S)-bimodule, and the formal triangular matrix ring $T := \begin{pmatrix} R & M \\ 0 & S \end{pmatrix}$ is a UNC ring by Corollary 2.6(3). One can verify that the central idempotents of T are trivial. So T is indecomposable. But since R is not a UU ring and S is not a nil-clean ring, T is neither a UU ring nor a nil-clean ring.

LEMMA 2.8. [**19, 20**] *An element $a \in R$ is strongly nil-clean if and only if $a - a^2$ is nilpotent.*

COROLLARY 2.9. *Let $C(R)$ be the center of R.*

1. *If R is nil-clean, then $C(R)$ is strongly nil-clean.*
2. *If R is a UNC ring, then $C(R)$ is a UU ring.*

PROOF. (1) Let $a \in C(R)$. Then $a \in R$ is nil-clean and central, so a is strongly nil-clean in R. Thus $a^2 - a$ is nilpotent by Lemma 2.8. Hence $C(R)$ is strongly nil-clean by Lemma 2.8.

(2) The proof is similar to (1). □

The proof of the next corollary actually shows that, if $2 \in J(R)$, then $a \in R$ is strongly nil-clean if and only if a can be written as $a = e + b$, where $e^2 = e \in R$, $b \in \mathrm{Nil}(R)$ and $eb = ebe$.

COROLLARY 2.10. *The following hold for a ring R:*

(1) *R is strongly nil-clean if and only if, for each $a \in R$, a can be written as $a = e + b$, where $e^2 = e \in R$, $b \in \mathrm{Nil}(R)$ and $eb = ebe$.*
(2) *R is a UU ring if and only if, for each $a \in U(R)$, a can be written as $a = e + b$, where $e^2 = e \in R$, $b \in \mathrm{Nil}(R)$ and $eb = ebe$.*

PROOF. (1) We show the sufficiency. For $a \in R$, let $a = e + b$ as given as in (1). Then $a^2 = e + eb + be + b^2$, so $a - a^2 = d - x$ where $d = b - b^2$ and $x = eb + be$. We have $ex = eb + ebe = 2eb \in J(R)$ as $2 \in J(R)$ (see [**8**]). It also follows from $eb = ebe$ that $eb^n = eb^n e$ for any $n \geq 0$, so $eb^n x = eb^n ex \in J(R)$. Hence, we deduce that $xd^n x \in J(R)$ for all $n \geq 0$. We now show that $a - a^2$ is a nilpotent. As $J(R)$ is nil by [**8**], it suffices to show that $\bar{a} - \bar{a}^2$ is nilpotent in $R/J(R)$. As $\bar{x}\bar{d}^n\bar{x} = \bar{0}$ in $R/J(R)$ for all $n \geq 0$, we have $(\bar{a} - \bar{a}^2)^{m+1} = (\bar{d} - \bar{x})^{m+1} = \bar{d}^{m+1} - \sum_{i+j=m} \bar{d}^i \bar{x} \bar{d}^j$ for any $m > 0$. So if $\bar{d}^m = \bar{0}$, then $(\bar{a} - \bar{a}^2)^{2m+1} = \bar{0}$.

(2) The proof is similar to (1). □

By [**3**], a commutative ring R is a UU ring if and only if so is $R[t]$. Next we present a generalization of this result. The prime radical $N(R)$ of a ring R is defined to be the intersection of the prime ideals of R. It is known that $N(R) = \mathrm{Nil}_*(R)$, the lower nilradical of R. A ring R is called a 2-primal ring if $N(R)$ coincides with $\mathrm{Nil}(R)$. For an endomorphism σ of R, R is called σ-compatible if, for any $a, b \in R$, $ab = 0 \Leftrightarrow a\sigma(b) = 0$ (see [**2**]), and in this case σ is clearly injective.

THEOREM 2.11. *Let R be a 2-primal ring and σ an endomorphism of R such that R is σ-compatible. The following are equivalent:*

(1) *$R[t;\sigma]$ is a UNC ring.*
(2) *$R[t;\sigma]$ is a UU ring.*
(3) *R is a UU ring.*
(4) *R is a UNC ring.*
(5) *$J(R) = \mathrm{Nil}(R)$ and $U(R) = 1 + J(R)$.*

PROOF. (2) ⇒ (1) and (3) ⇒ (4). The implications are obvious.

(1) ⇒ (4). As $R[t;\sigma]/(t) \cong R$ and units of $R[t;\sigma]/(t)$ are lifted to units of $R[t;\sigma]$, the implication holds.

(2) ⇒ (3). Argue as in proving (1) ⇒ (4).

(4) ⇒ (5). As R is 2-primal, $\mathrm{Nil}(R) \subseteq J(R)$, so $J(R) = \mathrm{Nil}(R)$ by Lemma 2.4. Hence $R/J(R)$ is a reduced ring that is a UNC ring. It follows that every unit of $R/J(R)$ is an idempotent, so $U(R/J(R)) = \{\bar{1}\}$. That is, $U(R) = 1 + J(R)$.

(5) ⇒ (2). As R is a 2-primal ring, we deduce from (5) that $J(R) = \mathrm{Nil}_*(R) = \mathrm{Nil}(R)$. So $R/J(R)$ is a reduced ring. As $\sigma(J(R)) \subseteq J(R)$, $\overline{\sigma} : R/J(R) \to R/J(R)$ defined by $\overline{\sigma}(\bar{a}) = \overline{\sigma(a)}$ is an endomorphism of $R/J(R)$. We next show that $R/J(R)$ is $\overline{\sigma}$-compatible. This is, for any $a, b \in R$, $ab \in \mathrm{Nil}(R) \Leftrightarrow a\sigma(b) \in \mathrm{Nil}(R)$. But this equivalence "⇔" has been established in the proof of Claims 1 and 2 of [**1**, Theorem 3.6] (also see [**5**, Lemma 2.1]). As $R/J(R)$ is a reduced ring that is $\overline{\sigma}$-compatible, by [**5**, Corollary 2.12] we have $U\big(\frac{R}{J(R)}[t;\overline{\sigma}]\big) = U\big(\frac{R}{J(R)}\big)$, which is equal to $\{\bar{1}\}$ by (5).

As $\frac{R[t;\sigma]}{J(R)[t;\sigma]} \cong \frac{R}{J(R)}[t;\overline{\sigma}]$, it follows that $\frac{R[t;\sigma]}{J(R)[t;\sigma]}$ is a UU ring. As R is a 2-primal ring that is σ-compatible, $\mathrm{Nil}_*(R[t;\sigma]) = \mathrm{Nil}_*(R)[t;\sigma]$ by [**5**, Lemma 2.2]. We infer that $J(R)[t;\sigma] = \mathrm{Nil}_*(R[t;\sigma])$, which is nil. Hence, by [**7**, Theorem 2.4] that $R[t;\sigma]$ is a UU ring. $\square$

COROLLARY 2.12. *A 2-primal ring R is a UNC ring if and only if $R[t]$ is a UNC ring, if and only if $J(R) = \mathrm{Nil}(R)$ and $U(R) = 1 + J(R)$.*

By [**7**], a ring R is strongly nil-clean if and only if R is a clean, UU ring. It is asked in [**7**] whether a clean, UNC ring is nil-clean (the converse holds clearly). We show that every semilocal UNC ring is nil-clean. In particular, a semiperfect, UNC ring is nil-clean. The following lemma is implicit in the proof of [**12**, Theorem 3].

LEMMA 2.13. [**12**] *Let D be a division ring. If $|D| \geq 3$ and $a \in D\backslash\{0,1\}$, then $\begin{pmatrix} a & \mathbf{0} \\ \mathbf{0} & \mathbf{0} \end{pmatrix} \in \mathbb{M}_n(D)$ is not nil-clean.*

The equivalence (2) $\Leftrightarrow$ (3) of the next theorem was proved in [**12**].

THEOREM 2.14. *The following are equivalent for a semilocal ring R:*
(1) *R is a UNC ring.*
(2) *R is a nil-clean ring.*
(3) *$J(R)$ is nil and $R/J(R)$ is a finite direct sum of matrix rings over $\mathbb{F}_2$.*

PROOF. (2) $\Leftrightarrow$ (3). The equivalence is [**12**, Corollary 5]. The implication (2) $\Rightarrow$ (1) is clear.

(1) $\Rightarrow$ (3). By Lemma 2.4, $J(R)$ is nil. As R is semilocal, $R/J(R) = R_1 \oplus R_2 \oplus \cdots \oplus R_n$ where each R_i is a matrix ring over a division ring D_i. If $|D_i| > 2$, let $a \in D_i\backslash\{0,1\}$. Then, for any $n \geq 1$, $A := \begin{pmatrix} a & \mathbf{0} \\ \mathbf{0} & \mathbf{0} \end{pmatrix} \in \mathbb{M}_n(D_i)$ is not nil-clean by Lemma 2.13, and this implies that $I_n - A$ is not nil-clean in $\mathbb{M}_n(D_i)$. But $I_n - A$ is a unit of $\mathbb{M}_n(D_i)$, we deduce that $\mathbb{M}_n(D_i)$ is not a UNC ring. Hence R_i is not a UNC ring, so R is not a UNC ring by Lemma 2.3 and Theorem 2.5. This contradiction shows that $|D_i| = 2$. $\square$

2.2. UNC matrix rings. Any proper matrix ring can not be a strongly nil-clean ring by [**8**] (indeed, can not be a UU ring by [**3**]). It is still unknown whether the matrix ring over a nil-clean ring is again nil-clean (see [**8**, Question 3]). Next we determine when the matrix ring over a commutative ring is a UNC ring. As a consequence, the matrix ring over a UNC ring need not be a UNC ring.

The (i,j)-cofactor of an $n \times n$ matrix A over a commutative ring, denoted by A_{ij}, is $(-1)^{i+j}$ times the determinant of the $(n-1) \times (n-1)$ submatrix obtained from A by deleting row i and column j. Let E_{ij} be the square matrix with (i,j)-entry 1 and all other entries 0.

LEMMA 2.15. [**13**] *Let R be a commutative ring and let $n \geq 1$. If $A \in \mathbb{M}_n(R)$ and $x \in R$, then $\det(xE_{ij} + A) = xA_{ij} + \det(A)$.*

THEOREM 2.16. *Let R be a commutative ring and $n \geq 2$. Then $\mathbb{M}_n(R)$ is a UNC ring if and only if $J(R)$ is nil and $R/J(R)$ is Boolean.*

PROOF. ($\Leftarrow$). This is by [**14**, Corollary 6.2].

($\Rightarrow$). By Lemma 2.4, $\mathbb{M}_n(J(R))$ is nil, so $J(R)$ is nil. Thus, it suffices to show that $R/J(R)$ is Boolean. As $\mathbb{M}_n(R/J(R)) \cong \mathbb{M}_n(R)/J(\mathbb{M}_n(R))$ is a UNC ring (by

Theorem 2.5) and $R/J(R)$ is reduced, we can assume without loss of generality that R is reduced. We next show that R is Boolean. As every commutative reduced ring is a subdirect product of integral domains, there exist a family of ideals $\{I_\alpha\}$ of R such that $\cap I_\alpha = 0$ and R/I_α is an integral domain for each α. To show that R is Boolean, it suffices to show that each $R/I_\alpha \cong \mathbb{Z}_2$.

Firstly, we show that R/I_α is a field. For $a \in R$, write $\overline{a} = a + I_\alpha \in R/I_\alpha$. For $A = (a_{ij}) \in \mathbb{M}_n(R)$, write $\overline{A} = (\overline{a}_{ij}) \in \mathbb{M}_n(R/J(R))$. Assume that R/I_α is not a field. Then there exists $\bar{0} \neq \overline{x} \in R/I_\alpha$ such that $\overline{x} \notin U(R/I_\alpha)$. The matrix

$$U := I_n + xE_{12} + xE_{21} + x^2E_{22}$$
$$= \begin{pmatrix} 1 & x & 0 & \cdots & 0 \\ x & 1+x^2 & 0 & \cdots & 0 \\ 0 & 0 & 1 & \cdots & 0 \\ \vdots & \vdots & \vdots & \ddots & \vdots \\ 0 & 0 & 0 & \cdots & 1 \end{pmatrix}$$

is a unit in $S := \mathbb{M}_n(R)$, so U is nil-clean in S. Hence $\overline{U}$ is nil-clean in $\overline{S} = \mathbb{M}_n(R/I_\alpha)$. Write $\overline{U} = \epsilon + \beta$ where $\epsilon^2 = \epsilon \in \overline{S}$ and $\beta \in \text{Nil}(\overline{S})$. One easily sees that $\epsilon \neq I_n$. So $\det(\epsilon) \neq \bar{1}$, and hence $\det(\epsilon) = \bar{0}$. Thus, $\bar{0} = \det(\epsilon) = \det(\overline{U} - \beta) = \det\big(\overline{x}E_{12} + \overline{x}E_{21} + \overline{x}^2E_{22} + (I_n - \beta)\big)$. By Lemma 2.15, there exist $a, b, c \in R/I_\alpha$ such that

$$\begin{aligned} \det\big(\overline{x}E_{12} + \overline{x}E_{21} + \overline{x}^2E_{22} + (I_n - \beta)\big) &= \overline{x}a + \det\big(\overline{x}E_{21} + \overline{x}^2E_{22} + (I_n - \beta)\big) \\ &= \overline{x}a + \overline{x}b + \det\big(\overline{x}^2E_{22} + (I_n - \beta)\big) \\ &= \overline{x}a + \overline{x}b + \overline{x}^2c + \det\big(I_n - \beta\big). \end{aligned}$$

It follows that $\overline{x}(a + b + \overline{x}c) = -\det(I_n - \beta)$. As β is nilpotent, $I_n - \beta$ is invertible, $\det(I_n - \beta)$ is a unit in R/I_α. Hence, we deduce that $\overline{x} \in U(R/I_\alpha)$, a contradiction. Thus, we have proved that R/I_α is a field.

Next we show that $R/I_\alpha \cong \mathbb{Z}_2$. Let $\overline{y} \in R/I_\alpha$. Then the matrix

$$V = \begin{pmatrix} 0 & 1 & 0 & \cdots & 0 \\ 1 & y & 0 & \cdots & 0 \\ 0 & 0 & 1 & \cdots & 0 \\ \vdots & \vdots & \vdots & \ddots & \vdots \\ 0 & 0 & 0 & \cdots & 1 \end{pmatrix}$$

is a unit in S, so V is nil-clean in S. Hence $\overline{V}$ is nil-clean in $\overline{S}$. Write $\overline{V} = \beta + \epsilon$ where $\epsilon^2 = \epsilon \in \overline{S}$ and $\beta \in \text{Nil}(\overline{S})$. By [**21**], ϵ is similar to a diagonal matrix, so it is similar to $\begin{pmatrix} I_k & \mathbf{0} \\ \mathbf{0} & \mathbf{0} \end{pmatrix}$ for some $0 < k < n$. Moreover, as β is nilpotent, it is similar to a strictly upper triangular matrix. Thus, as the trace is similarity-invariant, we obtain that $\text{trace}(\beta) = \bar{0}$ and $\text{trace}(\epsilon) = \bar{k}$. As $\bar{2} = \bar{0}$ in $\overline{S}$, for any $m \in \mathbb{Z}$, $\overline{m} = \bar{0}$ or $\overline{m} = \bar{1}$. So we see that $\text{trace}(\overline{V}) = \overline{y} + \overline{n-2}$, which is equal to $\overline{y}$ or $\overline{y} + \bar{1}$, and that $\text{trace}(\epsilon)$ is equal to $\bar{0}$ or $\bar{1}$. Therefore, from $\text{trace}(\overline{V}) = \text{trace}(\beta) + \text{trace}(\epsilon)$, we deduce that $\overline{y} = \bar{0}$ or $\overline{y} = \bar{1}$. Hence $R/I_\alpha \cong \mathbb{Z}_2$. □

The matrix ring over a UNC ring need not be a UNC ring.

EXAMPLE 2.17. *For any $n \geq 2$, $\mathbb{M}_n\big(\mathbb{Z}_2[t]\big)$ is not a UNC ring, while $\mathbb{Z}_2[t]$ is a UNC ring.*

In [**7**, Theorem 2.6], it is proved that any unital subring of a UU ring is again a UU ring. But a subring of a UNC ring may not be a UNC ring.

EXAMPLE 2.18. *Take* $u = \begin{pmatrix} 0 & 0 & 1 \\ 1 & 0 & 0 \\ 0 & 1 & 0 \end{pmatrix} \in \mathbb{M}_3(\mathbb{Z}_2)$.
Then $u^3 = 1$. *Let* R *be the unital subring of* $\mathbb{M}_3(\mathbb{Z}_2)$ *generated by* u. *That is,*

$$\begin{aligned} R &= \{a1 + bu + cu^2 : a, b, c \in \mathbb{Z}\} \\ &= \{0, 1, u, u^2, 1+u, 1+u^2, u+u^2, 1+u+u^2\}. \end{aligned}$$

One easily sees that R *is reduced. As* $u^2 \neq u$, u *is not nil-clean. So* R *is not a UNC ring, but* $\mathbb{M}_3(\mathbb{Z}_2)$ *is a UNC ring.*

2.3. UNC group rings. (Strongly) nil-clean group rings have been discussed in [**14, 17, 19**]. Here we consider when a group ring is a UNC or UU ring following the idea in [**19**]. A group G is called locally finite if every finitely generated subgroup of G is finite. Let p be a prime number. A group G is called a p-group if the order of each element of G is a power of p. The center of a group G is denoted by $\mathcal{Z}(G)$.

For the group ring RG of a group G over a ring R, the ring homomorphism $\omega : RG \to R$, $\Sigma r_g g \mapsto \Sigma r_g$ is called the augmentation map, and the kernel $\ker(\omega)$ is called the augmentation ideal of the group ring RG and is denoted by $\triangle(RG)$. Note that $\triangle(RG)$ is an ideal of RG generated by the set $\{1 - g : g \in G\}$.

PROPOSITION 2.19. *If* R *is a UNC ring and* G *is a locally finite* 2*-group, then* RG *is a UNC ring.*

PROOF. As G is locally finite, to show that RG is a UNC ring it suffices to show that GH is a UNC ring for any finite subgroup H of G. So, without loss of generality, one can assume that G is a finite 2-group. As R is a UNC ring, $2 \in J(R)$ is nilpotent by Lemma 2.4. So, by [**6**, Theorem 9], $\triangle(RG)$ is nilpotent. As $RG/\triangle(RG) \cong R$, it follows from Theorem 2.5 that RG is a UNC ring. □

The hypercenter of a group G, denoted by $H(G)$, is defined to be the union of the (transfinite) upper central series of the group G.

THEOREM 2.20. *Let* R *be a ring and* G *be a group. If* RG *is a UNC ring, then* R *is a UNC ring and* $H(G)$ *is a* 2*-group.*

PROOF. Let $1 = \mathcal{Z}_0(G) \subseteq \mathcal{Z}_1(G) \subseteq \mathcal{Z}_2(G) \subseteq \ldots \subseteq \mathcal{Z}_\alpha(G) = H(G)$ be the upper central series of length α for G. Clearly, $\mathcal{Z}_0(G)$ is a 2-group. Assume that, for some $\beta \leq \alpha$, $\mathcal{Z}_\sigma(G)$ is a 2-group for all $\sigma < \beta$. We next verify that $\mathcal{Z}_\beta(G)$ is a 2-group. This is certainly true if β is a limit ordinal. If β is not a limit ordinal, then $\mathcal{Z}_{\beta-1}(G)$ is a 2-group, and $\mathcal{Z}_\beta(G)/\mathcal{Z}_{\beta-1}(G) = \mathcal{Z}(\overline{G})$, where $\overline{G} = G/\mathcal{Z}_{\beta-1}(G)$. Let $\overline{R} = R/2R$. As $\overline{RG}$ is an image of RG, it is a UNC ring. So, for $g \in \mathcal{Z}(\overline{G})$, g is nil-clean in $\overline{RG}$. As g is central, it is strongly nil-clean. Hence, by Lemma 2.8, $g(1-g) = g - g^2$ is nilpotent. It follows that $1 - g \in \overline{RG}$ is nilpotent. So, for some $n > 0$, $(1-g)^{2^n} = 0$. That is, $g^{2^n} = 1$. Hence, $\mathcal{Z}_\beta(G)/\mathcal{Z}_{\beta-1}(G) = \mathcal{Z}(\overline{G})$ is a 2-group. As $\mathcal{Z}_{\beta-1}(G)$ is a 2-group, it follows that $\mathcal{Z}_\beta(G)$ is a 2-group. Therefore, by the Transfinite Induction, $H(G) = \mathcal{Z}_\alpha(G)$ is a 2-group. □

A nilpotent group is a group G such that $G = \mathcal{Z}_n(G)$ for a finite number n.

THEOREM 2.21. *Let R be a ring and G be a nilpotent group. Then RG is a UNC ring if and only if R is a UNC ring and G is a 2-group.*

PROOF. The claim follows from Proposition 2.19 and Theorem 2.20. □

THEOREM 2.22. *If RG is a UU ring, then R is a UU ring and G is a 2-group. The converse holds if G is locally finite.*

PROOF. If RG is a UU ring, then as an image of RG, R is certainly a UU ring. Moreover, G is a 2-group by the same argument as in the proof of Theorem 2.20 or by [**7**]. Conversely, as R is a UU ring, $2 \in J(R)$ is nilpotent by Lemma 2.4 or [**7**]. As G is a locally finite 2-group, $\triangle(RG)$ is locally nilpotent by [**6**, Corollary, p. 682]. Let $x \in U(RG)$. Then $\omega(x) \in U(R)$ is strongly nil-clean, so $\omega(x) - \omega(x)^2 \in R$ is nilpotent. Hence, for some $n > 0$, $\omega((x - x^2)^n) = (\omega(x - x^2))^n = 0$. So $(x - x^2)^n \in \triangle(RG)$. It follows that $x - x^2$ is nilpotent. So x is strongly nil-clean by Lemma 2.8. □

2.4. Semipotent UU rings. An element a in a ring R is called uniquely nil-clean if there exists a unique idempotent e in R such that $a - e$ is nilpotent, and the ring R is called uniquely nil-clean if each element of R is uniquely nil-clean. Uniquely nil-clean rings were characterized in Diesl [**8**]. Here, using a recent result of Ŝter in [**20**], we first show that UU rings are exactly those rings whose units are uniquely nil-clean.

THEOREM 2.23. *A ring R is a UU ring if and only if every unit of R is uniquely nil-clean.*

PROOF. ($\Rightarrow$). Let $u \in U(R)$. By the hypothesis, $u = 1 + b$ with $b \in \text{Nil}(R)$. Assume that $u = e + x$ where $e^2 = e$ and $x \in \text{Nil}(R)$. Then $1 - e = x - b$. By Ŝter [**20**, Corollary 2.13], $\text{Nil}(R)$ is closed under addition. So $x - b \in \text{Nil}(R)$, and hence $1 - e$ is nilpotent. It follows that $e = 1$ and $x = b$. Hence u is uniquely nil-clean.

($\Leftarrow$). Let $u \in U(R)$. Then there exist $e^2 = e \in R$ and $x \in \text{Nil}(R)$ such that $u = e + x$. Thus, $u = u^{-1}eu + u^{-1}xu$ is another nil-clean decomposition in R. So it follows that $e = u^{-1}eu$. This gives $eu = ue$. So u is strongly nil-clean. □

In contrast to Theorem 2.23, a unipotent unit need not be uniquely nil-clean, even in a nil-clean ring.

EXAMPLE 2.24. *Let $R = \mathbb{M}_3(\mathbb{Z}_2)$. Then R is a nil-clean ring. As observed in* [**9**]*, if A is the strictly upper triangular matrix in R whose all entries above the diagonal are equal to 1 and let B be the transpose of A, then A, B are nilpotent and $E := A + B = \begin{pmatrix} 0 & 1 & 1 \\ 1 & 0 & 1 \\ 1 & 1 & 0 \end{pmatrix}$ is a nonzero idempotent. Thus, $U := I_3 - A = (I_3 - E) + B$ is a unipotent unit that is not uniquely nil-clean.*

A ring R is strongly π-regular if, for each $a \in R$, $a^n \in a^{n+1}R \cap Ra^{n+1}$ for some $n \geq 1$. A ring is clean if every element is a sum of a unit and an idempotent, and a ring R is an exchange ring if, for each $a \in R$, $a - e \in (a - a^2)R$ for some $e^2 = e \in R$ (see [**18**]). A ring R is semipotent if every right ideal not contained in $J(R)$ contains a non-zero idempotent. We have the implications: strongly π-regular $\Rightarrow$ clean $\Rightarrow$ exchange $\Rightarrow$ semipotent; and none of the arrows is reversible.

Diesl [**8**, Corollary 3.11] proved that a ring R is strongly nil-clean if and only if R is a strongly π-regular, UU ring. Danchev and Lam [**7**, Theorem 4.3] generalized the direction "$\Leftarrow$" by showing that a ring R is strongly nil-clean if and only if R is an exchange (or clean), UU ring. We now generalize the direction "$\Leftarrow$" further by showing the following

THEOREM 2.25. *A ring R is a strongly nil-clean ring if and only if it is a semipotent, UU ring.*

PROOF. ($\Rightarrow$). The implication is clear.

($\Leftarrow$). Let R be a semipotent, UU ring. So $R/J(R)$ is a semipotent, UU ring (by [**7**, Theorem 2.4(2)]), and moreover, $J(R)$ is nil. So R is strongly nil-clean if and only if so is $R/J(R)$ by [**14**, Theorem 2.7]. Hence, to show the necessity, we can assume that $J(R) = 0$.

We now show that R is a reduced ring. Assume $a^2 = 0$ for some $0 \neq a \in R$. As R is a semipotent ring with $J(R) = 0$, there exists $e^2 = e \in R$ such that eRe is isomorphic to a 2×2 matrix ring over a non trivial ring by Levitzki [**16**, Theorem 2.1]. But, as eRe is again a UU ring by [**3**, Proposition 2.5], this gives a contradiction to [**3**, Corollary 3.3]. Hence R is reduced. It follows that $U(R) = \{1\}$.

We next show that R is a Boolean ring. Assume on the contrary that $a^2 \neq a$ for some $a \in R$. As R is semipotent with $J(R) = 0$, $(a - a^2)R$ contains a nonzero idempotent, say e. Write $e = (a - a^2)b$ with $b \in R$. Then $e = e(a - a^2)b = ea \cdot e(1 - a)b = e(1 - a) \cdot eab$. As eRe is reduced, ea and $e(1 - a)$ are units of eRe. As $U(R) = \{1\}$, we have $U(eRe) = \{e\}$. Hence $ea = e$ and $e(1 - a) = e$. It follows that $e = 0$, a contradiction. □

In [**10**, Theorem 2.9], Henriksen proved that a von Neumann regular ring R with $U(R) = \{1\}$ is Boolean. Danchev and Lam [**7**, Corollary 4.2] proved that a ring R is an exchange ring with $U(R) = \{1\}$ if and only if R is Boolean.

COROLLARY 2.26. *A ring R is semipotent with $U(R) = \{1\}$ if and only if R is Boolean.*

The question of Danchev and Lam whether a clean, UNC ring is nil-clean is still open. In view of Theorem 2.25 [**8**, Corollary 3.11] and [**8**, Question 4], the following questions arise.

QUESTIONS 2.27. *Is a semipotent, UNC ring a nil-clean ring? What can be said about strongly π-regular UNC rings?*

3. Units being sums of a nilpotent and two idempotents

As a generalization of a strongly nil-clean ring, a strongly 2-nil-clean ring was introduced by Chen and Sheibani [**4**] to be the ring for which every element is a sum of a nilpotent and two idempotents that commute with one another. The structure of these rings is obtained in [**4**]. In this section, we consider the "unit" version of strongly 2-nil-clean rings. That is, the rings for which every unit is a sum of a nilpotent and two idempotents that commute with one another. These rings extend UU rings, and will be completely characterized here. A special situation of a strongly 2-nil-clean ring is the property that every element of a ring is a sum of two commuting idempotents, first considered by Hirano and Tominaga [**11**]. Here the "unit" version of this property is also discussed.

3.1. Units being sums of a nilpotent and two idempotents.

DEFINITION 3.1. A ring R is called a UNII-ring if every unit of R is a sum of a nilpotent and two idempotents. The ring R is called a strong UNII-ring if for each $u \in U(R)$, $u = b + e + f$ where $b \in \mathrm{Nil}(R)$, $e, f \in \mathrm{idem}(R)$ such that b, e, f commute.

LEMMA 3.2. *A ring R is strong UNII-ring if and only if $R = A \oplus B$ where A, B are strong UNII-rings, $2 \in J(A)$ is nilpotent and $3 \in J(B)$ is nilpotent.*

PROOF. ($\Leftarrow$). The implication is clear.

($\Rightarrow$). Write $-1 = b + e + f$ where $b \in \mathrm{Nil}(R)$, $e, f \in \mathrm{idem}(R)$ and b, e, f all commute. Then

$$\begin{aligned} 1 + 2b + b^2 = (-1-b)^2 = (e+f)^2 &= e + f + 2ef = (-1-b) + 2e(-1-b-e) \\ &= -1 - b - 4e - 2eb, \end{aligned}$$

which gives $2 + 4e = -3b - b^2 - 2eb$. So $6e = (2+4e)e = (-3b - b^2 - 2eb)e = -(5e + be)b$ is nilpotent. Similarly, $6f$ is nilpotent. So there exists $n \geq 1$ such that $6^n e = 0$ and $6^n f = 0$. Thus $0 = 6^n(e+f) = 6^n(-1-b)$. As $-1-b \in U(R)$, $6^n = 0$, i.e., $2^n R \cap 3^n R = 0$. So $R = A \oplus B$ where $A \cong R/2^n R$ and $B \cong R/3^n R$, and A, B are strong UNII-rings. □

LEMMA 3.3. *The following are equivalent for a ring R:*

1. *A ring R is a strong UNII-ring with 2 nilpotent.*
2. *Each unit of R is a sum of a nilpotent and a tripotent that commute and 2 is nilpotent.*
3. *R is a UU ring.*

PROOF. (3) $\Rightarrow$ (1). The implication is clear.

(1) $\Rightarrow$ (2). Let $u \in U(R)$ and write $u = b + e + f$ where $b \in \mathrm{Nil}(R)$, $e, f \in \mathrm{idem}(R)$ and b, e, f all commute. Then $g := e + f - 2ef$ is an idempotent and $c := u - g = b + 2ef$ is nilpotent. Moreover, g, c commute. So $u = c + g$ is strongly nil-clean. Hence (2) holds.

(2) $\Rightarrow$ (3). Let $u \in U(R)$ and write $u = b + t$ where $b \in \mathrm{Nil}(R)$, $t^3 = t$ and $bt = tb$. Then $(t - t^2)^2 = 2(t^2 - t) \in \mathrm{Nil}(R)$, so $b + (t - t^2)$ is nilpotent. Thus, $u = (b + t - t^2) + t^2$ is strongly clean. So R is a UU ring. □

LEMMA 3.4. *Let R be a ring with $3 \in \mathrm{Nil}(R)$. The following are equivalent:*

1. *R is a strong UNII-ring.*
2. *Each unit of R is a sum of a nilpotent and a tripotent that commute.*
3. *$1 - u^2$ is nilpotent for every $u \in U(R)$.*

PROOF. (1) $\Rightarrow$ (2). Let $u \in U(R)$ and write $u = b + e + f$ where $b \in \mathrm{Nil}(R)$, $e, f \in \mathrm{idem}(R)$ and b, e, f all commute. Then $g := e + f - 3ef$ is a tripotent and $b + 3ef$ is nilpotent. Moreover, $g, b + 3ef$ commute. So $u = (b + 3ef) + g$ is a sum of a nilpotent and a tripotent that commute. Hence (2) holds.

(2) $\Rightarrow$ (3). Let $u \in U(R)$ and write $u = b + f$ where $b \in \mathrm{Nil}(R)$, f is a tripotent and b, f all commute. Then $u^3 \equiv b^3 + f \pmod{3R}$, so $u - u^3 \equiv b - b^3 \pmod{3R}$. As $3, b \in \mathrm{Nil}(R)$, $u - u^3 \in \mathrm{Nil}(R)$, so $1 - u^2$ is nilpotent.

(3) $\Rightarrow$ (1). Let $u \in U(R)$ and write $u = b + x$ where $b \in \mathrm{Nil}(R)$, $x^3 = x$ and $xb = bx$. Then $u = b + x^2 + (x - x^2)$, and $(x - x^2) - (x - x^2)^2 = 3(x - x^2) \in J(R)$ is nilpotent. By [**22**, Lemma 3.5], there exists a polynomial $\theta(t) \in \mathbb{Z}[t]$ such that

$\theta(x)^2 = \theta(x)$ and $j := (x - x^2) - \theta(x) \in J(R)$. Thus, $u = (b + j) + x^2 + \theta(x)$, where $x^2, \theta(x)$ are idempotents, and $b + j$ is nilpotent and they commute with each other. □

THEOREM 3.5. *The following are equivalent for a ring R:*

(1) *R is a strong UNII-ring.*
(2) *$1 - u^2$ is nilpotent for every $u \in U(R)$ and 6 is nilpotent.*
(3) *R is one of the following types:*
   (a) *R is a UU ring.*
   (b) *$1 - u^2$ is nilpotent for every $u \in U(R)$ and 3 is nilpotent.*
   (c) *$R = A \oplus B$, where A is a UU ring, $1 - u^2$ is nilpotent for every $u \in U(B)$ and $3 \in B$ is nilpotent.*

PROOF. (1) $\Leftrightarrow$ (3). The equivalence follows from Lemmas 3.2-3.4.

(3) $\Rightarrow$ (2). The implication is clear.

(2) $\Rightarrow$ (3). Let $6^n = 0$ with $n \geq 1$. Then $R = A \oplus B$ where $A \cong R/2^n R$ and $B \cong R/3^n R$. The hypothesis on R shows that $1 - u^2$ is nilpotent if $u \in U(A)$ or $u \in U(B)$. For $u \in U(A)$, $(1 - u)^2 = (1 - u^2) + 2(u^2 - u)$ is nilpotent, so $1 - u$ is nilpotent. This shows that A is a UU ring. □

The condition that 6 is nilpotent can not be removed in Theorem 3.5(2): As $U(\mathbb{Z}) = \{-1, 1\}$, $1 - u^2 = 0$ for all $u \in U(\mathbb{Z})$, but $6 \in \mathbb{Z}$ is not nilpotent. Theorem 3.6 below can be viewed as an extension of Theorem 2.25.

THEOREM 3.6. *A ring is a semipotent, strong UNII-ring if and only if $R = A \oplus B$, where A is zero or $A/J(A)$ is Boolean with $J(A)$ nil and B is zero or $B/J(B)$ is a subdirect product of $\mathbb{Z}_3$'s with $J(B)$ nil.*

PROOF. ($\Leftarrow$). As A is strongly nil-clean, it is a strong UNII-ring. For any $u \in U(B)$, $u^3 - u$ is nilpotent. So $(u^2)^2 - u^2 = u(u^3 - u)$ is nilpotent and $(u - u^2)^2 - (u - u^2) = (u - 2)(u^3 - u) + 3(u^2 - u)$ is nilpotent. By [**22**, lemma 3.5], there exist $\theta_1(t), \theta_2(t) \in \mathbb{Z}[t]$ such that $\theta_1(u)^2 = \theta_1(u)$, $\theta_2(u)^2 = \theta_2(u)$, and $u^2 - \theta_1(u)$, $(u - u^2) - \theta_2(u)$ are nilpotent. So $u = \theta(u) + \theta_1(u) + \theta_2(u)$, where $\theta(u) = (u^2 - \theta_1(u)) + [(u - u^2) - \theta_2(u)]$ is nilpotent. So B is a strong UNII-ring. Hence R is a strong UNII-ring.

($\Rightarrow$). By Theorem 3.5, $R = A \oplus B$, where A is zero or A is a UU ring, and B is zero or $1 - u^2 \in \text{Nil}(B)$ for every $u \in U(B)$ with $3 \in \text{Nil}(B)$. We can assume that $A \neq 0$ and $B \neq 0$. As R is semipotent, A, B are semipotent. So A is strongly nil-clean by Theorem 2.25, and hence $A/J(A)$ is Boolean with $J(A)$ nil. For $j \in J(B)$, $(1 + j)^2 - 1 = j(2 + j) \in \text{Nil}(B)$. As $2 + j \in U(B)$, $j \in \text{Nil}(B)$. So $J(B)$ is nil. We next show that $\overline{B} := B/J(B)$ is reduced. Assume $x^2 = 0$ for some $0 \neq x \in \overline{B}$. As $\overline{B}$ is semiprimitive semipotent, there exists $f^2 = f \in \overline{B}$ such that $f\overline{B}f \cong \mathbb{M}_2(S)$ for a non trivial ring S by Levitzki [**16**, Theorem 2.1]. In $\mathbb{M}_2(S)$, $\begin{pmatrix} 1 & 1 \\ 1 & 0 \end{pmatrix}^2 - \begin{pmatrix} 1 & 0 \\ 0 & 1 \end{pmatrix} = \begin{pmatrix} 1 & 1 \\ 1 & 0 \end{pmatrix}$ is not nilpotent. So, there exists a unit v of $f\overline{B}f$ such that $f - v^2$ is not nilpotent. Thus $u := v + 1 - f$ is a unit of $\overline{B}$ and $1 - u^2 = f - v^2$ is not nilpotent. This is a contradiction. Hence $\overline{B}$ is reduced. It follows that $w^2 = 1$ for any $w \in U(\overline{B})$. Next we show that $y^3 = y$ in $\overline{B}$ for every $y \in \overline{B}$. Assume that $y - y^3 \neq 0$ for some $y \in \overline{B}$. As $\overline{B}$ is semipotent, there exists $0 \neq e^2 = e \in (y - y^3)\overline{B}$. Write $e = (y - y^3)z$ with $z \in \overline{B}$. Then $e = ey \cdot e(1 - y^2)z = e(1 - y) \cdot ey(1 + y)z = e(1 + y) \cdot ey(1 - y)z$. As $e\overline{B}e$ is reduced,

$ey, e(1-y)$ and $e(1+y)$ all are units of $e\overline{B}e$. As $u^2 = 1$ for each $u \in \overline{B}$, $v^2 = e$ for all $v \in e\overline{B}e$. Hence, $e = (ey)^2 = (e(1+y))^2 = (e(1-y))^2$. It follows that $2e = 0$. As 2 is a unit of $e\overline{B}e$, $e = 0$. This contradiction shows that $y^3 = y$ in $\overline{B}$ for every $y \in \overline{B}$. As $3 = 0$ in $\overline{B}$, $\overline{B}$ is a subdirect product of $\mathbb{Z}_3$'s (see [**15**, Ex. 12.11]). □

COROLLARY 3.7. *A ring is a strongly 2-nil-clean ring if and only if it is a semipotent, strong UNII-ring.*

PROOF. This is by Theorem 3.6 and [**4**, Theorem 4.2]. □

COROLLARY 3.8. *If R is a semipotent strong UNII-ring, then every element of R is a sum of a nilpotent and two idempotents that commute with one another.*

PROOF. This is by Theorem 3.6 and [**4**, Theorem 4.5]. □

3.2. Units being sums of two idempotents.

Hirano and Tominaga [**11**] have characterized the rings in which every element is the sum of two commuting idempotents.

DEFINITION 3.9. A ring R is called a UII-ring if every unit of R is a sum of two idempotents, and R is called a strong UII-ring if every unit of R is a sum of two commuting idempotents.

LEMMA 3.10. *A finite direct sum $R_1 \oplus R_2 \oplus \cdots \oplus R_n$ of rings is a (strong) UII-ring if and only if each R_i is a (strong) UII-ring.*

LEMMA 3.11. *A ring R is a UII-ring if and only if $R = A \oplus B$ where A, B are UII-rings, $2 = 0$ in A and $3 = 0$ in B.*

PROOF. ($\Leftarrow$). The implication is clear.

($\Rightarrow$). Write $-1 = e + f$ where e, f are idempotents of R. Then $-1 - e = f = f^2 = (-1-e)^2 = 1+3e$, so $4e+2 = 0$. Thus, $0 = (4e+2)e = 6e$. Similarly, $6f = 0$. Hence $6 = -6(e+f) = -6e - 6f = 0$, so $2R \cap 3R = 0$. It follows that $R = A \oplus B$ where $A \cong R/2R$ and $B \cong R/3R$. Moreover, A, B are clearly UII-rings. □

LEMMA 3.12. *A ring R is strong UII with $ch(R) = 2$ if and only if $U(R) = \{1\}$.*

PROOF. If $U(R) = \{1\}$, then R is clearly strong UII, and $-1 = 1$; so $2 = 0$ in R. If R is strong UII with $2R = 0$, then, for any $u \in U(R)$, $u = e+f$ where e, f are commuting idempotents of R. Thus $u+e = f = f^2 = (u+e)^2 = u^2+2ue+e = u^2+e$. It follows that $u^2 = u$, i.e., $u = 1$. □

LEMMA 3.13. *A ring R is a strong UII-ring with $ch(R) = 3$ if and only if $U(R) = 1 + idem(R)$.*

PROOF. ($\Leftarrow$). Suppose $U(R) = 1 + idem(R)$. Then $2 = 1 + 1 \in U(R)$. As $-1 = 1+(-2)$, we infer that $(-2)^2 = -2$, so $-2 = 1$. That is, $3 = 0$ in R. Moreover, R is clearly a strong UII-ring.

($\Rightarrow$). For $e^2 = e \in R$, $1 + e = 1 - 2e \in U(R)$. Let $u \in U(R)$. Write $u = e + f$ where e, f are commuting idempotents. Then $u - e = f = f^2 = (u-e)^2 = u^2 - 2ue + e$, so

$$e = eu - u + u^2.$$

Multiplying the equality by e from the left gives $e = eu^2$. It follows that $eu^2 = eu-u+u^2$, and hence $eu = e-1+u$, i.e., $e = 1-u+eu$. We deduce that $1-u+eu = eu - u + u^2$, so $u^2 = 1$. This gives that $1 = u^2 = (e+f)^2 = e + f + 2ef = u + 2ef$. That is, $u = 1 + g$ with $g = ef$. □

THEOREM 3.14. *A ring R is a strong UII-ring if and only if R is one of the following types:*

(1) $U(R) = \{1\}$.
(2) $U(R) = 1 + \text{idem}(R)$.
(3) $R = A \oplus B$ *where* $U(A) = \{1\}$ *and* $U(B) = 1 + \text{idem}(B)$.

PROOF. It follows from Lemmas 3.10-3.13. □

COROLLARY 3.15. *If R is a strong UII-ring, then $J(R) = 0$ and $\text{Nil}(R) = 0$.*

PROOF. One easily sees that $U(R) = \{1\}$ or $U(R) = 1 + \text{idem}(R)$ implies that $J(R) = 0$ and $\text{Nil}(R) = 0$. □

In view of Corollary 3.15, we see that, for any ring R and $n \geq 2$, the rings $\mathbb{M}_n(R)$, $\mathbb{T}_n(R)$ and $R[[t]]$ are not strong UII-rings.

PROPOSITION 3.16. *Let σ be an endomorphism of a ring R. Then $R[t;\sigma]$ is a strong UII-ring if and only if R is a strong UII-ring and $\sigma(e) = e$ for all $e \in \text{idem}(R)$.*

PROOF. ($\Rightarrow$). For $u \in U(R)$, $u \in U(R[t;\sigma])$, so $u = \sum_{i\geq 0} a_i t^i + \sum_{i\geq 0} b_i t^i$ is a sum of two commuting idempotents in $R[t;\sigma]$. Thus $u = a_0 + b_0$ is a sum of two commuting idempotents in R. So R is a strong UII-ring. Since $R[t;\sigma]$ is a reduced ring, each $e \in \text{idem}(R)$ is central in $R[t;\sigma]$. So $et = te = \sigma(e)t$, showing that $\sigma(e) = e$.

($\Leftarrow$). As R is a strong UII-ring, R is a reduced ring. So idempotents of R are central, and hence units of R are central. Moreover, the assumption that $\sigma(e) = e$ for all $e^2 = e \in R$ implies that $\sigma(u) = u$ for all $u \in U(R)$. Now one can easily show that $\text{idem}(R[t;\sigma]) = \text{idem}(R)$ and $U(R[t;\sigma]) = U(R)$. It follows that $R[t;\sigma]$ is a strong UII-ring. □

PROPOSITION 3.17. *Let R be a ring and G a nontrivial group. Then RG is a strong UII-ring if and only if $U(R) = 1 + \text{idem}(R)$ and G is a group of exponent* 2.

PROOF. ($\Rightarrow$). Let $u \in U(R)$. Then $u = \alpha + \beta$, where α, β are commuting idempotents in RG. So $u = \omega(\alpha) + \omega(\beta)$ is a sum of two commuting idempotents in R. So R is a strong UII-ring. By Theorem 3.14, $R = A \oplus B$ where A, B are strong UII-rings, $2 = 0$ in A and $3 = 0$ in B. Thus $RG \cong AG \oplus BG$, so AG is a strong UII-ring. As $2 = 0$ in AG, $U(AG) = \{1\}$ by Lemma 3.12. Because G is a nontrivial group, it must be that $A = 0$. So $R = B$, and hence $U(R) = 1 + \text{idem}(R)$. As $3 = 0$ in RG, $U(RG) = 1 + \text{idem}(RG)$ by Lemma 3.13. It follows that $g^2 = 1$ for all $g \in G$.

($\Leftarrow$). Since G is a group of exponent 2, G is locally finite. So it suffices to show that for any finite subgroup H of G, RH is a strong UII-ring. But $2 \in U(R)$ by Lemma 3.13, so RH is a direct sum of finitely many copies of R. Hence RH is a strong UII-ring. □

By Hirano and Tominaga [**11**], the ring R must be commutative if every element of R is a sum of two commuting idempotents. However, a strong UII-ring need not be commutative. In fact, by Henriksen [**10**, pp. 86], the Weyl algebra over $\mathbb{Z}_2$ in the noncommuting indeterminants x, y subject to the relation $yx = xy + 1$ is a strong UII-ring that is not commutative.

Acknowledgments

The authors thank the referee for valuable comments and suggestions. Karimi-Mansoub acknowledges Tarbiat Modares University (Iran) to grant her a Ph.D scholarship for visiting Memorial University of Newfoundland. Zhou's research was supported by a Discovery Grant from NSERC of Canada.

References

[1] A. Alhevaz, A. Moussavi, and M. Habibi, *On rings having McCoy-like conditions*, Comm. Algebra **40** (2012), no. 4, 1195–1221. MR2912979

[2] S. Annin, *Associated primes over skew polynomial rings*, Comm. Algebra **30** (2002), no. 5, 2511–2528. MR1904650

[3] G. Călugăreanu, *UU rings*, Carpathian J. Math. **31** (2015), no. 2, 157–163. MR3408811

[4] H. Chen and M. Sheibani, *Strongly 2-nil-clean rings*, J. Algebra Appl. **16** (2017), no. 9, 1750178, 12. MR3661645

[5] W. Chen, *On constant products of elements in skew polynomial rings*, Bull. Iranian Math. Soc. **41** (2015), no. 2, 453–462. MR3345531

[6] I. G. Connell, *On the group ring*, Canad. J. Math. **15** (1963), 650–685. MR0153705

[7] P. V. Danchev and T.-Y. Lam, *Rings with unipotent units*, Publ. Math. Debrecen **88** (2016), no. 3-4, 449–466. MR3491753

[8] A. J. Diesl, *Nil clean rings*, J. Algebra **383** (2013), 197–211. MR3037975

[9] M. Ferrero, E. R. Puczyłowski, and S. Sidki, *On the representation of an idempotent as a sum of nilpotent elements*, Canad. Math. Bull. **39** (1996), no. 2, 178–185. MR1390353

[10] M. Henriksen, *Rings with a unique regular element*, Rings, modules and radicals (Hobart, 1987), Pitman Res. Notes Math. Ser., vol. 204, Longman Sci. Tech., Harlow, 1989, pp. 78–87. MR1035729

[11] Y. Hirano and H. Tominaga, *Rings in which every element is the sum of two idempotents*, Bull. Austral. Math. Soc. **37** (1988), no. 2, 161–164. MR930784

[12] M. T. Koşan, T.-K. Lee, and Y. Zhou, *When is every matrix over a division ring a sum of an idempotent and a nilpotent?*, Linear Algebra Appl. **450** (2014), 7–12. MR3192466

[13] T. Koşan, S. Sahinkaya, and Y. Zhou, *On weakly clean rings*, Comm. Algebra **45** (2017), no. 8, 3494–3502. MR3609355

[14] T. Koşan, Z. Wang, and Y. Zhou, *Nil-clean and strongly nil-clean rings*, J. Pure Appl. Algebra **220** (2016), no. 2, 633–646. MR3399382

[15] T. Y. Lam, *A first course in noncommutative rings*, Graduate Texts in Mathematics, vol. 131, Springer-Verlag, New York, 1991. MR1125071

[16] J. Levitzki, *On the structure of algebraic algebras and related rings*, Trans. Amer. Math. Soc. **74** (1953), 384–409. MR0053089

[17] W. Wm. McGovern, S. Raja, and A. Sharp, *Commutative nil clean group rings*, J. Algebra Appl. **14** (2015), no. 6, 1550094, 5. MR3338090

[18] W. K. Nicholson, *Lifting idempotents and exchange rings*, Trans. Amer. Math. Soc. **229** (1977), 269–278. MR0439876

[19] S. Sahinkaya, G. Tang, and Y. Zhou, *Nil-clean group rings*, J. Algebra Appl. **16** (2017), no. 7, 1750135, 7. MR3660418

[20] J. Šter, *Rings in which nilpotents form a subring*, Carpathian J. Math. **32** (2016), no. 2, 251–258. MR3587893

[21] A. Steger, *Diagonability of idempotent matrices*, Pacific J. Math. **19** (1966), 535–542. MR0202754

[22] Z. Ying, T. Koşan, and Y. Zhou, *Rings in which every element is a sum of two tripotents*, Canad. Math. Bull. **59** (2016), no. 3, 661–672. MR3563747

Department of Pure Mathematics, Faculty of Mathematical Sciences, Tarbiat, Modares University, Tehran, Iran, P.O. Box 14115-134
Email address: arezoukarimimansoub@gmail.com

Department of Mathematics, Gebze Technical University, Gebze/Kocaeli, Turkey
Email address: mtkosan@gtu.edu.tr

Department of Mathematics and Statistics, Memorial University of Newfoundland, St.John's, Nfld A1C 5S7, Canada
Email address: zhou@mun.ca

Contemporary Mathematics
Volume **715**, 2018
https://doi.org/10.1090/conm/715/14413

Commutators and Anti-Commutators of Idempotents in Rings

Dinesh Khurana and T. Y. Lam

Abstract. We show that a ring R has two idempotents e, e' with an invertible commutator $ee' - e'e$ if and only if $R \cong \mathbb{M}_2(S)$ for a ring S in which 1 is a sum of two units. In this case, the "anti-commutator" $ee' + e'e$ is automatically invertible, so we study also the broader class of rings having such an invertible anti-commutator. Simple artinian rings R (along with other related classes of matrix rings) with one of the above properties are completely determined. In this study, we also arrive at various new criteria for *general* 2×2 matrix rings. For instance, R is such a matrix ring if and only if it has an invertible commutator $er - re$ where $e^2 = e$.

1. Introduction

The work in this paper was inspired by an insightful exercise of Kaplansky in his 1968 book [**Ka**] on rings of operators. Given two idempotents e, e' in a ring R with the property that their commutator $ee' - e'e$ is invertible, "Exercise 6" in [**Ka**, p. 25] asked the readers to show that the idempotents $e, e', 1-e, 1-e'$ are pairwise isomorphic; in other words, the four principal right ideals generated by them are isomorphic as right$R$-modules. This exercise was intended for prospective use toward the proofs of Theorem 60 and Theorem 61 on Baer $*$-rings in [**Ka**, pp. 91–96]. For his Exercise 6 on p. 25, Kaplansky kindly offered his readers a "Hint". However, rendered with the author's trademark brevity, this "Hint" was itself no less than another substantial exercise. In all fairness, a full solution of Kaplansky's "Exercise 6" proved to be a considerable challenge by any yardstick.

That was the situation about fifty years ago, in 1967–68. Nowadays, with so much more known about idempotents and idempotent identities in rings, a rather natural new solution can be given for Kaplansky's "Exercise 6". More remarkably, the theme of this exercise can be further developed so as to give a full-fledged characterization theorem for the rings R that appeared in the exercise. To explain this from a more general point of view, we'll use the notation $[x, y]$ for the *commutator* $xy - yx$, and the notation $\langle x, y\rangle$ for the *anti-commutator* $xy + yx$, for any two elements x, y in a ring R. For the main purposes of this paper, it is convenient to

2010 *Mathematics Subject Classification.* 12E15, 15B33, 15B36, 16E50, 16N40, 16U60.
Key words and phrases. Units, idempotents, commutators, anti-commutators, Bott-Duffin inverse, matrix rings, simple artinian rings, exchange rings, commutative rings.

introduce the following two ring-theoretic properties, where $\mathrm{idem}(R)$ will henceforth denote the set of idempotents in the ring R, and $\mathrm{U}(R)$ will denote the group of units of R.

PROPERTY K. *There exist* $e, e' \in \mathrm{idem}(R)$ *such that* $[e, e'] \in \mathrm{U}(R)$.

PROPERTY $\overline{\mathbf{K}}$. *There exist* $e, e' \in \mathrm{idem}(R)$ *such that* $\langle e, e' \rangle \in \mathrm{U}(R)$.

The easiest class of rings to deal with in the investigation of these properties is the class of *abelian rings*; that is, rings in which all idempotents are central. It is easy to see that such a ring R has Property K iff $R = 0$, while it has Property $\overline{\mathrm{K}}$ iff $2 \in \mathrm{U}(R)$. (In fact, in view of the fact that $\langle 1, 1 \rangle = 2$, any ring with $2 \in \mathrm{U}(R)$ has Property $\overline{\mathrm{K}}$.) Another easy observation we can make is that, if $R \to R'$ is a (unital) ring homomorphism, then R having one of the two properties above implies that R' has the same property, whereby any matrix ring $\mathbb{M}_n(R)$ will also have the same property.

The first two main results in this paper are Theorem A and Theorem B below, where the former (to be proved in §2) explains our choice of the notation $\overline{\mathrm{K}}$ since it has the obvious consequence that the class of rings with Property K is (properly) contained in the class of rings with Property $\overline{\mathrm{K}}$.

THEOREM A. *For any* $e, e' \in \mathrm{idem}(R)$, $[e, e'] \in \mathrm{U}(R)$ *implies that* $\langle e, e' \rangle \in \mathrm{U}(R)$ (*though not conversely*). *In particular, if* R *has* Property K, *then it has* Property $\overline{\mathrm{K}}$.

THEOREM B. *A ring* R *has* Property K *iff* $R \cong \mathbb{M}_2(S)$ *for a ring* S *in which* $1 \in \mathrm{U}(S) + \mathrm{U}(S)$.

From our more general perspective, the solution of Kaplansky's Exercise 6 is just a part of the work needed for proving the "only if" part of Theorem B. In §3, this theorem is proved by using judiciously various idempotent identities of Nicholson [**Ni**], Kato [**Kt**], and Koliha-Rakočević [$\mathbf{KR}_1, \mathbf{KR}_2$], which we first develop *ad hoc* in §2. The classical notion of Bott-Duffin invertibility (relative to an idempotent) introduced in [**BD**] turns out to play a significant role in proving Theorem B, so the basic ingredients of the Bott-Duffin theory are briefly recalled at the beginning of §2 as well.

With Theorem A and Theorem B at our disposal, a natural question to ask is when would a matrix ring $\mathbb{M}_n(T)$ (over a given type of base rings T) have Property K or Property $\overline{\mathrm{K}}$. For instance, in the case where T is a division ring, this would amount to asking which simple artinian rings would have Property K or Property $\overline{\mathrm{K}}$. Indeed, taking T to be a local ring or a nonzero commutative ring, we have the following result in §3 on matrix rings $\mathbb{M}_n(T)$ with Property K.

THEOREM C. *Let* $R = \mathbb{M}_n(T)$, *where* T *is a local ring or a nonzero commutative ring. Then* R *has* Property K *iff either* $n \in \{4, 6, 8, \dots\}$, *or* $n = 2$ *and* $1_T \in \mathrm{U}(T) + \mathrm{U}(T)$. (*Note that in the case where* $(T, \mathfrak{m})$ *is local ring, the condition* $1_T \in \mathrm{U}(T) + \mathrm{U}(T)$ *would amount to the simpler statement that* $|T/\mathfrak{m}| > 2$.)

As for Property $\overline{\mathrm{K}}$, the corresponding result (to be proved in §4) is the following.

THEOREM D. *Let* T *be a local ring or a commutative ring. Then* $R = \mathbb{M}_n(T)$ *has* Property $\overline{\mathrm{K}}$ *iff we are in one of the following cases:*

(1) $n \in \{4, 6, 8, \dots\}$.
(2) $n = 2$ *and* $1_T \in \mathrm{U}(T) + \mathrm{U}(T)$.
(3) n *is odd and* $2 \in \mathrm{U}(T)$.

As a spin-off of our investigations on invertible commutators of the form $[e, e']$ where $e, e' \in \mathrm{idem}(R)$, we also consider the case where e' is allowed to be an arbitrary element of R, while e is replaced by an element with some specific property "comparable" to idempotency. Working with these more general commutators (and anti-commutators) and building on the work of Fuchs-Maxson-Pilz in [**FMP**, III.2], we obtain in §5 a number of new criteria for 2×2 matrix rings, three of which are summarized as follows.

THEOREM E. *A ring* R *is a* 2×2 *matrix ring over some other ring iff there is an invertible commutator* $[e, *]$ *where* $e^2 = e$, *iff there is an invertible commutator* $[p, *]$ *where* $p^2 = 0$, *iff there is an invertible commutator* $[u, *]$ *where* $u^2 = 1$.

The terminology and notations introduced so far in this Introduction will be used freely throughout the paper. For any ring R, $\mathrm{rad}(R)$ denotes the Jacobson radical of R, and the words "exchange ring" will be used in the sense of Warfield [**Wa**] and Nicholson [**Ni**]. By saying that a matrix $M \in \mathbb{M}_n(T)$ is *diagonalizable,* we'll mean that M is similar to a diagonal matrix in $\mathbb{M}_n(T)$. Other standard terminology and conventions in ring theory follow mainly those in [**Go**], [**La**$_1$], and [**La**$_3$].

2. Idempotent Identities of Kato and Koliha-Rakočević

One key ingredient used in this beginning section is the basic notion of Bott-Duffin invertibility introduced in the early paper [**BD**]. In order to give a relatively self-contained exposition of our results, we'll start by recalling some definitions and facts from [**BD**]. For any idempotent e in a ring R, an element $a \in R$ is said to be *Bott-Duffin invertible* relative to e if $eae \in \mathrm{U}(eRe)$. In this case, the inverse of eae in the corner ring eRe is said to be the Bott-Duffin inverse of a relative to e. For the sake of completeness, we state and prove the following classical characterization result for Bott-Duffin invertibility.

THEOREM 2.1. *For* $a \in R$ *and* $e \in \mathrm{idem}(R)$, *the following are equivalent:*

(1) a *is Bott-Duffin invertible relative to* e.
(2) $1 - e + ae \in \mathrm{U}(R)$. (3) $1 - e + ea \in \mathrm{U}(R)$. (4) $1 - e + eae \in \mathrm{U}(R)$.

Proof. For any ring R, Jacobson's Lemma (see, e.g. [**La**$_2$, Exercise 1.6]) states that, for any $x, y \in R$, $1 - xy \in \mathrm{U}(R)$ iff $1 - yx \in \mathrm{U}(R)$. Since $1 - e + ae = 1 - (1 - ae)\,e$, this Lemma gives (2) $\Leftrightarrow$ (4). By left-right symmetry, we have also (3) $\Leftrightarrow$ (4). To see that (4) $\Leftrightarrow$ (1), let $f = 1 - e$. With respect to the idempotent e, the element $f + eae$ has a diagonal Peirce decomposition matrix $\begin{pmatrix} eae & 0 \\ 0 & f \end{pmatrix}$. Such a matrix is invertible iff $eae \in \mathrm{U}(eRe)$, which is the defining condition for (1). This shows that (4) $\Leftrightarrow$ (1). □

Next, we state the following key result from Kato's book [**Kt**, (I.4.34), (I.4.44)].

PROPOSITION 2.2 (Kato's Identities). *For any* $e, e' \in \mathrm{idem}(R)$, *we have*

$$(e - e')^2 + (1 - e - e')^2 = 1, \tag{2.3}$$

which amounts to

$$(2.4)\qquad (e-e')^2 = (e+e')\,(2-e-e'),$$

where the two factors on the RHS *commute. Also, writing* $r = 1-e+e'e$ *and* $s = 1-e'+ee'$, *we have*

$$(2.5)\qquad rs = sr = (1+e-e')\,(1-e+e') = (1-e-e')^2.$$

Proof. The identity (2.3) is easily verified by a direct expansion of the LHS. By transposing the term $(1-e-e')^2$ to the RHS and using the factorization $1-x^2 = (1-x)\,(1+x)$, we get the identity (2.4). Similarly, by transposing the term $(e-e')^2$ in (2.3) to the RHS and using the factorization $1-y^2 = (1+y)\,(1-y)$, we get the last equality in (2.5). The first two equalities in (2.5) are easily verified by direct expansions. □

COROLLARY 2.6. *Two idempotents* $e, e' \in R$ *are Bott-Duffin invertible relative to each other iff* $1-e-e' \in \mathrm{U}(R)$.

Proof. This follows from (2.5), applied in conjunction with Theorem 2.1. □

REMARK 2.7. In general, e' being Bott-Duffin invertible relative to e alone does not imply that e is Bott-Duffin invertible relative to e'. For instance, if $e' = 1 \neq e$, then e' is Bott-Duffin invertible relative to e, but e is not Bott-Duffin invertible relative to e'. In this example, we have $1-e-e' = -e \notin \mathrm{U}(R)$.

In [**KR**$_1$, Theorem 3.5], it was proved that, for any idempotents e, e' in any ring, $e-e' \in \mathrm{U}(R)$ iff $e+e' \in \mathrm{U}(R)$ and $1-ee' \in \mathrm{U}(R)$. (The hypothesis $2 \in \mathrm{U}(R)$ was included in the statement of this theorem, but the result can be seen to be true without such a hypothesis.) Using here the crucial identity (2.4), we'll prove the following closely related result.

THEOREM 2.8. *For any* $e, e' \in \mathrm{idem}\,(R)$, *we have* $e-e' \in \mathrm{U}(R)$ *iff* $e+e' \in \mathrm{U}(R)$ *and* $f+f' \in \mathrm{U}(R)$, *where* $f = 1-e$ *and* $f' = 1-e'$. *In this case,* e *is similar to* $1-e'$.

Proof. The "iff" statement follows from the identity (2.4) (since $2-e-e' = f+f'$). The last statement is a special case of a result of Nicholson [**Ni**, Proposition 1.8] on the clean representations of elements in a ring. To make our exposition self-contained, we recall Nicholson's proof (in our special case). If $u := e-e' \in \mathrm{U}(R)$, then

$$(1-e')\,u = (1-e')\,(e-e') = (1-e')\,e = (e-e')\,e = ue.$$

Thus, $e = u^{-1}(1-e')\,u$ is similar to $1-e'$. □

REMARK 2.9. In general, $e+e' \in \mathrm{U}(R)$ alone need not imply that $e-e' \in \mathrm{U}(R)$. For instance, in any nonzero ring R in which $2 \in \mathrm{U}(R)$, the idempotents $e = e' = 1$ have the property that $e+e' = 2 \in \mathrm{U}(R)$, but $e-e' = 0 \notin \mathrm{U}(R)$.

Next we give a systematic derivation for two useful identities on the commutator $[\,e, e'\,] = ee' - e'e$ and the anti-commutator $\langle\, e, e'\,\rangle = ee' + e'e$ from [**KR**$_2$, p. 103] and [**KR**$_1$, p. 289].

PROPOSITION 2.10 (Koliha-Rakočević Identities). *For any* $e, e' \in \mathrm{idem}\,(R)$ *and* $f = 1-e$, *we have*

$$(2.11)\qquad [\,e, e'\,] = (e-e')\,(e'-f) = (f-e')\,(e-e');$$

$$(2.12)\qquad \langle\, e, e'\,\rangle = (e+e')\,(e'-f) = (e'-f)\,(e+e').$$

Proof. From $(e+e')^2 = e+e'+\langle e,e'\rangle$, transposition of the term $e+e'$ gives the two equalities in (2.12). Similarly, from $(e-e')(e+e') = e-e'+[e,e']$, transposition of the term $e-e'$ gives the first equality in (2.11), and the second equality follows by working instead with $(e+e')(e-e') = e-e'-[e,e']$. □

Kaplansky asserted in [**Ka**, p. 25, Exercise 6] that, under the assumption $[e,e'] \in \mathrm{U}(R)$, the idempotents $e, e' \in R$ are Bott-Duffin invertible relative to each other. With the help of Proposition 2.10, one can prove the more precise result in (1) below, and by the same token, prove also its complete analogue in (2) for anti-commutators.

THEOREM 2.13. *For any* $e, e' \in \mathrm{idem}(R)$ *and* $f = 1-e$, *the following hold.*

(1) $[e,e'] \in \mathrm{U}(R)$ *iff* $e-e' \in \mathrm{U}(R)$ *and* $f-e' \in \mathrm{U}(R)$, *iff* $e-e' \in \mathrm{U}(R)$ *and* e, e' *are Bott-Duffin invertible relative to each other.*

(2) $\langle e,e'\rangle \in \mathrm{U}(R)$ *iff* $e+e' \in \mathrm{U}(R)$ *and* $f-e' \in \mathrm{U}(R)$, *iff* $e+e' \in \mathrm{U}(R)$ *and* e, e' *are Bott-Duffin invertible relative to each other.*

(3) $[e,e'] \in \mathrm{U}(R)$ *implies that* $\langle e,e'\rangle \in \mathrm{U}(R)$.

Proof. For (1), the first "iff" statement follows from (2.11); see [**KR**$_2$, 3.6]. The second "iff" statement then follows from Corollary 2.6. (2) is proved similarly by using (2.11) and Corollary 2.6 (although the first "iff" statement in (2) was stated in [**KR**$_1$, Theorem 3.6] with an extra assumption that $2 \in \mathrm{U}(R)$). Finally, (3) follows from (1) and (2), since $e-e' \in \mathrm{U}(R) \Rightarrow e+e' \in \mathrm{U}(R)$ according to Theorem 2.8. □

Recalling from §1 that "Property K" (respectively, "Property $\overline{\mathrm{K}}$") on a ring R means the existence of $e, e' \in \mathrm{idem}(R)$ such that $[e,e'] \in \mathrm{U}(R)$ (respectively, $\langle e,e'\rangle \in \mathrm{U}(R)$), we can draw the following conclusions from Theorem 2.13.

COROLLARY 2.14. *If a ring* R *has* Property K, *then it has* Property $\overline{\mathrm{K}}$. *However, the converse of this statement fails in general.*

Proof. The first statement here follows from part (3) of Theorem 2.13. The second statement follows from the observation (made in §1) that any ring R with $2 \in \mathrm{U}(R)$ has Property $\overline{\mathrm{K}}$, while an abelian ring $R \neq 0$ cannot have Property K. □

For future reference, we record in the following a useful result on the Jacobson radical and an immediate consequence thereof.

PROPOSITION 2.15. *Let* $J \subseteq \mathrm{rad}(R)$ *be an ideal of* R *such that idempotents lift modulo* J. *Then* R *has* Property K *iff* R/J *does. The same statement holds for* Property $\overline{\mathrm{K}}$.

Proof. As we have observed in §1, the "only if" part holds without any assumptions on the ideal J. In the case where idempotents lift modulo $J \subseteq \mathrm{rad}(R)$, the "if" part follows from the observation that an element $s \in R$ is a unit iff $\overline{s}$ is a unit in $\overline{R} = R/J$. □.

COROLLARY 2.16. *Let* J *be an ideal of* R *such that one of the following holds:*

(1) J *is a nil ideal.*

(2) $J \subseteq \mathrm{rad}(R)$ *and* R *is an exchange ring.*

Then the conclusions of Proposition 2.15 *hold for the two rings* R *and* R/J.

Proof. (1) If J is a nil ideal, we have $J \subseteq \text{rad}(R)$ by [**La**$_1$, 4.11]. On the other hand, by [**La**$_1$, 21.28], idempotents lift modulo J. Thus, Proposition 2.15 applies.

(2) If R is an exchange ring, idempotents can be lifted modulo any ideal in R (according to [**Ni**]), so again Proposition 2.15 applies. □

We shall conclude the discussions in this section by using Corollary 2.16 to prove the following two easy results on upper triangular rings.

PROPOSITION 2.17. *Let S, T be two rings, and let M be any (S,T)-bimodule. Then the formal triangular ring* $R = \begin{pmatrix} S & M \\ 0 & T \end{pmatrix}$ *has* Property K (*respectively, Property* $\overline{\text{K}}$) *iff S and T both do.*

Proof. The "only if" part is clear, since S and T can both be thought of as factor rings of R. Conversely, assume that S and T have Property K. Then $S \times T$ also does. As $J := \begin{pmatrix} 0 & M \\ 0 & 0 \end{pmatrix}$ is a square-zero ideal in R, and $R/J \cong S \times T$, it follows from Case (1) of Corollary 2.16 that R also has Property K. The case where S and T both have Property $\overline{\text{K}}$ is similar. □

In the case of $n \times n$ upper triangular matrices over a given ring S, the ideal J of strictly upper triangular matrices has the property that $J^n = 0$. Thus, the same method of proof (using Case (1) of Corollary 2.16) yields the following similar result.

PROPOSITION 2.18. *Let $R = \mathbb{T}_n(S)$ be the ring of $n \times n$ upper triangular matrices over a ring S. Then R has* Property K (*respectively, Property* $\overline{\text{K}}$) *iff S does.*

3. Rings with Property K

To study rings with the Property K, we begin by giving a streamlined proof for the rest of Kaplansky's "Exercise 6" from [**Ka**, p. 25]; see also Theorem 60 in [**Ka**, p. 96]. The proof to be given below is in some sense much easier to grasp than that given by Kaplansky since it makes full use of the efficient tools developed in §2.

THEOREM 3.1. *If $e, e' \in \text{idem}(R)$ are such that $[e, e'] \in \text{U}(R)$, then the following hold.*

(1) *The four idempotents $e, e', 1-e, 1-e'$ are pairwise similar.*
(2) $R = eR \oplus e'R = Re \oplus Re'$.

Proof. (1) Let $f = 1 - e$. By Theorem 2.13(1), we have $e - e' \in \text{U}(R)$ and $f - e' \in \text{U}(R)$. Therefore, by the last part of Theorem 2.8, e' is similar to f as well as to e. By symmetry, it follows that $e, e', 1-e, 1-e'$ are pairwise similar.

(2) By symmetry, it suffices to prove that $R = eR \oplus e'R$. Since $e - e' \in \text{U}(R)$, clearly $eR + e'R = R$. Also, for any $x \in eR \cap e'R$, we have $x = ex = e'x$. Thus, $(e - e')\,x = 0$, and hence $x = 0$. This proves that $R = eR \oplus e'R$. (In conclusion, we note that (2) can also be deduced as a consequence of [**KR**$_2$, Theorem 3.2].) □

REMARK 3.2. In general, if two idempotents e, e' in a ring R are such that $e, e', 1-e, 1-e'$ are pairwise similar, the commutator $[e, e']$ need not be a unit in R, and e' need not be Bott-Duffin invertible relative to e. For instance, if $R = \mathbb{M}_2(S)$ over a nonzero ring S and E_{ij} are the matrix units, then $e = E_{11}$ and $e' = E_{22}$ certainly have the similarity properties mentioned above, but $[e, e'] = 0$, and $ee'e = 0$ (so e' is not Bott-Duffin invertible relative to e). On the other hand, if we have only $[e, r] \in \mathrm{U}(R)$ where $r \in R$ (or even $r \in \mathrm{U}(R)$), it also does not follow that r is Bott-Duffin invertible relative to e, as is shown by the example $e = E_{11}$ and $r = E_{12} + E_{21}$, for which $[e, r] = E_{12} - E_{21} \in \mathrm{U}(R)$, but again $ere = 0$.

Now we are in a position to prove Theorem B stated in the Introduction. The remarkable thing about this theorem is that a *single* commutator property on a ring turns out to be enough to determine the structure of the whole ring. Just for the record, we note that it was Kaplansky's "Exercise 6" in [**Ka**, p. 25] which had given us the impetus to prove the "only if" part of this theorem.

THEOREM 3.3. *A ring R has* Property K *iff $R \cong \mathbb{M}_2(S)$ for a ring S in which $1 \in \mathrm{U}(S) + \mathrm{U}(S)$.*

Proof. First assume that $1 = a + b \in S$ where $a, b \in \mathrm{U}(S)$. In the matrix ring $\mathbb{M}_2(S)$, we check easily that $e = \begin{pmatrix} a & b \\ a & b \end{pmatrix}$ is an idempotent, and that the commutator $[e, E_{11}] = \begin{pmatrix} 0 & -b \\ a & 0 \end{pmatrix}$ is a unit. Thus, $\mathbb{M}_2(S)$ has Property K. (In testimony to the implication statement in Theorem 2.13(3), the anti-commutator $\langle e, E_{11} \rangle = \begin{pmatrix} 2a & b \\ a & 0 \end{pmatrix}$ is indeed a unit too, so the ring $\mathbb{M}_2(S)$ also has Property $\overline{\mathrm{K}}$ as expected.)

Conversely, if a ring R has Property K, fix two idempotents $e, e' \in R$ such that $[e, e'] \in \mathrm{U}(R)$. By Theorem 3.1, $f := 1 - e$ is similar to e. This implies that $fR \cong eR$, and so $R_R = eR \oplus fR \cong eR \oplus eR$. Taking endomorphism rings, we have

$$R \cong \mathrm{End}_R(eR \oplus eR) \cong \mathbb{M}_2(S)$$

where $S := \mathrm{End}_R(eR)$ can be identified with the corner ring eRe. By Theorem 2.13(1), $ee'e \in \mathrm{U}(S)$. Similarly, for $f' := 1 - e'$, the fact that $[e, f'] = -[e, e'] \in \mathrm{U}(R)$ implies that $ef'e \in \mathrm{U}(S)$ too. Thus, $1_S = e = ee'e + ef'e \in \mathrm{U}(S) + \mathrm{U}(S)$, as desired. $\square$

In connection with the condition $1 \in \mathrm{U}(S) + \mathrm{U}(S)$ appearing in Theorem 3.3, we should point out that it is in fact equivalent to the following ostensibly stronger condition: for any integer $n \geq 1$, any unit in S is a sum of n units in S. In the case where S is a unit-regular ring, the condition $1_S \in \mathrm{U}(S) + \mathrm{U}(S)$ can also be self-strengthened into $S = \mathrm{U}(S) + \mathrm{U}(S)$, according to [**GW**, Theorem 3.8]. In the following, we will state two interesting consequences of Theorem 3.3.

COROLLARY 3.4. *A ring R has* Property K *with $2 \in \mathrm{U}(R)$ iff $R \cong \mathbb{M}_2(S)$ for some ring S with $2 \in \mathrm{U}(S)$.*

Proof. The "if" part follows from Theorem 3.3 since $2 \in \mathrm{U}(S)$ implies that $1 = 2^{-1} + 2^{-1} \in \mathrm{U}(S) + \mathrm{U}(S)$, as well as $2 \in \mathrm{U}(\mathbb{M}_2(S))$. Conversely, if a ring R

has Property K along with $2 \in \mathrm{U}(R)$, Theorem 3.3 implies that $R \cong \mathbb{M}_2(S)$ for some ring S. Clearly, the fact that $2 \in \mathrm{U}(R)$ implies that $2 \in \mathrm{U}(S)$. □

A second application of the "if" part of Theorem 3.3 is that it gives a good supply of examples of matrix rings having the Property K, as follows.

COROLLARY 3.5. *If* $R = \mathbb{M}_{2m}(T)$ *(for any ring* T*) where* $m \geq 2$*, or* $R = \mathrm{End}_D(V)$ *for some infinite-dimensional right vector space* V *over a division ring* D*, then* R *has* Property K.

Proof. First assume that $R = \mathbb{M}_{2m}(T)$ where $m \geq 2$. By a result of Henriksen [**He**], the identity matrix I_m is a sum of two units in $\mathbb{M}_m(T)$. Thus, by Theorem 3.3, $R \cong \mathbb{M}_2\big(\mathbb{M}_m(T)\big)$ has Property K. Finally, assume that $R = \mathrm{End}\,(V_D)$ as in the statement of the Corollary. By [**La**$_3$, Example 1.4], we have $R \cong \mathbb{M}_n(R)$ for every $n \geq 1$. Applying this for $n = 4$ (for instance), we are back to the case treated above. □

EXAMPLE 3.6. In general, if T is a ring such that $R := \mathbb{M}_2(T)$ has Property K, Theorem 3.3 only says that $R \cong \mathbb{M}_2(S)$ for some ring S in which 1 is a sum of two units. Since T may not be isomorphic to S, this *does not* imply that $1_T \in \mathrm{U}(T) + \mathrm{U}(T)$. An explicit example to illustrate the possible failure of $1_T \in \mathrm{U}(T) + \mathrm{U}(T)$ is as follows. For any field k, we construct after Leavitt [**Le**] and Cohn [**Co**$_1$] a k-algebra T with a generic invertible 3×2 matrix. This matrix defines a right T-module isomorphism $T^2 \cong T^3$, which induces another isomorphism $T^2 \cong T^4$. Taking T-endomorphism rings gives a ring isomorphism $R := \mathbb{M}_2(T) \cong \mathbb{M}_4(T)$. Thus, it follows from Corollary 3.5 that R has Property K. By a recent result of G. Bergman [**Be**$_2$] based on his earlier work [**Be**$_1$], $\mathrm{U}(T) = \mathrm{U}(k)$. In particular, if we take k to be the field of two elements, we'll have $\mathrm{U}(T) = \{1\}$, in which case clearly $1_T \notin \mathrm{U}(T) + \mathrm{U}(T)$.

Example 3.6 leads naturally to the following

QUESTION 3.7. *If* T *is a ring such that* $\mathbb{M}_2(T)$ *has* Property K, *under what additional assumptions on* T *can we conclude that* $1_T \in \mathrm{U}(T) + \mathrm{U}(T)$ *?*

The difficulty in dealing with this question stems mainly from the fact that $1_T \in \mathrm{U}(T) + \mathrm{U}(T)$ is *not* a Morita invariant property of rings. Nevertheless, there are many classes of rings T for which we can answer Question 3.7 in a satisfactory way. First, recall from [**La**$_3$, §17C] that a ring T is said to be **$\mathbb{M}_n$-unique** if, for any ring S, $\mathbb{M}_n(T) \cong \mathbb{M}_n(S)$ implies that $T \cong S$. Taking stock in this definition, we do have the following partial positive answer to Question 3.7.

THEOREM 3.8. *Let* $R = \mathbb{M}_2(T)$ *where* T *satisfies one of the following conditions:*

(1) T *is* $\mathbb{M}_2$*-unique.*

(2) T *is an abelian ring and every idempotent in* R *is diagonalizable.*

Then R *has* Property K *iff* $1_T \in \mathrm{U}(T) + \mathrm{U}(T)$.

Proof. The "if" part is true without any assumptions on T by Theorem 3.3. The "only if" part in the two cases (1) and (2) will be handled separately, as follows.

CASE (1). If R has Property K, Theorem 3.3 implies that $R \cong \mathbb{M}_2(S)$ for some ring S with $1_S \in \mathrm{U}(S) + \mathrm{U}(S)$. Given the $\mathbb{M}_2$-unique assumption in Case (1), we have $T \cong S$, and so $1_T \in \mathrm{U}(T) + \mathrm{U}(T)$.

CASE (2). If R has Property K, fix a commutator $[e, e'] \in \mathrm{U}(R)$ with $e, e' \in \mathrm{idem}(R)$. Given the hypothesis in this case, we may assume (after a conjugation) that $e' = \mathrm{diag}(s, t)$, where $s, t \in T$ are necessarily central idempotents. Writing $e = \begin{pmatrix} a & b \\ c & d \end{pmatrix}$, we have $[e, e'] = \begin{pmatrix} 0 & b(t-s) \\ c(s-t) & 0 \end{pmatrix} \in \mathrm{U}(R)$, which implies that $b, c \in \mathrm{U}(T)$. From the equation $e^2 = e$, we get $a^2 + bc = a$, so a, $1 - a$ are units in T, with sum 1. □

Having proved Theorem 3.8, we will mention in (3.9) – (3.13) below some of the more important classes of base rings T to which the theorem can be applied.

3.9. The most obvious class of examples of a ring T satisfying the hypothesis (2) of Theorem 3.8 is given by the projective-free rings defined by P. M. Cohn in [**Co**$_2$, **Co**$_3$]: a ring T is *projective-free* if every finitely generated projective right module over T is free of a unique rank. (This notion is known to be left-right symmetric; see [**Co**$_2$].) Clearly, such a ring T has only trivial idempotents (so it is abelian), and any idempotent in $\mathbb{M}_n(T)$ is similar to a diagonal matrix with 0's and 1's on the diagonal.

3.10. Another example of a ring satisfying the hypothesis (2) in Theorem 3.8 is an abelian (von Neumann) regular ring T. The fact that any idempotent matrix over a regular ring is diagonalizable can be shown in at least two ways: first by using the refinement theorem (for finitely generated projective modules) of Goodearl and Handelman in [**GH**, 3.8], and second, by using [**Go**, 2.6] in conjunction with [**SG**, Theorem 9].

3.11. In the formulation of "Open Problem (47)" in Goodearl's book [**Go**, p. 349], there is an extensive list of different classes of regular rings T that are classically known to be $\mathbb{M}_n$-unique for all n. Thus, Theorem 3.8 is applicable to all such regular rings, under the hypothesis (1). The two best known classes among those mentioned in [**Go**, p. 349] are right self-injective regular rings [**Go**, 10.35], and all regular rings whose primitive factors are artinian [**Go**, 6.12]. (The latter class includes all abelian regular rings, which we have already mentioned in (3.10).) But unfortunately, unit-regular rings are *not* known to be $\mathbb{M}_n$-unique (even for $n = 2$), so the answer to Question 3.7 has so far remained unknown in the case where the base ring T is unit-regular.

3.12. According to [**La**$_3$, 17.26], any semilocal ring T is $\mathbb{M}_n$-unique for all n. Thus, Theorem 3.8 is applicable to T again under the hypothesis (1). In the special case where $(T, \mathfrak{m})$ is a local ring, it would be projective-free as well. In this case, as we have noted in the statement of Theorem B in §1, the conclusion $1_T \in \mathrm{U}(T) + \mathrm{U}(T)$ in Theorem 3.8 would amount to the simpler statement that $|\,T/\mathfrak{m}\,| > 2$.

3.13. Theorem 3.8 is applicable to any commutative ring T (under hypothesis (1)) since T is also known to be $\mathbb{M}_n$-unique (for all n) by [**La**$_3$, 17.31]. In this case, the conclusion of Theorem 3.8 is capable of another more concrete derivation, as follows. If e, e' are idempotents in $R = \mathbb{M}_2(T)$ with $[e, e'] \in \mathrm{U}(R)$, it is easy to show using [**KLS**, Formula (1.1)] that $\mathrm{tr}(ee')$ and $1 - \mathrm{tr}(ee')$ are both units in T, with sum 1.

Using Theorem 3.3 and Corollary 3.5, it is now easy to determine all matrix rings over local rings and nonzero commutative rings that have Property K.

THEOREM 3.14. *Let* T *be a local ring or a nonzero commutative ring. Then* $R := \mathbb{M}_n(T)$ *has* Property K *iff either* $n \in \{4, 6, 8, \dots\}$, *or* $n = 2$ *and* $1_T \in \mathrm{U}(T) + \mathrm{U}(T)$.

Proof. The "if" part follows from Theorem 3.3 and Corollary 3.5. For the "only if" part, assume that $R = \mathbb{M}_n(T)$ has Property K. Since T is a local ring or a nonzero commutative ring, there exists an ideal $\mathfrak{m} \subseteq T$ such that $T/\mathfrak{m}$ is a division ring. Then $\mathbb{M}_n(T/\mathfrak{m}) \cong \mathbb{M}_n(T)/\mathbb{M}_n(\mathfrak{m})$ also has Property K. Applying Theorem 3.3, we have $\mathbb{M}_n(T/\mathfrak{m}) \cong \mathbb{M}_2(S)$ for some ring S with $1_S \in \mathrm{U}(S) + \mathrm{U}(S)$. By the uniqueness part of Wedderburn's theorem, we see that n must be even. Finally, in the special case $n = 2$, $R = \mathbb{M}_2(T)$ having Property K would imply that $1_T \in \mathrm{U}(T) + \mathrm{U}(T)$ by the remarks in (3.12) and (3.13). □

EXAMPLE 3.15. Let $R = \mathbb{M}_n(D)$ be a typical simple artinian ring, where D is a division ring. According to Theorem 3.14, R has Property K except precisely when n is odd or when $n = 2$ and $|D| = 2$.

EXAMPLE 3.16. Let $\mathbb{Z}_{(p)}$ denote the localization of $\mathbb{Z}$ at the prime ideal (p). If $T = \mathbb{Z}$ or $\mathbb{Z}_{(2)}$, $\mathbb{M}_n(T)$ has Property K iff $n \in \{4, 6, 8, \dots\}$. On the other hand, if $T = \mathbb{Z}_{(p)}$ where p is an odd prime, then $\mathbb{M}_n(T)$ has Property K iff n is even.

EXAMPLE 3.17. If the ring T in Theorem 3.14 is allowed to be noncommutative and not a local ring, the "only if" part in the theorem may no longer hold. For instance, we may take T to be any nonzero ring with Property K (so T is necessarily noncommutative). Then for *any* $n \geq 1$ (odd or even), $R = \mathbb{M}_n(T)$ also has Property K since T does.

4. Interplay Between Property K and Property $\overline{\mathrm{K}}$

In this section, we turn our attention to rings with Property $\overline{\mathrm{K}}$. Our goal is to understand more precisely the relationship and interaction between Property K and Property $\overline{\mathrm{K}}$. We begin by recalling that any ring R with $2 \in \mathrm{U}(R)$ has Property $\overline{\mathrm{K}}$, but not necessarily Property K. On the other hand, a 2×2 matrix ring such as $R = \mathbb{M}_2(\mathbb{F}_2)$ does not have Property K (e.g. by Theorem 3.3), and hence also does not have Property $\overline{\mathrm{K}}$ since $\mathrm{char}\,(R) = 2$. From this example, it follows for instance that if T is any ring with an ideal of index 2, then $\mathbb{M}_2(T)$ (with factor ring R) does not have Property $\overline{\mathrm{K}}$, and hence also not Property K.

Our first main result in this section is Theorem 4.1 below, where part (1) gives some necessary conditions on the invertibility of an anti-commutator $\langle e, e' \rangle \in R$, while part (2) offers a somewhat unexpected characterization for the rings with Property $\overline{\mathrm{K}}$ without the use of products of idempotents.

THEOREM 4.1. (1) *If* $e, e' \in \mathrm{idem}\,(R)$ *are such that* $\langle e, e' \rangle \in \mathrm{U}(R)$, *then* e' *is similar to* e, *and we have* $\langle e, w \rangle \in \mathrm{U}(R)$ *for some* $w \in \mathrm{U}(R)$.
(2) *A ring* R *has* Property $\overline{\mathrm{K}}$ *iff there exists a unit* $u \in \mathrm{idem}\,(R) + \mathrm{idem}\,(R)$ *such that* $1 - u \in \mathrm{U}(R)$.

Proof. (1) According to Theorem 2.13(2), $\langle e, e' \rangle \in \mathrm{U}(R)$ amounts to two conditions:

$$e + e' \in \mathrm{U}(R), \quad \text{and} \quad 1 - e - e' \in \mathrm{U}(R). \tag{4.2}$$

The latter implies (by Theorem 2.8) that $e' = wew^{-1}$ for some $w \in \mathrm{U}(R)$. Then $e + wew^{-1} = e + e' \in \mathrm{U}(R)$ implies that $\langle e, w \rangle = ew + we = uw \in \mathrm{U}(R)$.

(2) If R has Property $\overline{\mathrm{K}}$, there exist $e, e' \in \mathrm{idem}\,(R)$ satisfying (4.2). Adding the two elements in (4.2) shows that 1 is a sum of two units, the first one of which is a sum of two idempotents in R. Conversely, if $1 = u + u'$ where $u, u' \in \mathrm{U}(R)$ and $u = e + e'$ for some $e, e' \in \mathrm{idem}\,(R)$, then $1 - e - e' = 1 - u = u' \in \mathrm{U}(R)$. Thus, (4.2) is satisfied, so R has Property $\overline{\mathrm{K}}$. □

Theorem 4.1 for rings with Property $\overline{\mathrm{K}}$ has a complete analogue for rings with Property K too, which we shall present below. One good reason we have chosen to prove Theorem 4.1 for Property $\overline{\mathrm{K}}$ first is that the proof of its part (2) is easier and more intuitive, while it gives an impetus toward finding an analogous (but somewhat harder) characterization result in the case of Property K.

THEOREM 4.3. (1) *If* $e, e' \in \mathrm{idem}\,(R)$ *are such that* $[\,e, e'\,] \in \mathrm{U}(R)$, *then we have* $[\,e, v\,] \in \mathrm{U}(R)$ *for some* $v \in \mathrm{U}(R)$.
(2) *A ring* R *has* Property K *iff there exists a unit* $v \in \mathrm{idem}\,(R) - \mathrm{idem}\,(R)$ *such that* $1 \pm v \in \mathrm{U}(R)$ *for both signs.*

Proof. (1) By Theorem 2.13(1), $[\,e, e'\,] \in \mathrm{U}(R) \Rightarrow v := e - e' \in \mathrm{U}(R)$, and so

$$[\,e, v\,] = [\,e,\, e - e'\,] = -[\,e, e'\,] \in \mathrm{U}(R). \tag{4.4}$$

The fact that one can "replace" the idempotent e' by a unit v (while still achieving the property $[\,e, v\,] \in \mathrm{U}(R)$) reminds us somewhat of the theorem of de Sá as discussed in [**Sa**] and [**KL**], although our result here applies quite generally to any ring R, as long as $e, e' \in \mathrm{idem}\,(R)$. (Needless to say, the converse of (1) fails in general. For instance, taking $R = \mathbb{M}_2(T)$ for any ring T, $[e, v]$ is an invertible commutator for $e = E_{11} \in \mathrm{idem}\,(R)$ and $v \in \mathrm{U}(R)$. But for $S = \mathbb{Z}$ or $\mathbb{F}_2$ for instance, we cannot find *any* invertible $[e_1, e_2]$ with $e_1, e_2 \in \mathrm{idem}\,(R)$.)

(2) First assume R has Property K and fix an invertible commutator $[\,e, e'\,]$ with $e, e' \in \mathrm{idem}\,(R)$. We have $(1 - e) - e' \in \mathrm{U}(R)$ (by (2.13)(1)), so Theorem 2.8 implies that $(1 - e) + e' \in \mathrm{U}(R)$; that is, $1 - v \in \mathrm{U}(R)$ for $v := e - e' \in \mathrm{U}(R)$ as in the last paragraph. On the other hand, $1 + v = (1 - e') + e \in \mathrm{U}(R)$ too, again by Theorem 2.8 since $(1 - e') - e \in \mathrm{U}(R)$. This proves the "only if" part of (2). For the converse, assume that there is a unit $v = e - e'$ for suitable $e, e' \in \mathrm{idem}\,(R)$ such that $1 \pm v \in \mathrm{U}(R)$ for both signs. By Theorem 2.13(1), we will have Property K on R if we can show that $1 - e - e' \in \mathrm{U}(R)$. Appealing to Theorem 2.8 once more (with the idempotent e there replaced by $1 - e$), this would follow if $(1 - e) + e' \in \mathrm{U}(R)$ and $e + (1 - e') \in \mathrm{U}(R)$. The former amounts to $1 - v \in \mathrm{U}(R)$, while the latter amounts to $1 + v \in \mathrm{U}(R)$. Since both of these conditions were given, we are done. □

Our next goal is to study the possible presence of Property $\overline{\mathrm{K}}$ on $n \times n$ matrix rings over certain types of rings T. The first case we can treat without too much difficulty is when $(T, \mathfrak{m})$ is a local ring. Here, a judicious use of Corollary 2.16 enables us to (essentially) "replace" T by the division ring $T/\mathfrak{m}$. As we have pointed out before in (3.9), the condition $1_T \in \mathrm{U}(T) + \mathrm{U}(T)$ in the statement (2) below can be more simply expressed by $|\,T/\mathfrak{m}\,| > 2$. However, we still prefer to use the former condition because of its more general nature. Similarly, in the statement (3) below, we prefer the condition "$2 \in \mathrm{U}(T)$" to the equivalent condition "char $(T/\mathfrak{m}) \neq 2$".

THEOREM 4.5. *Let $(T, \mathfrak{m})$ be a local ring. Then $R = \mathbb{M}_n(T)$ has* Property $\overline{\mathrm{K}}$ *iff we are in one of the following cases:*

(1) $n \in \{4, 6, 8, \dots\}$.
(2) $n = 2$ *and* $1_T \in \mathrm{U}(T) + \mathrm{U}(T)$.
(3) n *is odd and* $2 \in \mathrm{U}(T)$.

Proof. If (1) or (2) holds, we know from Corollary 3.5 and Theorem 3.3 respectively that R has Property K, so of course R has Property $\overline{\mathrm{K}}$. If (3) holds instead, then $<I_n, I_n>= 2\, I_n \in \mathrm{U}(R)$ shows that R has Property $\overline{\mathrm{K}}$. *Conversely, assume in the following that R has* Property $\overline{\mathrm{K}}$. If $n \in \{4, 6, 8, \dots\}$, then (1) holds, so for the rest of the proof, we need only work with odd n and $n = 2$.

CASE A. *T is a division ring.* Fix an anti-commutator $\langle\, e, e' \,\rangle \in \mathrm{U}(R)$ with (necessarily nonzero) $e, e' \in \mathrm{idem}\,(R)$. *First assume that $n = 2$.* If $e = I_2$, $\langle\, e, e' \,\rangle = 2\, e' \in \mathrm{U}(R)$ implies that $2 \in \mathrm{U}(T)$, in which case $1_T = 2^{-1} + 2^{-1} \in \mathrm{U}(T) + \mathrm{U}(T)$. If $e \neq I_2$, we may assume (after a conjugation) that $e' = E_{11}$. Writing $e = \begin{pmatrix} a & b \\ c & d \end{pmatrix}$, we have $\langle\, e, e' \,\rangle = \begin{pmatrix} 2a & b \\ c & 0 \end{pmatrix} \in \mathrm{U}(R)$, so $b, c \in \mathrm{U}(T)$. As $e^2 = e$ implies that $a^2 + b\,c = a$, we see that a, $1 - a$ are in $\mathrm{U}(T)$ too, with sum 1, so (2) holds. *Finally, suppose n is odd.* Here we may assume that $e' = \mathrm{diag}\,(I_k, 0_{n-k})$ where $1 \le k \le n$. If $2 \notin \mathrm{U}(T)$, then $2 = 0 \in T$, and so $k < n$ (for otherwise $\langle\, e, e' \,\rangle = 2\, e' = 0$, which is impossible). Writing $e = \begin{pmatrix} A & B \\ C & D \end{pmatrix}$ with $A \in \mathbb{M}_k(T)$ and $D \in \mathbb{M}_{n-k}(T)$, we have $\langle\, e, e' \,\rangle = \begin{pmatrix} 0 & B \\ C & 0 \end{pmatrix}$. As B and C are *non-square matrices*, we can take a nonzero vector v such that either $B\,v = 0$ or $C\,v = 0$. Then we'll have either $\langle\, e, e' \,\rangle \begin{pmatrix} 0 \\ v \end{pmatrix} = 0$ or $\langle\, e, e' \,\rangle \begin{pmatrix} v \\ 0 \end{pmatrix} = 0$, in contradiction to the fact that $\langle\, e, e' \,\rangle \in \mathrm{U}(R)$. This completes the proof that $2 \in \mathrm{U}(T)$ when n is odd.

CASE B. *$(T, \mathfrak{m})$ is a local ring.* Let $\overline{T}$ be the division ring $T/\mathfrak{m}$. By [**La**$_1$, p. 57], $\mathrm{rad}\,(\mathbb{M}_n(T)) = \mathbb{M}_n(\mathrm{rad}\,(T)) = \mathbb{M}_n(\mathfrak{m})$. Therefore,

$$\mathbb{M}_n(T)/\mathrm{rad}\,(\mathbb{M}_n(T)) = \mathbb{M}_n(T)/\mathbb{M}_n(\mathfrak{m}) \cong \mathbb{M}_n(\overline{T}). \tag{4.6}$$

Since $\mathbb{M}_n(T)$ is an exchange ring (by [**Wa**] or [**Ni**]), Corollary 2.16 implies that it has Property $\overline{\mathrm{K}}$ iff $\mathbb{M}_n(\overline{T})$ does. Therefore, we are free to replace T by $\overline{T}$ to assume that T is a division ring, in which case we are fully covered by Case A above. □

In general, we do not know exactly when a general matrix ring $\mathbb{M}_n(T)$ will have Property $\overline{\mathrm{K}}$ if n is odd or $n = 2$. Aside from the case where T is local (as treated above in Theorem 4.5), a second manageable case is where T is a commutative ring. Working under this assumption, it turns out that the more substantial case to consider is $n = 2$. In this case, we were pleasantly surprised to find that Property K and Property $\overline{\mathrm{K}}$ are *equivalent* on $\mathbb{M}_2(T)$! To prove this, we start by first working out some easy facts about determinants of 2×2 matrices over a commutative ring.

PROPOSITION 4.7. *For $A, B \in R = \mathbb{M}_2(T)$ where T is a commutative ring, we have*

$$\det\,(A + B) + \det\,(A - B) = 2\,[\,\det\,(A) + \det\,(B)\,] \in T. \tag{4.8}$$

Proof. Writing $A = \begin{pmatrix} p & q \\ r & s \end{pmatrix}$ and $B = \begin{pmatrix} w & x \\ y & z \end{pmatrix}$, a quick computation shows that

$$\det(A+B) = \det(A) + \det(B) + (pz + sw - qy - rx), \text{ and}$$
$$\det(A-B) = \det(A) + \det(B) - (pz + sw - qy - rx).$$

Adding these two formulas gives the desired equation (4.8). □

COROLLARY 4.9. *In the notations of* Proposition 4.7, *if* $2\,[\det(A)+\det(B)] \in \operatorname{rad}(T)$, *then* $A+B \in \mathrm{U}(R)$ *iff* $A-B \in \mathrm{U}(R)$. *In particular, this "iff" statement holds in case* $\det(A) + \det(B) = 0$; *or more specifically, if* $\det(A) = \det(B) = 0$.

Proof. The first "iff" statement holds on account of (4.8) since a matrix $C \in R$ is invertible iff $\det(C) \in \mathrm{U}(T)$, while for any $t, t' \in T$, $t + t' \in \operatorname{rad}(T)$ implies that $t \in \mathrm{U}(T)$ iff $t' \in \mathrm{U}(T)$. The rest of the Corollary is clear. □

With the above Corollary providing a crucial link between the invertibility of $A+B$ and $A-B$, we are now ready to prove the following result.

THEOREM 4.10. *Let* $R = \mathbb{M}_2(T)$ *where* T *is a commutative ring. Then* R *has* Property K *iff it has* Property $\overline{\mathrm{K}}$.

Proof. Of course, only the "if" part is at stake. Before beginning its proof, we first point out that the "if" part does not mean that for any $e, e' \in R$, $\langle e, e' \rangle \in \mathrm{U}(R)$ implies that $[e, e'] \in \mathrm{U}(R)$. Indeed, in the case where $2 \in \mathrm{U}(T)$, choosing $e = e' = I_2$ gives $\langle e, e' \rangle = 2I_2 \in \mathrm{U}(R)$, but $[e, e'] = 0 \notin \mathrm{U}(R)$ if $T \neq 0$. (Nevertheless, under the assumption that $2 \in \mathrm{U}(T)$, R does have Property K by taking $a = b = 2^{-1}$ in the proof of Theorem 3.1.) *In the following, we assume that* R *has Property* $\overline{\mathrm{K}}$. We'll show that R has Property K in two steps.

STEP 1. *The desired conclusion is true if* T *is a connected ring*; that is, if $\operatorname{idem}(T) = \{0, 1\}$. Indeed, taking an invertible $\langle e, e' \rangle$ (for suitable $e, e' \in \operatorname{idem}(R)$), $\det(e)$ is either 1 or 0. In the former case, e is invertible, so $e = I_2$. Here, $\langle e, e' \rangle = 2e' \in \mathrm{U}(R)$ implies that $2 \in \mathrm{U}(T)$, so of course R has Property K. We may thus assume that $\det(e) = 0$, in which case $\det(ee') = \det(e'e) = 0$. As $ee' + e'e \in \mathrm{U}(R)$, the last part of Corollary 4.9 implies that $ee' - e'e \in \mathrm{U}(R)$, so R has Property K.

STEP 2. *Assume that* R *does not have* Property K. By applying Zorn's Lemma to the family $\mathcal{F}$ of ideals $J_i \subseteq T$ such that $\mathbb{M}_2(T/J_i)$ does not have Property K, we see that $\mathcal{F}$ has a maximal member (with respect to inclusion), say J. The maximal choice of J implies that T/J is a connected ring, for otherwise $\mathbb{M}_2(T/J)$ would have been a direct product of a pair of 2×2 matrix rings each of which has Property K, in contradiction to the fact that $J \in \mathcal{F}$. On the other hand, $\mathbb{M}_2(T/J)$ is isomorphic to a factor ring of $\mathbb{M}_2(T)$, so it has Property $\overline{\mathrm{K}}$. As T/J is a connected ring, what we have done in Step 1 implies that $\mathbb{M}_2(T/J)$ has Property K. This is a contradiction. □

With the help of Theorem 4.10, we can now decide exactly when a matrix ring $\mathbb{M}_n(T)$ over a commutative ring T will have Property $\overline{\mathrm{K}}$. It is of interest to note that the conclusions in the following theorem happen to be *identical* to those of Theorem 4.5 over a (possibly noncommutative) local ring T, although the proofs are rather different in the key case where $n = 2$.

THEOREM 4.11. *For any commutative ring T, $R = \mathbb{M}_n(T)$ has* Property $\overline{\mathrm{K}}$ *iff we are in one of the following cases:*

(1) $n \in \{4, 6, 8, \dots\}$.
(2) $n = 2$ *and* $1_T \in \mathrm{U}(T) + \mathrm{U}(T)$.
(3) n *is odd and* $2 \in \mathrm{U}(T)$.

Proof. If (1) or (2) holds, R would have Property K, and therefore Property $\overline{\mathrm{K}}$. If (3) holds, of course R has trivially Property $\overline{\mathrm{K}}$. *Conversely, assume R has* Property $\overline{\mathrm{K}}$. If $n \in \{4, 6, 8, \dots\}$, we are in Case (1). If $n = 2$, we know from Theorem 4.10 that R has Property K, so Theorem 3.14 implies that $1_T \in \mathrm{U}(T) + \mathrm{U}(T)$. Finally, let n be odd. If $2 \notin \mathrm{U}(T)$, then 2 lies in some maximal ideal $\mathfrak{m} \subseteq T$. Since $\mathbb{M}_n(T/\mathfrak{m})$ is isomorphic to a factor ring of R, it also has Property $\overline{\mathrm{K}}$. As $T/\mathfrak{m}$ is a field of characteristic 2, $\mathbb{M}_n(T/\mathfrak{m})$ would also have Property K, which would contradict Theorem 3.14. Therefore, we must have $2 \in \mathrm{U}(T)$, so we are in Case (3). □

The result above suggests that the case of $\mathbb{M}_2(T)$ for commutative rings T is of special interest in the treatment of Property K (or equivalently, Property $\overline{\mathrm{K}}$ by Theorem 4.10). We'll now conclude this paper by giving some more characterizations for these properties by using commutators and anti-commutators whose second entries are diagonalizable matrices instead of idempotent matrices. Such a replacement is by no means automatic since in general these two classes of matrices are logically independent.

THEOREM 4.12. *For $R = \mathbb{M}_2(T)$ where T is a commutative ring, the following statements are equivalent:*

(1) *R has Property K (or equivalently, $1_T \in \mathrm{U}(T) + \mathrm{U}(T)$).*
(2) *$[\,e, \delta\,] \in \mathrm{U}(R)$ for some $e \in \mathrm{idem}\,(R)$ and some diagonalizable matrix $\delta \in R$.*
(3) *$\langle\, e, \delta\,\rangle \in \mathrm{U}(R)$ for some $e \in \mathrm{idem}\,(R)$ and some diagonalizable matrix $\delta \in R$.*

In the case where T is an exchange ring, these statements are also equivalent to:

(4) *T has no ideal of index 2.*

Proof. First assume (1) holds; say $1 = a + b$ for some $a, b \in \mathrm{U}(T)$. Taking $e = \begin{pmatrix} a & b \\ a & b \end{pmatrix} \in \mathrm{idem}\,(R)$ and $\delta = \mathrm{diag}\,(1, 0)$, we have $[\,e, \delta\,] \in \begin{pmatrix} 0 & b \\ -a & 0 \end{pmatrix} \in \mathrm{U}(R)$, and $\langle\, e, \delta\,\rangle \in \begin{pmatrix} 2\,a & b \\ a & 0 \end{pmatrix} \in \mathrm{U}(R)$, so (2) and (3) both hold. Next, (2) $\Rightarrow$ (1) follows from the calculation in the proof of Case (2) in Theorem 3.8, after a reduction to the case where δ is diagonal. Now we come to the harder implication (3) $\Rightarrow$ (1). Given e, δ as in (3), we may again assume that $\delta = \mathrm{diag}\,(s, t)$ for some $s, t \in T$. Letting $e = \begin{pmatrix} a & b \\ c & d \end{pmatrix}$, the determinant of $\langle\, e, \delta\,\rangle$ is easily computed to be

$$4\,adst - b\,c\,(s + t)^2 \in \mathrm{U}(T). \tag{4.13}$$

Let $g = \det\,(e) = ad - bc$, which is an idempotent in T. We have projection maps from T onto gT and onto $(1 - g)\,T$. Taken together, these define a natural ring isomorphism from T onto $gT \times (1 - g)\,T$. Therefore, our job is reduced to showing that 1 is a sum of two units in both gT and $(1 - g)\,T$. Under the two projections,

g projects to the identity and to zero respectively. Thus, it suffices to argue in the following two cases.

CASE 1. $g = 1$. In this case, $e = I_2$, so $\langle e, \delta \rangle = 2\delta \in \mathrm{U}(R)$ implies that $2 \in \mathrm{U}(R)$. Thus, $2 \in \mathrm{U}(T)$, and so $1 = 2^{-1} + 2^{-1} \in \mathrm{U}(T) + \mathrm{U}(T)$.

CASE 2. $g = 0$. In this case, $ad = bc$, so (4.13) shows that $bc \in \mathrm{U}(T)$. But $e \in \mathrm{idem}(R)$ implies that $a^2 + bc = a$, so a, $1 - a$ are units in T, with sum 1.

Finally, (1) $\Rightarrow$ (4) is always true, since the existence of an ideal $I \subseteq T$ of index two would imply that R has a factor ring isomorphic to $\mathbb{M}_2(\mathbb{F}_2)$, which does not have Property K. Conversely, assume that T is an exchange ring satisfying (4). To prove (1), we assume instead that $1_T \notin \mathrm{U}(T) + \mathrm{U}(T)$. Consider the nonempty family $\mathcal{G}$ of ideals $I_i \subseteq T$ for which 1 is not a sum of two units in T/I_i. Applying Zorn's Lemma to the family $\mathcal{G}$, we see that $\mathcal{G}$ has a maximal member I (with respect to inclusion). The maximal choice of I implies (as in the proof of (4.10)) that the commutative factor ring T/I is connected. Since T/I remains to be an exchange ring (and $T/I \neq 0$), it must be a local ring (by [**Wa**, Proposition 1]). As 1 is not a sum of two units in T/I, the local ring T/I must have residue field $\mathbb{F}_2$. This implies that T has a factor ring isomorphic to $\mathbb{F}_2$, which contradicts (4). □

5. Commutator Characterizations for 2×2 Matrix Rings

According to Theorem 3.3, rings with Property K are precisely 2×2 matrix rings over base rings with a specific unit property. One may naturally ask: how about characterizations of the *most general* 2×2 matrix rings? In the literature, there is a simple criterion for these ((A) $\Leftrightarrow$ (H) in Theorem 5.1), that is due to Fuchs, Maxson and Pilz [**FMP**, III.2]. To make our exposition more useful and yet completely self-contained, we'll reprove the Fuchs-Maxson-Pilz theorem in a much expanded form below, following the idea of using anti-commutators (as well as commutators) in [**La**$_4$, p. 349] but not assuming any general matrix ring recognition theorems in [**La**$_3$, §17]. In the following result, the statements (B) through (G) are expressed in terms of anti-commutators, while (I), (J) and (K) are expressed in terms of commutators.

THEOREM 5.1. *For any ring R, the following statements are equivalent:*
(*A*) $R \cong \mathbb{M}_2(S)$ *for some ring* S.
(*B*) *There exist* $p, q \in R$ *with* $p^2 = q^2 = 0$ *such that* $\langle p, q \rangle = 1$.
(*C*) *There exist* $p, q \in R$ *with* $p^2 = q^2 = 0$ *such that* $\langle p, q \rangle \in \mathrm{U}(R)$.
(*D*) *There exist* $v \in \mathrm{U}(R)$ *and* $p \in R$ *with* $p^2 = 0$ *such that* $\langle p, v \rangle = 1$.
(*E*) *There exist* $v \in \mathrm{U}(R)$ *and* $p \in R$ *with* $p^2 = 0$ *such that* $\langle p, v \rangle \in \mathrm{U}(R)$.
(*F*) *There exist* $r, p \in R$ *with* $p^2 = 0$ *such that* $\langle p, r \rangle = 1$.
(*G*) *There exist* $r, p \in R$ *with* $p^2 = 0$ *such that* $\langle p, r \rangle \in \mathrm{U}(R)$.
(*H*) *There exist* $p, q \in R$ *with* $p^2 = q^2 = 0$ *and* $p + q \in \mathrm{U}(R)$.
(*I*) *There exist* $p, q \in R$ *with* $p^2 = q^2 = 0$ *such that* $[p, q] \in \mathrm{U}(R)$.
(*J*) *There exist* $v \in \mathrm{U}(R)$ *and* $p \in R$ *with* $p^2 = 0$ *such that* $[p, v] \in \mathrm{U}(R)$.
(*K*) *There exist* $r, p \in R$ *with* $p^2 = 0$ *such that* $[p, r] \in \mathrm{U}(R)$.

Proof. (A) $\Rightarrow$ (B). If (say) $R = \mathbb{M}_2(S)$, then (B) holds for $p = E_{12}$ and $q = E_{21}$.

(B) $\Rightarrow$ (C) is a tautology.

(C) $\Rightarrow$ (H). For p, q as in (C), we have $(p + q)^2 = \langle p, q \rangle \in \mathrm{U}(R)$, so $p + q \in \mathrm{U}(R)$.

(H) $\Rightarrow$ (D). For $p, q \in R$ as in (H), let v be the inverse of $u := p + q \in \mathrm{U}(R)$. Left multiplying by q and right multiplying by p give $qu = qp$ and $up = qp$ respectively. Equating these gives $vq = pv$, so $\langle p, v \rangle = pv + vp = vq + vp = vu = 1$.

(D) $\Rightarrow$ (G) is a tautology.

(G) $\Rightarrow$ (F). For r, p as in (G), let $w = pr + rp \in \mathrm{U}(R)$. Left multiplying by p gives $pw = prp$, and right multiplying by p gives $wp = prp$. Thus, $pw = wp$, so left multiplying by w^{-1} yields $1 = w^{-1}pr + w^{-1}rp = p\,(w^{-1}r) + (w^{-1}r)\,p$, which verifies (F).

(F) $\Rightarrow$ (A). For r, p as in (F), left multiplying $pr + rp = 1$ by $e := rp$ shows that $e \in \mathrm{idem}\,(R)$, with complementary idempotent $f := pr$. By [**La**$_1$, 21.20], e, f are *isomorphic* idempotents. Thus, we have $R_R = eR \oplus fR \cong eR \oplus eR$ as right R-modules. Taking endomorphism rings gives $R \cong \mathrm{End}_R(eR) \cong \mathbb{M}_2(eRe)$.

(D) $\Leftrightarrow$ (E). We need only show that (E) $\Rightarrow$ (D), so let $w := pv + vp \in \mathrm{U}(R)$ where $p^2 = 0$ and $v \in \mathrm{U}(R)$. Applying the same proof for (G) $\Rightarrow$ (F) here gives $1 = p\,(w^{-1}v) + (w^{-1}v)\,p$. Since $w^{-1}v \in \mathrm{U}(R)$, this proves (D).

(A) $\Rightarrow$ {(I), (J) and (K)}. If (say) $R = \mathbb{M}_2(S)$, then (I) holds for $p = E_{12}$ and $q = E_{21}$, and (J), (K) both hold for $p = E_{12}$ and $v = r = E_{12} + E_{21}$.

(I) $\Rightarrow$ (C). Given p, q in (I), let $w := [\,p, q\,] \in \mathrm{U}(R)$. Repeating the first steps in the proof of (G) $\Rightarrow$ (F), we can show here that w *anti-commutes* with both p and q. Thus,

$$(5.2) \qquad 1 = w^{-1}pq - w^{-1}qp = -p\,(w^{-1}q) - (w^{-1}q)\,p = -\langle\, p, w^{-1}q \,\rangle\,.$$

Since $(w^{-1}q)^2 = w^{-1}qw^{-1}q = -w^{-2}q^2 = 0$, this verifies (C).

(J) $\Rightarrow$ (K) $\Rightarrow$ (G). It suffices to prove the latter implication. Given $w := pr - rp \in \mathrm{U}(R)$ as in (K), we have again $wp = -pw$. Thus, by repeating the calculations in (5.2) (with q replacing r), we get $1 = -\langle\, p,\, w^{-1}r \,\rangle$. Since $p^2 = 0$, this verifies (G). $\square$

REMARK 5.3. It would be tempting to think that in the statements (B), (D) and (F), we can replace the anti-commutators by commutators and get three more equivalent statements in terms of commutators. However, this is not the case! In fact, given (A), we may not be able to get an equation $[\,p, q\,] = 1$ with $p^2 = q^2 = 0$. Indeed, if $R = \mathbb{M}_2(S)$ for a commutative ring S, *any* commutator in R will have zero trace, while the identity matrix I_2 has trace 2. Similarly, we must refrain from trying to change the anti-commutator into a commutator in the statements (D) and (F). As for statement (H) (the Fuchs-Maxson-Pilz characterization for (A)), it is easy to see that we also cannot change the requirement $p + q \in \mathrm{U}(R)$ into $p + q = 1 \in R$ (or $p + q$ being a *central* unit), since this would have implied that $p = q = 0$, and hence $R = 0$!

REMARK 5.4. As we have pointed out in the paragraph preceding Theorem 5.1, the proof of that theorem was designed to be entirely self-contained so that it can be read independently of the standard recognition theorems for matrix rings as developed, for instance, in [**La**$_3$]. In the simplest case, according to [**La**$_3$, 17.10], a standard characterization for a ring R to be a 2×2 matrix ring is that there exist $p, r, r' \in R$ with $p^2 = 0$ such that $pr + r'p = 1$. This condition can thus be added to all the other equivalent conditions listed in Theorem 5.1. We note that it is ostensibly "weaker" (as a sufficient condition for (A)) than the three conditions (B), (D), and (F).

At this point, it would be helpful to recall a key definition introduced in our earlier work [**KL**] on invertible commutators. In that paper, we defined a ring element $a \in R$ to be *completable* if there exists $r \in R$ such that $[a, r] \in \mathrm{U}(R)$. Using this terminology, the equivalence of (A) and (K) in Theorem 5.1 says precisely that R is a 2×2 matrix ring (over some other ring) iff R has a square-zero completable element. Inspired by this fact (as well as by our earlier Theorem 3.3), we stumbled upon the idea of looking for *completable idempotents* in R. It was a pleasant surprise to us that the existence of a completable idempotent in a ring R turns out to be *also* equivalent to R being a 2×2 matrix ring. This fact and some of its variations are collected in the new characterization theorem below for 2×2 matrix rings. Recall from Corollary 3.4 that, in the special case where $2 \in \mathrm{U}(R)$, we did know that R is a 2×2 matrix ring iff R has Property K (that is, $[e, e'] \in \mathrm{U}(R)$ for some $e, e' \in \mathrm{idem}\,(R)$). However, the following new equivalence result, like Theorem 5.1, is valid for all rings.

THEOREM 5.5. *For any ring R, the following statements are equivalent:*
(1) $R \cong \mathbb{M}_2(S)$ *for some ring* S.
(2) *There exist* $e \in \mathrm{idem}\,(R)$ *and* $r \in R$ *with* $r^2 = 1$ *such that* $[e, r] \in \mathrm{U}(R)$.
(3) *There exist* $e \in \mathrm{idem}\,(R)$ *and* $r \in R$ *with* $r^2 = -1$ *such that* $[e, r] \in \mathrm{U}(R)$.
(4) *There exist* $e \in \mathrm{idem}\,(R)$ *and* $r \in R$ *with* $r^3 = 1$ *such that* $[e, r] \in \mathrm{U}(R)$.
(5) *There exist* $e \in \mathrm{idem}\,(R)$ *and* $r \in \mathrm{U}(R)$ *such that* $[e, r] \in \mathrm{U}(R)$.
(6) R *has a completable idempotent.*
(7) *There exist* $e \in \mathrm{idem}\,(R)$ *and* $r \in R$ *with* $r^2 = 0$ *such that* $[e, r] \in \mathrm{U}(R)$.
(8) *There exist* $e \in \mathrm{idem}\,(R)$ *and a nilpotent* $r \in R$ *such that* $[e, r] \in \mathrm{U}(R)$.

Proof. We first prove that (1) implies (2), (3), (4) and (7). Indeed, if $R = \mathbb{M}_2(S)$ for some ring S, we can take $e = \begin{pmatrix} 1 & 0 \\ 0 & 0 \end{pmatrix}$, $r_2 = \begin{pmatrix} 0 & 1 \\ 1 & 0 \end{pmatrix}$, $r_3 = \begin{pmatrix} 0 & 1 \\ -1 & 0 \end{pmatrix}$, $r_4 = \begin{pmatrix} -1 & 1 \\ -1 & 0 \end{pmatrix}$, and $r_7 = \begin{pmatrix} 1 & 1 \\ -1 & -1 \end{pmatrix}$. Then $e^2 = e$, $r_2^2 = r_4^3 = 1$, $r_3^2 = -1$, and $r_7^2 = 0$, and we check easily that $[e, r_i] \in \mathrm{U}(R)$ for all i.

(7) $\Rightarrow$ (8) $\Rightarrow$ (6) are tautologies. [We note on the side that (7) also trivially implies (K) (and hence (A)) in Theorem 5.1, although this information is not needed here.]

(6) $\Rightarrow$ (1). Let $e = e^2 \in R$ and $r \in R$ be such that the following commutator is a unit:

$$[e, r] = er - re = er\,(1 - e) - (1 - e)\,re. \tag{5.6}$$

Letting $p = er\,(1 - e)$ and $q = -(1 - e)\,re$, we have clearly $p^2 = q^2 = 0$, and $p + q \in \mathrm{U}(R)$ according to (5.6). Thus, (H) $\Rightarrow$ (A) in Theorem 5.1 gives (1).

The proof is now complete, since $\{$(2) or (3) or (4)$\}$ $\Rightarrow$ (5) $\Rightarrow$ (6) are tautologies. $\square$

It is of interest to point out that, in the various criteria above, r can be chosen to be a unit (respectively, a square root of ± 1 or a cubic root of 1), or a nilpotent element (respectively, a square-zero element), *but not necessarily an idempotent.* We note also that, in the criterion (5) above, the commutator $[e, r]$ cannot be replaced by the anti-commutator $\langle e, r\rangle$. Indeed, if we take R to be any nonzero commutative ring with $2 \in \mathrm{U}(R)$, and choose $e = r = 1$, then $\langle e, r\rangle \in \mathrm{U}(R)$. Here, R cannot possibly be a 2×2 matrix ring. Nevertheless, we do have the following result for anti-commutators.

COROLLARY 5.7. *If R is a ring with an idempotent e and an element r such that $\langle e, r\rangle \in \mathrm{U}(R)$, then $R/2R \cong \mathbb{M}_2(S)$ for some ring S.*

Proof. In $\overline{R} = R/2R$, we have $\overline{er - re} = \overline{er + re} \in \mathrm{U}(\overline{R})$, so we can apply (5) $\Rightarrow$ (1) in Theorem 5.5 to the factor ring $\overline{R}$. □

Emboldened by the fact that 2×2 matrix rings are characterized by the existence of completable idempotents as well as the existence of completable square-zero elements, we were led to the consideration of *completable involutions* as well. (By an involution, we simply mean an element $u \in R$ with $u^2 = 1$.) In ring theory, it is rather rare that idempotents, involutions and square-zero elements would play parallel roles in the treatment of a certain problem or property. But for the problem of characterizing 2×2 matrix rings, this does turn out to be the case, as the following result shows.

THEOREM 5.8. *A ring R is a 2×2 matrix ring iff it has a completable involution.*

Proof. If $R = \mathbb{M}_2(S)$ for some ring S, we have already pointed out earlier that the involution $E_{12} + E_{21} \in R$ is completable. Conversely, assume that there exist $u, r \in R$ with $u^2 = 1$ such that $v := [u, r] \in \mathrm{U}(R)$. Repeating the idea (in the proof of (5.1)) of left and right multiplying by u, and keeping in mind the equation $u^2 = 1$, we see here that $uv = -vu$. Thus,

$$(5.9) \qquad 1 = v^{-1}ur - v^{-1}ru = (1 - u)\, v^{-1}r - v^{-1}r\,(1 + u);$$

that is, $1 = e + f$, where $e := (1 - u)\, v^{-1}r$ and $f := -v^{-1}r\,(1 + u)$. Since $fe = 0$, we have $f \in \mathrm{idem}\,(R)$, with complementary idempotent e. *It suffices to show that e and f are isomorphic idempotents,* as that will show (as in (F) $\Rightarrow$ (A) in the proof of Theorem 5.1) that $R \cong \mathbb{M}_2(eRe)$. Right multiplying (5.9) by $1-u$ and left multiplying it by $1+u$ give the following two "von Neumann regularity" equations:

$$(1 - u)\, v^{-1}r\,(1 - u) = 1 - u, \quad \text{and} \quad (1 + u)\,(-v^{-1}r)\,(1 + u) = 1 + u.$$

From the former, we have $eR = (1-u)\, R$. From the latter, we have $Rf = R\,(1+u)$, which is well known to imply that $fR \cong (1 + u)\, R$ as right R-modules. (For a full proof of this implication, see [**La**$_2$, Exercise 1.17].) Noting that

$$(1 - u)\, R = (1 - u)\, vR = v\,(1 + u)\, R \cong (1 + u)\, R,$$

we conclude that $eR \cong fR$, as desired. □

REMARK 5.10. In view Theorem 5.8 (and parts of Theorem 5.5), it might be tempting to surmise that the existence of a completable element $u \in R$ with $u^n = 1$ for some $n \geq 3$ might also imply that R is a 2×2 matrix ring (over some other ring). However, this is not the case. For instance, let R be Hamilton's quaternion division algebra generated over $\mathbb{R}$ by i, j, with the relations $i^2 = j^2 = -1$ and $ij = -ji$. Let $u = a + b\,i$ $(a, b \in \mathbb{R})$ be a primitive n-th of unity in $\mathbb{R}\,[\,i\,] \cong \mathbb{C}$. If $n \geq 3$, then $b \neq 0$, and $[u, j] = [a + b\,i, j] = b\,[i, j] = 2\,b\,ij \in \mathrm{U}(R)$. This shows that $u \in R$ is completable. However, being a division algebra, R is not a 2×2 matrix ring over any ring. In a similar vein, if a ring R has a completable element w such that $w^n \in \{0, w\}$ for some $n \geq 3$, R need not be a 2×2 matrix ring either. For instance, let $R = \mathbb{M}_3(S)$ over a ring S. For the two matrices

$$w = \begin{pmatrix} 1 & 1 & 0 \\ 0 & 0 & 1 \\ 0 & 0 & -1 \end{pmatrix} \text{ and } w_1 = \begin{pmatrix} 1 & 0 & 0 \\ 0 & 0 & 0 \\ 1 & 1 & 0 \end{pmatrix}, \text{ we have } [w, w_1] = \begin{pmatrix} 0 & -1 & 0 \\ 1 & 1 & 0 \\ -2 & -2 & -1 \end{pmatrix} \in \mathrm{U}(R).$$

Thus, w is completable in R, and we can check easily that $w^3 = w$. However, if we choose S to be a division ring (or $S = \mathbb{Z}$), R is *not* a 2×2 matrix ring over any ring. The case where $w^n = 0$ for some $n \geq 3$ can be handled similarly by taking instead $w = \begin{pmatrix} 0 & 1 & 0 \\ 0 & 0 & 1 \\ 0 & 0 & 0 \end{pmatrix}$ and $w_2 = \begin{pmatrix} 0 & 0 & 0 \\ 1 & 0 & 1 \\ 1 & 0 & 0 \end{pmatrix}$, with $w^3 = 0$ and $[w, w_2] = \begin{pmatrix} 1 & 0 & 1 \\ 1 & -1 & 0 \\ 0 & -1 & 0 \end{pmatrix} \in \mathrm{U}(R)$.

Acknowledgments. We are grateful to Professor S. K. Berberian who kindly communicated to us his detailed personal notes on Kaplansky's "Exercise 6" from [**Ka**, p. 25]. These notes helped us come to a better understanding of Kaplansky's proposed solution of his exercise, which eventually led to the present work. We thank Professor G. M. Bergman for his valuable contribution [**Be**$_2$] toward the validation of Example 3.6, and we also thank the referee of this paper for his/her many thoughtful comments.

References

[Be$_1$] G. M. Bergman, *Modules over coproducts of rings*, Trans. Amer. Math. Soc. **200** (1974), 1–32. MR0357502

[Be$_2$] G. M. Bergman: Email communication, July 3, 2017.

[BD] R. Bott and R. J. Duffin, *On the algebra of networks*, Trans. Amer. Math. Soc. **74** (1953), 99–109. MR0056573

[Co$_1$] P. M. Cohn, *Some remarks on the invariant basis property*, Topology **5** (1966), 215–228. MR0197511

[Co$_2$] P. M. Cohn, *Some remarks on projective-free rings*, Algebra Universalis **49** (2003), no. 2, 159–164. Dedicated to the memory of Gian-Carlo Rota. MR2015350

[Co$_3$] P. M. Cohn, *Another criterion for a ring to be projective-free*, Bull. London Math. Soc. **37** (2005), no. 6, 857–859. MR2186718

[FMP] P. Fuchs, C. J. Maxson, and G. Pilz, *On rings for which homogeneous maps are linear*, Proc. Amer. Math. Soc. **112** (1991), no. 1, 1–7. MR1042265

[Go] K. R. Goodearl, *von Neumann regular rings*, 2nd ed., Robert E. Krieger Publishing Co., Inc., Malabar, FL, 1991. MR1150975

[GH] K. R. Goodearl and D. Handelman, *Simple self-injective rings*, Comm. Algebra **3** (1975), no. 9, 797–834. MR0379593

[GW] H. K. Grover, Z. Wang, D. Khurana, J. Chen, and T. Y. Lam, *Sums of units in rings*, J. Algebra Appl. **13** (2014), no. 1, 1350072, 10 pp. MR3096853

[He] M. Henriksen, *Two classes of rings generated by their units*, J. Algebra **31** (1974), 182–193. MR0349745

[Ka] I. Kaplansky, *Rings of operators*, W. A. Benjamin, Inc., New York-Amsterdam, 1968. MR0244778

[Kt] T. Kato, *Perturbation theory for linear operators*, Classics in Mathematics, Springer-Verlag, Berlin, 1995. Reprint of the 1980 edition. MR1335452

[KL] D. Khurana and T. Y. Lam, *Invertible commutators in matrix rings*, J. Algebra Appl. **10** (2011), no. 1, 51–71. MR2784754

[KLS] D. Khurana, T. Y. Lam, and N. Shomron, *A quantum-trace determinantal formula for matrix commutators, and applications*, Linear Algebra Appl. **436** (2012), no. 7, 2380–2397. MR2889999

[KR$_1$] J. J. Koliha and V. Rakočević, *Invertibility of the sum of idempotents*, Linear Multilinear Algebra **50** (2002), no. 4, 285–292. MR1936951

[KR$_2$] J. J. Koliha and V. Rakočević, *Invertibility of the difference of idempotents*, Linear Multilinear Algebra **51** (2003), no. 1, 97–110. MR1950415

[La$_1$] T. Y. Lam, *A first course in noncommutative rings*, 2nd ed., Graduate Texts in Mathematics, vol. 131, Springer-Verlag, New York, 2001. MR1838439

[La$_2$] T. Y. Lam, *Exercises in classical ring theory*, 2nd ed., Problem Books in Mathematics, Springer-Verlag, New York, 2003. MR2003255

[La$_3$] T. Y. Lam, *Lectures on modules and rings*, Graduate Texts in Mathematics, vol. 189, Springer-Verlag, New York, 1999. MR1653294

[La$_4$] T. Y. Lam, *Exercises in modules and rings*, Problem Books in Mathematics, Springer, New York, 2007. MR2278849

[Le] W. G. Leavitt, *Modules without invariant basis number*, Proc. Amer. Math. Soc. **8** (1957), 322–328. MR0083986

[Ni] W. K. Nicholson, *Lifting idempotents and exchange rings*, Trans. Amer. Math. Soc. **229** (1977), 269–278. MR0439876

[Sa] E. Marques de Sá, *The rank of the difference of similar matrices*, Portugal. Math. **46** (1989), no. 2, 177–187. MR1020965

[SG] G. Song and X. Guo, *Diagonability of idempotent matrices over noncommutative rings*, Linear Algebra Appl. **297** (1999), no. 1-3, 1–7. MR1723833

[Wa] R. B. Warfield Jr., *Exchange rings and decompositions of modules*, Math. Ann. **199** (1972), 31–36. MR0332893

Department of Mathematics, Panjab University, Chandigarh 160 014, India
Email address: `dkhurana@pu.ac.in`

Department of Mathematics, University of California, Berkeley, CA 94720, USA
Email address: `lam@math.berkeley.edu`

Contemporary Mathematics
Volume **715**, 2018
https://doi.org/10.1090/conm/715/14414

Distributive hierarchies of binary operations

Sergio López-Permouth and Louis H. Rowen

Abstract. This paper is about sets equipped with two operations, one of which distributes over the other (a type of structure that we call a "distributive magma" to emphasize that often no other properties are assumed for the two operations involved). Upon closer scrutiny, one should notice that the paper actually centers on a set with a single associative binary operation (usually a group) with the intention to determine whether a suitable companion operation may be found so that the resulting structure is a distributive magma. The motivation here comes, for example, from the opposite roles that semigroups and abelian groups play in the theories of rings and semirings (such as the max-plus algebra). As there are two underlying operations in a distributive magma, the distributed operation and the distributor, the problem we pose here necessarily comes in two flavors depending on the role one expects from the operation with which one begins. Additional layers arise as one endows the beginning operation with specific algebraic structures or demands additional structure from the companion sought. The type of question explored includes whether there are group-theoretic differences between the two abelian group structures associated with a field and, if so, what are the inherent properties of addition and multiplication that entitles one of them to be a distributor and the other one to be the distributed one?

1. Introduction, motivation and philosophy

Algebraic structures with two binary operations on a single set, one of them distributive over the other, abound in Modern Algebra. Rings and fields are the foremost examples of such a structure. The importance of semirings (cf. [**Go**] for a standard reference), such as the max-plus algebra[1], keeps growing; in particular, semirings serve as an algebraic underpinning for tropical geometry and control theory, and have many applications. As algebraic concepts often are enriched with additional structure (topological semigroups, topological vector spaces, ordered groups, algebraic groups, etc), structures with two operations may be understood by studying one operation via its interaction with another. In this case,

2010 *Mathematics Subject Classification.* Primary 03C07, 08A02, 08A05, 08A99; Secondary 08A62.

Key words and phrases. Max-plus algebra, distributivity, semiring.

The second author would like to thank the Ohio University Center of Ring Theory and its applications in Athens, Ohio, for their hospitality.

We thank D. Saltman for a helpful conversation.

[1]The max-plus algebra [**ABG**] is defined by an ordered abelian group G, where multiplication is the group operation of G and addition is defined by taking the maximum, i.e., $a + b$ is defined as $\max\{a, b\}$ in G.

each operation is the source of enrichment for the other one and the (left, right or two-sided) **distributive properties** provide the connection between the two operations. An operation $\circ$ distributes over $*$ on the left if :

$$a \circ (b * c) = (a \circ b) * (a \circ c), \ \forall a, b, c \in A, \tag{1}$$

or on the right, if

$$(b * c) \circ a = (b \circ a) * (c \circ a), \ \forall a, b, c \in A. \tag{2}$$

Distributivity is a relation between operations. If the operation $\circ$ distributes over $*$, we say that $\circ$ is **distributed** and $*$ is the **distributor**. Alternatively, we say that $\circ$ is at the **bottom** (and $*$ is at the **top**) of the distributive relation. The purpose of this paper is to seek implications from the existence of such relations, as well as conditions on operations that allow them to be, respectively, distributed or distributors.

It is commonplace in the literature to refer to a pair $(A, \circ)$ where $\circ$ is a (binary) operation on A as a **magma**. We also refer to magmas having two operations, one distributing over the other, the intention being to have an expression to refer to operations without explicit assumptions of any other properties (such as associativity, commutativity, identity or inverses).

This kind of question was addressed for skew fields and rings by Cohn [**Co**] and Dicker [**D**]. Distributivity in magmas was also considered by Przytycki [**P**]. Our approach can be put into a general framework in which, in some sense, semifields (an abelian group which distributes over a monoid) are opposite to rings (a monoid which distributes over an abelian group). The connection is formalized by using the terminology from Definition 11. Addition of reals plays the distributed role in the max-plus algebra, but the max operation does not carry a group structure; what other operations can act as distributors for $+$, and how different can they be from (or how similar must they be to) "max"?

Our program: Given a set A and two classes of magmas X and Y, and an operation $\circ$, $(A, \circ)$ belonging to Y, characterize those binary operations $*$ on A such that $(A, *, \circ)$ is an $(X;Y)$-magma, cf. Definition 8 (or, alternatively, $(A, \circ, *)$ is a (Y, X) magma). Does any such $*$ even exist? For instance, for $X =$ Sgrp and $Y =$ AbGrp, $A = \mathbb{R}$ (or more generally any ordered group), and $\circ = +$, it is known that **max** and **min** are solutions to the problem. The so-called **max-plus algebra** is the basic setting for tropical geometry. This paper concerns the case of an Abelian group $(A, +)$ and seeing how close it is to the max-plus algebra, given various natural conditions on the operators.

In Section 2, we introduce the necessary terminology as we present the notion of a distributive magma as a mechanism to talk about having two binary operations on a single set, one of which distributes over the other, without making any other assumptions on the two operations. Distributive actions and hierarchy graphs, two alternative perspectives which motivate our study and point to further directions for research, are presented in the closing appendix.

The following question has a long history, cf. [**Fu**, Chapter XVII, especially the notes on p. 311].

QUESTION 1. *Which groups can be the additive group of a ring, an integral domain, or a field?*

REMARK. It is well-known in the theory of finite fields that a finite group G is the multiplicative group of a finite field if and only if it is cyclic of order $p^n - 1$, for some prime p. Likewise, the only finite abelian groups which are the additive group of a field are those isomorphic to $(\mathbb{Z}_p)^m$; this observation settles Question 1 in the case of a finite integral domain, and one turns to the infinite case.

A general answer has been given in [**Co**] and [**D**, Theorem 4], via a function on $(G, \cdot)$ which is required to have certain properties (and take on the role of $g \mapsto 1 - g$, [**D**, §3]). Szele and Fuchs [**Fu**, Theorem 122.4] characterize additive groups of Artinian rings.

In this paper we only highlight those aspects directly pertaining to magmas. Explicit questions include whether it is possible to have an operation over the reals that gives them the structure of a field with the usual addition playing instead the role of multiplication. (The answer is positive: see Example 42.)

Similarly, is it possible to have an associative operation over the reals that distributes over the usual multiplication? Can the resulting structure be a division ring? The answer to the last part is negative: see Corollary 35.

Our main objective is to turn Question 1 around and ask,

QUESTION 2. *How can we construct a semifield from a multiplicative abelian group, with various specified properties?*

This question is handled for fields in [**Fu**, Chapter XVIII]. In Section 3 we study the kinds of distributive magmas that may be constructed with an Abelian group at the bottom. We find that, in this case, the top operation is always characterized by an arbitrary **defining function** (Subsection 3.1) and we study how that defining function characterizes various properties of the top operation. The goal of Subsection 3.2 is to analyze how one obtains various monoid properties for the top operation in terms of the defining function. Section 4 considers when one can obtain field or division ring structures beginning with a specific group, with the goal of it being either the additive part or the multiplicative part (Subsection 4.2).

Thinking about the typical algebraic structures of tropical geometry, where the addition on a field becomes the multiplication of a semifield whose addition is a very modest monoid (associative and with an identity but without additive inverses), led us to ponder the possibility of placing richer structure in the additive role. Aside from the issue of modeling the desired geometric properties, is it feasible, from a purely algebraic point of view, to put other properties on addition? Following this thread has yielded the results we present here and, in fact, raised even more questions. The questions themselves seem valuable and we believe are an important contribution.

After writing the paper, we realize that it is hard to figure out the extent to which some of our results constitute new contributions. Some of the questions considered here even have been discussed in mathematical chat groups online. Statements and proofs in online chats wind up blending together, making it hard to determine precisely what is known, and who deserves the credit. We feel that in addition to providing new ideas and results, this paper will serve, like a survey, as a record of some of ideas that may already be floating around and as a pointer for new directions for future research.

1.1. Summary of the main results. In Section 3 we consider intrinsic properties needed for a distributive magma to be isomorphic to a max-plus algebra: Commutativity (Proposition 15), idempotence (Proposition 16), alternativity (Theorem 22), associativity (Lemma 19), "upper bound" (Proposition 24), and bipotence (Proposition 29). These are synchronized in Proposition 30, to provide a criterion, in terms of "positive" and "negative" elements, for the distributor to provide a total order.

Other results: Proposition 34 points out that if a group G has exactly one element of order 2, then G cannot be the additive group of an integral domain. It follows that the multiplicative group of a field of characteristic $\neq 2$ cannot also be the additive group of an integral domain. In the other direction, to get a field from a multiplicative group, it is customary to adjoin the element 0. This raises the question of whether a group can already be the multiplicative group of a field, without adjoining an extra element. We see in Proposition 32 that this is impossible.

Finally the degenerate case where $* = \circ$, (the characterization of vertices in the hierarchy graph that host loops, described in the Appendix), is shown to reduce to the theory of monoids; in particular, by Proposition 47, a commutative monoid $(A, \circ)$ is a distributive magma $(A, \circ, \circ)$ iff A is idempotent.

As an indication of the spirit of our investigation, consider the following result, which is part of the folklore. It comes up when one asks whether addition must always be a commutative operation in ring theory; as the result shows, for rings with a multiplicative identity, there is no other option. The result illustrates the power of distributivity; nothing is required about $\circ$ other than distributivity over $*$; in particular associativity is not assumed. Furthermore, the only places in the proof where distributivity is used are for e over arbitrary $*$ products of elements of A, and for elements of A over the single element $e * e$.

PROPOSITION 3. *If a set A is equipped with two operations $*$ and $\circ$ such that $(A, \circ)$ has an identity element e, $(A, *)$ is a cancellative semigroup, and $\circ$ distributes over $*$, then $*$ is commutative.*

PROOF. Let $a, b \in A$. Then $(a * b) * (a * b) = ((a * b) \circ e) * ((a * b) \circ e) = (a * b) \circ (e * e) = [(a \circ (e * e)] * [(b \circ (e * e)] = a * a * b * b$, and the result follows from cancellation in $*$. $\square$

Structures such as hyperfields may also be viewed in a similar way as those studied here, except with the caveat that the extra operation in this case is multi-valued (see, e.g. [**Ba**]).

2. Distributive magmas

We present the notion of distributive magma as a mechanism to frame our deliberations. In the Appendix, we present the notions of distributive actions and hierarchy graphs as alternative perspectives for the problems discussed here.

We lay out some terminology here for structures with two operations.

DEFINITION 4.

(1) A triple $(A, *, \circ)$ where $(A, \circ)$ and $(A, *)$ are magmas is also called a magma. We will often be interested in the case when $\circ$ has an identity element and then we will usually denote it e.

(2) A magma $(A, *, \circ)$ is called a **left (resp. right) distributive magma** if $\circ$ left (resp. right) distributes over $*$. A magma is simply called a **distributive magma** if it is both a left and right distributive magma. If emphasis is needed, we say that the magma is **two-sided distributive**.
(3) If $(A, *, \circ)$ is a (left, right, two-sided) distributive magma, we say that the pair $(*, \circ)$ is a (left, right) distributive pair, and that $*$ is at the **top** and $\circ$ is at the **bottom** of $(A, *, \circ)$ (or $(*, \circ)$).

REMARK 5. Clearly, if the operation at the bottom of a one-sided distributive magma is commutative then the magma is indeed two-sided distributive.

In particular, one gets the following corollary to Proposition 3.

COROLLARY 6. If an abelian group A is at the bottom of a left distributive magma $(A, *, \circ)$, and $(A, *)$ is a cancellative semigroup, then $(A, *)$ is commutative.

DEFINITION 7. Let $(A, \circ)$ be a magma.
(1) An element e of A is **absorbing** if $e \circ a = a \circ e = e$ for all $a \in A$.
(2) Let 0 be an element not belonging to A. We write $A^{\#}$ for $A \cup \{0\}$, and we extend $\circ$ to $A^{\#}$ in such a way that 0 is absorbing in $A^{\#}$. We refer to $A^{\#}$ as A **buffered** (by 0). When A is a group, we will say that $A^{\#}$ is a buffered group (even though $A^{\#}$ is no longer a group). Similar expressions may be used for the buffered versions of other structures with a given operation.

The definition above is motivated by the fact that the additive identity in a field absorbs its multiplication. In that case, the non-zero elements are a group and the underlying set for the field is the buffered group, since the non-zero elements must be extended by 0 as above. We will see in Theorem 32 that this situation is inevitable when a group operation distributes over another operation which has its own identity. While studying ring theory, the separation between non-zero elements and the additive identity of the ring happens precisely when one is dealing with a domain (a ring where the product of non-zero elements is non-zero). Otherwise, zero-divisors force non-trivial mingling between non-zero elements and zero.

Mostly, our distributive pair will be denoted as $(*, \circ)$. Often the symbol $\circ$ will be dropped in computations, and we will write, say, ab for $a \circ b$. When dealing with ring-like structures, the distributive pair will be $(+, \cdot)$ and $\cdot$ will likewise be dropped. One exception is that, when dealing with semirings, inspired by the max-plus algebra we will favor the notation $(*, +)$. In those cases, $+$ will never be tacit.

DEFINITION 8. Let X and Y denote classes of algebraic structures with a single operation. An (X, Y)**-magma** is a distributive magma $(A, *, \circ)$, for which $(A, *)$ is in X and $(A, \circ)$ is in Y. We denote the class of (X, Y)-magmas simply by (X, Y).

The following are initial examples of possible *values* for X and Y in the above definition:

- Grp, the class of groups.
- AbGrp, the class of Abelian groups.
- Sgrp, the class of semigroups, cf. [**LN**].
- AbSgp, the class of Abelian semigroups.
- Mon, the class of monoids, and AbMon, the class of Abelian monoids.

QUESTION 9. *For an arbitrary set S, does there exist an operation $*$ that is a universal distributor; that is such that for any $\circ$, $(S, *, \circ)$ is a distributive magma? One famous example from semigroup theory is the right zero band, cf.* [**Cl**], *given by $a \circ b = b$ for all a, b. The right zero band is a universal left distributor since* (14) *reduces to the tautology $s_1 * s_2 = s_1 * s_2$.*

QUESTION 10. *For any algebraic structures X and Y, can a given magma in Y (resp. in X) be at the bottom (resp. top) of an (X, Y)-magma?*

Finding an answer for Question 10 in general does not seem to be very easy; however, extra structure comes in handy. Frequently in the sequel, in a distributive magma $(A, *, \circ)$, $(A, \circ)$ will be a group.

A semiring without an additive identity element will be called a **semiring**†. We will be interested mainly in semirings† whose multiplicative monoid is an Abelian group. With the notation introduced here:

- (Sgrp, Mon) is the class of semirings†.
- (Mon, Mon) is the class of semirings $(S, *, \circ)$ with top identity 0. Note, however, that the definition of a semiring found in [**Go**] requires, in addition, the **absorbing property of the top identity** (i.e. $0*a = 0 = a*0$, for all $a \in S$).
- (AbGrp, Mon) is the class of rings.
- (AbGrp, AbMon) is the class of commutative rings.
- (AbGrp, Sgrp) is the class of rings without 1.

For any class X of monoids, we write $X^{\#}$ for the class of $M^{\#}$ for $M \in X$. Of course, we do not suggest abandoning the standard terminology such as "division ring" in favor of "(AbGrp, Grp$^{\#}$)." However, the reader might concede that the alternate nomenclature proposed here is more self-explanatory, and provides a convenient category ((x, y)-magmas) with which to work, via the forgetful functor. For instance, expressions such as **quasi-magma, semiring, pseudoring, hemiring, hyperring** are used with various conflicting meanings in the literature and, as the interests of researchers (addressing questions like the ones posed in this manuscript) grows in many directions, the approach of adding modifiers to the word **ring** will clearly not suffice, but rather add to the confusion.

We are interested in the following notion:

DEFINITION 11. The **opposite** (or **flip**) of the class of (X, Y) is the class (Y, X).

With this terminology, we get that:

(i) The opposite of the class of rings (AbGrp, AbMon) is (AbMon, AbGrp), the class of semifields.
(ii) The opposite of (AbGrp, AbSgp) is (AbSgp, AbGrp), the class of semifields†.

The kind of results one seeks are, for example, contrasts or similarities between a class of distributive magmas and its opposite. Lemma 31 and Theorem 32 illustrate a somewhat surprising connection between (AbGrp, Magma) and its opposite.

3. Building Magmas from Groups: The defining function and its uses

If $(A, *, \circ)$ is a distributive magma with a group $(A, \circ)$ at the bottom, there exists a function φ from A to A which determines completely the distributor operation $*$. Making that function explicit simplifies many computations in the sequel.

3.1. The defining function.

DEFINITION 12. Given a magma $(A, *, \circ)$, if $(A, \circ)$ has an identity element e, for all $a \in A$, we define

$$\varphi(a) = e * a. \tag{3}$$

PROPOSITION 13. *Given a magma $(A, *, \circ)$ such that $\circ$ has an identity element e, if $g \in A$ is invertible with respect to $\circ$ then, for all $a \in A$ (using our notational convention to make the distributed operation $\circ$ tacit in equations) we have*

$$g * a = g\varphi(g^{-1}a). \tag{4}$$

PROOF. $g * a = g(e * g^{-1}a) = g\varphi(g^{-1}a)$. □

Since the function $\varphi : A \to A$ determines $*$ in terms of $\circ$, we refer to it as **the defining function** of the magma. Although there is an analogy with the function of [**Co**], the defining function here is taken at the opposite end (and thus its properties are quite different)!

Pertaining to Proposition 13, the result can be extended by considering a submonoid $(G, \circ, e)$ of $(A, \circ, e)$. In the case when G is indeed a subgroup, one need only assume that A left distributes over $(G, \circ, e)$, to obtain similarly as above that $g * a = g\varphi(g^{-1}a)$ for all $g \in G$, $a \in A$.

From now on, we start with the function φ and define $*$ by means of the formula (4). It turns out that, when $(A, \circ)$ is a group then any arbitrary function $\varphi : A \to A$ will indeed be the defining function for a distributive magma. It seems surprising at first that no other constraints on φ are needed to assure distributivity. That is the subject of our next Theorem.

THEOREM 14. *If $(A, \circ)$ is a group then, for every function $\varphi : A \to A$, there exists a binary operation $*$ on A such that $(A, *, \circ)$ is a left distributive magma having φ as its defining function.*

PROOF. Define $g * a = g\varphi(g^{-1}a)$ for all $g, a \in A$. Then, for any $a, c, d \in A$,

$$(ac) * (ad) = (ac)\varphi((ac)^{-1}(ad)) = a(c\varphi(c^{-1}d)) = a(c * d).$$ □

Arbitrary functions $\varphi : A \to A$ may be used in a similar way to create right distributors. Furthermore, if $(A, \circ)$ is an abelian group then the resulting magma $(A, *, \circ)$ is indeed two-sided distributive.

As can be seen in Theorem 14, for any group $(G, \circ)$, every function $\varphi : G \to G$ is the defining function of a left distributive magma having $\circ$ at the bottom. It is interesting to investigate the additional conditions needed to guarantee two-sided distributivity or additional properties at the top (such as associativity and commutativity). These concerns are also considered in Section 3.

The feasibility of answering any instance of Question 10 depends on extra assumptions being placed on the given $(A, \circ)$, on our expectations for $(A, *)$, and on whether $(*, \circ)$ or $(\circ, *)$ is expected to be a distributivity pair.

3.2. Commutativity and idempotence, plus associativity and some generalizations.

We turn to Question 2. Reversing the situation of Question 1, our results highlight the role in determining additional properties of the distributor in a magma with a group $(A, *, \circ)$ at the bottom. We superimpose an operation $*$ on A, over which $\circ$ is distributed. Our basic concern is, "How close must $*$ be to a sup operation on A?" In particular, "When do we get a max-plus algebra?" We are hardly in the customary ring-theoretic situation, since e will not be the identity element for $*$.

Writing $+$ instead of $\circ$ (and 0 instead of e) to emphasize that we are aiming for the max-plus algebra, we recall the defining function $\varphi : A \to A$ given by $\varphi(c) = 0 * c$, and now we write

$$a * b = a + \varphi(b - a).$$

Surprisingly, to assure commutativity of $*$, it suffices to show that all elements commute with 0.

PROPOSITION 15. *Suppose a group $(G, +)$ with identity e is at the bottom of a left distributive magma $(G, *, +)$. Then:*

(1) *$*$ is commutative if and only, for every $a \in G$,*

$$a = \varphi(a) - \varphi(-a). \tag{5}$$

(2) *$*$ is commutative if and only if every element commutes with* 0.

PROOF. Given two elements $b, c \in G$, $b*c = c*b$ if and only if $b+\varphi(c-b) = c+\varphi(b-c)$. So $*$ is commutative if and only if, for all $b, c \in G$, $c-b = \varphi(c-b)-\varphi(b-c)$. The result follows when one substitutes a for $c - b$. On the other hand, an element $a \in G$ commutes with 0 if and only if $a = \varphi(a) - \varphi(-a)$. □

3.2.1. *Idempotence.*

Since the max-plus algebra is idempotent under sup, idempotence is characterized next.

PROPOSITION 16. *The operation $*$ is idempotent iff $\varphi(0) = 0$.*

PROOF. Applying Equation (4) for $b = a$ yields $a * a = a + \varphi(0)$. □

Here is a non-idempotent example.

EXAMPLE 17. Define $a*b = \log(\exp(a)+\exp(b))$. Then $(\mathbb{R}, *, +)$ is a semiring†, which is not idempotent. Its defining function $\varphi(a) = 0 * a = \log(1 + \exp(a))$ resembles what is known as the **Zech logarithm** in coding theory; it is as a mechanism to facilitate addition in finite fields when the elements are written as powers of a primitive element. In fact, Zech logarithms were earlier used by Jacobi, cf. [**LN**] for number theoretic considerations.

3.2.2. *Associativity.*

We require two somewhat technical lemmas for associativity.

LEMMA 18. $((b - a) * a) * (c - a) = (b - a) * (a * (c - a))$ *iff*

$$\varphi(a + \varphi(b - a)) = \varphi(a) + \varphi(b - \varphi(a)). \tag{6}$$

PROOF.

$$(a * b) * c = (a + \varphi(b - a)) * c = (a + \varphi(b - a)) + \varphi(c - a - \varphi(b - a)). \tag{7}$$

$$a * (b * c) = a + \varphi((b * c) - a) = a + (\varphi(b + \varphi(c - b) - a)). \tag{8}$$

We get equality iff

$$\varphi(b - a) + \varphi(c - a - \varphi(b - a)) = \varphi(b + \varphi(c - b) - a) \tag{9}$$

Replacing a by $b-a$ yields

$$\varphi(a)+\varphi(a+c-b-\varphi(a))=\varphi(a+\varphi(c-b)). \tag{10}$$

Now replacing b by $a+c-b$ yields

$$\varphi(a)+\varphi(b-\varphi(a))=\varphi(a+\varphi(b-a)). \tag{11}$$

□

LEMMA 19. *The following conditions are equivalent:*

(i) *The operation $*$ is associative;*
(ii) *Equation* (6) *holds for all $a,b\in A$;*
(iii) $(b*a)*(-a)=b*(a*(-a))$ *for all $a,b\in A$.*

PROOF. (i) $\Leftrightarrow$ (ii). Apply the lemma, noting that $b-a$ and $c-a$ run through A, for any given $a\in A$.

(ii) $\Leftrightarrow$ (iii). Now take $c=0$ in the lemma, to get

$$((b-a)*a)*(-a)=(b-a)*(a*(-a)),$$

noting again that $b-a$ runs through A. □

The following yields a nice necessary condition for associativity.

PROPOSITION 20. *If $*$ is associative, then $\varphi(a+\varphi(a))=\varphi(a)+\varphi(a)$, $\forall a\in A$. In particular,*

$$\varphi(\varphi(0))=\varphi(0)+\varphi(0).$$

PROOF. Take $b=a+\varphi(a)$ in (6), and then $a=0$. □

3.2.3. *Alternativity.*

The criterion for associativity in Lemma 19 is considerably more technical than those encountered before; it involves a condition reminiscent of a cocycle, as given in Lemma 18. However, a slightly weaker condition, that of alternativity (cf. [**Ro1**, Definition 21B.9]), has a rather nice and simpler characterization.

DEFINITION 21. An operation $*$ is **alternative** when $(a*a)*b=a*(a*b)$. The class of sets with an alternative operation is denoted Alt.

Thus (AbGrp, Alt) is simply the class of **alternative** algebras, and its opposite (Alt, AbGrp) plays an important role in the subsequent discussion.

PROPOSITION 22. *Suppose that $*$ is commutative. Then $*$ is alternative iff*

$$\varphi^2=\varphi+c_0,$$

where c_0 is the constant $\varphi(0)+\varphi(-\varphi(0))$.

PROOF.

$$(a*a)*b=(a+\varphi(0))*b=a+\varphi(b-a)+\varphi(0)*b,$$

whereas

$$a*(a*b)=a+\varphi(a*b-a)=a+\varphi(a+\varphi(b-a)-a)=a+\varphi^2(b-a).$$

Thus the alternative condition $(a*a)*b=a*(a*b)$ holds iff

$$\varphi^2(b-a)=\varphi(b-a)+\varphi(0)*b$$

for all a,b. Replacing b by $b-b=0$ and a by $-a$ yields $\varphi^2(a)-\varphi(a)=\varphi(0)*0=\varphi(0)+\varphi(-\varphi(0))$. □

COROLLARY 23. Suppose that the operation $*$ is commutative and idempotent. Then $*$ is alternative iff $\varphi^2 = \varphi$.

PROOF. $\varphi(0) = 0$, implying $c_0 = 0$. □

This result opens the way to studying algebras whose *addition* (rather than multiplication) is alternative. For example, by Artin's theorem [**Ro1**, Exercise 21B.11], every subalgebra of a classical alternative algebra generated by two elements is associative. However, we do not see here that the condition $(a * a) * b = a * (a * b)$ for all a, b should imply $(a * (-a)) * b = a * ((-a) * b)$, and in fact the latter is equivalent to associativity by Lemma 19.

3.2.4. *The ub property.*

Another property of the max-plus algebra which comes up in various interesting contexts [**IKR**] is the rule

$$\text{ub property}: \quad a * (b * c) = a \quad \Rightarrow \quad a * b = a. \tag{12}$$

LEMMA 24. *The ub property holds iff* $\varphi(b + \varphi(c)) = 0$ *implies* $\varphi(b) = 0$, *for all* $b, c \in A$.

PROOF. $a * (b * c) = a + \varphi((b * c) - a) = a + \varphi(b + \varphi(c - b) - a)$. Equating this with a yields $\varphi(b - a + \varphi(c - b)) = 0$. Likewise, matching with $a = a * b = a + \varphi(b - a)$, and replacing b by $b - a$ and c by $c - b$ yields $[\varphi(b + \varphi(c)) = 0$ implies $\varphi(b) = 0]$, as desired. □

3.3. The role of positive and negative elements of $(A, *)$.

Assume throughout this subsection that A is commutative, so that we can apply Equation (5).

DEFINITION 25. $A^+ = \{a \in A : \varphi(a) = a\}$ is the set of **positive** elements, and $A^- = \{a \in A : \varphi(a) = 0\}$ is the set of **negative** elements.

0 is both positive and negative. Not all elements need be positive or negative, and in a pathological situation one might have no nonzero positive or negative elements.

LEMMA 26.
(i) $b - a \in A^+$ *iff* $a * b = b$.
(ii) $b - a \in A^-$ *iff* $a * b = a$.

PROOF. (i) $a * b = b$ iff $a + \varphi(b - a) = b$, iff $\varphi(b - a) = b - a$.
(ii) $a * b = a$ iff $a + \varphi(b - a) = a$, iff $\varphi(b - a) = 0$. □

LEMMA 27. $a \in A$ *is positive iff* $-a$ *is negative.* $A^+ \cap A^- \subseteq \{0\}$, *equality holding iff the semigroup* $(A, *)$ *is idempotent.*

PROOF. The first assertion is immediate from (5); likewise $a \in A^+ \cap A^-$ iff $a = \varphi(a) = 0$. Hence $0 \in A^+ \cap A^-$ iff $\varphi(0) = 0$, which means that $*$ is idempotent, by Proposition 16. □

The next result is easy, but provides a nice description of $*$ on a certain part of A.

LEMMA 28. *If* $a \in A$ *is negative, then* $\varphi(b) = \varphi(a + \varphi(b - a)), \quad \forall b \in A$.

PROOF. Substitute $\varphi(a) = 0$ into (6). □

The operation $*$ is called **bipotent** iff $a * b \in \{a, b\}$.

PROPOSITION 29.

(i) *A^+ and A^- are submonoids of $(A, +)$.*
(ii) *A is a bipotent semiring, iff $A = A^+ \cup A^-$ and $\varphi(0) = 0$, then*

PROOF. (i) If a, $b-a \in A^-$, then $\varphi(b) = \varphi(a+\varphi(b-a)) = \varphi(a+0) = \varphi(a) = 0$. If a, $b - a \in A^+$, then $-a$, $a - b \in A^-$, so $-b \in A^-$ and $b \in A^+$.

(ii) ($\Rightarrow$) We need to verify associativity for $*$. By Lemma 19, it is enough to verify (6) for $a \in A^-$ and $b - a \in A^+$. The associativity criterion (6) reduces to the tautology

$$\varphi(b) = \varphi(a + \varphi(b - a)) = \varphi(a + b - a) = \varphi(b).$$

Bipotence follows from Lemma 26 and Proposition 16, since for any $a, b \in A$, $b - a \in A = A^+ \cup A^-$.

($\Leftarrow$) $\varphi(0) = 0$ by Proposition 16, since bipotence implies idempotence. Furthermore, $b + 0 \in \{b, 0\}$ for any $b \in A$, implying $b \in A^+ \cup A^-$ by Lemma 26. $\square$

In a sense, we can now achieve our original goal of characterizing the max-plus algebra. We recall the **Green relation** $a \leq b$ iff $a * b = b$.

PROPOSITION 30. *(For A commutative) $*$ yields a total order with respect to the Green relation iff $A = A^+ \cup A^-$ with $\varphi(0) = 0$, in which case $*$ is bipotent, i.e., A is a max-plus algebra.*

PROOF. If $a \leq b$ and $b \leq a$ then $a = b * a = a * b = b$. Thus the Green relation is a partial order. Furthermore $(*)$ is bipotent by Proposition 29, which translates to the Green relation being total. $\square$

4. Building fields from abelian groups

We now change gears, and consider how to build $(X, X^\#)$ magmas when both $X = $ AbGrp. The possibility of studying (X, X) instead is ruled out by the next lemma.

4.1. The absorbing property of the top identity.

The next observation should sound familiar from ring theory. We include it here to emphasize that its proper setting is (AbGrp, Magma) and not Rings. No restrictions are placed on the bottom operation.

LEMMA 31. *If the top $(A, +)$ of a distributive magma is an Abelian group with identity 0, then 0 absorbs all elements with respect to the distributed operation $\circ$ (i.e., $a0 = 0 = 0a$, for all $a \in R$).*

PROOF. $a0 = a(0 + 0) = a0 + a0$, implying $0 = a0$ and likewise $0 = 0a$. $\square$

The following Proposition 32 could be stated to say that in a (Magma, AbGrp), if the top operation has an identity e then e absorbs the bottom operation. That formulation is vacuous at the outset – The family (Magma, AbGrp) would consist of a single structure (a singleton with trivial operations). However, expressing it that way highlights that (AbGrp, Magma) and its opposite (Magma, AbGrp) (as in Definition 11) share a certain behavior.

The main thrust of Proposition 32 is that if one wants a group at the bottom, one must actually settle for AbGrp# or Grp# for bottom structures when the top structure has an identity. We will consider the multiplicative group of non-zero elements of a field as **the group at the bottom of the structure.** See Remark 33 below to see how that impacts our use of a defining function.

PROPOSITION 32. *The distributor of a distributive magma* $(A, *, +)$ *with a non-trivial group at the bottom cannot have an identity element with respect to* $*$.

PROOF. Let $(A, *, \circ)$ be a distributive magma such that $(A, \circ, e)$ is a non-trivial group. Suppose that f is an identity element for $*$. The equality $a = a * f$ is equivalent to $a = a(e * (a^{-1}f))$. Thus $e * (a^{-1}f) = e$ for all a. But then, $e * a = e$ for all $a \in A$. Likewise $a * e = e$, implying $a^{-1} = a^{-1}e = a^{-1}(a * e) = e * a^{-1} = e$, proving the contrapositive. □

The above considerations do not represent an impediment for rings in general to be considered as (AbGrp, Mon) magmas. Monoids are not required in general be cancellative, and therefore the absorbing behavior of 0 does not present a challenge to this designation. Every ring $(R, +, \cdot)$ is in (AbGrp, Mon). Since the 0 element is always absorbing, it is tempting to think that this may also be written as $(\mathrm{AbGrp}, \mathrm{Mon}^{\#})$, but one must notice that, in general, $R \setminus \{0\}$ is not closed under multiplication, so the inclusion $(\mathrm{AbGrp}, \mathrm{Mon}^{\#}) \subset (\mathrm{AbGrp}, \mathrm{Mon})$ is proper. In fact, a magma $(R, +, \cdot)$ in (AbGrp, Mon) is in $(\mathrm{AbGrp}, \mathrm{Mon}^{\#})$ if and only if R is a domain, i.e. if and only if $R \setminus \{0\}$ is closed under multiplication (and hence multiplicatively cancellative). In particular, $(\mathrm{AbGrp}, \mathrm{AbGrp}^{\#})$ is indeed the class of fields.

Semirings ((Mon, Mon)) do not share any of these concerns: the identity of the monoid at the top need not absorb all elements with respect to the distributed operation. In fact, [**Go**] has an example (positive integers with *lcm* and multiplication) where the top and bottom monoids share the same identity element 1.

REMARK 33. We have seen before that distributive magmas with a group at the bottom are determined by the so-called defining function. It is easy to see that this notion can be extended also for cases when the bottom structure is a buffered group. The only case when φ is not already defined is $\varphi(0)$, but the purpose of φ is to determine the top operation and the role of 0 there is already established.

4.2. How to place specific Abelian groups at the top or at the bottom of a field.

REMARK 34. All non-zero torsion elements of the additive group of an integral domain have the same order, which is prime.

PROOF. For all a, $a + \cdots + a = 0$ if and only if $a(1 + \cdots + 1) = 0$. □

The two-element group is the additive group of the field with two elements, and the multiplicative group of the field with three elements. A straightforward argument shows that this is the only instance of such a dual role.

PROPOSITION 35. *The only multiplicative group of a field of characteristic* $\neq 2$ *that is also the additive group of an integral domain, is the group with two elements.*

PROOF. There is a unique element $a \neq 1$ (namely, $a = -1$) which has order 2. So, by Remark 34, the group has exactly two elements 1 and -1. □

THEOREM 36. *If some nonzero element in a* $(\text{Mon}, \text{Grp}^\#)$*-magma* $(A, +, \cdot)$ *is invertible with respect to the top operation, then the top is actually a group (i.e. the magma is* $(\text{Grp}, \text{Grp}^\#)$*). In particular, if some non-zero element of a* $(\text{Mon}, \text{AbGrp}^\#)$*-magma has a negative, then the magma is a field.*

PROOF. Write $A^\times = A \setminus \{0\}$, where 0 is the top identity and thus the absorbing element. A has non-trivial bottom group $(A^\times, \circ, 1)$. Suppose $a \in A^\times$ has a negative $-a$. Then, from the fact that $0 = a + (-a)$, multiplying both sides by a^{-1} and using distributivity yields $1 + (-a)a^{-1} = 0$ and therefore 1 has a negative $-1 \in A^\times$. Multiplying the expression $1 + (-1) = 0$ by an arbitrary element $b \in A^\times$ yields that b has a negative $-b = (-1)b$. Thus every element of A has a negative. The last assertion follows from Corollary 6 and Proposition 3. □

Obviously the multiplicative group of a field is in $(\text{Grp}, \text{Grp}^\#)$, raising the following question:

Which Abelian groups are in $(\text{AbGrp}, \text{AbGrp}^\#)$?

Since the structure of multiplicative groups of fields has been traversed rather thoroughly, cf. [**Fu**, Chapter XIII], we give a taste just of this aspect.

The torsion part is shown in [**Fu**, Theorem 127.1] to be the subgroups of $\mathbb{Q}/\mathbb{Z}$ with nontrivial 2-component. Skolem [**Sk**] described the multiplicative group of a finite algebraic extension of a prime field. Thus, we focus on torsion-free groups.

Fuchs [**Fu**, Theorems 127.2 and 127.3] describes the multiplicative subgroups of algebraically closed fields and real-closed fields, and in [**Fu**, Theorem 127.5] gives Hensel's description of the multiplicative group of the p-adic number field.

LEMMA 37. *If a torsion-free group* G *is the multiplicative group of a field* F*, then every element of* F *is transcendental over the ground field* $\mathbf{F}_2 = \{0, 1\}$.

PROOF. The characteristic subfield is $\mathbf{F}_2$, since otherwise -1 is a 2-torsion element of G. If F contains an algebraic element a, then $\mathbf{F}_2(a)$ is a finite field and thus torsion as a group, contrary to hypothesis. □

REMARK 38. It is easy to see that if R is a unique factorization domain, then the multiplicative group of its field of fractions is free, based on the irreducible elements.

This applies in particular to the multiplicative group of the purely transcendental field $\mathbf{F}_2(\lambda)$ of degree 1, yielding

LEMMA 39. *If a torsion-free group* A *is the multiplicative group of a field, then* A *contains a free Abelian group of infinite countable rank.*

EXAMPLE 40. The free Abelian group of rank $\aleph$ is the multiplicative group of the purely transcendental field $\mathbf{F}_2(\Lambda)$, where $|\Lambda| = \aleph$.

We can view any torsion-free group G as a module over $\mathbb{Z}$, and take its divisible closure, which is the localization $Q(G) := \mathbb{Q} \otimes_{\mathbb{Z}} G$.

LEMMA 41. *If a torsion-free group* G *is the multiplicative group of a field closed under taking radicals, then* G *is a vector space over* $\mathbb{Q}$ *under the action* $\frac{m}{n}a = \sqrt[n]{a^m}$*, and so is a direct product of copies of* $(\mathbb{Q}, +)$.

PROOF. For any $a \in G$ the element $\frac{m}{n}a$ is written in the multiplicative structure as $\sqrt[n]{a^m}$. Thus, $\sqrt[n]{a^m} \in G$ for all $a \in G$ and all m, n. When G itself is divisible, this happens iff $\sqrt[n]{a} \in G$ for all $a \in G$ and all n. □

(We thank D. Saltman for pointing this out.) This means that, up to isomorphism, there is only one divisible torsion-free Abelian group of each infinite cardinality, which enables us to match multiplicative subgroups.

EXAMPLE 42.

(i) $(\mathbb{R},+)$ is divisible. On the other hand, let F be the purely transcendental extension of $\mathbf{F}_2$ with a transcendence base of $\aleph$ elements, and let $\hat{F}$ be its radical closure, i.e., adjoining n-th roots of all elements of F, and continuing via transfinite induction. Then by Lemma 41 the multiplicative group of $\hat{F}$ is $\mathbb{Q}^{(\aleph)}$. In particular, this is isomorphic to $(\mathbb{R},+)$ when $\aleph = |\mathbb{R}|$, and is isomorphic to $(\bar{\mathbb{Q}},+)$ (the algebraic closure of $\mathbb{Q}$), when $\aleph = \aleph_0$.

(ii) The group $(\mathbb{Q},+)$ is uniform, i.e., any two cyclic subgroups intersect nontrivially. For example, the cyclic groups generated by any two elements $\frac{m_1}{n_1}$ and $\frac{m_2}{n_2}$ both contain $m_1 m_2$. Hence $(\mathbb{Q},+)$ cannot be isomorphic to the multiplicative group of a field F, in view of Lemma 41. ($F^\times$ would have to be torsion-free and thus F would have characteristic 0, but the groups generated by 2 and 3 would have to intersect.)

(iii) By Remark 38, the free Abelian group of rank $\aleph$ is the multiplicative group of the purely transcendental field extension $\mathbf{F}_2(\Lambda)$ of degree $\aleph$.

(iv) Since Lemma 37 rules out characteristic 0, one is led to consider proper semigroups of $\mathbb{Q}$ and $\mathbb{R}$. For example, here is a well known fact: The multiplicative group of $\mathbb{Q}^+$ is free, and thus also isomorphic to the multiplicative group of the purely transcendental field extension $\mathbf{F}_2(\Lambda)$ of rank $\aleph_0$, in view of (iii).

(v) Recall the **domain of Puiseux series** $F((t))_{\mathcal{G}}$, cf. [**IMS**], which is

$$\left\{ p(t) := \sum_{i\in\mathcal{G},\ c_i\in F} c_i t^i \right\},$$

where $\mathcal{G}$ is a given subgroup of $(\mathbb{R},+)$, and the powers of t are taken over well-ordered subsets of $\mathcal{G}$ (depending on p). Addition is obtained by adding coefficients, and multiplication is given by the usual formula

$$\left(\sum_i c_i t^i\right)\left(\sum_j d_j t^j\right) = \sum_{u\in\mathcal{G}} \left(\sum_{i+j=u} c_i d_j\right) t^u. \tag{13}$$

The product is defined because each sum is finite. Then $\mathcal{G}$ is isomorphic to $\{t^i : i \in \mathcal{G}\} \subset F((t))_{\mathcal{G}}$, but we do not have equality.

5. Appendix: Distributive Actions and Hierarchy graphs

5.1. Actions. The familiar notion of groups acting on sets can be generalized substantially by stripping the algebraic structure of the acting agent. For instance, our next definition illustrates how things would look for a monoid action on a set.

DEFINITION 43. A monoid $(A,\cdot,1)$ **acts** on a set $\mathcal{S}$ if there is a map $\circ : A\times\mathcal{S} \to \mathcal{S}$ satisfying $1\circ s = s$ and $(a_1\circ a_2)\circ s = a_1\circ(a_2\circ s)$ for all $a_i \in A$ and $s \in S$.

On the other hand, one often considers an algebraic structure on the acted-upon set. In these cases, we assume that A acts on a set S via actions $\circ : A \times S \to S$

and $\circ' : S \times A \to S$, and there is a binary operation $* : S \times S \to S$, over which we would like $\circ$ and $(\circ')$ to distribute, in the sense of Equations (1), (2), rewritten a bit more generally as

$$a \circ (s_1 * s_2) = (a \circ s_1) * (a \circ s_2) \tag{14}$$

and

$$(a_1 \cdot a_2) \circ s = (a_1 \circ s) * (a_2 \circ s), \tag{15}$$

for all $a, a_i \in A,\ s, s_i \in S$.

In light of this, the notion of a distributive magma $(A, *, \circ)$ can be viewed from the perspective that $(A, \circ)$ acts as endomorphisms of $(A, *)$.

Another way of asserting Equation (14) is that we want the action $\circ$ to make $(S, *)$ into a (left) A-module. For Equation (15) to make sense, we also must endow A with a binary operation $* : A \times A \to A$, for which $(A, *)$ is a semigroup, and say that $(S, *)$ is an $(A, *)$-**module** if (15) holds. How can we describe properties of $(S, *)$ (commutativity, idempotence, bipotence, alternativity, associativity) in terms of the two operations?

Another way of writing (1) is to define the **left map** $\ell_a : A \to A$ by $\ell_a(b) = a \circ b$, and require ℓ_a to be a $*$-homomorphism in the sense that

$$\ell_a(b * c) = \ell_a(b) * \ell_a(c), \ \forall b, c \in S. \tag{16}$$

(2) says

$$\ell_{a_1 * a_2} = \ell_{a_1} * \ell_{a_2}. \tag{17}$$

We propose the following more general terminology. Given a distributive magma $(A, *_A, \circ_A)$, a **right modular magma** $(M, *_M, \circ_M)$ over A is a magma $(M, *_M)$ with a map $\circ_M : M \times A \to M$ (which we would call scalar multiplication) satisfying the following distributivity conditions:

(1) $(m_1 *_M m_2) \circ_M a = (m_1 \circ_M a) *_M (m_2 \circ_M a),\ \forall m_1, m_2 \in M,\ a \in A$,
(2) $m \circ_M (a_1 *_A a_2) = (m \circ_M a_1) *_M (m \circ_M a_2),\ \forall m \in M,\ a_1, a_2 \in A$, and
(3) $m \circ_M (a_1 \circ_A a_2) = (m \circ_M a_1) \circ_M a_2,\ \forall m \in M,\ a_1, a_2 \in A$,

One could delete the subscripts in the notation; then the ambiguity in the notation of $*$ and $\circ$ would be consistent with the ambiguity that arises with the usage of $+$ and $\cdot$ in the theory of modules over rings.

REMARK 44. These questions also could be phrased in the language of blueprints [**Lo**], which are monoids with extra structure. More generally, we could formulate all of this in terms of universal algebra, cf. [**Ja**, Chapter 2], and then make appropriate modifications. An advantage of the universal algebra approach to questions like Questions 1 and 2 is that one has the **relatively free** algebra satisfying an identity such as in Proposition 13, and then all of its homomorphic images also automatically satisfy it.

5.2. Hierarchy graphs. Yet another mechanism that may help to organize future research is the following graph-theoretic paradigm.

DEFINITION 45. Let S be a set and V be a set of binary operations on S (for example, V could consist of all associative operations, all group operations, all cancellative operations, etc.). The (left, right, two-sided) **distributive hierarchy graph** is the directed graph $\mathcal{H}(V) = (V, E)$ having V as its vertex set and such that there is an edge from $\circ$ to $*$ if $\circ$ (left, right, left and right) distributes over $*$. In other words, $\circ \to *$ means that $(S, *, \circ)$ is a (left, right, two-sided) distributive magma. Referring to the set of edges in the graph as E and keeping in mind that there is at most one arrow from one point to another, we will simply write (as is customary in graph theory) $(\circ, *) \in E$ when $\circ \to *$. We will call the graph $\mathcal{H}(V)$ the **hierarchy graph** when the additional qualifications are clear from the context.

By Theorem 14, every group structure has nonzero out-degree in the left and right hierarchy graphs, and by Proposition 15, Abelian group structures do so in the two-sided hierarchy graph. Many natural questions come to mind; here is a sample collection.

REMARK 46.

(1) The overriding question in this paper (Question 10) is of the form: Given $\circ \in V$, does there exist $* \in V$ such that $(*, \circ) \in E$ (or $(\circ, *) \in E$)? Furthermore, one can consider specific subsets V_1 and V_2 of V and ask whether for $* \in V_1$ one can find a suitable $\circ \in V_2$.
(2) From that point of view, Proposition 13 states that the answer for the above question is affirmative for the one-sided hierarchy graphs when V_1 consists of groups and $V_2 = V$, and also for the two-sided hierarchy graph when V_1 consists of Abelian groups; Section 3 deals with the complications involved with tightening the target family V_2.
(3) Another example of the scenario proposed in (1) is the question of whether a given buffered abelian group $S^\#$ (Definition 7) can be the multiplicative monoid of a semifield (or of a field), with V_1 consisting of buffered abelian group structures $S^\#$ and V_2 consisting of semigroup (abelian group) structures on that same set, Proposition 32.
(4) Basic examples can be analyzed in this framework. The remark that the **degenerate case** of Proposition 47 consists of idempotent operations, is addressing the question of what vertices in $\mathcal{H}(V)$ host loops.
(5) The remark that all operations left distribute over the right zero band operation is a statement about a terminal vertex (such that all other vertices, including itself, connect directly to it) in the appropriate graph. This fact was pointed out to us by A. Rafieipour and M. Mazrooei.

What if we want $*$ and $\circ$ to be the same operation, i.e., we want $(A, \circ, \circ)$ to be a distributive magma? When the two operations are the same, we are talking about semigroups that satisfy the identity

$$a(bc) = (ab)(ac), \tag{18}$$

and in the associative case, every time some element repeats in the middle, we can erase it. This is a strong version of idempotence (at the expense of requiring one extra element in both sides of the product). By Proposition 3, $\circ$ is commutative in the cancellative case.

PROPOSITION 47. *(i) A commutative monoid $(A, \circ, \circ)$ is a distributive magma iff A is idempotent under $\circ$.*

(ii) More generally, a commutative semigroup $(A, \circ, \circ)$ is a distributive magma iff A satisfies the identity $abc = a^2bc$ for all $a, b, c \in A$.

PROOF. (i) Taking $b = c = e$ shows $a = a^2$, and conversely

$$abc = aabc = abac,$$

yielding (18).

(ii) Analogous to (i). ☐

EXAMPLE 48. (1) Any semilattice (i.e., set with a sup operation) is a distributive magma, where $* = \circ = \sup$. In particular, this holds for $\{\mathcal{P}(S), \cup\}$, the power set of a set.

(2) It is easy to construct the generic monoid for (i), which is precisely the set of commuting words in which no letter repeats (because of idempotence). But this is clearly isomorphic to $\{\mathcal{P}(S), \cup\}$, where S is the set of letters.

(3) Likewise the generic semigroup for (ii) is the set of commuting words in which no letter repeats in a word of length 3 or more.

EXAMPLE 49. The notion of a "quandle," important in knot theory, reduces to a set A with a non-associative binary operation $\triangle$ which, among other properties, right distributes over itself. In analogy to Proposition 47, all elements of a quandle are idempotent. A good reference for quandles is [**EN**].

References

[ABG] M. Akian, R. Bapat, and S. Gaubert, Max-plus algebra, In: Hogben, L., Brualdi, R., Greenbaum, A., Mathias, R. (eds.) *Handbook of Linear Algebra*. Chapman and Hall, London, 2006.

[Ba] M. Baker. Matroids over hyperfields, arXiv:1601.01204v2 [math.CO], 2016.

[Cl] A. H. Clifford and G. B. Preston, *The algebraic theory of semigroups. Vol. I*, Mathematical Surveys, No. 7, American Mathematical Society, Providence, R.I., 1961. MR0132791

[Co] P. M. Cohn, *On the embedding of rings in skew fields*, Proc. London Math. Soc. (3) **11** (1961), 511–530. MR0136632

[D] R. M. Dicker, *A set of independent axioms for a field and a condition for a group to be the multiplicative group of a field*, Proc. London Math. Soc. (3) **18** (1968), 114–124. MR0219519

[EN] M. Elhamdadi and S. Nelson, *Quandles—an introduction to the algebra of knots*, Student Mathematical Library, vol. 74, American Mathematical Society, Providence, RI, 2015. MR3379534

[Fu] L. Fuchs, *Infinite abelian groups. Vol. II*, Academic Press, New York-London, 1973. Pure and Applied Mathematics. Vol. 36-II. MR0349869

[Go] J. S. Golan, *Semirings and their applications*, Kluwer Academic Publishers, Dordrecht, 1999. Updated and expanded version of *The theory of semirings, with applications to mathematics and theoretical computer science* [Longman Sci. Tech., Harlow, 1992; MR1163371 (93b:16085)]. MR1746739

[Ja] N. Jacobson, *Basic algebra. II*, W. H. Freeman and Co., San Francisco, Calif., 1980. MR571884

[IMS] I. Itenberg, G. Mikhalkin, and E. Shustin, *Tropical algebraic geometry*, Oberwolfach Seminars, vol. 35, Birkhäuser Verlag, Basel, 2007. MR2292729

[IKR] Z. Izhakian, M. Knebusch, and L. Rowen, *Quadratic and symmetric bilinear forms on modules with unique base over a semiring*, Doc. Math. **21** (2016), 773–808. MR3548133

[LN] R. Lidl and H. Niederreiter, *Finite fields*, 2nd ed., Encyclopedia of Mathematics and its Applications, vol. 20, Cambridge University Press, Cambridge, 1997. MR1429394

[Lo] O. Lorscheid, *A blueprint view on $\mathbb{F}_1$-geometry*, Absolute arithmetic and $\mathbb{F}_1$-geometry, Eur. Math. Soc., Zürich, 2016, pp. 161–219. MR3701901

[P] J. H. Przytycki, *Distributivity versus associativity in the homology theory of algebraic structures*, Demonstratio Math. **44** (2011), no. 4, 823–869. MR2906433

[Ro1] L. H. Rowen, *Graduate algebra: noncommutative view*, Graduate Studies in Mathematics, vol. 91, American Mathematical Society, Providence, RI, 2008. MR2462400

[Ro2] L.H. Rowen, *Algebras with a negation map*, arXiv:1602.00353 [math.RA], 72 pages (2016).

[Sk] Th. Skolem, *On the existence of a multiplicative basis for an arbitrary algebraic field*, Norske Vid. Selsk. Forh., Trondhjem **20** (1947), no. 2, 4–7. MR0026080

Department of Mathematics, Ohio University, Athens, Ohio
Email address: lopez@ohio.edu

Department of Mathematics, Bar-Ilan University, Ramat Gan, Israel
Email address: rowen@math.biu.ac.il

Contemporary Mathematics
Volume **715**, 2018
https://doi.org/10.1090/conm/715/14415

Idealizers in Differential Polynomial Rings and Generalized HNP Rings

Hidetoshi Marubayashi and Akira Ueda

ABSTRACT. Polynomial rings over hereditary Noetherian prime rings have the following properties: The global dimension is two, and any ideal with left v-ideal or right v-ideal is a projective ideal, that is, left and right projective. By using latter property we defined a concept of generalized hereditary Noetherian prime rings (G-HNP rings) to study the structure of projective ideals. In this paper, let $K[z][y][x;\delta]$ be a differential polynomial ring over a field K with char $K = 0$, which is a generalization of the first Weyl algebras from the view-point of arithmetic ideal theory and $R := \mathbf{I}_S(xS)$, the idealizer of xS in S.

We completely describe all the projective ideals of R and obtain that R is a G-HNP ring with $2 \leq$ gl dim $R \leq 4$. Furthermore, we show that there is an infinite descending chain of not G-HNP rings R_n with rt gl dim $R_n = 2$ such that $R = \cap R_n$. Finally, we raise several open questions about G-HNP rings.

1. Introduction

Let R be a hereditary Noetherian prime ring, σ be an automorphism of R and δ be a derivation of R. The skew polynomial ring $R[x;\sigma]$ and differential polynomial ring $R[x;\delta]$ have the following properties ([**2**], [**7**]):

(1) Global dimensions are two.

(2) Any ideal which is left v-ideal or right v-ideal is of the form;

$$X\mathfrak{a}[x;\sigma] \ (X\mathfrak{b}[x;\delta]),$$

where X is an invertible ideal and $\mathfrak{a}$ is a σ-invariant ideal of R ($\mathfrak{b}$ is a δ-ideal of R), that is, they are all the projective ideals (see the paragraph after Proposition 2.10 for the definition of v-ideals).

By using the second property, we have defined a concept of generalized hereditary Noetherian prime rings which is a natural generalization of HNP rings from the view-point of arithmetic ideal theory.

DEFINITION. A prime Goldie ring R is a generalized hereditary Noetherian prime ring (a G-HNP ring for short) if

(i) Any ideal of R which is a left v-ideal or right v-ideal is a projective ideal, that is, left and right projective.

(ii) R is τ-Noetherian (see [**8**] for the definition of τ-Noetherian).

2010 *Mathematics Subject Classification.* 16S32, 16D25.

Key words and phrases. projective ideal, hereditary Noetherian prime ring, differential polynomial ring, generalized hereditary Noetherian prime ring.

A G-HNP ring is *strongly* if any left (right) v-ideal is left (right) projective.

In [**3**], we have completely described the structure of projective ideals in G-HNP rings under a mild condition. In Section 2, in order to give an example of G-HNP rings which have enough projective ideals and global dimension ≥ 2, we generalize the first Weyl algebras, adding a variable, as follows:

Let K be a field with char $K = 0$, $K[z, y] = K[z][y]$ be a polynomial ring over K in variables z, y and δ be a derivation of $K[z, y]$ such that $\delta(y) = 1$ and $\delta(a) = 0$ for all $a \in K[z]$.

Let $S = K[z, y][x; \delta]$ be a differential polynomial ring which is a generalization of the first Weyl algebras and $R = \mathbf{I}_S(xS)$, the idealizer of xS in S.

We explicitly describe projective ideals of R (Theorem 2.15) which is one of the purposes in the paper, and show that R is a G-HNP ring with $2 \leq \text{gl dim } R \leq 4$.

As an application, we provide, in Section 3, an infinite descending chain of not G-HNP rings R_n containing R such that $R = \cap R_n$ and rt gl dim $R_n = 2$ for all n. Section 4 contains several open questions about G-HNP rings.

This paper is basically a continuation of [**3**] and the results in Section 2 were announced in [**3**] without the proofs. Note that the class of G-HNP rings ranges from rings with global dimension two to rings with infinite global dimension, though HNP rings are just global dimension one ([**3**]). We refer the readers to [**9**] and [**8**] for order theory and global theory.

2. Idealizers in differential polynomial rings

Throughout this paper, K is a field with char $K = 0$, $K[z, y] = K[z][y]$ is a polynomial ring over K in variables z, y, $S = K[z, y][x; \delta]$ is the differential polynomial ring over $K[z][y]$ in a variable x, that is, $xr = rx + \delta(r)$ for all $r \in K[z][y]$.

The derivation δ on R is extended to an derivation of δ of S by $\delta(s) = xs - sx$ for all $s \in S$, that is, an inner derivation induced by x, which is equivalent to; for $s = r_0 + r_1x + \cdots + r_nx^n$, where $r_i \in K[z][y], \delta(s) = \delta(r_0) + \delta(r_1)x + \cdots + \delta(r_n)x^n$. It is sometimes convenient to write the coefficients on the right side, that is, for $s \in S,\ \ s = r_0 + xr_1 + \cdots + x^n r_n$.

We frequently use the following lemma without reference, which is easily proved by induction on natural numbers.

LEMMA 2.1. For any natural number ℓ, $y^\ell x = xy^\ell - \ell y^{\ell-1}$ and $yx^\ell = x^\ell y - \ell x^{\ell-1}$.

We denote by $\mathbb{Z}(R)$ the center of a ring R.

LEMMA 2.2. $\mathbb{Z}(S) = K[z]$.

Proof. It is clear that $\mathbb{Z}(S) \supseteq K[z]$. Let $s = r_0 + r_1x + \cdots + r_\ell x^\ell \in \mathbb{Z}(S)$, where $r_i \in K[z][y]$. From $xs = sx$, $0 = xs - sx = \delta(r_0) + \delta(r_1)x + \cdots + \delta(r_\ell)x^\ell$ and so $\delta(r_i) = 0$ for all i $(0 \leq i \leq \ell)$. Thus $r_i \in K[z]$. Next,

$$ys = yr_0 + yr_1x + \cdots + yr_\ell y^\ell$$

and

$$\begin{aligned} sy &= r_0y + r_1xy + r_2x^2y + \cdots + r_\ell x^\ell y \\ &= r_0y + r_1(yx + 1) + r_2(yx^2 + 2x) + \cdots + r_\ell(yx^\ell + \ell x^{\ell-1}). \end{aligned}$$

Since $ys = sy$, we have $r_1 + 2r_2x + \cdots + \ell r_\ell x^{\ell-1} = 0$ and $r_i = 0$ for all i $(1 \leq i \leq \ell)$. Hence $s = r_0 \in K[z]$. □

Throughout the paper,

$$R = \mathbf{I}_S(xS) = \{s \in S \mid sxS \subseteq xS\},$$

idealizer of xS in S. Note that δ induces a derivation of R.

LEMMA 2.3. (1) $SxS = S$.
(2) $R = K[z] + xS$.

Proof. (1) is clear since $xy = yx + 1$.

(2) Clearly $R \supseteq K[z] + xS$, because any element in $K[z]$ is central. To prove the converse inclusion, let $s = r_0 + xs_1 \in R$, where $r_0 \in K[z][y]$ and $s_1 \in S$. Since $xS \ni sx = r_0x + xs_1x = xr_0 - \delta(r_0) + xs_1x = -\delta(r_0) + xs_2$ for some $s_2 \in S$, we have $\delta(r_0) = 0$, that is, $r_0 \in K[z]$ and hence $s \in K[z] + xS$. □

In order to show that R is a right Noetherian ring, we describe all right R-submodules of S containing xS (see [**10**, Proposition 2.3]). Note that any right R-submodule I of S containing xS is δ-stable, that is, $\delta(I) \subseteq I$ since δ is an inner derivation induced by x and that any ideal of S (resp., R) is also δ-stable.

LEMMA 2.4. Let I be a right R-submodule of S with $I \supset xS$ and set $\mathfrak{a} = K[z] \cap I$. If $r = a_0 + a_1y \in I$, where $a_0, a_1 \in K[z]$, then $a_0, a_1 \in \mathfrak{a}$ and $a_1y \in I$.

Proof. We note that $I \ni \delta(r) = a_1$, $a_1 \in \mathfrak{a}$, and

$$I \ni ryx = a_0yx + a_1y^2x = a_0(xy - 1) + a_1(xy^2 - 2y) = xs - (a_0 + 2a_1y),$$

where $s \in S$. Thus $a_0 + 2a_1y \in I$ and so $I \ni 2r - (a_0 + 2a_1y) = a_0$. Hence $a_0 \in \mathfrak{a}$ and $a_1y \in I$. □

LEMMA 2.5. Let I be a right R-submodule of S with $I \supset xS$. Then

(1) $\mathfrak{a} = I \cap K[z] \neq (0)$.

(2) $\mathfrak{a}_i = \{a \in K[z] \mid \exists r = a_0 + a_1y + \cdots + a_iy^i + \cdots + a_\ell y^\ell \in I\}$. Then $\mathfrak{a}_i = \mathfrak{a}_{i+1}$ for all $i \geq 1, \mathfrak{a}_i \subseteq \mathfrak{a}$ and $\mathfrak{a}_iy^i \subseteq I$.

Proof. We let $r = a_0 + a_1y + \cdots + a_\ell y^\ell \in I \setminus xS$, where $a_i \in K[z]$ and $a_\ell \neq 0$.

(1) Because $\delta^\ell(r) = \ell!\,a_\ell \in I$, $a_\ell \in I \cap K[z] = \mathfrak{a}$ and $\mathfrak{a} \neq (0)$.

(2) Since $I \ni \delta^{\ell-1}(r) = (\ell-1)(\ell-2)\cdots 1a_{\ell-1} + \ell(\ell-1)\cdots 2a_\ell y$, we have $a_{\ell-1} \in \mathfrak{a}$ and $a_\ell y \in I$ by Lemma 2.4. Further, since $I \ni a_\ell yxy^\ell = a_\ell(xy-1)y^\ell$, $a_\ell y^\ell \in I$ and so $a_0 + a_1 + \cdots + a_{\ell-1}y^{\ell-1} \in I$. Continuing this process, we have $a_i \in \mathfrak{a}$, that is, $\mathfrak{a}_i \subseteq \mathfrak{a}$ and $a_iy^i \in I$ for any i, $0 \leq i \leq \ell$. It also follows from the proof above that $\mathfrak{a}_i = \{a \in K[z] \mid ay^i \in I\}$. For each $a \in \mathfrak{a}_i, ay^i \in I$. Thus

$$I \ni ay^ixy^2 = a(xy^i - iy^{i-1})y^2$$

and so $ay^{i+1} \in I$, that is, $a \in \mathfrak{a}_{i+1}$.

Conversely, if $ay^{i+1} \in I$, then $I \ni ay^{i+1}x = a(xy^{i+1} - (i+1)y^i)$ and so $ay^i \in I$, that is, $a \in \mathfrak{a}_i$. Hence $\mathfrak{a}_i = \mathfrak{a}_{i+1}$ for all $i \geq 1$. □

PROPOSITION 2.6. Any right R-submodule properly containing xS is one the following types:

$$I_1 = \mathfrak{a} + xS,\ I_2 = \mathfrak{a}[y] + xS \text{ and } I_3 = \mathfrak{a} + \mathfrak{b}[y] + xS,$$

where $\mathfrak{a}$ and $\mathfrak{b}$ are ideals of $K[z]$ with $\mathfrak{a} \supset \mathfrak{b} \supset (0)$. In particular, these are all left R-submodules of S.

Proof. It is clear that $I_1 = \mathfrak{a} + xS$ is a right R-submodule of S, where $\mathfrak{a}$ is a nonzero ideal of $K[z]$. Let $I = \mathfrak{a} + \mathfrak{b}[y] + xS$, where $\mathfrak{a}$ and $\mathfrak{b}$ are ideals of $K[z]$ with $\mathfrak{a} \supseteq \mathfrak{b} \supset (0)$. We prove that I is a right R-submodule of S. By Lemmas 2.2 and 2.3, we have $IK[z] \subseteq I$, $\mathfrak{a}xS \subseteq I$, $xSxS = xS \subseteq I$. Thus it suffices to prove:

$$\mathfrak{b}[y]xS \subseteq I. \qquad (*)$$

For any $r = b_0 + b_1y + \cdots + b_ly^l \in \mathfrak{b}[y]$,

$$\begin{aligned} rx &= (b_0 + b_1y + \cdots + b_\ell ly^\ell)x \\ &= b_0x + b_1(xy-1) + b_2(xy^2 - 2y) + \cdots + b_\ell(xy^\ell - \ell y^{\ell-1}) \\ &= xr - (b_1 + 2b_2y + \cdots + \ell b_\ell y^{\ell-1}) \\ &\in xS + \mathfrak{b}[y] \subseteq I \end{aligned}$$

and so $Ix \subseteq I$.

Let $s = r_0 + xr_1 + \cdots + x^mr_m \in S$, where $r_i \in K[z][y]$. Since

$$rxs = \left(\sum_{i=0}^{\ell} b_iy^ix\right)\left(\sum_{j=0}^{m} x^jr_j\right) = \sum_{i,j} b_iy^ix^{j+1}r_j,$$

to prove (*), it is enough to prove $by^kx^nt \in I$, where $k \geq 0, n \geq 1, b \in \mathfrak{b}$ and $t \in K[z][y]$. This is proved by induction on n. If $n = 1$, then

$$by^kxt = b(xy^k - ky^{k-1})t = xby^kt - kby^{k-1}t \in xS + \mathfrak{b}[y] \subseteq I.$$

We may assume that $by^kx^nt \in I$. Then

$$by^kx^{n+1}t = by^kx^n(tx + \delta(t)) = by^kx^ntx + by^kx^n\delta(t) \subseteq Ix + I \subseteq I.$$

Hence I is a right R-submodule of S.

Conversely, let I be a right R-submodule of S with $I \supset xS$. Then , by Lemma 2.5, $I \supseteq \mathfrak{a} + \mathfrak{b}[y] + xS$, where $\mathfrak{a} = I \cap K[z], \mathfrak{b} = \mathfrak{a}_i$ for any $i \geq 1$. Let

$$r = a_0 + a_1y + \cdots + a_\ell y^\ell \in I \setminus xS,$$

where $a_i \in \mathfrak{b}(1 \leq i \leq \ell)$. By Lemma 2.5, $a_iy^i \in I$, so $a_0 = r - (a_1y + \cdots + a_\ell y^\ell) \in I$, which implies $a_0 \in \mathfrak{a}$ and hence $I = \mathfrak{a} + \mathfrak{b}[y] + xS$.

It is easy to see that I_i are all left R-submodules. □

REMARK. In Proposition 2.6, I_2 is a right ideal of S. However, neither I_1 nor I_3 is a right ideal of S.

LEMMA 2.7. *R is left and right Noetherian.*

Proof. Let $I = \mathfrak{a} + \mathfrak{b}[y] + xS$, where $\mathfrak{a}$ and $\mathfrak{b}$ are ideals of $K[z]$ such that either $\mathfrak{a} = \mathfrak{b}$ or $\mathfrak{a} \supset \mathfrak{b} \supseteq (0)$. If $\mathfrak{b} = (0)$, that is, $I = I_1$, then I is a finitely generated right R-submodule by [**10**, Lemma 2.1]. If $\mathfrak{b}$ is non-zero, then for each $b \in \mathfrak{b}$ and $\ell \geq 1$,

$$\mathfrak{b}yR \ni byxy^\ell = b(xy-1)y^\ell = bxy^{\ell+1} - by^\ell$$

and so $\mathfrak{a} + \mathfrak{b}yR + xS \supseteq \mathfrak{b}[y]$. Hence $I = \mathfrak{a}R + \mathfrak{b}yR + xS$ and it is a finitely generated right R-module. Since S is a left and right Noetherian ring ([**9**, (1.2.9)]), R is right Noeherian by [**10**, Proposition 2.3].

In order to show that R is left Noetherian, we consider $S' = xSx^{-1}$. It is easily obtained that $S' = xSx^{-1} = K[z][y'][x;\delta]$, where $y' = xyx^{-1}$, that $S'x = xS$ and $S'xS' = S'$.

We now consider the idealizer $\mathbf{I}_{S'}(S'x)$. It is obvious that $\mathbf{I}_{S'}(S'x) \supseteq R$ because $S'xR = xSR = xS = S'x$. Let $xsx^{-1} \in S'$ such that $S'xxsx^{-1} \subseteq S'x$. Then $xSxsx^{-1} \subseteq xS$ and so $xsx^{-1} \in S$ and $xsx^{-1}xS = xsS \subseteq xS$. Thus $xsx^{-1} \in R$, that is, $\mathbf{I}_{S'}(S'x) = R$. Hence R is left Noetherian by the left sided version. □

LEMMA 2.8. Let P' be an ideal of S. Then P' is a prime ideal if and only if $P' = \mathfrak{m}[y][x;\delta]$ for some maximal ideal $\mathfrak{m}$ of $K[z]$.

Proof. Let $P' = \mathfrak{m}[y][x;\delta]$, where $\mathfrak{m}$ is a maximal ideal of $K[z]$. As $\delta(\mathfrak{m}[y]) = \mathfrak{m}[y]$,

$$P' = \sum_{i=0}^{\infty} x^i \mathfrak{m}[y]$$

and so P' is an ideal of S. δ naturally induces a derivation δ on $(K[z][y]/\mathfrak{m}[y])$ and $S/\mathfrak{m}[y][x;\delta] \cong (K[z]/\mathfrak{m})[y][x;\delta]$. Since $\operatorname{char}(K[z]/\mathfrak{m}) = 0$, $(K[z]/\mathfrak{m})[y][x;\delta]$ is the first Weyl algebra and so it is a simple ring ([**9**, (1.3.5)]). Hence P' is a prime ideal (in fact it is maximal).

Conversely, let P' be a prime ideal of S. First we prove that $P' \cap K[z][y] \neq (0)$. Suppose, on the contrary, $P' \cap K[z][y] = (0)$ and let $s \in P'$. Write

$$s = r_0 + r_1 x + \cdots + r_\ell x^\ell$$

and

$$K[z][y] \ni r_i = a_{i0} + a_{i1}y + \cdots + a_{in_i} y^{n_i},$$

where $a_{ij} \in K[z], a_{in_i} \neq 0$ if $r_i \neq 0$. We may assume that ℓ is minimal. Set $n = \max\{n_i \mid 0 \leq i \leq \ell\}$. If $n > n_\ell$, then

$$P' \ni \delta^n(s) = \delta^n(r_0) + \delta^n(r_1)x + \cdots + \delta^n(r_{n-1})x^{\ell-1} + 0x^\ell.$$

Thus, by the choice of ℓ, $\delta^n(r_i) = 0$ for all i $(0 \leq i \leq n-1)$. Hence $n = n_\ell$ and $\delta^n(s) = b_0 + b_1 x + \cdots + b_\ell x^\ell$, where $b_i \in K[z]$ for all i and $b_\ell = n!\, a_{\ell n_\ell}$. Now note that

$$\begin{aligned} P' &\ni \delta^n(s)y - y\delta^n(s) \\ &= b_0 y + b_1 xy + \cdots + b_\ell x^\ell y - (yb_0 + b_1 yx + \cdots + b_\ell y x^\ell) \\ &= b_1 + 2b_2 x + \cdots + \ell b_\ell x^{\ell-1}. \end{aligned}$$

Again, the minimality of ℓ implies $b_j = 0$ for all j, $1 \leq j \leq \ell$, which is a contradiction. Hence $P' \cap K[z][y] \neq (0)$. Let $r = a_0 + a_1 y + \cdots + a_n y^n \in P' \cap K[z][y]$ be non-zero, where $a_i \in K[z]$ and $a_n \neq 0$. Then $P' \ni \delta^n(r) = n!\, a_n$. Hence $\mathfrak{m} = K[z] \cap P' \neq 0$.

It is clear that $\mathfrak{m}$ is a prime ideal of $K[z]$ since $K[z] = \mathbb{Z}(S)$ and it is a maximal ideal of $K[z]$. Since $P' \supseteq \mathfrak{m}[y][x;\delta]$, we have $P' = \mathfrak{m}[y][x;\delta]$. □

REMARK. Any prime ideal of S is a maximal ideal of S.

In order to study prime ideals of R, we use the ring of Morita context. Set

$$T = \begin{pmatrix} S & S \\ xS & R \end{pmatrix}.$$

LEMMA 2.9. There exists an one to one correspondence between $\operatorname{Spec}(S)$ and $\operatorname{Spec}^*(R) = \{P \in \operatorname{Spec}(R) \mid P \not\supseteq xS\}$, given by $P' \to P = \{r \in R \mid SrxS \subseteq P'\}$, where $P' \in \operatorname{Spec}(S)$. In particular, $P = P' \cap R$.

Proof. By [**9**, (3.6.2)], the correspondence is one-to-one since $SxS = S$. We only need to prove that $P = P' \cap R$. It is clear that $P' \cap R \subseteq P$ since $S(P' \cap R)xS \subseteq P'$. To prove the converse inclusion, let $P' = \mathfrak{m}[y][x;\delta]$, where $\mathfrak{m}$ is a maximal ideal of $K[z]$ by Lemma 2.8. Let $r = r_0 + xr_1 + \cdots + x^\ell r_\ell \in P$, where $r_0 \in K[z]$ and $r_i \in K[z][y]$. Now

$$\begin{aligned} P' &\ni rx = xr - \delta(r) \\ &= x(r_0 - \delta(r_1)) + x^2(r_1 - \delta(r_2)) + \cdots + x^\ell(r_{\ell-1} - \delta(r_\ell)) + x^{\ell+1} r_\ell. \end{aligned}$$

So $r_l \in \mathfrak{m}[y]$ and $r_i - \delta(r_{i+1}) \in \mathfrak{m}[y]$ for all i, $0 \le i \le \ell - 1$. Thus $r \in P'$ and hence $P = P' \cap R$. □

PROPOSITION 2.10. *The set of all prime ideals of R is $\{\mathfrak{m}R, \mathfrak{m} + xS$ and xS, where $\mathfrak{m}$ runs over all maximal ideals of $K[z]\}$.*

Proof. Let P be a prime ideal of R. Suppose $P \not\supseteq xS$, then $P = P' \cap R$ by Lemma 2.9, where P' is a maximal ideal of S. Write $P' = \mathfrak{m}[y][x;\delta]$ for some maximal ideal $\mathfrak{m}$ of $K[z]$. It is clear that $P \supseteq \mathfrak{m}R$. Conversely, let $r = r_0 + xr_1 + \cdots + x^l r_l \in P$. Then $r_0 \in \mathfrak{m}$ and $x^i r_i \in x^i \mathfrak{m}[y]$ (see the proof of Lemma 2.9) and so $r \in \mathfrak{m}R$. As a consequence $P = \mathfrak{m}R$.

Suppose $P \supseteq xS$. Then either $P = xS$ or $P = \mathfrak{m} + xS$ by Proposition 2.6. □

We use the following notation: For a right R-ideal I, we put

$$(R : I)_\ell = \{q \in Q \mid qI \subseteq R\},$$

where Q is the quotient ring of R, which is a left R-ideal. Similarly, for a left R-ideal J, we let

$$(R : J)_r = \{q \in Q \mid Jq \subseteq R\}.$$

Put

$$I_v = (R : (R : I)_\ell)_r,$$

which contains I. If $I = I_v$, then it is called a *right v-ideal*. Similarly, we take ${}_vJ = (R : (R : J)_r)_\ell$. Then J is called a *left v-ideal* if ${}_vJ = J$.

Note that if I is a right projective R-module, then I is a right v-ideal, which is easily obtained by the dual basis lemma ([**6**, Lemma 1.5]). By *projective ideals* we mean it is right and left projective.

In the remainder of this section, we describe all the projective ideals by using the notion of v-ideals.

LEMMA 2.11. *Let A be a proper ideal of R such that $A \supset xS$. Then $A_v = R = {}_vA$.*

Proof. By Proposition 2.6, $A = \mathfrak{a} + xS$, where $\mathfrak{a}$ is a non-zero ideal of $K[z]$. It is clear that $(R : A)_\ell \supseteq R$. To prove the converse inclusion, note that

$$xSx^{-1} = O_\ell(xS) = \{q \in Q \mid qxS \subseteq xS\} = (R : xS)_\ell \supseteq (R : A)_\ell$$

since xS is idempotent.

For any $xsx^{-1} \in (R : A)_\ell$, where $s \in S$ and $xsx^{-1}(\mathfrak{a} + xS) \subseteq R$. In particular, $xsx^{-1}\mathfrak{a} \subseteq R$. Thus $xs\mathfrak{a} \subseteq Rx = xK[z] + xSx$ and so $s\mathfrak{a} \subseteq K[z] + Sx$. Write $s = r_0 + r_1 x + \cdots + r_l x^l$, where $r_i \in K[z][y]$ for all i. For any $a \in \mathfrak{a}$,

$$K[z] + Sx \ni sa = r_0 a + r_1 ax + \cdots + r_\ell ax^\ell$$

and so $r_0 a \in K[z]$. Thus $r_0 \in K[z]$ and

$$xsx^{-1} = r_0 + xr_1 + \cdots + xr_\ell x^{\ell-1} \in K[z] + xS = R.$$

Thus $(R : A)_\ell \subseteq R$ and so $(R : A)_\ell = R$. Hence $A_v = R$.

Since $(R : A)_r \subseteq O_r(xS) = \{q \in Q \mid xSq \subseteq xS\} = S$, we have $R =_v A$ by the similar method. □

LEMMA 2.12. Let A be an ideal of R with $A = A_v$ or ${}_vA = A$. Assume that $A \not\supseteq xS$ and $xS \not\supseteq A$. Then $A = \mathfrak{a}R$ for some ideal $\mathfrak{a}$ of $K[z]$.

Proof. We only prove the lemma for the case when $A = A_v$ and we assume, on the contrary, that there is an ideal A not satisfying the conclusion and also assume A is maximal with this property. By Proposition 2.10 and Lemma 2.11, A is not maximal in the set

$$\mathcal{S} = \{B: \text{ ideals of } R \mid B = B_v \text{ and } B \supseteq A\}$$

since the maximal element in $\mathcal{S}$ is a prime ideal (see [**2**, Lemma 2.1]). So there is a prime ideal P in $\mathcal{S}$ which is of the form $\mathfrak{m}R$ for some maximal ideal $\mathfrak{m}$ of $K[z]$ by Proposition 2.10 and Lemma 2.11.

If $P^{-1}A \supseteq xS$, then $P^{-1}A = xS$ since $(P^{-1}A)_v = P^{-1}A$ by [**2**, Lemma 2.1], and so $A \subseteq xS$, a contradiction. Thus $P^{-1}A \not\supseteq xS$ and, of course, $xS \not\supseteq P^{-1}A$. It follows from the choice of A that $P^{-1}A = \mathfrak{b}R$ for some ideal $\mathfrak{b}$ of $K[z]$. Hence $A = P\mathfrak{b} = \mathfrak{m}\mathfrak{b}R$, a contradiction. Therefore $A = \mathfrak{a}R$. Similarly, $A = \mathfrak{a}R$ in case ${}_vA = A$. □

LEMMA 2.13. Let A be an ideal of R.

(1) If $x^\ell a \in A$ for some $a \in K[z]$ and $\ell > 1$, then $x^{\ell-1}a \in A$.

(2) If $x \in A$, then $A \supseteq xS$.

Proof. (1) Put $r = xy^{\ell+1}x^\ell a \in A$. Then $A \ni \delta^{\ell-1}(r) = (\ell+1)l \cdots 3axy^2x^\ell$ and $A \ni axy^2x^\ell$ follow. Because $y^2x^\ell = x^\ell y^2 - 2\,\ell x^{\ell-1}y + \ell\,(\ell-1)x^{\ell-2}$, we have that

$$\begin{aligned} axy^2x^\ell &= ax(x^\ell y^2 - 2\,\ell x^{\ell-1}y + \ell\,(\ell-1)x^{\ell-2}) \\ &= ax^{\ell+1}y^2 - 2\,\ell\,ax^\ell y + \ell\,(\ell-1)ax^{\ell-1}. \end{aligned}$$

Since $ax^{\ell+1}y^2 \in A$, $s = 2ax^\ell y - (\ell-1)ax^{\ell-1} \in A$. Furthermore,

$$A \ni yxs = 2ax^{\ell+1}y^2 - (3\,\ell+1)ax^\ell y + \ell\,(\ell-1)ax^{\ell-1}.$$

As $ax^{\ell+1}y^2 = ax^\ell xy^2 \in A$, $u = (3\,\ell+1)ax^\ell y - \ell\,(\ell-1)ax^{\ell-1} \in A$. Therefore it follows that $(3\,\ell+1)s - 2u = -(\ell-1)(\ell+1)x^{\ell-1}a \in A$ and hence $x^{\ell-1}a \in A$.

(2) Since $yx^2 = x^2y - 2x$, we have $Sx^2S \ni x$ and so $Sx^2S = SxS = S$. If $x \in A$, then $A \supseteq RxR \supseteq xSxxS = xS$. □

LEMMA 2.14. Let A be a non-zero ideal of R such that $xS \supset A$. Then $A = \mathfrak{a}xS$ for some proper ideal $\mathfrak{a}$ of $K[z]$.

Proof. Let $r = xr_1 + \cdots + x^\ell r_\ell \in A$, where $r_i \in K[z][y]$ and $r_\ell \neq 0$. We may assume that ℓ is the minimal degree. We first prove that $x^\ell a \in A$ for some nonzero $a \in K[z]$. Set $r_i = a_{i0} + a_{i1}y + \cdots + a_{in_i}y^{n_i}$, where $a_{ij} \in K[z]$, $a_{in_i} \neq 0$ if $r_i \neq 0$ and set $n = \max\{n_i \mid 1 \leq i \leq \ell\}$. As in the proof of Lemma 2.8, we have $n = n_\ell$ and hence

$$A \ni s = \delta^n(r) = xa_1 + \cdots + x^\ell a_\ell,$$

where $a_i \in K[z]$ and $a_\ell = n!\,a_{\ell n_\ell}$.

If $n > n_i$ for all i $(1 \leq i < \ell)$, then $s = x^\ell a_\ell$ as desired. If $n = n_i$ for some i $(1 \leq i < \ell)$, that is, $a_i \neq 0$, then, since $yx \in R$, we have

$$A \ni u = yxs = x^2ya_1 + x^3ya_2 + \cdots + x^{\ell+1}ya_\ell - (2xa_1 + 3x^2a_2 + \cdots + (\ell+1)x^\ell a_\ell)$$

and

$$A \ni v = syx = x^2ya_1 + x^3ya_2 + \cdots + x^{\ell+1}ya_\ell - (xa_1 + x^2a_2 + \cdots + x^\ell a_\ell).$$

Thus, $v - u = xa_1 + 2x^2a_2 + \cdots + \ell x^\ell a_\ell \in A$ and

$$\ell\, s - (v-u) = (\ell-1)xa_1 + (\ell-2)x^2a_2 + \cdots + x^{\ell-1}a_{\ell-1} \in A. \tag{**}$$

Thus, by the choice of ℓ, $a_i = 0$ for all i $(1 \leq i < \ell)$. Hence $s = x^\ell a$ and by Lemma 2.13, $\ell = 1$.

Set $\mathfrak{a} = \{a \in K[z] \mid xa \in A\}$, which is a nonzero, proper ideal of $K[z]$ by Lemma 2.13. Finally we prove that $A = \mathfrak{a}xS$. It is clear that $\mathfrak{a}xS$ is an ideal of R. For any $a \in \mathfrak{a}$ and any $n \geq 1$, $A \ni xy^{n+1}ax = axy^{n+1}x = ax(xy^{n+1} - (n+1)y^n)$. Thus $\mathfrak{a}xy^n \subseteq A$. Hence $A \supseteq \mathfrak{a}xS$ since $\mathfrak{a}xS = x\mathfrak{a}[y] + \mathfrak{a}x^2S$.

Let $r = xr_1 + \cdots + x^\ell r_\ell \in A \setminus \mathfrak{a}xS$, where $r_i \in K[z][y], r_\ell \notin \mathfrak{a}[y]$ and we may assume that ℓ is the minimal degree with this property and so $\ell > 1$. Then as in before, $A \ni s = \delta^n(r) = xa_1 + \cdots + x^\ell a_\ell$, where $a_i \in K[z], a_\ell \notin \mathfrak{a}$ and $A \ni (\ell-1)xa_1 + (\ell-2)x^2a_2 + \cdots + x^{\ell-1}a_{\ell-1}$ (see (**)). By the choice of ℓ, $a_i \in \mathfrak{a}$ for all i $(1 \leq i < \ell)$ and so $x^\ell a_\ell \in A$. By Lemma 2.13, $x^{\ell-1}a \in A$, a contradiction. Hence $A = \mathfrak{a}xS$. □

THEOREM 2.15. *Let A be an ideal of R with $A = A_v$ or ${}_vA = A$. Then A is one of the following:*

$$A = xS,\ A = \mathfrak{a}R,\ A = \mathfrak{a}xS \text{ for some nonzero ideal } \mathfrak{a} \text{ of } K[z].$$

They are all the projective ideals.

Proof. If $A \not\supseteq xS$ and $xS \not\supseteq A$, then $A = \mathfrak{a}R$ for some nonzero ideal $\mathfrak{a}$ of $K[z]$ by Lemma 2.12. If $A \supseteq xS$, then $A = xS$ by Lemma 2.11 since xS is right R-projective by [**10**, Lemma 2.1], and if $A \subseteq xS$, then $A = \mathfrak{a}xS$ for some nonzero ideal $\mathfrak{a}$ of $K[z]$. They are all the left R-projective by the left sided version (we observe that $xS = S'x$). □

We denote the (right) global dimension of R by (rt) gl dim R.

PROPOSITION 2.16. *R is a G-HNP ring with $2 \leq$ gl dim $R \leq 4$.*

Proof. R is a G-HNP ring by Theroem 2.15 and the definition of G-HNP rings and $2 \leq$ gld $R \leq 4$ follows from [**9**, (7.5.13)]. □

PROPOSITION 2.17. *The following are equivalent:*

(1) *Maximal projective ideals of R.*
(2) *Prime projective ideals of R.*
(3) *Minimal prime ideals of R.*

Proof. Let P be a maximal projective ideal of R. To prove that P is prime, let $AB \subseteq P$, where A and B are ideals of R with $B \subset P$. Then $B_v = R$ since B_v is a Projective ideal by Theorem 2.15. So $P \supseteq (AB)_v = (AB_v)_v = A_v \supseteq A$ by [**5**, lemma 1.1]. Hence P is a prime ideal. Now the proposition is easily proved by Proposition 2.10 and Lemma 2.11. □

3. Some overrings of $R = K[z] + xS$

In this section, we consider some overrings of $R = K[z] + xS$ to get more properties on R by using idealizer theory and also give differences between HNP rings and G-HNP rings.

LEMMA 3.1. gl dim $S = 2$ and S is a strongly G-HNP ring.

Poof. Since gl dim $K[z][y] = 2$, we have that $2 \leq$ gl dim $S \leq 3$ by [**9**, (7.5.3)]. By Lemma 2.9, any prime ideal P' of S is height one and char $S/P' = 0$ as char $K = 0$. Hence gl dim $S = 2$ by [**9**, (7.10.3)]. Since S is a Noetherian maximal order ([**8**, (2.3.19)] and [**9**, (1.2.9)]), it follows that $A_v = {}_vA$ for any ideal A of S ([**8**, p.110]). Thus S is a strongly G-HNP ring by [**4**, Proposition 5.2]. □

Let $\mathfrak{m}_1, \ldots, \mathfrak{m}_n$ be distinct maximal ideals of $K[z]$, $\mathfrak{a}_n = \mathfrak{m}_1 \cap \cdots \cap \mathfrak{m}_n$ and $R_n = K[z] + \mathfrak{a}_n[y] + xS$, overrings of $R = K[z] + xS$.

LEMMA 3.2. Let $I_n = \mathfrak{m}_n + \mathfrak{a}_n[y] + xS$. Then

(1) I_n is a maximal right ideal of R_{n-1} with $R_{n-1}I_n = R_{n-1}$ (in case $n = 1$, $R_0 = S$).

(2) $R_n = \mathbf{I}_{R_{n-1}}(I_n)$, the idealizer of I_n in R_{n-1}, and R_n is right Noetherian.

(3) rt gl dim $R_n = 2$ and R_n is not a G-HNP ring.

Proof. For simplicity, put $I = I_n$, $\mathfrak{a} = \mathfrak{a}_n$ and $\mathfrak{b} = \mathfrak{a}_{n-1}$ (in case $n = 1$, $\mathfrak{b} = K[z]$).

(1) It is clear that I is a right ideal of R_{n-1}. Let J be a right ideal of R_{n-1} with $J \supset I$. Since J is a right R-submodule of S it follows from Proposition 2.6 that $J = \mathfrak{c} + \mathfrak{c}_1[y] + xS$, where $\mathfrak{c}$ and $\mathfrak{c}_1$ are ideals of $K[z]$ with $\mathfrak{c} \supseteq \mathfrak{c}_1 \supseteq (0)$ and so $\mathfrak{c} \supseteq \mathfrak{m}_n$. If $\mathfrak{c} \supset \mathfrak{m}_n$, then $\mathfrak{c} = K[z]$ and $J = R_{n-1}$. If $\mathfrak{c} = \mathfrak{m}_n$, then $\mathfrak{c}_1 \subseteq \mathfrak{m}_n \cap \mathfrak{b} = \mathfrak{a}$ and hence $J = I$, a contradiction. Therefore I is a maximal right ideal of R_{n-1}.

To prove $R_{n-1}I = R_{n-1}$, it suffices to show that I is not a left ideal of R_{n-1}. We assume, on the contrary, that I is a left ideal of R_{n-1}. Then for any $by \in R_{n-1}$, where $b \in \mathfrak{b} \setminus \mathfrak{a}$, $I \ni byx = b(xy - 1) = xby - b$. Thus $b \in I$ and $\mathfrak{b} \subseteq \mathfrak{m}_n$, a contradiction.

(2) It is obvious that $\mathbf{I}_{R_{n-1}}(I) \supseteq R_n = K[z] + \mathfrak{a}[y] + xS$. Conversely, take $r \in \mathbf{I}_{R_{n-1}}(I)$. Then we may assume that $r = b_1y + \cdots + b_\ell y^\ell$, where $b_i \in \mathfrak{b}$. Because $I \ni rx = xr - \delta(r)$, we have that $I \ni \delta(r) = b_1 + \cdots + \ell\, b_\ell y^{\ell-1}$ and thus $b_1 \in \mathfrak{m}_n, b_i \in \mathfrak{a}$ $(2 \leq i \leq \ell)$. In particular, $b_1 \in \mathfrak{m}_n \cap \mathfrak{b} = \mathfrak{a}$, that is, $r \in \mathfrak{a}[y]$. So

$$\mathbf{I}_{R_{n-1}}(I) = K[z] + \mathfrak{a}[y] + xS,$$

that is, $R_n = \mathbf{I}_{R_{n-1}}(I)$. It is clear that R_n is right Noetherian by [**10**, Proposition 2.3] and the proof of Lemma 2.7.

(3) Since $R_0 = S$, rt gl dim $R_1 = 2$ by [**10**, Theorem 2.9], Lemma 3.1 and (1). Inductively, assume that rt gl dim $R_{n-1} = 2$, then rt gl dim $R_n = 2$ by [**10**, Theorem 2.9] and (1).

In order to show that R_n is not a G-HNP ring, it is enough to prove that I is left projective and $I_v = R_n$. Note that I is idempotent since $I^2 = IR_{n-1}I = IR_{n-1} = I$ and $O_r(I) \supseteq R_{n-1}$ since I is a right ideal of R_{n-1}. Let $q \in O_r(I)$, that is, $Iq \subseteq I$. Then $q \in R_{n-1}q = R_{n-1}Iq \subseteq R_{n-1}I = R_{n-1}$ and $O_r(I) = R_{n-1}$. Thus

$$(R_n : I)_r I = O_r(I)I = R_{n-1}I = R_{n-1} = O_r(I).$$

Therefore I is left projective by [**6**, Lemma 1.5]. To prove $I_v = R_n$, it is enough to show $(R_n : I)_\ell = R_n$. It is clear that $(R_n : I)_\ell \supseteq R_n$ since I is an ideal of R_n.

Conversely, let $r \in (R_n : I)_\ell \subseteq (S : xS)_\ell = Sx^{-1}$. So $t = xr = xsx^{-1}$ for some $s \in S$. Write $s = s_0 + s_1 x$, where $s_0 \in K[z][y]$ and $s_1 \in S$. Then we claim that

$$r = x^{-1}s_0 + s_1.$$

We observe that $t = x(s_0 + s_1 x)x^{-1} = xs_0x^{-1} + xs_1$ and $t \in (R_n : I)_\ell$ imply $xs_0x^{-1} \in (R_n : I)_\ell$. Write $s_0 = \sum_{i=0}^k b_i y^i$, where $b_i \in K[z]$. Then

$$\begin{aligned} R_n &\supseteq \mathfrak{a}x(\sum_{i=0}^k b_i y^i)x^{-1} = \mathfrak{a}(\sum_{i=0}^k b_i x y^i)x^{-1} \\ &= \mathfrak{a}(\sum_{i=0}^k b_i(y^i x + i y^{i-1})x^{-1}) \\ &= \mathfrak{a}(s_0 + (\sum_{i=0}^k i b_i y^{i-1})x^{-1}). \end{aligned}$$

Thus $\mathfrak{a}(\sum_{i=0}^k i b_i y^{i-1})x^{-1} \subseteq R_n$ as $\mathfrak{a}s_0 \subseteq R_n$, and so

$$\mathfrak{a}(\sum_{i=0}^k i b_i y^{i-1}) \subseteq R_n x \subseteq Sx,$$

which implies $b_1 = \cdots = b_k = 0$, that is, $s_0 = b_0 \in K[z]$. Hence it follows that $t = xr = xb_0x^{-1} + xs_1$ and $r = x^{-1}s_0 + s_1$ as claimed.

Furthermore, as $(x^{-1}s_0 + s_1)\mathfrak{a} = \mathfrak{a}r \subseteq R_n \subseteq S$, $s_0 = 0$, that is, $r = s_1 \in S$. Write $r = r_0 + xs_2$, where $r_0 = a_0 + a_1 y + \cdots + a_\ell y^\ell$ $(a_i \in K[z])$ and $s_2 \in S$. Since $xs_2 \in R_n$, we have $r_0 \in (R_n : I)_\ell$. Therefore

$$R_n \ni r_0 x = xr_0 - (a_1 + 2a_2 y + \cdots + l a_l y^{\ell-1}),$$

which implies $a_2, \ldots, a_\ell \in \mathfrak{a}$ and so $a_1 y = r_0 - (a_0 + a_2 y^2 + \cdots + a_\ell y^\ell) \in (R_n : I)_\ell$. Thus $R_n \ni a_1 yxy = a_1(xy - 1)y = xa_1y^2 - a_1 y$ and $a_1 \in \mathfrak{a}$, that is, $r_0 \in R_n$ and $r \in R_n$. Hence $(R_n : I)_\ell = R_n$ and $I_v = R_n$, that is, I is not right projective. This completes the proof. □

PROPOSITION 3.3. Let $I_n = \mathfrak{m}_n + \mathfrak{a}_n[y] + xS$ and $R_n = K[z] + \mathfrak{a}_n[y] + xS$ as in Lemma 3.2. Then we have an infinite descending chain: $S = R_0 \supset R_1 \supset \cdots \supset R_n \supset \cdots \supset R$ such that:

(i) R_n is right Noetherian, rt gl dim $R_n = 2$ and R_n is not a G-HNP ring;
(ii) $R = \cap_{n \in \mathbb{N}} R_n$, where $\mathbb{N}$ is the set of natural numbers.

Proof. (i) follows from Lemma 3.2 and (ii) is clear since $\cap \mathfrak{a}_n = (0)$. □

We end this section with the following remark which exhibiting differences between HNP rings and G-HNP rings.

REMARK. (1) As it was noted in introduction, HNP rings are just global dimension one. However, the class of G-HNP rings ranges from rings with global dimension two to rings with infinite global dimension ([**3**]).

(2) In case of HNP rings, the idealizer of a generative, isomaximal right ideal is again an HNP ring. However R_1 in Lemma 3.2 is not a G-HNP ring, though R_0 is a strongly G-HNP ring with gl dim $R_0 = 2$. Moreover, there are infinite descending chain

$$S = R_0 \supset R_1 \supset \cdots \supset R_n \supset \cdots \supset R = K[z] + xS$$

such that rt gl dim $R_n = 2, R_n$ is not a G-HNP ring and $R = \cap_{n\in\mathbb{N}} R_n$, a G-HNP ring.

(3) xS is a maximal projective right ideal of S (right ideals maximal amongst the projective right ideals) since $S = J_v$ for any right ideal J of S with $J \supset xS$, which is proved in the same way as in Lemma 3.2.

These phenomena happened in the preceding remark illustrate that the structures of G-HNP rings are more complicated and include deeper properties than ones of HNP rings.

4. Some open questions about G-HNP rings

In this section, we provide several open questions about G-HNP rings.

QUESTION 4.1. (i) Let $R = K[z] + xS$ as in Section 3. What is the exact value of gl dim R? (We expect either gl dim $R = 2$ or gl dim $R = 3$.)

(ii) Is R_n in Lemma 3.2 left Noetherian? If this is true, then gl dim $R_n = 2$ ([**11**]).

QUESTION 4.2. Let R be a G-HNP ring, σ be an automorphism of R and δ be a left σ-derivation of R. Then is the Ore extension $R[x;\sigma,\delta]$ a G-HNP ring? If this is true, then describe all the projective ideals of $R[x;\sigma,\delta]$ (see [**2**] and [**7**] in case R is an HNP ring, $R[x;\sigma]$ and $R[x;\delta]$).

QUESTION 4.3. Let $S = \bigoplus_{n\in\mathbb{Z}} R_n$ be a strongly graded ring of type $\mathbb{Z}$. If R_0, the part of degree 0, is a G-HNP ring, then is S a G-HNP ring? If this is true, then describe all the projective ideals of S. In particular, if R_0 is an HNP ring, then is S is a strongly G-HNP ring with gl dim $S = 2$?

QUESTION 4.4. Let $T = \begin{pmatrix} R & V \\ W & S \end{pmatrix}$ be a ring of Morita context which is a prime Goldie ring. Then T is a (strongly) G-HNP ring if and only if so are R and S?

Following [**1**], define notions of G-HNP (R, S)-modules V and of G-HNP (S, R)-modules W, respectively. T is a (strongly) G-HNP ring if and only if V is a (strongly) G-HNP (R, S)-module and W is a (strongly) G-HNP (S, R)-module?

QUESTION 4.5. Let S be a G-HNP ring and I be a generative right ideal of S. Then $R = \mathbf{I}_S(I)$ is a G-HNP ring if and only if I is a finite intersection of maximal projective right ideals (see [**10**, Theorem 5.3] in case of HNP rings and aslo [Remark (3) in Section 3]).

QUESTION 4.6. Let R be a prime Goldie ring satisfying a PI condition with the center D. If R is a G-HNP ring, then study relations between R and D from the view-point of arithmetic ideal theory.

References

[1] B. Saraç, E. Akalan, P. Aydoğdu, and H. Marubayashi, *Rings of Morita contexts which are maximal orders*, J. Algebra Appl. **15** (2016), no. 7, 1650129, 13 pp. MR3528557

[2] E. Akalan, P. Aydoğdu, H. Marubayashi, B. Saraç, and A. Ueda, *Projective ideals of skew polynomial rings over HNP rings*, Comm. Algebra **45** (2017), no. 6, 2546–2556. MR3594537

[3] E. Akalan, H. Marubayashi and A. Ueda, *Generalized hereditary Noetherian prime rings*, Preprint.

[4] H. Bass, *Finitistic dimension and a homological generalization of semi-primary rings*, Trans. Amer. Math. Soc. **95** (1960), 466–488. MR0157984

[5] H. Marubayashi, *A Krull type generalization of HNP rings with enough invertible ideals*, Comm. Algebra **11** (1983), no. 5, 469–499. MR693797

[6] H. Marubayashi, H. Miyamoto, and A. Ueda, *Non-commutative valuation rings and semi-hereditary orders*, K-Monographs in Mathematics, vol. 3, Kluwer Academic Publishers, Dordrecht, 1997. MR1461811

[7] H. Marubayashi and A. Ueda, *Projective ideals of differential polynomial rings over HNP rings*, Preprint.

[8] H. Marubayashi and F. Van Oystaeyen, *Prime divisors and noncommutative valuation theory*, Lecture Notes in Mathematics, vol. 2059, Springer, Heidelberg, 2012. MR3185163

[9] J. C. McConnell and J. C. Robson, *Noncommutative Noetherian rings*, Pure and Applied Mathematics (New York), John Wiley & Sons, Ltd., Chichester, 1987. MR934572

[10] J. C. Robson, *Idealizers and hereditary Noetherian prime rings*, J. Algebra **22** (1972), 45–81. MR0299639

[11] J. J. Rotman, *Notes on homological algebras*, Van Nostrand Reinhold Mathematical Studies, No. 26. Van Nostrand Reinhold Co., New York-Toronto, Ont.-London, 1970. MR409590

Department of Mathematics, Naruto University of Education, Naruto, Tokushima 772-8502, Japan

Email address: marubaya@naruto-u.ac.jp

Department of Mathematics, Shimane University, Matsue, Shimane 690-8504, Japan

Email address: ueda@riko.shimane-u.ac.jp

Contemporary Mathematics
Volume **715**, 2018
https://doi.org/10.1090/conm/715/14416

Module hulls—similarities and contrasts

Jae Keol Park and S. Tariq Rizvi

Dedicated to the memory of Professor Bruno J. Müller (1934–2014)

ABSTRACT. While the notions of Baer, quasi-Baer and extending modules are related to each other, it is well known that these are distinct notions in general. In this paper, we investigate a number of instances under which these notions coincide with each other and other instances when these are distinct and disparate from each other. As a consequence, for a certain module M over a commutative ring and a fixed injective hull $E(M)$, we study the coincidence, similarities and contrasts between the Baer module hull, the quasi-Baer module hull, and the extending module hull of M. We show that for an essential extension V_R of a nonsingular cyclic module over a commutative ring R, the notions of V_R being Baer, quasi-Baer, quasi-continuous, extending, FI-extending, or strongly FI-extending, coincide with each other. Using this result, we prove the existence and coincidence of the Baer, the quasi-Baer, the extending, and the FI-extending hulls of any nonsingular cyclic module over a commutative ring. As an application, we explicitly describe such module hulls and show that each of these is precisely given by ΩM, where Ω is the subring of $H = \text{End}(E(M))$ generated by all idempotents of H, namely, the quasi-continuous hull of the nonsingular cyclic module. As a byproduct, when R is a commutative semiprime ring, we characterize all intermediate Baer (extending, quasi-continuous) rings between R and $Q(R)$, where $Q(R)$ is the maximal (right) ring of quotients of R. A characterization for a commutative domain to be Prüfer is obtained by using extending module hulls of certain modules. A number of examples which exhibit the disparities and differences between Baer hulls, quasi-Baer hulls, extending hulls, and FI-extending hulls of a module are provided.

1. Introduction

Shoda [**32**] and Eckmann and Schopf [**10**] showed independently that for every module M, there exists a unique (up to isomorphism) minimal injective overmodule of M called an injective hull of M, denoted by $E(M)$. The module M is called *quasi-injective* if for any submodule N of M, any $f \in \text{Hom}(N, M)$ can be extended to a homomorphism of M. Thereby, M is quasi-injective if and only if M is a fully invariant submodule of $E(M)$.

Consider the following conditions for a module M:

(C_1) Every submodule of M is essential in a direct summand of M.

2010 *Mathematics Subject Classification.* 16D25, 16D40, 16D80, 16E60, 16R20, 16U10.
Key words and phrases. Baer module, Baer hull, quasi-Baer module, quasi-Baer hull, extending module, extending hull, quasi-retractable, $\mathcal{K}$-cononsingular.

(C_2) If $V \leq M$ and $V \cong N \leq^{\oplus} M$, then $V \leq^{\oplus} M$.

(C_3) If M_1 and M_2 are direct summands of M with $M_1 \cap M_2 = 0$, then $M_1 \oplus M_2$ is a direct summand of M.

(FI) Every fully invariant submodule of M is essential in a direct summand of M.

M is called *continuous* if it satisfies (C_1) and (C_2) conditions; M is called *quasi-continuous* if it satisfies (C_1) and (C_3) conditions; M is called *extending* if it satisfies (C_1) condition; and M is called *FI-extending* if it satisfies (FI) condition.

It is well known that the following implications hold true for modules while the reverse implications do not hold true (see [**21**], [**9**] and [**6**]):

$$\text{injective} \Rightarrow \text{quasi-injective} \Rightarrow \text{continuous} \Rightarrow$$
$$\text{quasi-continuous} \Rightarrow \text{extending} \Rightarrow \text{FI-extending}.$$

For a given module M, let $H = \text{End}_R(E(M))$ denote the endomorphism ring of its injective hull $E(M)$. Then by Johnson and Wong [**13**], the quasi-injective hull of the module M (i.e., the smallest quasi-injective overmodule of M in $E(M)$) is given by HM. In [**12**] Goel and Jain showed that there always exists a unique smallest quasi-continuous overmodule for every module M within a fixed injective hull $E(M)$. They termed it as *quasi-continuous hull* of M and explicitly described it by ΩM, where Ω is the subring of $H = \text{End}(E(M))$ generated by all idempotents of H (see [**21**] for more details on quasi-injective, continuous, and quasi-continuous modules). In contrast to the quasi-injective hulls and the quasi-continuous hulls, it was shown by Müller and Rizvi in [**22**] that continuous hulls do not always exist in general. However, they did prove the existence of continuous hulls of certain classes of modules over a commutative ring (such as nonsingular cyclic ones) and provided a description of these continuous hulls (see [**22**, Theorem 8]). Similar to the case of continuous hulls, it is also known that extending hulls do not always exist (for example see [**6**, Example 8.4.13, p. 319]). Closely linked to these notions are the notions of (quasi-)Baer modules and (quasi-)Baer rings.

Kaplansky introduced the notion of a Baer ring in [**15**]. He and many others obtained a number of interesting results on Baer rings which have their roots in Functional Analysis. Recall that a ring R is called a Baer ring if the right annihilator of every nonempty subset of R is eR with $e^2 = e \in R$. It can be routinely checked that a ring R is a Baer ring if and only if the left annihilator of any nonempty subset of R is Rf with $f^2 = f \in R$.

More recently, the notion of a Baer ring was extended to an analogous module theoretic notion via the endomorphism ring of a module by Rizvi and Roman [**28**] as follows. A module M_R is called a *Baer* (resp., *quasi-Baer*) *module* if, for any $N_R \leq M_R$ (resp., $N \trianglelefteq M$), there is $e^2 = e \in S := \text{End}(M_R)$ with $\ell_S(N) = Se$, where $\ell_S(N) = \{f \in S \mid f(N) = 0\}$.

We note that a ring R is a (quasi-)Baer ring if and only if R_R is a (quasi-)Baer module. There are close connections between an extending ring and a Baer ring (see [**7**]). In [**28**] it was shown that every $\mathcal{K}$-nonsingular extending module is Baer and that every Baer module under a dual condition becomes extending (see Theorem 2.5).

The notions of Baer and quasi-Baer modules are very different in general. However, when M is a finitely generated module over a Dedekind domain, it was shown in [**16**] that M is a Baer module if and only if M is a quasi-Baer module (see Theorem 2.19). Motivated by this result, in this paper we investigate similarity and contrasts mainly between Baer modules, quasi-Baer modules, and extending modules. Also, for a module M of a certain type over a commutative ring, and a fixed injective hull $E(M)$, we study the similarities and contrasts between the Baer *module* hull, the quasi-Baer *module* hull, and the extending *module* hull of M. We obtain certain classes of modules where Baer module hulls do exist, while extending module hulls do not exist, and vice versa. Explicit description of each type of hulls of M for these classes of modules is provided.

We show that any essential extension of a nonsingular cyclic module over a commutative ring is Baer if and only if it is quasi-Baer if and only if it is quasi-continuous if and only if it is extending if and only if it is FI-extending if and only if it is strongly FI-extending. We also exhibit a number of examples where Baer, extending and quasi-continuous hulls of a module differ from each other. Furthermore, Baer hulls and extending hulls of certain classes of modules are also compared and differentiated.

All rings are assumed to have identity and all modules are assumed to be unitary. For right R-modules M_R and N_R, we use $\text{Hom}(M_R, N_R)$, $\text{Hom}_R(M, N)$, or $\text{Hom}(M, N)$ to denote the set of all R-module homomorphisms from M_R to N_R. Likewise, $\text{End}(M_R), \text{End}_R(M)$, or $\text{End}(M)$ denotes the endomorphism ring of an R-module M. For an R-module homomorphism $f \in \text{Hom}_R(M, N)$, $\text{Ker}(f)$ denotes the kernel of f.

We use $E(M_R)$ or $E(M)$ to denote an injective hull of a module M_R. For a module M, $K \leq M$, $L \trianglelefteq M$, $N \leq^{\text{ess}} M$, and $U \leq^{\oplus} M$ denote that K is a submodule of M, L is a fully invariant submodule of M, N is an essential submodule of M, and U is a direct summand of M, respectively. If M is an R-module, $\text{Ann}_R(M)$ stands for the annihilator of M in R.

For a module M and a nonempty set Λ, let $M^{(\Lambda)}$ denote the direct sum of $|\Lambda|$ copies of M. When Λ is finite with $|\Lambda| = n$ and $n > 1$, $M^{(n)}$ stands for $M^{(\Lambda)}$. We use $\text{CFM}_\Lambda(R)$ to denote the $\Lambda \times \Lambda$ column finite matrix ring over a ring R. Let $\text{Mat}_n(R)$ denote the $n \times n$ matrix ring over a ring R.

For a ring R, we let $Q(R)$ and $Q^\ell(R)$ stand for the maximal right ring of quotients of R and the maximal left ring of quotients of R, respectively. The symbols $\mathbb{C}$, $\mathbb{R}$, $\mathbb{Q}$, $\mathbb{Z}$, and $\mathbb{Z}_n (n > 1)$ denote the field of complex numbers, the field of real numbers, the field of rational numbers, the ring of integers, and the ring of integers modulo n, respectively. Ideals of a ring without the adjective "left" or "right" mean two-sided ideals. Finally, we note that for a ring R, the notation $I \trianglelefteq R$ denotes that I is an ideal of R.

2. Similarities and Contrasts

Our focus in this section, is to investigate the similarities and contrasts between Baer modules, extending modules, and related classes of modules. We also compare Baer module hulls and extending module hulls over a commutative domain.

For an essential extension V_R of a nonsingular cyclic module over a commutative ring R, we prove that V_R is Baer if and only if V_R is quasi-Baer if and only if V_R is quasi-continuous if and only if V_R is extending if and only if V_R is FI-extending. Using this result, we prove that the Baer hull, the quasi-Baer hull, the extending hull, and the FI-extending hull of a nonsingular cyclic module M over a commutative ring coincide with each other and are precisely equal to the quasi-continuous hull of M. As a byproduct, when R is a commutative semiprime ring, we characterize all possible intermediate Baer (extending, quasi-continuous) rings between R and $Q(R)$. So we obtain the smallest intermediate Baer (extending, quasi-continuous) ring between a commutative semiprime ring R and $Q(R)$.

For Baer modules properties, and the relationship between Baer modules and extending modules, we start with the following.

DEFINITION 2.1. A module M_R is called a *Baer module* if, for any $N_R \leq M_R$, there is $e^2 = e \in S := \mathrm{End}(M_R)$ with $\ell_S(N) = Se$, where $\ell_S(N) = \{f \in S \mid f(N) = 0\}$. Equivalently, a module M is Baer if and only if for any left ideal I of S, $r_M(I) = fM$ with $f^2 = f \in S$, where $r_M(I) = \{m \in M \mid Im = 0\}$.

THEOREM 2.2 ([**28**, Theorem 2.17]). Let M be a Baer module. Then every direct summand of M is also a Baer module.

DEFINITION 2.3 ([**28**, Definition 2.5]). A module M_R is said to be $\mathcal{K}$*-nonsingular* if, for all $\varphi \in \mathrm{End}_R(M)$, $\mathrm{Ker}(\varphi)_R \leq^{\text{ess}} M_R$ implies $\varphi = 0$.

Every Baer module is $\mathcal{K}$-nonsingular [**28**, Lemma 2.15]. Also in [**28**], it is proved that every nonsingular module is $\mathcal{K}$-nonsingular, but not conversely in general. For more details on $\mathcal{K}$-nonsingular modules, see [**28**] and [**29**].

DEFINITION 2.4 ([**28**, Definition 2.7]). A module M_R is called $\mathcal{K}$*-cononsingular* if, for $N_R \leq M_R$, $\ell_S(N) = 0$ implies $N_R \leq^{\text{ess}} M_R$, where $S = \mathrm{End}_R(M)$.

There are close connections between an extending ring and a Baer ring. Indeed, it is proved in [**7**] that a ring R is right extending and right nonsingular if and only if R is a Baer ring which is right cononsingular. The following result shows a strong bond between extending modules and Baer modules.

THEOREM 2.5 ([**28**, Theorem 2.12]). A module M_R is extending and $\mathcal{K}$-nonsingular if and only if M_R is Baer and $\mathcal{K}$-cononsingular.

DEFINITION 2.6 ([**30**, Definition 2.3]). Let M_R be an R-module and $S = \mathrm{End}_R(M)$. Then M_R is called *quasi-retractable* if $\mathrm{Hom}_R(M, r_M(I)) \neq 0$ for every left ideal I of S with $r_M(I) \neq 0$ (or, equivalently, if $r_S(I) \neq 0$ for every left ideal I with $r_M(I) \neq 0$).

THEOREM 2.7 ([**30**, Theorem 2.5]). A module M_R is Baer if and only if $\mathrm{End}_R(M)$ is a Baer ring and M_R is quasi-retractable.

Recall from [**8**] that a ring R is called *quasi-Baer* if the left annihilator of every ideal of R is generated, as a left ideal, by an idempotent. It is routine to show that a ring R is quasi-Baer if and only if the right annihilator of every ideal of R is generated, as a right ideal, by an idempotent.

DEFINITION 2.8 ([**28**, Definition 3.2]). A right R-module M is said to be a *quasi-Baer module* if for each $N \trianglelefteq M$, $\ell_S(N) = Se$ for some $e^2 = e \in S := \mathrm{End}_R(M)$, where $\ell_S(N) = \{f \in S \mid f(N) = 0\}$. Thus for a right R-module M, M is a quasi-Baer module if and only if, for any ideal J of $S, r_M(J) = fM$ for some $f^2 = f \in S$, where $r_M(J) = \{m \in M \mid Jm = 0\}$.

We note that a ring R is quasi-Baer if and only if R_R is a quasi-Baer module. It is shown in [**28**] that there are close connections between a quasi-Baer module and an FI-extending module (see Theorem 2.16). As a consequence, any direct sum of nonsingular uniform modules is quasi-Baer but may not be Baer.

THEOREM 2.9. (i) ([**28**, Theorem 4.1]) Let M_R be a quasi-Baer module. Then $\mathrm{End}_R(M)$ is a quasi-Baer ring.

(ii) ([**17**, Proposition 2.11]) Let M_R be a quasi-retractable and $\mathrm{End}_R(M)$ is a quasi-Baer ring. Then M_R is a quasi-Baer module.

(iii) ([**28**, Theorem 3.17]) Any direct summand of a quasi-Baer module is a quasi-Baer module.

(iv) ([**28**, Corollary 3.20]) A free module over a quasi-Baer ring is a quasi-Baer module.

DEFINITION 2.10 ([**18**, Definition 2.2]). Let M be a right R-module and $S = \mathrm{End}(M)$. Then M is called a *Rickart module* if for each $\phi \in S$, $r_M(\phi) = \mathrm{Ker}(\phi) = eM$ for some $e^2 = e \in S$.

Recall that a ring R is called right (resp., left) Rickart if the right (resp., left) annihilator of any element of R is generated by an idempotent, as a right (resp., left) ideal of R. Thus R is a right Rickart ring if and only if R_R is a Rickart module. By [**19**, Theorem 2.2], every direct summand of a Rickart module is a Rickart module. The following example comes from [**16**, Example 1.4].

EXAMPLE 2.11. (i) Every Baer module is a quasi-Baer and a Rickart module.

(ii) All semisimple modules are (quasi-)Baer and Rickart modules. For a von Neumann regular ring R, R_R and ${}_RR$ are Rickart modules.

(iii) Every nonsingular injective (even extending) module is a Baer module (see Theorem 2.5).

(iv) Any free R-module $R_R^{(\Lambda)}$ with Λ a nonempty set, over a quasi-Baer ring R which is not a Baer ring (e.g., $R = T_2(\mathbb{Z})$, the 2×2 upper triangular matrix ring over $\mathbb{Z}$), is a quasi-Baer module but it is not a Baer module.

(v) There exist Rickart modules which are not (quasi-)Baer. For example, let F be a field and R be the ring of infinite sequences with elements from F which are eventually constant. Then R is a von Neumann regular ring so R_R is a Rickart module. But R is not a (quasi-)Baer ring. Thus R_R is not a (quasi-)Baer module.

(vi) There exist quasi-Baer rings R which are not right Rickart rings (e.g., $R = \mathrm{Mat}_n(\mathbb{Z}[x])$, where $n > 1$). So R_R is a quasi-Baer module, but R_R is neither a Rickart nor a Baer module.

(vii) Let R be a right hereditary ring which is not Baer. Then every free R-module is Rickart but not a Baer module (see [**18**, Proposition 2.29]).

DEFINITION 2.12 ([**28**, Definition 3.6]). A right R-module M is said to be *FI-$\mathcal{K}$-nonsingular* if for any $I \trianglelefteq S$, $r_M(I) \leq^{\mathrm{ess}} eM$ with $e^2 = e \in S$ implies $Ie = 0$, where $S = \mathrm{End}_R(M)$.

EXAMPLE 2.13. (i) Let R be a ring. Then R_R is $\mathcal{K}$-nonsingular if and only if R_R is nonsingular (see [**28**, Remark 2.9]).

(ii) Every $\mathcal{K}$-nonsingular module is FI-$\mathcal{K}$-nonsingular (see [**28**, Corollary 3.9]).

(iii) If R is a semiprime ring, then R_R is FI-$\mathcal{K}$-nonsingular. In fact, say $I \trianglelefteq R$ such that $r_R(I) \leq^{\mathrm{ess}} eR_R$ with $e^2 = e \in R$. By contradiction assume that $Ie \neq 0$. Since R is semiprime and $I \trianglelefteq R$, $eI \neq 0$ and hence there exists $x \in I$ such that $0 \neq ex \in eI \subseteq eR$. Therefore there is $r \in R$ with $0 \neq exr \in r_R(I)$ because $r_R(I) \leq^{\mathrm{ess}} eR_R$. Thus $0 \neq exr \in r_R(I) \cap I = 0$, a contradiction. Hence $Ie = 0$. Therefore, R_R is FI-$\mathcal{K}$-nonsingular (see [**6**, Example 4.6.7, p. 132]).

(iv) From [**24**], there exists a prime ring R with $Z(R_R) \neq 0$, where $Z(R_R)$ is the singular submodule of R_R. By part (i), R_R is not $\mathcal{K}$-nonsingular. But note that R_R is FI-$\mathcal{K}$-nonsingular by part (iii) (see [**6**, Example 4.6.7, p.132]).

DEFINITION 2.14 ([**28**, Definition 3.7]). A right R-module M is said to be *FI-$\mathcal{K}$-cononsingular* if for any $N \trianglelefteq M$, $r_M(\ell_S(N)) \leq^{\oplus} M$ implies that $N \leq^{\text{ess}} r_M(\ell_S(N))$, where $S = \text{End}_R(M)$.

EXAMPLE 2.15. (i) Any $\mathcal{K}$-cononsingular module is an FI-$\mathcal{K}$-cononsingular module (see [**28**, Corollary 3.9]).

(ii) There exists a ring R such that R_R is FI-$\mathcal{K}$-cononsingular, but R_R is not $\mathcal{K}$-cononsingular. For example, take

$$R = \begin{bmatrix} \mathbb{C} & \mathbb{C} \\ 0 & \mathbb{R} \end{bmatrix}.$$

Then R is a Baer ring, but R is not right extending. Thus R_R is not $\mathcal{K}$-cononsingular by Theorem 2.5. We note that R_R is FI-extending, hence R_R is FI-$\mathcal{K}$-cononsingular by [**28**, Lemma 3.11]. For more details, see [**6**, Example 4.6.7, p. 132].

THEOREM 2.16 ([**28**, Theorem 3.10]). A module M is FI-extending and FI-$\mathcal{K}$-nonsingular if and only if M is quasi-Baer and FI-$\mathcal{K}$-cononsingular.

Recall from [**4**] that a module M is said to be *strongly FI-extending* if every fully invariant submodule of M is essential in a fully invariant direct summand of M. Clearly, a strongly FI-extending module is FI-extending. In the following example, there exists a right self-injective ring R such that R_R is not strongly FI-extending. Thus the notion of a strongly FI-extending module does not generalize that of an injective module.

EXAMPLE 2.17 ([**4**, Example 1.1]). Let $R = \mathbb{Z}_3[S_3]$ be the group algebra of the symmetric group S_3 on $\{1, 2, 3\}$ over $\mathbb{Z}_3$. Then R is a right self-injective ring. But R_R is not strongly FI-extending.

LEMMA 2.18 ([**4**, Proposition 1.5]). Let M be a nonsingular module. Then M is FI-extending if and only if M is strongly FI-extending.

Recall that a commutative domain R is called *Dedekind* if R is hereditary (i.e., every ideal of R is projective as an R-module).

In the following, there is coincidence of being a Baer module, being a quasi-Baer module, and being a Rickart module for the class of finitely generated modules over a Dedekind domain.

THEOREM 2.19 ([**16**, Theorem 4.10]). Let M be a finitely generated module over a Dedekind domain. Then the following are equivalent.

(i) M is a Baer module.

(ii) M is a quasi-Baer module.

(iii) M is a Rickart module.

(iv) M is semisimple or torsion-free.

Motivated by Theorem 2.19, in the following, we provide a class of modules V for which the notions of being Baer, being quasi-Baer, being quasi-continuous, being extending, being FI-extending, and being strongly FI-extending all coincide.

We use $\mathbf{I}(R)$ and $\mathcal{B}(R)$ for the set of idempotents and the set of central idempotents of a ring R, respectively.

THEOREM 2.20. Let M_R be a nonsingular cyclic module over a commutative ring R, and let V_R be an essential extension of M_R. Then the following are equivalent.

(i) V_R is Baer.
(ii) V_R is quasi-Baer.
(iii) V_R is quasi-continuous.
(iv) V_R is extending.
(v) V_R is FI-extending.
(vi) V_R is strongly FI-extending.

Proof. (i)$\Rightarrow$(ii) is obvious.

(ii)$\Rightarrow$(iii) Say $M = R/I := \overline{R}$, where I is an ideal of R. Since M is nonsingular, $\overline{R}$ is a nonsingular ring (see the proof of [**22**, Lemma 2]) and hence the ring $\overline{R}$ is semiprime. We put $E = E(M_R)$. By the proof of [**22**, Lemma 2], $Q(\overline{R}) = E$ and $\text{End}(E_R) = E$. Thus E is a commutative von Neumann regular ring.

Now assume that V is quasi-Baer. Consider

$$\theta : \text{End}(V_R) \to \text{End}(E_R),$$

where $\theta(\varphi)$ is an extension of $\varphi \in \text{End}(V_R)$. If $\varphi_1 = \varphi_2$ with $\varphi_1, \varphi_2 \in \text{End}(V_R)$, then

$$(\theta(\varphi_1) - \theta(\varphi_2))(v) = \varphi_1(v) - \varphi_2(v) = 0 \text{ for each } v \in V.$$

Therefore $(\theta(\varphi_1) - \theta(\varphi_2))(V) = 0$, and hence

$$\theta(\varphi_1) - \theta(\varphi_2) \in \Delta := \{f \in \text{End}(E_R) \mid \text{Ker}(f)_R \leq^{\text{ess}} E_R\}.$$

As E_R is nonsingular, E_R is $\mathcal{K}$-nonsingular by [**28**], and so $\Delta = 0$. Thus we have that $\theta(\varphi_1) - \theta(\varphi_2) = 0$, so $\theta(\varphi_1) = \theta(\varphi_2)$. Hence θ is well-defined.

It is routine to verify that θ is a ring monomorphism, so $\text{End}(V_R)$ is a subring of a commutative von Neumann regular ring $\text{End}(E_R) = E$. Hence $\text{End}(V_R)$ is a commutative (semiprime) ring. As V_R is quasi-Baer, $\text{End}(V_R)$ is a quasi-Baer ring by Theorem 2.9(i). Thus $\text{End}(V_R)$ is a Baer ring.

Furthermore, V is an $\overline{R}$-module. For this, we need to show that $VI = 0$. Assume on the contrary that $va \neq 0$ for some $v \in V$ and $a \in I$. As M is nonsingular, so is V. Hence any essential submodule of V, in particular M, is dense in V. Thus there exists $r \in R$ such that $vr \in M$ and $var \neq 0$. Since $0 \neq var = (vr)a \in MI = 0$, we get a contradiction. Therefore $VI = 0$. For $v \in V$ and $\overline{r} \in \overline{R}$, define $v\overline{r} = vr$. Then $v\overline{r}$ is well-defined, and hence V is an $\overline{R}$-module, which is induced from the R-module structure of V.

Let $\lambda : \overline{R} \to \text{End}(V_R)$ be defined by $\lambda(\overline{r}) \in \text{End}(V_R)$ such that $\lambda(\overline{r})(v) = v\overline{r}$ for all $v \in V$. Since $\overline{R} = M \leq V$, we see that λ is a ring monomorphism. Therefore $\text{End}(V_R)$ is a commutative Baer intermediate ring between the semiprime ring $\overline{R}$ and $Q(\overline{R})$. From [**5**, Theorem 3.3] (or [**20**]), $\mathbf{I}(Q(\overline{R})) = \mathcal{B}(Q(\overline{R})) \subseteq \text{End}(V_R)$. Thus it follows that each $e \in \mathbf{I}(Q(\overline{R})) = \mathbf{I}(\text{End}(E_R))$ is in $\text{End}(V_R)$, thus $eV \subseteq V$. Therefore V is quasi-continuous by [**12**].

(iii)$\Rightarrow$(iv)$\Rightarrow$(v) is evident.

(v)$\Rightarrow$(i) Assume that V is FI-extending. Because V is nonsingular, V is FI-$\mathcal{K}$-nonsingular. Therefore V is quasi-Baer by Theorem 2.16, so V is extending from the preceding implications. Hence Theorem 2.5 yields that V is Baer since V is nonsingular (hence $\mathcal{K}$-nonsingular).

(v)$\Leftrightarrow$(vi) follows from Lemma 2.18. □

In view of Theorem 2.19, we may expect Theorem 2.20(i)–(vi) are equivalent to the fact that V_R is Rickart. However, the following example eliminates this expectation.

EXAMPLE 2.21. Let F be a field and $F_n = F$ for $n = 1, 2, \ldots$. We put

$$R = \left\{ (a_n) \in \prod_{n=1}^{\infty} F_n \mid a_n \text{ is eventually constant} \right\},$$

which is a subring of $\prod_{n=1}^{\infty} F_n$. Then R is a commutative (von Neumann) regular ring. Hence R_R is a Rickart module. Take $M_R = V_R = R_R$. Then V_R is a nonsingular cyclic module over the commutative ring R. However, since R is not a Baer ring, V_R is not (quasi-)Baer, not extending, not quasi-continuous, not FI-extending, and not strongly FI-extending by Theorem 2.20.

DEFINITION 2.22 ([**6**, Definition 8.4.1, p.310]). Let M_R be a module. We fix an injective hull $E(M_R)$ of M_R. Let $\mathfrak{M}$ be a class of modules. We call, when it exists, a module H_R the $\mathfrak{M}$ *hull* of M_R if H_R is the smallest extension of M_R in $E(M_R)$ that belongs to $\mathfrak{M}$.

Let M be a module. By Definition 2.22, the *Baer* (resp., *extending,* etc.) *hull* of M is the smallest Baer (resp., extending, etc.) essential extension of M in $E(M)$.

COROLLARY 2.23. Every nonsingular cyclic module M over a commutative ring R has a Baer hull, which is given by ΩM, where Ω is the subring of $\text{End}(E(M))$ generated by $\mathbf{I}(\text{End}(E(M)))$. Moreover, the Baer hull, the quasi-Baer hull, the quasi-continuous hull, the extending hull, the FI-extending hull, and the strongly FI-extending hull of M coincide.

Proof. By [**12**], M is quasi-continuous if and only if $\Omega M \subseteq M$. So ΩM is the quasi-continuous hull of M because $\Omega(\Omega M) = \Omega M$. Thus ΩM is the Baer hull, the quasi-Baer hull, the extending hull as well as the FI-extending hull of M by Theorem 2.20. □

THEOREM 2.24 ([**27**, Corollary 5]). Let R be a commutative noetherian ring. Then every continuous R-module is quasi-injective.

While the description of continuous hulls over noncommutative rings in general is open, it is known from [**23**] that every continuous module M over an arbitrary ring decomposes as $M = U \oplus V$, where U is directly finite and V is purely infinite. Since V is quasi-injective being purely infinite, the question can be reduced to the description of directly finite continuous modules.

For a given module M, the quasi-injective hull of M exists by [**13**], while the quasi-continuous hull of M exists by [**12**]. However, continuous hull of M does not exists in general as in the following example.

EXAMPLE 2.25 ([**22**, Example 3]). Let V be a vector space over a field F with basis elements $v_m, w_k (m, k = 0, 1, 2, \ldots)$. We denote by V_n the subspace generated by the $v_m (m \geq n)$ and all the w_k. Also we denote by W_n the subspace generated by the $w_k (k \geq n)$. We write $\mathcal{S}$ for the shift operator such that $\mathcal{S}(w_k) = w_{k+1}$ and $\mathcal{S}(v_i) = 0$ for all k, i.

Let R be the set of all $\rho \in \text{End}_F(V)$ such that $\rho(v_m) \in V_m, \rho(w_0) \in W_0$ and $\rho(w_k) = \mathcal{S}^k \rho(w_0)$, for m, $k = 0, 1, 2, \ldots$. It is routine to check that R is a subring of $\mathrm{End}_F(V)$. Consider the left R-module $M = W_0$. Then M has no continuous hull. For more details, see [**6**, Example 8.4.4, p.311].

In contrast to Example 2.25, however, when the base ring R is commutative noetherian, any R-module M_R has a continuous hull, which is same as the quasi-injective hull of M_R.

THEOREM 2.26. Let R be a commutative noetherian ring and M be an R-module. Then M has a continuous module hull. In this case, the continuous module hull of M is precisely the quasi-injective module hull.

Proof. Let $\Lambda = \mathrm{End}_R(E(M))$. Then by Johnson and Wong [**13**], $V := \Lambda M$ is the quasi-injective hull of M. Thus V is continuous. To show that V is the continuous hull of M, assume that U is a continuous module such that $M \leq U \leq E(M)$. By Theorem 2.24, U is quasi-injective. Hence $V \leq U$ because V is the quasi-injective hull of M. Therefore $V = \Lambda M$ is the continuous hull of M. □

Examples 2.27 and 2.28 illustrate Theorem 2.20 and Corollary 2.23.

EXAMPLE 2.27. In view of Theorems 2.19 and 2.20, one may expect that part (iii) of Theorem 2.20 can be sharpened to the property that V is continuous. But it is impossible. For example, let R be a commutative domain which is not a field. Take $M = V = R_R$. Then V_R is Baer, extending, and quasi-continuous, however V_R is not continuous. Further, in this case V_R itself is the Baer, the extending, and the quasi-continuous hull of V_R itself.

Let F be the field of fractions of R. Then $\Lambda = \mathrm{End}_R(E(V)) = \mathrm{End}_R(F) = F$. By Johnson and Wong [**13**], $\Lambda V = F_R$ is the quasi-injective hull of V_R. Moreover, by [**22**] (see also [**6**, Theorem 8.4.11, p.319]) F_R is the continuous hull of V_R.

Recall that a commutative domain R is called *Prüfer* if R is semihereditary (i.e., every finitely generated ideal of R is projective).

EXAMPLE 2.28. (i) Let $M_R = \mathbb{Z}_{p^n}$, where $R = \mathbb{Z}$, p is a prime integer, and n is an integer such that $n > 1$. Then M is *cyclic but not nonsingular* over the commutative ring R. Note that M is quasi-injective, so the quasi-continuous hull of M and the extending hull of M are M itself.

Further, M is the FI-extending hull of M. However, M has no Baer hull. In fact, assume on the contrary that M has a Baer hull, say V. Then $V = \mathbb{Z}_{p^k}$ with $k \geq n$ or $V = \mathbb{Z}_{p^\infty}$, the Prüfer p-group. But V is not a Baer $\mathbb{Z}$-module. Also M is not a Baer module. Similarly M has no quasi-Baer hull.

(ii) ([**6**, Example 8.4.13, p.319]) Let $R = \mathbb{Z}[x, y]$, the polynomial ring over $\mathbb{Z}$. We put $M_R = R \oplus R$. Then M is a *nonsingular module which is not cyclic* over the commutative domain R. However, M_R is a direct sum of two copies of the indecomposable Baer module R_R.

By Theorem 2.9(iv), M is a quasi-Baer module, so M itself is the quasi-Baer module hull of M. Since R_R is FI-extending, M is FI-extending by [**6**, Theorem 2.3.5, p.48]. Thus M itself is also the FI-extending hull of M. However, M has neither Baer hull nor extending hull (hence M is neither Baer nor extending). Also note that M is not quasi-continuous.

In general, for a commutative domain A, $(A \oplus A)_A$ has a Baer hull if and only if $(A \oplus A)_A$ has an extending hull if and only if A is a Prüfer domain (cf. Theorem 2.38). In this case, $(A \oplus A)_A$ itself is the Baer and the extending hull of $(A \oplus A)_A$.

(iii) Let $R = \mathrm{Mat}_n(\mathbb{Z})$, where n is an integer such that $n > 1$. Put $M = R_R$. Then M is a cyclic nonsingular R-module, but R is *not commutative*. We see that M is Baer and extending, hence M itself is the Baer and the extending hull of M. However, M is *not quasi-continuous*. Furthermore, the Baer hull of M is not the quasi-continuous hull of M. Also the extending hull of M is not the quasi-continuous hull of M.

We recall from [**2**] that the ring $R\mathcal{B}(Q(R))$, the subring of $Q(R)$ generated by R and $\mathcal{B}(Q(R))$, is called the *idempotent closure* of a ring R.

THEOREM 2.29. Assume that R is a commutative semiprime ring and $e^2 = e \in R$. Then $eR\mathcal{B}(Q(R))_R$ is the Baer, the quasi-Baer, the quasi-continuous, the extending, the FI-extending, and the strongly FI-extending hull of eR_R.

Proof. Note that R is nonsingular and so eR_R is nonsingular cyclic R-module. Also $E(eR_R) = eQ(R)_R$. So $\mathbf{I}(\mathrm{End}_R(E(eR_R)) = \mathbf{I}(eQ(R)) = \mathcal{B}(eQ(R)) = e\mathcal{B}(Q(R))$. Let $\langle e\mathcal{B}(Q(R))\rangle$ be the subring of $\mathrm{End}_R(E(eR_R)) = \mathrm{End}_R(eQ(R)) = eQ(R)e$ generated by $e\mathcal{B}(Q(R))$. Note that $\langle e\mathcal{B}(Q(R))\rangle = e\mathcal{B}(Q(R))$. From Corollary 2.23, $\langle e\mathcal{B}(Q(R))\rangle eR_R = eR\mathcal{B}(Q(R))_R$ is the Baer hull, the quasi-Baer hull, the quasi-continuous, the extending hull, the FI-extending hull, and the strongly FI-extending hull of eR_R. □

As a consequence, we obtain the next corollary (see [**6**, Theorem 8.2.4, p.276]). In particular Corollary 2.30(i) is [**20**, Proposition 2.5].

COROLLARY 2.30. Let R be a commutative semiprime ring and let A be an intermediate ring between R and $Q(R)$. Then:

(i) A is a Baer ring if and only if $\mathcal{B}(Q(R)) \subseteq A$.

(ii) A is an extending ring if and only if $\mathcal{B}(Q(R)) \subseteq A$.

(iii) A is a quasi-continuous ring if and only if $\mathcal{B}(Q(R)) \subseteq A$.

Thereby, A is a Baer ring if and only if A is an extending ring if and only if A is a quasi-continuous ring. Also $R\mathcal{B}(Q(R))$ is the smallest Baer (extending, quasi-continuous) intermediate ring between R and $Q(R)$.

Proof. (i) From Theorem 2.29, $R\mathcal{B}(Q(R))_R$ is the Baer hull of R_R. Hence by Theorem 2.7, $\mathrm{End}_R(R\mathcal{B}(Q(R))) = R\mathcal{B}(Q(R))$ is a Baer ring. Let A be a ring of quotients of R which is Baer. Then $\mathrm{End}_R(A) = A$. Say I is an ideal of A. Since A is a Baer ring, $r_A(I) = eA$ with $e^2 = e \in A$, where $r_A(I)$ is the right annihilator of I in A. Hence A_R is Baer by Definition 2.1, thus $R\mathcal{B}(Q(R)) \subseteq A$ because $R\mathcal{B}(Q(R))_R$ is the Baer hull of R_R from Theorem 2.29. Conversely, assume $\mathcal{B}(Q(R)) \subseteq A$. By [**6**, Corollary 8.3.19, p.293], A is a quasi-Baer ring. Hence A is a Baer ring since A is commutative.

(ii) Say A is an extending ring of quotients of R. As A is nonsingular, A is a Baer ring by [**7**] or Theorem 2.5. Thus $R\mathcal{B}(Q(R)) \subseteq A$ by part (i). For the converse, suppose $\mathcal{B}(Q(R)) \subseteq A$. Then by part (i), A is a Baer ring. Since A is a commutative semiprime ring, A is cononsingular. Thus A is an extending ring from [**7**] or Theorem 2.5.

(iii) Let A be a quasi-continuous ring of quotients of R. Then A is an extending ring of quotients of R, and so $\mathcal{B}(Q(R)) \subseteq A$ by part (ii). Conversely, suppose

$\mathcal{B}(Q(R)) \subseteq A$. Then by part (ii), A is an extending ring. Since A is commutative, A is a quasi-continuous ring. □

EXAMPLE 2.31. Let R be a commutative semiprime ring and n an integer such that $n > 1$. Assume that $e^2 = e \in \mathrm{End}_R(R^{(n)})$. In view of Theorem 2.29, one may expect that $eR\mathcal{B}(Q(R))_R^{(n)}$ is the Baer hull of $eR_R^{(n)}$. But this does not hold true. For a commutative domain R, we note that $R\mathcal{B}(Q(R))_R^{(n)} = R_R^{(n)}$ has a Baer hull if and only if R is Prüfer. So if R is a non-Prüfer commutative domain, then $R\mathcal{B}(Q(R))_R^{(n)} = R_R^{(n)}$ has no Baer hull.

When R is a *commutative nonsingular* ring (hence R is semiprime), $R\mathcal{B}(Q(R))_R$ is the Baer hull, the extending hull, and the quasi-continuous hull of R_R by Theorem 2.29. But as in the following example, there is a *noncommutative right (and left) nonsingular* semiprime ring R such that neither the Baer hull nor the extending hull of R_R exist. Also $R\mathcal{B}(Q(R))_R$ is not the quasi-continuous hull of R_R.

EXAMPLE 2.32. Let $R = \mathrm{Mat}_n(F[x, y])$, where n is an integer such that $n > 1$ and $F[x, y]$ is the polynomial ring over a field F with commuting indeterminates x and y. Then:

(i) The Baer hull of R_R does not exist. For this, put $U = \mathrm{Mat}_n(F(x)[y])$ and $V = \mathrm{Mat}_n(F(y)[x])$. Then $\mathrm{End}(U_R) = U$, $\mathrm{End}(V_R) = V$. Further, U and V are Baer rings. We can check that U_R and V_R are quasi-retractable, and so U_R and V_R are Baer modules by Theorem 2.7.

Now say B_R is the Baer hull of R_R. Then $B \subseteq U \cap V = R$ (see [**6**, Example 8.3.34, p.300]). Hence $B = R$. Now $\mathrm{End}(R_R) = R$ is not a Baer ring because $F[x, y]$ is not a Prüfer domain. Thus $B_R = R_R$ is not a Baer module by Theorem 2.7, a contradiction. Hence the Baer hull of R_R does not exist.

(ii) The extending hull of R_R does not exist. Assume C_R is the extending hull of R_R. Let U and V be as in part (i). Then U_R and V_R are $\mathcal{K}$-cononsingular by direct computation. By Theorem 2.5, U_R and V_R are extending. So $C \subseteq U \cap V = R$. Thus $C_R = R_R$ and so R_R is extending. Since R_R is nonsingular, R_R is Baer by Theorem 2.5, which is a contradiction by the preceding argument. Thus the extending hull of R_R does not exist.

(iii) $R\mathcal{B}(Q(R))_R$ is not the quasi-continuous hull of R_R. Because R is a prime ring, $\mathcal{B}(Q(R)) = \{0, 1\}$, so $R = R\mathcal{B}(Q(R))$. Hence $R\mathcal{B}(Q(R))_R$ is not quasi-continuous from [**12**] as it does not contain all idempotents of $Q(R)$. Therefore $R\mathcal{B}(Q(R))_R$ is not the quasi-continuous hull of R_R.

We recall that a ring R is called *semiprimary* if $R/J(R)$ is artinian and $J(R)$ is nilpotent, where $J(R)$ is the Jacobson radical of R. It is well-known that if R is a semiprimary ring, then R is right hereditary if and only if R is left hereditary.

LEMMA 2.33 ([**33**, Theorem 2] and [**30**, Theorem 3.3]). *Let R be a ring. Then the following are equivalent.*

(i) *R is a semiprimary right (and left) hereditary ring.*

(ii) *$\mathrm{CFM}_\Lambda(R)$ is a Baer ring for any nonempty set Λ.*

EXAMPLE 2.34. Let Γ be an uncountable set. Then from [**34**, Remark, p.32], $\mathbb{Z}^{(\Gamma)}$ is not a Baer $\mathbb{Z}$-module (see also [**6**, Example 4.5.2(v), p.125]). Furthermore, $\mathbb{Z}^{(\Gamma)}$ has no Baer hull. For this, assume on the contrary that $\mathbb{Z}^{(\Gamma)}$ has a Baer hull as a $\mathbb{Z}$-module, say V. From [**19**, Theorem 2.16] for each $\alpha \in \Gamma$, $\mathbb{Z} \oplus \mathbb{Q}^{(\Gamma\backslash\{\alpha\})}$ is a Baer

$\mathbb{Z}$-module (see also [**6**, Theorem 4.2.18, p.107]). So $V \subseteq \cap_{\alpha\in\Gamma}(\mathbb{Z}\oplus\mathbb{Q}^{(\Gamma\setminus\{\alpha\})}) = \mathbb{Z}^{(\Gamma)}$. Hence $V = \mathbb{Z}^{(\Gamma)}$. Thus we get a contradiction.

On the other hand, assume on the contrary that $\mathbb{Z}^{(\Gamma)}$ has an extending hull as a $\mathbb{Z}$-module, say W. By [**1**, Proposition 1.8(ii) and Corollary 3.3(i)], $\mathbb{Z}\oplus\mathbb{Q}^{(\Gamma\setminus\{\alpha\})}$ is an extending $\mathbb{Z}$-module, for each $\alpha \in \Gamma$. Therefore $W \subseteq \cap_{\alpha\in\Gamma}(\mathbb{Z}\oplus\mathbb{Q}^{(\Gamma\setminus\{\alpha\})}) = \mathbb{Z}^{(\Gamma)}$. Hence $W = \mathbb{Z}^{(\Gamma)}$. So $\mathbb{Z}^{(\Gamma)}$ is an extending $\mathbb{Z}$-module, which is a contradiction as Γ is infinite (see [**9**, p.56]). Thus $\mathbb{Z}^{(\Gamma)}$ has no extending hull.

Recall from [**11**, p.112] that a module M is said to be Σ-*injective* if $M^{(\Lambda)}$ is injective for any set Λ. By [**11**, Exercise 20.4C(d), pp.114–115], a ring R is right Goldie if and only if $E(R_R)$ is Σ-injective.

Motivated by Example 2.34, in the following, we study the existence of a Baer hull and that of an extending hull of $R_R^{(\Lambda)}$ for any nonempty set Λ for the case when R is a right extending semiprime right Goldie ring. Note that a semiprime right Goldie ring is right nonsingular.

PROPOSITION 2.35. Let R be a right extending semiprime right Goldie ring. Then the following are equivalent.

(i) $R_R^{(\Lambda)}$ has a Baer hull for any nonempty set Λ.

(ii) $R_R^{(\Gamma)}$ has an extending hull for any nonempty set Γ.

(iii) R is semismimple artinian.

Proof. Note that $E(R_R)$ is Σ-injective (see [**11**, Exercise 20.4C(d), pp.114–115]) since R is right Goldie. Therefore $E(R_R^{(\Omega)}) = E(R_R)^{(\Omega)}$ for any set Ω.

(i)$\Rightarrow$(iii) Since R is semiprime right Goldie, R is right nonsingular. Let Λ be a nonempty set. First, suppose $|\Lambda| = 1$. Since R is right extending and right nonsingular, R is a Baer ring by Theorem 2.5 or [**7**].

Next suppose $|\Lambda| \geq 2$. By assumption, $R_R^{(\Lambda)}$ has a Baer hull, say V_R. Take $\alpha \in \Lambda$. As R_R is nonsingular extending (hence R_R is Baer by Theorem 2.5 or [**7**]) and $E(R_R^{(\Lambda\setminus\{\alpha\})}) = E(R_R)^{(\Lambda\setminus\{\alpha\})}$ is nonsingular injective, by [**19**, Theorem 2.16] (or [**6**, Theorem 4.2.18, p.107]) $R \oplus E(R_R)^{(\Lambda\setminus\{\alpha\})}$ is Baer as a right R-module. Thus $V \subseteq R \oplus E(R_R)^{(\Lambda\setminus\{\alpha\})}$ for all $\alpha \in \Lambda$, and so

$$V \subseteq \cap_{\alpha\in\Lambda}(R \oplus E(R_R)^{(\Lambda\setminus\{\alpha\})}) = R^{(\Lambda)}.$$

Hence $V = R^{(\Lambda)}$, and thus $R_R^{(\Lambda)}$ is Baer for any set Λ with $|\Lambda| \geq 2$.

Consequently, $R_R^{(\Lambda)}$ is a Baer module for any nonempty set Λ. By Theorem 2.7, $\mathrm{CFM}_\Lambda(R)$ is a Baer ring for any nonempty set Λ because $\mathrm{End}_R(R^{(\Lambda)}) = \mathrm{CFM}_\Lambda(R)$. By Lemma 2.33, R is semiprimary, and thus R is semisimple artinian because R is semiprime.

(ii)$\Rightarrow$(iii) Note that $E(R_R)$ is Σ-injective. Let Γ be a nonempty set. First, suppose $|\Gamma| = 1$. Since R is right extending and right nonsingular, R is a Baer ring by [**7**] or Theorem 2.5. Thus R_R is a Baer module.

Next we assume $|\Gamma| \geq 2$. Take $\beta \in \Gamma$. By hypothesis, $R_R^{(\Gamma)}$ has an extending hull, say U_R. As R_R is nonsingular extending and $E(R_R^{(\Gamma\setminus\{\beta\})}) = E(R_R)^{(\Gamma\setminus\{\beta\})}$ is nonsingular injective, it follows that $R \oplus E(R_R)^{(\Gamma\setminus\{\beta\})}$ is extending as a right R-module from [**1**, Proposition 1.8(ii) and Corollary 3.3(i)]. Thus we have that $U \subseteq R \oplus E(R_R)^{(\Gamma\setminus\{\beta\})}$ for all $\beta \in \Gamma$, and so

$$U \subseteq \cap_{\beta\in\Gamma}(R \oplus E(R_R)^{(\Gamma\setminus\{\beta\})}) = R^{(\Gamma)}.$$

Hence $U = R^{(\Gamma)}$, so $R_R^{(\Gamma)}$ is extending for any nonempty set Γ with $|\Gamma| \geq 2$. Since $R_R^{(\Gamma)}$ is nonsingular, $R_R^{(\Gamma)}$ is a Baer module by Theorem 2.5 for any nonempty set Γ with $|\Gamma| \geq 2$.

As a consequence, $R_R^{(\Gamma)}$ is a Baer module for any nonempty set Γ. Therefore R is semisimple artinian by the argument used in the proof of (i)⇒(iii).

(iii)⇒(i) and (iii)⇒(ii) are obvious. □

When R is a commutative domain, Proposition 2.35 yields that $R_R^{(\Lambda)}$ has a Baer hull (or an extending hull) for any nonempty set Λ if and only if R is a field.

REMARK 2.36. Let R be a right FI-extending semiprime ring. Then R_R is quasi-Baer by [**3**, Theorem 4.7] (also see [**6**, Theorem 3.2.37, p.83]). Therefore $R_R^{(\Lambda)}$ is quasi-Baer for any nonempty set Λ by Theorem 2.9(iv). Thus $R_R^{(\Lambda)}$ itself is the quasi-Baer hull of $R_R^{(\Lambda)}$. Also we note that $R_R^{(\Lambda)}$ is FI-extending by [**3**, Theorem 1.3] (also see [**6**, Theorem 2.3.5, p.48]). Hence $R_R^{(\Lambda)}$ itself is the FI-extending hull of $R_R^{(\Lambda)}$.

Assume that R is a commutative domain and N is an R-module. Let $t(N) = \{n \in N \mid na = 0 \text{ for some } 0 \neq a \in R\}$. Then $t(N)$ is a submodule of N and $t(N)$ is called the torsion submodule of N. When $N = t(N)$, N is called a torsion module.

LEMMA 2.37 ([**14**, Corollary 2]). *Let R be a commutative domain, and let N be an R-module which is not a torsion module. Then N is extending if and only if its torsion submodule $t(N)$ is injective and $N/t(N)$ is extending.*

We exhibit the disparity between Baer hulls and extending hulls for a certain class of modules. The following theorem describes the extending hull of certain modules over a commutative domain.

THEOREM 2.38. *Assume that R is a commutative domain with the field of fractions F. Let A be an intermediate domain between R and F. Then the following are equivalent.*

(i) A is a Prüfer domain.

(ii) $E(M_R) \oplus A_R^{(n)}$ is the extending hull of $M_R \oplus A_R^{(n)}$ for any R-module M with $\text{Ann}_R(M) \neq 0$ and for any positive integer n.

(iii) $A_R^{(2)}$ is an extending module.

(iv) $A_R^{(2)}$ is a Baer module.

Proof. (i)⇒(ii) Let V_R be an extending essential extension of $M_R \oplus A_R^{(n)}$. We put $I = \text{Ann}_R(M)$. Since V is extending, $\ell_V(I) \leq^{\text{ess}} U$ for some $U \leq^{\oplus} V$. Therefore $V = U \oplus W$ with $W \leq V$. By assumption $I \neq 0$, so $\ell_V(I)$ is a torsion module. We put $Y = E(M_R) \oplus E(A_R^{(n)})$. Then $M_R \oplus A_R^{(n)} \leq V_R \leq Y_R$.

Let $t(V)$ (resp., $t(Y)$) be the torsion submodule of V (resp., Y). As M is a torsion module and R is a commutative domain, $E(M_R)$ is also a torsion module. Thus $E(M_R) \subseteq t(Y)$. Next, let $y \in t(Y)$. Write $y = y_1 + y_2 \in E(M_R) \oplus E(A_R^{(n)})$ with $y_1 \in E(M_R)$ and $y_2 \in E(A_R^{(n)})$. Then there exists $0 \neq r \in R$ such that $0 = yr = (y_1 + y_2)r = y_1 r + y_2 r$, so $y_1 r = 0$ and $y_2 r = 0$. As $E(A_R^{(n)}) = E(A_R)^{(n)}$ and $E(A_R)$ is the field of fractions of R, $y_2 = 0$ and so $y = y_1 \in E(M_R)$. Thus $t(Y) \subseteq E(M_R)$. Hence $t(Y) = E(M_R)$.

Now $M \subseteq \ell_V(I) \subseteq t(V) \subseteq t(Y) = E(M)$, and hence $M \leq^{\text{ess}} \ell_V(I)$. Further, since $\ell_V(I) \leq^{\text{ess}} U$ and $\ell_V(I)$ is a torsion module, U is also a torsion module. Therefore $U \subseteq t(V)$. Because $U \cap W = 0$, $M \cap W = 0$ and hence $E(M) \cap W = 0$.

We show that W is torsion-free. For this, say $w \in W \leq E(M) \oplus E(A_R)^{(n)}$. Then $w = x_w + y_w \in E(M) \oplus E(A_R)^{(n)}$, where $x_w \in E(M)$ and $y_w \in E(A_R)^{(n)}$. Consider

$$f : W \to E(A_R)^{(n)}$$

which is defined by $f(w) = y_w$ for $w \in W$. Since $E(M) \cap W = 0$, we see that f is an R-module monomorphism. Therefore W is torsion-free because $E(A_R)$ is the field of fractions of R (note $E(A_R)^{(n)}$ is torsion-free).

Next, take $v \in t(V)$. Write $v = x_v + y_v \in U \oplus W$ with $x_v \in U$ and $y_v \in W$. There exists $0 \neq b \in R$ such that $vb = 0$, hence $0 = vb = x_v b + y_v b$. So $x_v b = 0$ and $y_v b = 0$. But as W is torsion-free, $y_v = 0$ and thus $v = x_v \in U$. Hence $t(V) \subseteq U$. Consequently, $t(V) = U$.

From the preceding argument, we obtain that $V = U \oplus W$ and $U = t(V)$. Furthermore, $U \leq^{\text{ess}} E(M)$ because $M \leq^{\text{ess}} \ell_V(I) \leq^{\text{ess}} U \leq t(Y) = E(M_R)$. Note that $V = U \oplus W$ is extending, $t(V) = U$, and $V/t(V) \cong W$ is torsion-free. Thus U is injective from Lemma 2.37. So $U = E(M)$, and thus $V = E(M) \oplus W$. Therefore $E(M_R) \oplus A_R^{(n)} \leq V_R$. We put $N = E(M_R) \oplus A_R^{(n)}$. Then $t(N) = E(M)$ is injective.

Further, as A is a Prüfer domain, $\text{Mat}_n(A)$ is a Baer ring by [**19**, Corollary 3.15] (also see [**6**, Theorem 6.1.4, p.191]). Since $\text{End}(A_R^{(n)}) = \text{Mat}_n(A)$ is a Baer ring and $A_R^{(n)}$ is quasi-retractable by [**25**, Lemma 2.10], $A_R^{(n)}$ is a Baer module from Theorem 2.7. Theorem 2.5 yields that $A_R^{(n)}$ is extending because $A_R^{(n)}$ is $\mathcal{K}$-cononsingular by [**26**, Theorem 3.5]. So $N = E(M_R) \oplus A_R^{(n)}$ is extending by Lemma 2.37. Because $E(M_R) \oplus A_R^{(n)} \leq V_R$, $E(M_R) \oplus A_R^{(n)}$ is the extending hull of $M_R \oplus A_R^{(n)}$.

(ii)⇒(iii) In (ii), put $M = 0$ and $n = 2$. By (ii), $A_R^{(2)}$ is an extending module.

(iii)⇒(iv) By assumption $A_R^{(2)}$ is extending. Since $A_R^{(2)}$ is nonsingular, $A_R^{(2)}$ is a Baer module by Theorem 2.5.

(iv)⇒(i) As $A_R^{(2)}$ is a Baer module, $\text{End}_R(A^{(2)}) = \text{Mat}_2(\text{End}_R(A)) = \text{Mat}_2(A)$ is a Baer ring by Theorem 2.7. Hence from [**19**, Corollary 3.15] (also see [**6**, Theorem 6.1.4, p.191]), A is a Prüfer domain. □

THEOREM 2.39. Assume that R is a commutative domain, which is not a field. Let $M = R/I$, where I is a nonzero proper ideal of R. Then $E(M_R)$ is not $\mathcal{K}$-nonsingular.

Proof. Take $0 \neq a \in I$. Define $f : aR \to E(M)$ by $f(ar) = r + I \in M \subseteq E(M)$ for $r \in R$. Then $f \in \text{Hom}_R(aR, E(M))$. Say $g \in \text{Hom}_R(R, E(M))$ is an extension of f. Put $g(1) = \alpha \in E(M)$. Then $1 + I = f(a) = g(a) = g(1)a = \alpha a$.

Consider $\varphi : E(M) \to E(M)$ defined by $\varphi(x) = xa$ for $x \in E(M)$. Then we see that $\varphi \in \text{End}_R(E(M))$ and $\varphi(\alpha) = \alpha a = 1 + I \neq 0$. Now $\varphi(M) = Ma = 0$ because $M = R/I$ and $a \in I$. So $M \subseteq \text{Ker}(\varphi)$ and hence $\text{Ker}(\varphi) \leq^{\text{ess}} E(M)$. But $\varphi \neq 0$. So $E(M_R)$ is not $\mathcal{K}$-nonsingular. □

Corollary 2.23 and Theorem 2.29 show that Baer hulls and extending hulls coincide for certain classes of modules. But in the following example, we exhibit the disparity between Baer hulls and extending hulls for some other classes of modules.

EXAMPLE 2.40. (i) Assume that R is a Prüfer domain. Let M be a semisimple module with only a finite number of homogeneous components. Say n is a positive integer. Because $\mathrm{Ann}_R(M) \neq 0$, $E(M_R) \oplus R_R^{(n)}$ is the extending hull of $M_R \oplus R_R^{(n)}$ by Theorem 2.38.

(ii) Let R, M, and n be as in part(i). We note that $M_R \oplus R_R^{(n)}$ is FI-extending by [**6**, Theorem 2.3.5, p.48]. Hence $M_R \oplus R_R^{(n)}$ itself is the FI-extending hull of $M_R \oplus R_R^{(n)}$.

(iii) Let R be a Prüfer domain, which is not a field. Say I is a nonzero proper ideal of R. From Theorem 2.38, $E((R/I)_R) \oplus R_R^{(n)}$ is the extending hull of $(R/I)_R \oplus R_R^{(n)}$ for any positive integer n. Since $E((R/I)_R)$ is not $\mathcal{K}$-nonsingular from Theorem 2.39, [**28**, Lemma 2.15] yields that $E((R/I)_R)$ is not Baer. By [**28**, Theorem 2.17], any direct summand of a Baer module is a Baer module. Thus for each positive integer n, $E((R/I)_R) \oplus R_R^{(n)}$ is not Baer.

Let R be a commutative domain with the field of fractions F. A submodule K of F_R is called a *fractional ideal* of R if $rK \subseteq R$ for some $0 \neq r \in R$. Thus $K_R \cong (rK)_R$ and rK is an ideal of R.

For a fractional ideal K of R, we put $K^{-1} = \{q \in F \mid qK \subseteq R\}$, which is called the *inverse* of K. We say that a fractional ideal K is *invertible* if $KK^{-1} = R$. It is well-known that for a nonzero ideal I of a commutative domain R, I_R is projective if and only if $II^{-1} = R$. In this case, I_R is finitely generated and I^{-1} is a fractional ideal of R. Thus for each nonzero ideal I of a Dedekind domain R, it follows that $II^{-1} = R$ because I_R is projective. We note that a Dedekind domain is noetherian because every ideal is projective (hence every ideal is finitely generated). See [**31**, Chapter 6] for more details on Dedekind domains.

If I is an invertible ideal of a commutative domain R, then we let

$$I^{-2} = I^{-1}I^{-1}, \ I^{-3} = I^{-1}I^{-1}I^{-1}, \quad \text{and so on.}$$

For convenience, we put $I^0 = R$.

Assume that R is a Dedekind domain. Then for nonzero ideals $I_1, I_2, \ldots, I_n$ of R, it can be checked that $(I_1 I_2 \cdots I_n)^{-1} = I_n^{-1} \cdots I_2^{-1} I_1^{-1}$ (see [**25**, Lemma 2.8]).

We denote the Baer hull of a module M by $\mathbf{B}(M)$ when it exists.

THEOREM 2.41 ([**25**, Theorem 2.13]). *Let R be a Dedekind domain. Assume that M is an R-module with nonzero annihilator in R, and $\{K_1, K_2, \ldots, K_m\}$ is a finite set of nonzero fractional ideals of R. Then the following are equivalent.*

(i) *$M_R \oplus (\oplus_{j=1}^m K_j)_R$ has a Baer hull.*

(ii) *M_R is semisimple.*

(iii) *$M_R \oplus (\oplus_{j=1}^m K_j)_R$ has a Baer essential extension.*

In this case, $\mathbf{B}(M_R \oplus (\oplus_{j=1}^m K_j)_R) = M_R \oplus (\oplus_{j=1}^m K_j A)_R$, where $A = \sum_{\ell \geq 0} I^{-\ell}$ with $I = \mathrm{Ann}_R(M)$. Furthermore, $A = R[q_1, q_2, \ldots, q_n]$, where $1 = \sum_{u=1}^n r_u q_u$ with $r_u \in I$ and $q_u \in I^{-1}$, $1 \leq u \leq n$.

The conclusion of this section exhibits explicit examples of different types of hulls and how they differ from each other in the following.

EXAMPLE 2.42. Let p be a prime integer in $\mathbb{Z}$. Then:

(i) $\mathbb{Z}_{p^\infty} \oplus \mathbb{Z}$ is the extending hull of $\mathbb{Z}_p \oplus \mathbb{Z}$ by Theorem 2.38.

(ii) $\mathbb{Z}_p \oplus \mathbb{Z}[1/p]$ is the Baer hull of $\mathbb{Z}_p \oplus \mathbb{Z}$ from Theorem 2.41.

(iii) $\mathbb{Z}_{p^2} \oplus \mathbb{Z}$ has extending hull, which is $\mathbb{Z}_{p^\infty} \oplus \mathbb{Z}$ by Theorem 2.38.

(iv) From Theorem 2.41, the $\mathbb{Z}$-module $\mathbb{Z}_{p^2} \oplus \mathbb{Z}$ has no Baer hull because $\mathbb{Z}_{p^2}$ is not semisimple.

(v) $\mathbb{Z}_p \oplus \mathbb{Z}$ itself is the FI-extending hull of $\mathbb{Z}_p \oplus \mathbb{Z}$ because $\mathbb{Z}_p \oplus \mathbb{Z}$ is FI-extending as a $\mathbb{Z}$-module by [**3**, Theorem 1.3] (see also [**6**, Theorem 2.3.5, p.48]).

Acknowledgments. The authors are very thankful to the referee for his/her nice comments and helpful suggestions for the improvement of this work. The authors are grateful for the partial grant support received from The Ohio State University at Lima, the Office of International Affairs, Mathematics Research Institute and the ASC College, The Ohio State University. The authors also thank each others' institutions for the hospitality received during the research work on this paper.

References

[1] E. Akalan, G.F. Birkenmeier, and A. Tercan,*Goldie extending modules*, Comm. Algebra **37** (2009), 663–683, Corrigendum **38** (2010), 4747–4748, Corrigendum **41** (2013), 2005.

[1] E. Akalan, G. F. Birkenmeier, and A. Tercan, *Corrigendum to: Goldie extending modules [MR2493810]*, Comm. Algebra **38** (2010), no. 12, 4747–4748. MR2805141

[2] K. Beidar and R. Wisbauer, *Strongly and properly semiprime modules and rings*, Ring theory (Granville, OH, 1992), World Sci. Publ., River Edge, NJ, 1993, pp. 58–94. MR1344223

[3] G. F. Birkenmeier, B. J. Müller, and S. T. Rizvi, *Modules in which every fully invariant submodule is essential in a direct summand*, Comm. Algebra **30** (2002), no. 3, 1395–1415. MR1892606

[4] G. F. Birkenmeier, J. K. Park, and S. T. Rizvi, *Modules with fully invariant submodules essential in fully invariant summands*, Comm. Algebra **30** (2002), no. 4, 1833–1852. MR1894048

[5] G. F. Birkenmeier, J. K. Park, and S. T. Rizvi, *Hulls of semiprime rings with applications to C^*-algebras*, J. Algebra **322** (2009), no. 2, 327–352. MR2529092

[6] G. F. Birkenmeier, J. K. Park, and S. T. Rizvi, *Extensions of rings and modules*, Birkhäuser/Springer, New York, 2013. MR3099829

[7] A. W. Chatters and S. M. Khuri, *Endomorphism rings of modules over nonsingular CS rings*, J. London Math. Soc. (2) **21** (1980), no. 3, 434–444. MR577719

[8] W. E. Clark, *Twisted matrix units semigroup algebras*, Duke Math. J. **34** (1967), 417–423. MR0214626

[9] Nguyen Viet Dung, Dinh Van Huynh, P. F. Smith, and R. Wisbauer, *Extending modules*, Pitman Research Notes in Mathematics Series, vol. 313, Longman Scientific & Technical, Harlow; copublished in the United States with John Wiley & Sons, Inc., New York, 1994. MR1312366

[10] B. Eckmann and A. Schopf, *Über injektive Moduln* (German), Arch. Math. (Basel) **4** (1953), 75–78. MR0055978

[11] C. Faith, *Algebra II, Ring Theory*, Springer, Berlin-Heidelberg-New York, 1976.

[12] V. K. Goel and S. K. Jain, *π-injective modules and rings whose cyclics are π-injective*, Comm. Algebra **6** (1978), no. 1, 59–73. MR0491819

[13] R. E. Johnson and E. T. Wong, *Quasi-injective modules and irreducible rings*, J. London Math. Soc. **36** (1961), 260–268. MR0131445

[14] M. A. Kamal and B. J. Müller, *Extending modules over commutative domains*, Osaka J. Math. **25** (1988), no. 3, 531–538. MR969016

[15] I. Kaplansky, *Rings of operators*, W. A. Benjamin, Inc., New York-Amsterdam, 1968. MR0244778

[16] G. Lee, J.K. Park, S.T. Rizvi, and C.S. Roman, *Quasi-Baer module hulls and applications*, J. Pure Appl. Algebra **222** (2018), 2427–2455.

[17] G. Lee and S. Tariq Rizvi, *Direct sums of quasi-Baer modules*, J. Algebra **456** (2016), 76–92. MR3484136

[18] G. Lee, S. T. Rizvi, and C. S. Roman, *Rickart modules*, Comm. Algebra **38** (2010), no. 11, 4005–4027. MR2764847

[19] G. Lee, S. T. Rizvi, and C. S. Roman, *Direct sums of Rickart modules*, J. Algebra **353** (2012), 62–78. MR2872436

[20] A. C. Mewborn, *Regular rings and Baer rings*, Math. Z. **121** (1971), 211–219. MR0292894

[21] S. H. Mohamed and B. J. Müller, *Continuous and discrete modules*, London Mathematical Society Lecture Note Series, vol. 147, Cambridge University Press, Cambridge, 1990. MR1084376

[22] B. J. Müller and S. T. Rizvi, *On the existence of continuous hulls*, Comm. Algebra **10** (1982), no. 17, 1819–1838. MR674693

[23] B. J. Müller and S. T. Rizvi, *On injective and quasicontinuous modules*, J. Pure Appl. Algebra **28** (1983), no. 2, 197–210. MR697666

[24] B. L. Osofsky, *A non-trivial ring with non-rational injective hull*, Canad. Math. Bull. **10** (1967), 275–282. MR0212055

[25] J. K. Park and S. T. Rizvi, *Baer module hulls of certain modules over a Dedekind domain*, J. Algebra Appl. **15** (2016), no. 8, 1650141, 24 pp. MR3528763

[26] J.K. Park and S.T. Rizvi, *Properties inherited by direct sums of copies of a module*, Manuscript.

[27] S. T. Rizvi, *Commutative rings for which every continuous module is quasi-injective*, Arch. Math. (Basel) **50** (1988), no. 5, 435–442. MR942540

[28] S. T. Rizvi and C. S. Roman, *Baer and quasi-Baer modules*, Comm. Algebra **32** (2004), no. 1, 103–123. MR2036224

[29] S. T. Rizvi and C. S. Roman, *On $\mathcal{K}$-nonsingular modules and applications*, Comm. Algebra **35** (2007), no. 9, 2960–2982. MR2356310

[30] S. T. Rizvi and C. S. Roman, *On direct sums of Baer modules*, J. Algebra **321** (2009), no. 2, 682–696. MR2483287

[31] D. W. Sharpe and P. Vámos, *Injective modules*, Cambridge Tracts in Mathematics and Mathematical Physics, No. 62, Cambridge University Press, London-New York, 1972. MR0360706

[32] K. Shoda, *Zur Theorie der algebraischen Erweiterungen* (German), Osaka Math. J. **4** (1952), 133–143. MR0052403

[33] W. Stephenson and G.M. Tsukerman, *Rings of endomorphisms of projective modules*, Siberian Math. J. **11** (1970), 181–184 (English translation).

[34] G. V. Wilson, *Modules with the summand intersection property*, Comm. Algebra **14** (1986), no. 1, 21–38. MR814137

Department of Mathematics, Busan National University, Busan 46241, South Korea
Email address: jkp1128@yahoo.com

Department of Mathematics, The Ohio State University, Lima, OH 45804-3576, U.S.A.
Email address: rizvi.1@osu.edu

Contemporary Mathematics
Volume **715**, 2018
https://doi.org/10.1090/conm/715/14417

Direct sums of completely almost self-injective modules

Surjeet Singh

Dedicated to the memory of Professor Bruno J. Müller

ABSTRACT. For any two modules M and N over a ring R, the concept of M being *almost N-injective* was introduced by Baba in 1989. A module M is said to be *completely almost self-injective*, if for any two subfactors A, B of M, A is amost B-injective. The structure theory of completely almost self-injective modules has been developed by Singh (2017) and he has given a characterization of these modules. A direct sum of two completely almost self-injective modules need not be completely almost self-injective. In this paper, direct sums of completely almost self-injective modules are studied. Among other results, it has been proved that if a ring R is completely almost self-injective, then any finitely generated R-modules is completely almost self-injective.

Introduction

The concept of a module M_R being almost N_R-injective was defined by Baba [**4**], and it study was continued by Baba and Harada [**5**, **8**, **9**, **10**]. This concept has also been studied by many authors including Alahmadi, Jain and Singh in [**1**, **2**, **11**, **12**]. As defined in [**12**], a module M is said to be *completely almost self-injective*, if for any two subfactors A, B of M, A is almost B-injective. A characterization for a module to be completely almost self-injective has been given in [**12**, Theorem 2.11]. Thereafter in [**12**, Corollary 2.13], it is proved that a a von Neumann regular ring R is right completely almost self-injective if and only if $R/soc(R_R)$ is semi-simple and every minimal right ideal of R is injective. In this paper, we study properties of direct sums of completely almost self-injective modules and prove some basic properties of these direct sums. A direct sum of completely almost self-injective modules need not be completely almost self-injective.

Let $K = N_1 \oplus N_2$, where N_i are completely almost self-injective modules over a ring R. For K to be completely almost self-injective, two necessary conditions are: (a) If S is a simple submodule of a subfactor A of K, then any complement of S in A is a summand of A, (b) any non-simple uniserial subfactor of K is K-injective and has length 2, and any two non-simple uniserial subfactors U, V of K are isomorphic, whenever they have isomorphic socles. Conditions under which

2010 *Mathematics Subject Classification.* primary 16D50, secondary 16E50.
Key words and phrases. M-injectives, M-projectives, uniform modules, uniserial modules, Von-Neumann regular rings.

K satisfies these conditions are determined. Thereafter necessary sufficient conditions are determined under which K is completely almost self-injective, whenever $N_i/soc(N_i)$ are finitely generated (Theorem 13 and Theorem 14). From this it follows that if a ring R is such that R_R is completely almost self-injective, then any finitely generated module over R is completely almost self-injevtive (Theorem 15). At the end it is proved that a ring R is right completely almost self-injective if and only if $R/soc(R_R)$ is semi-simple and every non-simple uniserial subfactor of R_R is injective and projective (Theorem 16).

Preliminaries

All rings R considered here are with unity and all modules are unital right modules, unless otherwise stated. Let M_R be any module. Then $N \leq M$, $N < M$, $N \subseteq_e M$ denote that N is a submodule of M, a proper submodule of M, an essential submodule of M respectively. $E(M)$, $J(M)$ denote the injective hull, radical of M respectively. If M is of finite length, then $d(M)$ denotes its length.

A module M_R is said to be *almost N_R-injective*, if for any homomorphism $f : L \to M$, $L < N$, either f extends to a homomorphism from N to M, or there exist a decomposition $N = N_1 \oplus N_2$ with $N_1 \neq 0$ and a homomorphism $h : M \to N_1$ such that $hf(x) = \pi(x)$ for any $x \in L$, where $\pi : N \to N_1$ is a projection with kernel N_2. As observed in [**11**], a module M_R is almost N_R-injective if and only if for any maximal homomorphism $f : L \to M$, $L < N$ either f extends to a homomorphism from N to M or there exist a decomposition $N = N_1 \oplus N_2$ with $N_1 \neq 0$, and a homomorphism $h : M \to N_1$ such that $hf(x) = \pi(x)$ for any $x \in L$. Here $f : L \to M$, $L < N$ is *maximal* in the sense that there exists no extension of f, $g : L' \to M$, $L < L' < N$. A module N_R is called a *subfactor* of M, if it is isomorphic to a submodule of a homomorphic image of M. A module M_R is said to be *completely almost self-injective*, if given any two subfactors A, B of M, A is almost B-injective. For other concepts one may refer to [**3, 6**]. For von Neumann regular rings, one may refer to [**7**].

Completely almost self-injective modules

We start with following two results proved in [**12**].

THEOREM A ([**12**, Theorem 2.11]). Let M_R be any module. Then M is completely almost self-injective if and only if it satisfies the following conditions.

(i) $M/soc(M)$ is semi-simple.

(ii) If N is a subfactor of M, then any complement in N of a simple submodule S of N is a summand of N and if $S \subseteq J(N)$, then there exists a uniserial summand N_1 of N of length 2 that contains S.

(iii) Any two non-simple uniserial subfactors U, V of M are quasi-injective, and if they have isomorphic socles, they are isomorphic.

(iv) If A, B are two subfactors of M such that $J(A) = 0$, then any monomorphism $f : L \to B$, $L \subseteq_e A$ such that $f(soc(A)) \subseteq J(B)$ can be extended to a homomorphism from A to B. □

THEOREM B ([**12**, Corollary 2.12]). Let M_R be any module such that $M/soc(M)$ is finitely generated. Then M is completely almost self-injective if and only if it satisfies the following conditions.

(i) $M/soc(M)$ is semi-simple.

(ii) If N is a subfactor of M, then any complement in N of a simple submodule S of N is a summand of N and if $S \subseteq J(N)$, then there exists a uniserial summand N_1 of N of length 2 that contains S.

(iii) Any two non-simple uniserial subfactors U, V of M are quas-injective, and if they have isomorphic socles, they are isomorphic. □

It follows from [**12**, Lemma 2.7] that any subfactor of a completely almost self-injective module M satisfies condition (i) in the above theorem, so any subfactor of M is also completely almost self-injective, we also have $soc(M) \subseteq_e M$.

EXAMPLE. Let A be any uniserial module of length 2, which is quasi-injective but not quasi-projective. Obviously, A is completely almost self-injective. Consider $N = A \oplus A$. We have an automorphism σ of $A/J(A)$ which cannot be lifted to an endomorphism of A. Now $K = \{(a, b) \in N \mid b + J(A) = \sigma(a + J(A))\}$. Then $soc(N) \subseteq K$, $d(K) = 3$, and K is indecomposable. It follows that a complement in K of a simple submodule of K is not a summand of K. Hence N is not completely almost self-injective. □

DEFINITION 1. Let G_R, H_R be any two completely almost self-injective modules. They are said to satisfy condition $(*)$, if any non-simple uniserial subfactor A of G is H-injective and any non-simple uniserial subfactor of H is G-injective. □

Any non-simple uniserial subfactor of a completely almost self-injective module G is isomorphic to a factor of G and is G-injective [**12**, Lemma 2.10(b)]. Therefore a pair G, H of completely almost self-injective modules satisfy $(*)$ if and only if any non-simple uniserial subfactor of G or H is $G \oplus H$-injective.

LEMMA 2 ([**12**, Lemma 2.10]). *Let K_R be a module satisfying the following conditions.*

(i) $K/soc(K)$ is semi-simple.

(ii) Any complement in K of a minimal submodule S of K is a summand of K. If a minimal submodule $S \subseteq J(K)$, then there exists a uniserial submodule U of K of length 2 containing S.

(iii) Any uniserial subfactor of K of length 2 is quasi-injective. Any two non-simple uniserial subfactors U, V of K are isomorphic, whenever they have isomorphic socles.

Then the following hold.

(a) Any simple module S_R is either K-injective or there exists a uniserial summand U of K of length 2 such that $S \cong soc(U)$.

(b) Any uniserial subfactor of K of length 2 is K-injective. For any $V < K$ such that K/V is uniserial and of length 2, V is a summand of K. □

LEMMA 3. *Let B_1, B_2 be two uniserial modules of length not more than 2. Let $K = B_1 \oplus B_2$ and T be a proper submodule of K. Then K/T is uniform if and only if one of the following holds.*

(i) If at least one of B_i is simple, then T is isomorphic to one of B_i under natural projections or is equal to $soc(K)$.

(ii) If each of B_i is of length 2, then T maps on one of the B_i under natural projections.

Proof. Let K/T be uniform. Let $\pi_i : B_1 \oplus B_2 \to B_i$ be associated projections. If both B_i are simple, T is simple, and K/T is simple. Clearly, T is isomorphic to B_1 or B_2 under the associated projections. Suppose at least one of B_i, say B_1 has length 2. Let B_2 be simple. If $T \cap B_1 = 0$, then $T \cong B_2$ and $K/T \cong B_1$. If $T \cap B_1 \neq 0$, then $T \cong B_1$ under π_1 or $T = soc(K)$, thus (i) holds. Suppose both B_i's are of length 2. Then $d(T) \geq 2$. If $T \cap B_1 = 0$ or $T \cap B_2 = 0$, then T maps onto B_2 or B_1 under associated projections, so (ii) holds. Suppose $T \cap B_1 \neq 0$, $T \cap B_2 \neq 0$. If $d(T) = 2$, $T = soc(K)$ and K/T is not uniform. So $d(T) = 3$, and (ii) hold.

Conversely, if condition (i) or (ii) holds, it can be easily seen that K/T is uniserial. □

LEMMA 4. Let N_1, N_2 be a pair of completely almost self-injective modules over a ring R. Let $K = N_1 \oplus N_2$.

(i) Let S be a simple subfactor of K. Then S is a subfactor of N_1 or N_2.

(ii) Let A be a subfactor of K which is non-simple and uniserial. Then $d(A) = 2$ and A is a subfactor of N_1 or N_2.

Proof. By the hypothesis, there exist $x_i \in N_i$ such that for $x = x_1 + x_2 \in K$, $S = xR/T$ for some maximal submodule T of xR. We have associated projections $\pi_i : K \to N_i$. If $x_1 \notin \pi_1(T)$, then $S \cong x_1R/\pi_1(T)$. Suppose $x_1 \in \pi_1(T)$. Then there exists $y = x_1 + z_2 \in T$, where $z_2 \in \pi_2(T)$, and $x - y = x_2 - z_2 \neq 0$. As $x - y \notin T$, it generates S modulo T, so S is a subfactor of N_2 .

Let A be a non-simple uniserial subfactor of K. By [**12**, Lemma 2.7], $A/soc(A)$ is simple, therefore $d(A) = 2$. Now there exists $x = x_1 + x_2 \in K$, where $x_i \in N_i$, and a submodule T of xR such that $A \cong xR/T$. For some $r \in R$, $xr + T$ generates $B = soc(xR/T)$. As in (i), $\pi_1(xrR + T)/\pi_1(T)$ is simple or $\pi_2(xrR + T)/\pi_2(T)$ is simple. This completes the proof. □

LEMMA 5. Let N_1, N_2 be any pair of completely almost self-injective modules satisfying condition $(*)$. Let $K = N_1 \oplus N_2$.

(i) Any uniform subfactor of K is simple or of length 2. If a uniform subfactor of K is not simple, then it is quasi-injective.

(ii) Any non-simple uniserial subfactor of K is K-injective.

Proof. Let H be a uniform subfactor of K. By [**12**, Lemma 2.7], $H/soc(H)$ is semi-simple. Thus $S = soc(H)$ is simple. Let H be not simple. Suppose $d(H) > 2$. We get two distinct uniserial submodules A, B of H of length 2. By Lemma 4, each of A, B is a subfactor of N_1 or N_2. Therefore by Theorem A, A and B are quasi-injective. If A, B both are subfactors of say N_1, by Theorem A, they are isomorphic. Suppose A is a subfactor of N_1 and B is a subfactor of N_2. By [**12**, Lemma 2.6], A is isomorphic to a summand of N_1 and B is isomorphic to a summand of N_2. It follows from Definition 1 that A is B-injective, therefore $A \cong B$. Which gives a contradiction. Hence $d(H) = 2$.

Let A be any non-simple uniserial subfactor of K. By Lemma 4, we take A a subfactor of N_1. Then, by Lemma 2(b), $N_1 = A \oplus F$ for some $F < N_1$. Then Theorem A gives (i). By Theorem A, A is N_1-injective. As condition $(*)$ holds, A is also N_2-injective. Hence A is K-injective. □

PROPOSITION 6. Let N_1, N_2 be a pair of completely almost self-injective modules satisfying condition $(*)$. Let $K = N_1 \oplus N_2$. Let A be any summand of K. Then any complement of a minimal submodule of A in A is a summand of A. If S is a simple submodule of K contained in $J(K)$, then there exists a uniserial summand L of length 2 such that $S \subseteq L$.

Proof. Let S be a simple submodule of K and L be a complement of S in K. We can take $S = (s_1+s_2)R$ with $s_i \in N_i$ and $s_1 \neq 0$. Then K/L is uniform and its socle is isomorphic to S and its length can be taken 2. Now $s = s_1+s_2$ is such that $s+L$ generates $soc(K/L)$, $K/L = \overline{N_1} + \overline{N_2}$, where $\overline{N_i} = (N_i + L)/L \cong N_i/L_i$, where $L_i = L \cap N_i$. Then at least one of $s_i + L$ is non-zero. We can take $s_1 + L \neq 0$. We get $L_1 \cap s_1R = 0$. As N_1/L_1 is uniform, L_1 is a complement of s_1R. By Theorem A, $N_1 = L_1 \oplus B_1$, where $B_1$is a uniserial submodule. Then $L = L_1 \oplus (L \cap (B_1 \oplus N_2))$. Thus $K/L \cong (B_1 \oplus N_2)/L'$, where $L' = L \cap (B_1 \oplus N_2)$. But $L' \cap B_1 = 0$, we get L' is a complement of B_1 in $B_1 \oplus N_2$. In addition, N_2/L_2 is uniform, where $L_2 = L' \cap N_2 = L \cap N_2$.

Case 1. L_2 is not essential in N_2. Then, by Lemma 2(b), $N_2 = L_2 \oplus B_2$, where B_2 is uniserial. So, $K/L \cong (B_1 \oplus B_2)/(L \cap (B_1 \oplus B_2))$, which is uniserial. As $(L \cap (B_1 \oplus B_2)) \cap B_1 = 0$, by Lemma 3, $L \cap (B_1 \oplus B_2) \cong B_2$. We get $B_1 \oplus B_2 = B_1 \oplus L \cap (B_1 \oplus B_2)$, therefore $B_1 \oplus N_2 = L_2 \oplus B_1 \oplus (L \cap (B_1 \oplus B_2))$. Hence $N = L \oplus B_1$.

Case 2. L_2 is essential in N_2. If $L_2 = N_2$. Then $L= L_1 \oplus N_2$, which is a summand of K. So we take $L_2 < N_2$. Then, by Theorem A, $\overline{N_2} = N_2/L_2$ is simple. We have natural projections $\eta_1 : B_1 \oplus N_2 \to B_1$, $\eta_2 : B_1 \oplus N_2 \to N_2$. Therefore $L'/L_2 = L'/(L' \cap N_2) \cong \eta_1(L')$. Suppose $\eta_1(L') = 0$, then $L' = L_2$. But $(B_1 \oplus N_2)/L'$ is uniform. Therefore $L' = N_2$, $L_2 = N_2$, which gives a contradiction. Hence $\eta_1(L') \neq 0$. But $B_1 \cap L' = 0$. Therefore, by Lemma 3, $\eta_2(L') = N_2$, $B_1 \oplus N_2 = B_1 \oplus L'$. Thus $K = L_1 \oplus B_1 \oplus L' = B_1 \oplus L$.

Now $K = A \oplus B$. Let S be a simple submodule of A and V be a complement of S in A. Then $V \oplus B$ is a complement of S in K. Therefore $K = T \oplus V \oplus B$, where T is a uniserial submodule of K. Then $A = V \oplus (A \cap (T \oplus B))$. □

PROPOSITION 7. Let N_1, N_2 be any pair of completely almost self-injective modules satisfying $(*)$ and $K = N_1 \oplus N_2$. Let S_R be any simple module. Then either S is K-injective or there exists a uniserial summand B of K of length 2 such that $S \cong soc(B)$.

Proof. Suppose S is not K-injective. So there exists a maximal homomorphism $f : L \to S$, $L < K$, which cannot be extended from K to S. Let $T = \ker f$. As $K/soc(K)$ is semi-simple, and $L \subseteq_e K$, we get T is not essential in L. Therefore $L = S' \oplus T$, where $S' \cong S$. We get a complement H in K of S' that contains T. The maximality of f gives that $H \subseteq L$, so that $T = H$. By Proposition 6, $K = B \oplus H$, where B is some uniform submodule with $soc(B) \cong S$. As $L < K$, B is not simple. By Lemma 5, $d(B) = 2$. □

LEMMA 8. Let $T = zR$, $N = yR$ be two non-simple cyclic modules such that T is uniserial and $d(T) = 2$. Let $\pi_1 : T \oplus N \to T$, $\pi_2 : T \oplus N \to N$ be associated projections. Let K be a submodule of $T \oplus N$ such that $soc(T) \subseteq K$ but $T \nsubseteq K$, $soc(N) \subseteq K$ but $N \nsubseteq K$, and $z + y \in K$. Then the following hold.

(i) There exists a maximal submodule L of yR containing $V = N \cap K$ such that $\theta : T/soc(T) \to yR/L$, $\theta(zr + soc(T)) = yr + L$ for any $r \in R$, defines an isomorphism.

(ii) The mapping θ can be lifted to a homomorphism from T to yR, if and if only if there exists a uniserial submodule U of K such that $U \cap V = 0$ and it has length 2.

(iii) If $T \oplus N$ is completely almost self-injective, then θ can be lifted from T to N.

Proof. (i) As $V = K \cap N$. Then $K/V \cong \pi_1(K) = T$, it is uniserial and has length 2. Let $soc(K/V) = L/V$. We have mapping $f : yR \to T/soc(T)$defined as follows. Let $yr \in yR$. If for some $u \in T$, $u + yr \in K$, set $f(yr) = u + soc(T)$. Suppose for some u_1, $u_2 \in T$, we have $u_1 + yr$, $u_2 + yr \in K$, we get $u_1 - u_2 \in K \cap T = soc(T)$. This shows that f is well defined. We see that $f(yr) = zr + soc(T)$ for any $r \in R$. Its kernel L is a maximal submodule of yR containing V. Which gives the desired isomorphism θ. This proves (i).

(ii) Suppose, we have a uniserial submodule $H = (zr_1 + yr_2)R$ of K, which is disjoint from V and has length 2. If $d(zr_1R) \leq 1$, then $d(yr_2R) = 2$ and we have an epimorphism $\lambda : yr_2R \to zr_1R$ with $\lambda(yr_2) = zr_1$, its kernel is non-zero. Let $yr_2r \in \ker\lambda$, then $zr_1r = 0$, therefore $yr_2r \in H \cap N \subseteq H \cap K \cap N = H \cap V = 0$, which gives a contradiction. Hence $d(zr_1R) = 2$, which gives $zr_1R = T$ and homomorphism $\mu : T \to yr_2R$ such that $\mu(zr_1) = yr_2$. Thus $z = zr_1s$ for some $s \in R$. Therefore $\mu(z) = yr_2s$. Now $zr_1s + yr_2s \in H$ gives

$$z + yr_2s \in K \text{ and } (z + y) - (z + yr_2s) = (y - yr_2s) \in V \subseteq L.$$

Thus $yr_2s + L = y + L$. This proves that μ lifts θ.

Conversely, let there exists a homomorphism $\lambda : T \to yR$ lifting θ. Then

$$U = \{x + \lambda(x) : x \in T\}$$

is uniserial and has length 2. Now $\theta(z + soc(T)) = y + L$. Therefore we have that $y + L = \lambda(z) + L$.

We show that $U \subseteq K$. Observe that, if $yr \in L$, then $zr \in soc(T) \subseteq K$, therefore $yr \in K \cap N = V$. Which gives $L \subseteq V$. But L is maximal in N, we get $L = V$, which gives $y + V = \lambda(z) + V$. Thus $\lambda(z) = y + v$ for some $v \in V$. We get $z + \lambda(z) = (z + y) + (0 + v) \in K$. Hence $U \subseteq K$. As $U \cap N = 0$, $U \cap V = 0$.

(iii) Let $T \oplus N$ be completely almost self-injective. Suppose θ cannot be lifted from T to N. Let $H = \{u + v : u \in T,\ v \in N \text{ such that } \theta(u + soc(T)) = v + L\}$. Then $soc(T) \subseteq H$, $L \subseteq H$, $T \not\subseteq H$, $N \not\subseteq H$. For $W = H \cap N$, $H/W \cong T$, therefore by Lemma 2(b), $H = U \oplus W$ for some uniserial submodule U of length 2. This contradicts (ii). □

PROPOSITION 9. Let A_R be a uniserial quasi-injective module of length 2 and N_R be any completely almost self-injective module.

(i) If $A \oplus N$ is completely almost self-injective, then A is N-projective.

(ii) If A is not quasi-projective and is N-projective, then $Hom(A, N/soc(N)) = 0$.

(iii) If $N \oplus N$ is completely almost self-injective. then any non-simple uniserial subfactor A of N is quasi-projective.

(iv) A is N-projective if and only if it is C-projective for every cyclic submodule C of N.

Proof. (i) Let $f : A \to N/L$ be a non-zero homomorphism, where $L < N$. If f is one-to one, by Theorem A, $N/L = K/L \oplus T/L$, where $K/L = f(A)$. As K is completely almost self-injective, $K = K_1 \oplus L$ for some uniserial submodule K_1 of length 2, therefore $N = K_1 \oplus T$ and we finish. Suppose $f(A)$ is simple. Now $f(A) = K/L$ for some $K \leq N$. If L is not essential in K, then $K = S \oplus L$, where S is a simple submodule, and we finish. Suppose $L \subseteq_e K$. Now $A = zR$. So $K = yR + L$, where $f(z) = y + L$. Thus, without loss of generality, we take $K = yR$ a cyclic submodule. Set $H = \{u+v \in A \oplus K : f(u) = v + L\}$. Then $soc(A) \oplus L \subseteq H$. But H does not contain A or yR. However, $A \oplus K$ being a submodule of a completely almost self-injective module $A \oplus N$, is completely almost self-injective. Therefore by Lemma 8, f can be lifted from A to K . Hence A is N-projective.

(ii) Let $f : A \to N/soc(N)$ be a non-zero homomorphism. It lifts to a homomorphism $g : A \to N$. If $A \cong g(A)$, then the hypothesis gives that A is $g(A)$-projective, and hence A is quasi-projective. Which is a contradiction. Hence $g(A)$ is a simple module, so $g(A) \subseteq soc(A)$, and that gives $f(A) = 0$, which is a contradiction. Hence $Hom(A, N/soc(N)) = 0$.

(iii) It is immediate from (i).

(iv) Let A be C-projective for every cyclic submodule C of N. Let $f : A \to N/L$ be a non-zero homomorphism. Now $f(A) = K/L$ with $K \leq N$. If $d(f(A)) = 2$, by Lemma 2(b), $K = B \oplus L$,for some $B \cong f(A)$ and we finish.

Suppose that $d(f(A)) = 1$. If L is not essential in K, trivially $K = S \oplus L$, for some simple submodule $S \cong f(A)$. Suppose $L \subseteq_e K$. Then $K = xR + L$ with $xR/(xR \cap L) \cong f(A)$. As A is xR-projective, $xR = S \oplus (xR \cap L)$, $K = S \oplus L$ for some simple submodule $S \cong f(A)$. Hence A is N-projective. □

Corollary 10. Let M_R be a completely almost self-injective module. Let A, B be two disjoint non-simple uniserial summands of M.

(i) $A \cong B$ if and only if $A/soc(A) \cong B/soc(B)$.

(ii) If A is not quasi-projective, A is not isomorphic to B.

Proof. Now $A \oplus B$ is isomorphic to a summand of M, therefore it is completely almost self-injective. Therefore the result follows Proposition 9. □

Theorem 11. Let N_1, N_2 be any two completely almost self-injective modules satisfying $(*)$. Let $K = N_1 \oplus N_2$.

(i) If K is completely almost self-injective, then the following hold.

(a) Any non-simple uniserial subfactor A of N_1 is N_2-projective and any non-simple uniserial subfactor of N_2 is N_1-projective.

(b) If a non-simple uniserial subfactor A of N_1 is not quasi-projective, then N_2 has no uniserial subfactor isomorphic to A and $Hom(A, N_2/soc(N_2)) = 0$; for no uniserial subfactor B of N, $A/soc(A) \cong B/soc(B)$.

(ii) If any non-simple uniserial subfactor of N_1 is C-projective, where C is any cyclic subfactor of N_2 and any non-simple uniserial subfactor of N_2 is C-projective, where C is any cyclic subfactor of N_1, then the following hold.

(c) For any subfactor A of K and any simple submodule S of A, any complement in A of S is a summand of A.

(d) Any non-simple uniserial subfactor of K is quasi-injective.

(e) Any two non-simple uniserial subfactors U, V of K are isomorphic, whenever $soc(U) \cong soc(V)$.

Proof. (i) Suppose K is completely almost self-injective. Let A be a non-simple uniserial subfactor of N_1 and C be any subfactor of N_2 . Then $A \oplus C$ being a subfactor of K, is completely almost self-injective. By Proposition 9, A is C-projective. Similarly, any non-simple uniserial subfactor A' of N_2 is C-projective, where C is a subfactor of N_1. This proves (i)(a). Suppose there exists a non-zero homomorphism $\theta : A \to N_2/soc(N_2)$. We have a lifting of θ, $\eta : A \to N_2$. If

$A \cong \eta(A)$, then A being $\eta(A)$-projective gives that A is quasi-projective, which is a contradiction. Thus $\eta(A)$ is simple, $\eta(A) \subseteq soc(N_2)$, therefore $\theta = 0$.

(ii) Firstly, let A be any submodule of K and S a simple submodule of A. Let V be a complement of S in A. We prove that V is a summand of A. By Lemma 5, $d(A/V) \leq 2$. If $d(A/V) = 1$, trivially, $A = S \oplus V$. Suppose $d(A/V) = 2$. Let $\pi_i : K \to N_i$ be associated projections. By Lemma 4, A/V is a subfactor of N_1 or N_2. For some $y \in A$, $A = yR + V$. Let $W = yR \cap V$. Then $A/V \cong yR/W$. Therefore $d(yR/W) = 2$. It is enough to prove that W is summand of yR. Now $y = y_1 + y_2$ with $y_1 \in N_1$, $y_2 \in N_2$.

Because of Lemma 4, we take $yR/W \cong y_1R/\pi_1(W)$, hence $y_1R = B_1 \oplus \pi_1(W)$ for some uniserial submodule B_1 isomorphic to yR/W. By the hypothesis, B_1 is N_2-injective, which gives that B_1 is $\pi_2(yR)$-injective. By Lemma 2, B_1 is $\pi_1(W)$-injective. So, B_1 is $\pi_1(W) \oplus y_2R$-injective.

Case 1. B_1 is quasi-projective. Proposition 9 gives, B_1 is $y_1R \oplus y_2R$-projective. As $yR \subseteq y_1R \oplus y_2R$, we get B_1 is yR-projective. Thus yR/W is yR-projective. Hence $yR = B \oplus W$, $A = B \oplus V$, where B is uniserial and has length 2.

Case 2. B_1 is not quasi-projective. Therefore $Hom(B_1, N_2/soc(N_2)) = 0$ and N_2 has no submodule isomorphic to B_1. Now $B_1 = z_1R$. For some $z_2 \in y_2R$, $z = z_1 + z_2 \in yR$. As $yR = zR + W$, we take $y = z$, so $z_1 = y_1$. If $y_1R \subseteq yR$, then $yR = B_1 \oplus W$, and we finish. Suppose $y_2 \in \pi_2(W)$. Then there exists $w = w_1 + y_2 \in W$, where $w_1 \in \pi_1(W)$. Then $y - w = y_1 - w_1$, we can replace y by $y - w$, and we finish.

Thus we are left with the situation that $y_2 \notin \pi_2(W)$ and $y_1 \notin yR$. It follows that $y_2 \notin yR$. If $yR \cap y_1R = 0$, then yR embeds in the completely almost self-injective module y_2R, and we finish. Therefore $soc(y_1R) \subseteq yR$. Now $y_2R/\pi_2(W)$ is uniserial and $d(y_2R/\pi_2(W)) \leq 2$. There exists a unique maximal submodule L of y_2R containing $\pi_2(W)$ and $y_1R/soc(y_1R) \cong y_2R/L$. Suppose $d(y_2R/\pi_2(W)) = 2$. Then $y_2R = C \oplus \pi_2(W)$, where C is uniserial and has length 2. Then $y_1R \cong C$, which is a contradiction, as $Hom(y_1R, N_2/soc(N_2)) = 0$. Thus it follows that $d(y_2R/\pi_2(W)) = 1$ and $L = \pi_2(W)$. As $Hom(B_1, N_2/soc(N_2)) = 0$, and $y_2R/soc(y_2R)$ embeds in $N_2/soc(N_2)$, it follows that $Hom(B_1, y_2R/soc(y_2R)) = 0$. But we have an isomorphism

$$\theta : y_1R/soc(y_1R) \to y_2R/L \text{ defined by } \theta(y_1 + soc(y_1R)) = y_2 + L.$$

It has a lifting $\eta : y_1R \to y_2R$. It follows that $\eta(A)$ is simple. Which gives that

$$y_2R = \eta(y_1R) \oplus L, \; y_2 - \eta(y_1) \in L.$$

If $y_1r \in soc(y_1R)$, then $y_2r \in yR$, hence $L = yR \cap N_2$. Therefore we have that $y_1 + \eta(y_1) = y + (y_2 - \eta(y_1)) \in yR$. Then $yR = (y_1 + \eta(y_1))R \oplus W$, and we finish.

Let X be any subfactor of K. Then $X = C/V$ for some $C \leq K$ and $V < C$. We prove that any complement of a simple submodule of X is a summand of X. Let L/V be a simple submodule of X and G/V be a complement of L/V in X. We consider the case when $d(X/G') = 2$, where $G' = G/V$. Then $d(C/G) = 2$. By the first part $C = C' \oplus G$, where $C' \cong X/G'$. Then $X = \overline{C'} \oplus G'$, where $\overline{C'} = (C' + G)/G$. Let U, V be two subfactors of K such that $soc(U) \cong soc(V)$. The condition $(*)$ gives that U is V-injective, that gives $U \cong V$. □

THEOREM 12. Let $M_R = A \oplus N$, where A is a non-simple, quasi-injective uniserial module of length 2, and is not quasi-projective.

If M is completely almost self-injective, then the following hold.

(a) N is completely almost self-injective.

(b) $Hom(A, N/soc(N)) = 0$

(c) For any non-simple uniserial summand B of M which is disjoint from A, $soc(A) \ncong soc(B)$ and $A/soc(A) \ncong B/soc(B)$.

Conversely, if the above three conditions hold, then:

(d) A is N-injective and any non-simple uniserial summand B of N is A-injective.

(e) A is N-projective and every non-simple uniserial subfactor of N is A-projective.

(f) For any simple submodule S of a subfactor K of M, any complement V of S in K is a summand of K. If $S \subseteq J(K)$, then $K = B \oplus V$ for some non-simple uniserial submodule containing S.

(g) Any non-simple uniserial subfactor of M is quasi-injective, and any two non-simple uniserial subfactors of M are isomorphic if and only if they have isomorphic socles.

Proof. Let $M = A \oplus N$ be completely almost self-injective. By Theorem 11, conditions (a) and (b) hold. Let B be a non-simple uniserial subfactor of M. Because of Lemma 2(b), we take B a summand of M. If $B \cap A = 0$, then B embeds in N, therefore $A \ncong B$. By Theorem 11, A is B-projective, which gives $A/soc(A) \ncong B/soc(B)$. By Theorem A, $soc(A) \ncong soc(B)$.

Conversely, let M satisfy the given conditions. We show that the pair A, N satisfy condition $(*)$. As $N/soc(N)$ is semi-simple, $M/soc(M)$ is semi-simple. Suppose A is not N-injective. We get a maximal homomorphism $f : L \to A$, $L < N$, which cannot be extended from N to A. Let $K = \ker f$.

Assume that f is onto. Then $A \cong L/K$. By Lemma 2, $L = B \oplus K$ with $B \cong A$, which contradicts (c). Thus $f(L) = soc(A)$. We get $L = S \oplus K$ and K is a complement of S in N. Therefore, because N is completely almost self-injective, it follows that $N = B \oplus K$, where B is non-simple and uniserial and $soc(B) \cong soc(A)$. This contradicts (c). Hence A is N-injective. It follows from (c) that every non-simple uniserial subfactor B of N is A-injective.

Suppose we have a non-zero homomorphism $g : A \to N/H$ for some $H < N$. We get $g(A) = G/H$ for some $G \leq N$. If $g(A)$ is not simple, by Lemma 2(b), H is a summand of G and that contradicts (c). Therefore $g(A)$ is simple. If H is not essential in G, we get H a summand of G, and we finish. Therefore $H \subseteq_e G$ and $soc(G) \subseteq H$. But $G/soc(G)$ embeds in $N/soc(N)$, which contradicts (b). Hence A is N-projective. It follows from (c) that any non-simple uniserial subfactor of N is A-projective. Thus the pair A, N satisfy condition $(*)$ and the conditions in Theorem 11(ii). This gives the other parts of the theorem. □

THEOREM 13. Let N_1 and N_2 be two completely almost self-injective modules over a ring R such that $N_1/soc(N_1)$ and $N_2/soc(N_2)$ are finitely generated. Then $K = N_1 \oplus N_2$ is completely almost self-injective if and only the following hold.

(i) Any non-simple uniserial subfactor of N_1 is N_2-injective, and every non-simple uniserial subfactor of N_2 is N_1-injective.

(ii) Any non-simple uniserial subfactor A of N_1 is C-projective for any cyclic subfactor C of N_2, and any non-simple uniserial subfactor B of N_2 is C'-projective for any cyclic subfactor C' of N_1.

Proof. Suppose K is completely almost self-injective. Theorem B gives (i) and Theorem 11 gives (ii).

Conversely, let conditions (i) and (ii) hold. Thus K satisfies $(*)$. By Theorems 11 and 12, all the conditions in Theorem B are satisfied. Hence K is completely almost self-injective. □

THEOREM 14. Let N_R be any completely almost self-injective module such that $N/soc(N)$ is finitely generated, and A_R be a quasi-injective uniserial module of length 2, which is not quasi-projective. Then $M = A \oplus N$ is completely almost self-injective if and only if A is N-projective.

Proof. Suppose M is completely almost self-injective. By Theorem 11, A is N-projective.

Conversely, let A be N-projective. As in Theorem 11, $Hom(A, N/soc(N)) = 0$. It follows that all the conditions (a), (b) and (c) in Theorem 12 hold. Thereafter Theorem B gives that M is completely almost self-injective. □

THEOREM 15. Let M_R be a completely almost self-injective module such that $M/soc(M)$ is finitely generated.

(i) $M \oplus M$ is completely almost self-injective if and only if any non-simple uniserial subfactor of M is M-projective.

(ii) If $M \oplus M$ is completely almost self-injective, then any finite direct sum of copies of M is completely almost self-injective.

(iii) If a ring R is right completely almost self-injective, then any finitely generated module K_R is completely almost self-injective.

Proof. Observe that the pair M, M satisfies condition $(*)$.

(i) Let $M \oplus M$ be completely almost self-injective. Let A be a non-simple uniserial subfactor of M, then A being a summand of M, gives that $A \oplus M$ is completely almost self-injective. Therefore, by Proposition 9, A is M-projective. Conversely, let any non-simple uniserial subfactor A of M be M-projective, therefore A is also C-projective for any cyclic subfactor C of M. In addition any non-simple uniserial subfactor of M is M-injective. Hence , by Theorem 13, $M \oplus M$ is completely almost self-injective.

(ii) Let N be a finite direct sum of copies of M. Let A be a non-simple uniserial subfactor of N. By Lemma 4, A is a subfactor of M, therefore by (i), A is M-projective, hence N-projective. As in (i), by Theorem 13, N is completely almost self-injective.

(iii) Let R_R be completely almost self-injective. Let A be a non-simple uniserial subfactor of R_R. As A is a summand of R_R, it is projective. Therefore by (ii), any finite direct sum of copies of R_R is completely almost self-injective. This gives that any finitely generated R-module being a homomorphic image of a finite direct sum of copies of R_R, is completely almost self-injective. □

We end this paper by giving another characterization for a ring R to be right completely almost self-injective.

THEOREM 16. Let R be any ring. Then R is right completely almost self-injective if and only if it satisfies the following conditions.

(i) $R/soc(R_R)$ is semi-simple.

(ii) Any non-simple uniserial subfactor of R_R is of length 2 and is injective and projective.

Proof. Let R be a right completely almost self-injective ring. Let U be a non-simple uniserial subfactor of R_R. By Lemma 2, U isomorphic to a summand of R_R, therefore it is projective. Also U being R-injective, is injective.

Conversely, let the above two conditions hold. Observe that the above two conditions are satisfied by any subfactor of R. Let $\overline{S} = S/L$ be a simple submodule of a subfactor $\overline{A} = A/L$ of R and $\overline{K} = K/L$ be a complement of S/L in $\overline{A}$. Thus S/L embeds in $A/K \cong \overline{A}/\overline{K}$. Therefore A/K is uniform with simple socle. However, by [**12**, Lemma 2.7], $(A/K)/soc(A/K)$ is semi-simple. But, by the hypothesis, any non-simple uniserial subfactor of R is injective and has length 2. Which gives $d(A/K) \leq 2$.

If $d(A/K) = 1$, trivially $\overline{A} = \overline{S} \oplus \overline{K}$. If $d(A/K) = 2$, then $\overline{A}/\overline{K}$ is projective, therefore $\overline{A} = \overline{B} \oplus \overline{K}$ for some uniserial submodule $\overline{B}$ of length 2. If $\overline{S} \subseteq J(\overline{A})$, by [**12**, Lemma 2.8], we get a uniserial submodule U of $\overline{A}$ of length 2 containing $\overline{S}$.

Any two non-simple uniserial subfactors U, V of R with isomorphic socles are isomorphic, as they are injective. Hence, by Theorem B, R_R is completely almost self-injective. □

Acknowledgment. The author is thankful to the referee for his/her various suggestions.

References

[1] A. Alahmadi and S. K. Jain, *A note on almost injective modules*, Math. J. Okayama Univ. **51** (2009), 101–109. MR2482408
Characterizations of almost injective modules In: Non-commutative rings and

[2] A. Alahmadi, S. K. Jain, and S. Singh, *Characterizations of almost injective modules*, Non-commutative rings and their applications, Contemp. Math., vol. 634, Amer. Math. Soc., Providence, RI, 2015, pp. 11–17. MR3307386

[3] F. W. Anderson and K. R. Fuller, *Rings and categories of modules*, Graduate Texts in Mathematics, vol. 13, Springer-Verlag, New York-Heidelberg, 1974. MR0417223

[4] Y. Baba, *Note on almost M-injectives*, Osaka J. Math. **26** (1989), no. 3, 687–698. MR1021440

[5] Y. Baba and M. Harada, *On almost M-projectives and almost M-injectives*, Tsukuba J. Math. **14** (1990), no. 1, 53–69. MR1063835

[6] C. Faith, *Algebra. II. Ring theory*, Grundlehren der Mathematischen Wissenschaften, vol. 191, Springer-Verlag, Berlin-New York, 1976. MR0427349

[7] K. R. Goodearl, *von Neumann regular rings*, Monographs and Studies in Mathematics, vol. 4, Pitman (Advanced Publishing Program), Boston, Mass.-London, 1979. MR533669

[8] M. Harada, *On almost relative injectives on Artinian modules*, Osaka J. Math. **27** (1990), no. 4, 963–971. MR1088193

[9] M. Harada, *Direct sums of almost relative injective modules*, Osaka J. Math. **28** (1991), no. 3, 751–758. MR1144484

[10] M. Harada, *Note on almost relative projectives and almost relative injectives*, Osaka J. Math. **29** (1992), no. 3, 435–446. MR1181113

[11] S. Singh, *Almost relative injective modules*, Osaka J. Math. **53** (2016), no. 2, 425–438. MR3492807

[12] S. Singh, *Uniform almost relative injective modules*, J. Algebra **478** (2017), 353–366. MR3621678

House No.424, Sector 35 A, Chandigarh–160022, India
Email address: ossinghpal@yahoo.co.in

Selected Published Titles in This Series

715 **Sergio R. López-Permouth, Jae Keol Park, S. Tariq Rizvi, and Cosmin S. Roman, Editors,** Advances in Rings and Modules, 2018

712 **Nero Budur, Tommaso de Fernex, Roi Docampo, and Kevin Tucker, Editors,** Local and Global Methods in Algebraic Geometry, 2018

711 **Thomas Creutzig and Andrew R. Linshaw, Editors,** Vertex Algebras and Geometry, 2018

710 **Raphaël Danchin, Reinhard Farwig, Jiří Neustupa, and Patrick Penel, Editors,** Mathematical Analysis in Fluid Mechanics, 2018

709 **Fernando Galaz-García, Juan Carlos Pardo Millán, and Pedro Solórzano, Editors,** Contributions of Mexican Mathematicians Abroad in Pure and Applied Mathematics, 2018

708 **Christian Ausoni, Kathryn Hess, Brenda Johnson, Ieke Moerdijk, and Jérôme Scherer, Editors,** An Alpine Bouquet of Algebraic Topology, 2018

707 **Nitya Kitchloo, Mona Merling, Jack Morava, Emily Riehl, and W. Stephen Wilson, Editors,** New Directions in Homotopy Theory, 2018

706 **Yeonhyang Kim, Sivaram K. Narayan, Gabriel Picioroaga, and Eric S. Weber, Editors,** Frames and Harmonic Analysis, 2018

705 **Graham J. Leuschke, Frauke Bleher, Ralf Schiffler, and Dan Zacharia, Editors,** Representations of Algebras, 2018

704 **Alain Escassut, Cristina Perez-Garcia, and Khodr Shamseddine, Editors,** Advances in Ultrametric Analysis, 2018

703 **Andreas Malmendier and Tony Shaska, Editors,** Higher Genus Curves in Mathematical Physics and Arithmetic Geometry, 2018

702 **Mark Grant, Gregory Lupton, and Lucile Vandembroucq, Editors,** Topological Complexity and Related Topics, 2018

701 **Joan-Carles Lario and V. Kumar Murty, Editors,** Number Theory Related to Modular Curves, 2018

700 **Alexandre Girouard, Dmitry Jakobson, Michael Levitin, Nilima Nigam, Iosif Polterovich, and Frédéric Rochon, Editors,** Geometric and Computational Spectral Theory, 2017

699 **Mark L. Agranovsky, Matania Ben-Artzi, Catherine Bénéteau, Lavi Karp, Dmitry Khavinson, Simeon Reich, David Shoikhet, Gilbert Weinstein, and Lawrence Zalcman, Editors,** Complex Analysis and Dynamical Systems VII, 2017

698 **Alexander M. Blokh, Leonid A. Bunimovich, Paul H. Jung, Lex G. Oversteegen, and Yakov G. Sinai, Editors,** Dynamical Systems, Ergodic Theory, and Probability: in Memory of Kolya Chernov, 2017

697 **Fabrizio Broglia, Françoise Delon, Max Dickmann, Danielle Gondard-Cozette, and Victoria Ann Powers, Editors,** Ordered Algebraic Structures and Related Topics, 2017

696 **Ara S. Basmajian, Yair N. Minsky, and Alan W. Reid, Editors,** In the Tradition of Ahlfors–Bers, VII, 2017

695 **Katrina Barron, Elizabeth Jurisich, Antun Milas, and Kailash Misra, Editors,** Lie Algebras, Vertex Operator Algebras, and Related Topics, 2017

694 **Manjul Bhargava, Robert Guralnick, Gerhard Hiss, Klaus Lux, and Pham Huu Tiep, Editors,** Finite Simple Groups: Thirty Years of the Atlas and Beyond, 2017

693 **Michael Cwikel and Mario Milman, Editors,** Functional Analysis, Harmonic Analysis, and Image Processing, 2017

692 **Anatole Katok, Yakov Pesin, and Federico Rodriguez Hertz, Editors,** Modern Theory of Dynamical Systems, 2017

For a complete list of titles in this series, visit the AMS Bookstore at **www.ams.org/bookstore/conmseries/**.